W9-CNG-634

RELATIVISTIC POINT DYNAMICS

RELATIVISTIC POINT DYNAMICS

HENRI ARZELIÈS

General equations; constant proper masses; interactions between electric charges; variable proper masses; collisions

Translated by P. W. Hawkes from a thoroughly revised text, brought up to date to take account of recent developments

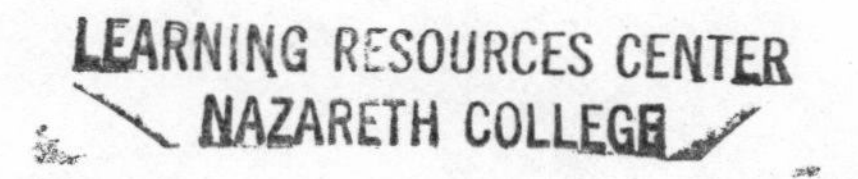

PERGAMON PRESS
OXFORD · NEW YORK · TORONTO
SYDNEY · BRAUNSCHWEIG

Pergamon Press Ltd., Headington Hill Hall, Oxford
Pergamon Press Inc., Maxwell House, Fairview Park, Elmsford, New York 10523
Pergamon of Canada Ltd., 207 Queen's Quay West, Toronto 1
Pergamon Press (Aust.) Pty. Ltd., 19a Boundary Street, Rushcutters Bay, N.S.W. 2011, Australia
Vieweg & Sohn GmbH, Burgplatz 1, Braunschweig

First edition 1971
Library of Congress Catalog Card No. 72-142173

This translation is based upon material drawn from both volumes of *Dynamique Relativiste*, published by Gauthier-Villars, Paris, in 1957 and 1958; Chapters XII and XIII are original.

Printed in Hungary
08 015842 0

The first volume of these studies in relativity was a tribute to the memory of Albert Einstein. It is with pleasure and a sense of fitness that I dedicate this volume to

J. C. Maxwell and H. A. Lorentz

to whom we owe the fundamental equations of the electromagnetic field and of particle dynamics.

CONTENTS

PREFACE

PRELIMINARY REMARKS ON THE IMPORTANCE OF EPISTEMOLOGICAL, HISTORICAL AND BIBLIOGRAPHICAL STUDIES

The presentation adopted in my *Kinematics*, which united in a single work an account suitable for beginners and the facts necessary for research, has met with a generally sympathetic reception. In practice, the formula seems a good one; it is therefore worth while to analyse it further, and thus stimulate useful discussion.

* * *

The important place that I accord epistemological considerations has provoked very varied reactions. Some critics have warmly approved my adoption of this attitude even in the expository part of my book. I must, however, admit that these critics were, for the most part, already among the converted. Far too many physicists, and the majority of mathematicians, still regard this as time wasted.

"Your first four chapters", said one of them to me, "are just padding. Your book really begins at Chapter V, at § [46] to be exact." The epistemological passages remind such readers of the celebrated discussions on the sex of the angels, in honour of Byzantium when Mahomet II was under the ramparts.†

In my opinion, no falser position could be held. The questions with which the first four chapters of *Kinematics* are concerned are basic. The choice of standards has twice been the cornerstone of decisive steps forward in the development of natural philosophy. The work of Copernicus, the founder of modern astronomy, reduces to a suitable choice of reference system. The essential feature of the doctrine of relativity lies in the choice of length and time standards. It is a commonplace remark that practically all the formulae of special relativity were in existence before Einstein; nevertheless, no one disputes the importance of the year 1905 in the evolution of physics. The difficulties

† I am far from following the historians in their indignant censure. A civilized society, worthy of the name, is characterized by the harmonious equilibrium of its various activities; the "sex of the angels" can and should be studied, even during critical periods, for the defence of civilization would otherwise be pointless. Obviously, the vital activities must not, however, be neglected in favour of purely intellectual pursuits. Consider the life of Marcus Aurelius or Julian, for example.

encountered today in the theory of elementary particles also seem closely connected with the definition of measuring procedures and standards.

* * *

The purpose of the notes at the end of each chapter is to shed light on the present state of each problem by situating it in the general framework, and to replace the digests of physics by texts, more alive and comprehensive. Their object is to give the reader *the taste for and the opportunity of referring to the sources*. The work necessary to prepare these notes cannot easily be undertaken by the physicist who is a member of a specialized team, to which he must devote all his energies, still less by an engineer. It is, on the contrary, perfectly suitable for a university professor. The latter can, if he so desires (although it is true that he does desire this less and less), withdraw from the feverish activity which agitates most scientific communities. He can take his time, spend weeks and months reading past articles or repeating experiments, reflect on the very meaning of his work and on the value of physics, unhampered by an exact schedule.

Nevertheless, the main centre of interest and the essential part of my book is of course the account of the present state of the subject (and the hurried reader may limit himself to this account). I have not undertaken the work of a historian of physics, and I do not pretend, therefore, to give a full historical study. Despite appearances, I have no desire to make a show of erudition which informed readers will quickly find defective.

My intention is to show that relativistic ideas fit smoothly into the general evolution without a break and that they were even present in the earliest stages of certain sciences. This belief, which I have already stated in *Kinematics*, is particularly strikingly illustrated in *Electricity*. In the present book, the reader will find such physicists as Oersted and Ampère quoted, whose writings appeared long before the birth of the principle of relativity. I shall have achieved my goal if, after studying the text, the presence of these names in a book on relativistic dynamics seems perfectly natural. Perhaps some historian—reversing my point of view and going more deeply into it—will take pleasure in elucidating the papers of the last century with the aid of relativity, and reveal new evolutionary threads.

The impression that a break had occurred, in 1905, and which still remains for most physicists, brings the following image to my mind. The development of a science may be compared with a long path which humanity must follow. Sometimes this path joins another and two sciences fuse (optics and electricity, with Maxwell); for a period, it is easy to advance. Then the path narrows again, and here and there it is difficult to trace: undergrowth chokes it and it breaks up into tracks which fade out altogether. In 1905 physicists had reached such a point; they had unwittingly strayed from the path and were following side-tracks with no future in them. Einstein set them back on to the right path; to do this, however, he had to break away from the habits acquired along the little side paths and return to the real path, that is, to the general principles which physicists more or less consciously imposed upon themselves (reduction and exact definition of the concepts used, suppression of absolute entities which are inaccessible, . . .).

* * *

In this period of aggressive modernity, young research workers need to be taught to acquire a feeling and taste for intellectual continuity. I should like to protect them from the ridicule attracted by the attitudes of a visionary prophet or a burning iconoclast, so common among philosophers, poets and artists. Quite apart from their excesses of language (goal: épater les "bourgeois"), many of them limit themselves to repeating, century after century: "everything done before me was just dabbling; just you wait and see what's in store for you". And what do we see? In exceptional cases, when the man has genius, when he is a Nietsche, a Rimbaud or a Gauguin, he finds a place in the history of philosophy, literature or art, alongside those who have gone before him. More frequently, however, it is but "much ado about nothing". It is amusing to watch them straining their brains to think up strange shapes, at least with an appearance of novelty. Before they came upon the scene, words had meanings which were practically the same for everyone; Boileau even asserted that "ce qui se conçoit bien s'énonce clairement".† What an old-world, out of date opinion! The ultimate end is, on the contrary, to put oneself into a state of poetic trance and write down disconnected words at the dictates of one's subconscious (so, at least, those authors pretend to do . . . and we are credulous enough to believe them). And why stop there? Use words, words which have meaning? Enough of that! Let us just retain the letters, and pour them out wholesale (an inspired pouring out wholesale, naturally . . .). Why not therefore just publish blank pages (of different colours, perhaps) with a few vague blotches and marks here and there? I'm sorry—it was done long ago in Sterne's *Tristram Shandy*.

Let us be serious, however. Between La Bruyère's "Tout est dit, et l'on vient trop tard, depuis plus de sept mille ans qu'il y a des hommes, et qui pensent",‡ and the contempt for the past which is periodically fashionable, there is room for a reasonable central attitude. I realize that for some, a reasonable being is the worst thing to be. However, it is not because one raves that one is a genius, or even talented.

Scientific milieux too have been tainted by the epidemic of modernity. I am myself an active partisan of a more extensive use of the theories said to be modern, as my books adequately prove; it would not be fair to ascribe to me feelings of narrow conservatism. The attraction of relativity and, above all, quantum theories is due all too often to the strange and even occult appearance which these theories have in the eyes of the young physicist, and which is eagerly opposed to that of the classical theories. It is this opposition that I hope to remove or lessen in my books. Perhaps I shall repel some potential candidates away from "modern" physics, but I hope to draw attention to the permanent principles of which I spoke earlier.

It is also necessary to prevent research physicists from connecting the value of a scientific paper with its date, a practice that is, alas, all too frequent. Quite apart from their purely historical interest (this is not the point of view that I adopt), relatively old texts are sometimes more useful than recent textbooks and articles. To treat the writings of ten or twenty years ago as historical documents is permissible in ladies' fashions, but not in physics.

† Well conceived is clearly stated.

‡ All is said, we have come too late; for more than seven thousand years there have been men, thinking.

The situation is certainly not the same as in literature: no one questions the value of the *results* obtained by our predecessors, but one is rebuffed by the characteristics of the period (different theoretical terminology, for example). There would be nothing to criticize in this were it not that this disdain for the original writings all too often goes with ignorance of their contents. By blazing the trail, I hope to make it easier to read and use certain memoirs which are now fifty years or more old, but where remarkable solutions are to be found ... and which are periodically rediscovered, more or less correctly. The captious reader, comparing what I hope to do with what he finds in the present work, will perhaps recall that the road to hell is paved with good intentions. I shall not mind such a criticism at all, if my intentions are accepted as good.

* * *

The historical effort is not everything. One must also, and above all, set out the present state of the subject faithfully, *taking the latest publications into consideration.* However, if the old articles are found difficult reading because of the difficulty of adapting oneself to unfamiliar, out-of-date concepts, the recent papers really horrify many by their quantity and by the variety of languages employed. A general treatise has a duty to guide the physicist through such a jungle, and to show him the lie of the land if not to provide a complete chart. The number of references is in general greater for the modern texts, which is natural since my book is centred on the present time; a historian of physics would select a different perspective. So far as the old articles are concerned, I quote only the great names or those references which offer a good bibliographical starting-point. The work produced today is richer than that of fifty years ago, not because we have more genius or are cleverer (this scarcely varies over millenia), but because the amount of research performed is incomparably greater.

If I dare risk yet another metaphor, bibliography is a tree, the roots and trunk of which are the classic papers, while recent articles form the branches and foliage. At what moment we change from trunk to branch is hard to decide. Nevertheless, we must ensure that the outermost twig, just because it is the outermost, does not forget that it is not the trunk, that without the latter it would be nothing and that tomorrow perhaps it will have withered away. By returning to the sources one learns modesty, a quality inseparable from every kind of culture.

* * *

The training of a good scientist must provide familiarity with certain foreign languages. It is essential nowadays to be able to read articles in English, French, German, Italian and Russian. This does not imply a thorough knowledge of these languages, but a rapid general course (one year is enough) after which a specialized vocabulary will be acquired, which is virtually the same in structure in all these languages. The vast majority of the scientific literature is written in the first four. The Russians only began to produce serious work towards 1930, but during the last twenty years especially, they have been taking giant strides. There had certainly always been some scientific activity in Russia, but their principal journals were published in German, French or English; the national tongue has now replaced these. Conclusion: young physicists, my friends, acquire some knowledge of Russian while your brains are still supple—it will be increasingly more useful to you. Indeed, I am really afraid that Chinese and, consequently, Japanese will soon have to be added to this list. Spanish and Portuguese, despite their human importance, are less useful in a relativistic bibliography.

PHYSICS AND PHILOSOPHY

The words used by the physicist can today be classified into two categories, which despite inevitable interpenetration remain quite distinct: one constitutes the technical vocabulary, the other the epistemological vocabulary.

The technical vocabulary consists of words like velocity, acceleration, force, electrical force and so on. All treatises worthy of the title define them carefully and draw attention to the difference which there often is between the physical meaning and the everyday meaning (thus a racing car completes a circuit at a steady speed according to the speedometer; in the physical sense, it is travelling with an acceleration, but not in the everyday sense).

The epistemological vocabulary consists of the following sorts of terms: truth (true, correct, false, . . .), existence (reality, matter, to annihilate, create or materialize, subjective, objective, . . .), cause (causality, finality, determinism, . . .), time (time-interval, simultaneity, co-existence, . . .), etc. On these, the authors of physics textbooks are practically mute.

Do they trust to the professional philosophers, since the terms are epistemological? It does not seem very likely, for they never refer their readers to philosophical works, which they themselves have in most cases never read.

The opinion is nearly universally held that these words in the common vocabulary are clear in themselves, and that there is no point in defining them therefore. In support of this opinion, and if memories of his schooldays are not too distant, the reader will be thinking of Pascal's remark about definitions in his pamphlet *De l'esprit géométrique:*

"Elle (la géométrie) ne définit aucune de ces choses, espace, temps, mouvement, nombre, égalité, ni les semblables qui sont en si grand nombre, parce que ces termes désignent si naturellement les choses qu'ils signifient à ceux qui entendent la langue, que l'éclaircissement qu'on en voudrait faire apporterait plus d'obscurité que d'instruction."†

Although he is dealing with geometry, the context shows that Pascal applied these ideas to physics.‡

† It (geometry) defines none of these things—space, time, motion, number, equality, nor similar things of which there are so many—because these terms evoke the things they signify so naturally for those who understand the language that any elucidation that one might undertake would lead to more obscurity than instruction.

‡ In the same passage, Pascal states propositions which he considers "parfaitement démontrées . . . par la lumière naturelle" [perfectly proven . . . by the light of common day], but which the modern physicist cannot accept: ". . . quelque mouvement, quelque espace, quelque nombre, quelque temps que ce soit, il y en a toujours un plus grand et un moindre . . ." [for any motion, any space, any number, or any time whatsoever, there is always a greater and a lesser one].

This shows just how much "the light of common day" varies at different periods, and doubtless among different individuals as well.

Since 1905 especially, however, physicists have taken great pains to define these epistemological terms and in certain cases, the resulting progress has been so vast that the whole of physics seems to have been newly conceived (for instance, the definition of the concept of simultaneity). In contradiction to Pascal's thesis, the modern physicist defines time and discards, among the everyday quantities, "what everybody means by such and such a word". He feels, more or less confusedly, that this method should be extended to the whole of the "primitive" vocabulary which Pascal did not wish to touch.

* * *

Let us first examine the meaning of the word "definition" without letting ourselves be distracted by the paradoxical nature of this suggestion (of defining definition). According to Pascal, definitions are "impositions de noms aux choses qu'on a clairement désignées en termes parfaitement connus".† We must therefore stop somewhere as we make our way back through the vocabulary, whence certain "taboo" words. Furthermore, a word is completely defined as soon as it is introduced once and for all.

This point of view is at the root of ordinary mathematics. The physicist, however, must ascribe a much wider meaning to the concept of definition. For him, there is often only an *analysis in several stages, and the terms employed shed light on one another with a gradual overall improvement*. (We notice that in certain cases, the modern mathematician too seems to be coming round to this point of view.) Physical definitions are different in kind from mathematical definitions in Pascal's sense.

A physics treatise begins with a certain number of words which are taken to have their ordinary meaning, for want of any better. These words are not particularly privileged, however, and their appearance of clarity is often a trap. One has constantly to go back and sharpen the meaning of these primitive words which had at first been treated as implicit. The semantic progress of physics is far from being a continuous ascent. It is more like the work of the settler reclaiming virgin forest who first cuts tracks to give access to the interior. He then uses the materials to hand to widen these tracks, to transform them into roads, and so on. We shall meet this idea again in connection with the different theories.

In conclusion, then, I believe that epistemological terms must be treated like any others. There is no taboo vocabulary in physics; little by little, the epistemological vocabulary should disappear.

* * *

The definition of a new physical quantity X has today two essential features. The quantity X can and in general must be defined mathematically by means of a formula containing X and other quantities already known. In addition, X must be defined operationally, which gives it physical meaning, by describing a physical procedure by means of which X could be measured in terms of a standard. The primary quantity can of course be defined only by the second procedure, but it *must be so defined*. In the present state of science, this primary quantity is time. Every other quantity must be defined by means of both procedures. Nevertheless, these two procedures (chiefly the second one) use a language which must be continually modified as science and technology progress.

* * *

† The giving of names to things that have been clearly indicated in perfectly understood terms.

I do not make the absurd claim of studying the subject in its full generality, and exhaustively. My goal is more modest: to prompt the reader to ponder, and to suggest various criticisms to him. It may be—it is likely even—that I am sometimes wrong (not too often, I venture to hope), but mistakes clearly expressed, and hence readily corrected, are far more useful than vague ideas.

A healthy understanding of the theory of relativity and quantum theory requires reflection on the concepts of truth, existence, objectivity, ... By making their vocabulary more exact, physicists can appreciate and eliminate the *pseudo-problems which an insufficiently exact use of language engenders.* Many arguments—the most futile are the most virulent—can in fact be reduced to exchanges of words, to pure verbiage.

It is very important today that the physicist should study the epistemological vocabulary; *a course on this should form part of a technical training.*†

With extremely rare exceptions, such questions are considered only—and all too often inadequately—in books on the philosophy of science which physicists are unlikely to read. Whence the necessity to include them in books on physics: the reader will then consider them naturally, if only by way of relaxation.

The majority of the remarks in this preface will be found banal by physicists who have pondered these questions; however, numerous conversations and wide reading have taught me that this is not true of the majority of physicists, *nor above all, of professional philosophers.* Intercourse between physicists and philosophers should overcome such divergent attitudes.

* * *

The following reflections tend to resolve themselves into a philosophy. This doctrine, which is constantly being modified in the same way as the physical theories, finds no general or absolute statement, however; it arises from experience and does not claim to apply to anything except experience. It is *essentially independent of the metaphysical systems*; if, at certain stages of its development, it *seems* to violate some of these, the physicist must remain unaffected.

It is a matter of establishing the rules of behaviour which will enable us to organize our terrestrial existence in the least bad way possible, and of not letting ourselves be inveigled by some metaphysical absolute. Our philosophy is therefore formed after physics, or at least, in company with physics and not before. It is taught by physics and in general by the natural sciences, and does not teach them (by the natural sciences, we also understand the psychological and social sciences, *in so far as they employ experimental methods*).

It is therefore vital to leave this philosophy "open", to maintain its *inferior* rank. We notice, however, that if it has nothing to teach physics, it can instruct certain physicists who are too fond of the concepts prevalent before 1905.

* * *

† If we exploit only the *technical* possibilities of the knowledge acquired, then epistemology, whether good or bad, has no role to play; only the dictates of economics and finance are involved. Fundamental physical research, on the other hand, is nowadays becoming more and more strikingly epistemological in nature; thought is being given to the basic concepts: time, space, identity, reality and so on. Higher education in all branches of the sciences ought therefore to devote a few hours to epistemology.

The only way of reaching agreement between all the physicists upon the epistemological problems created by their profession is to free physics from metaphysics. This would be the prelude to a genuine scientific humanism, the absence of which weighs heavily upon our period.

I must insist: *there is—there can be—no such thing as Buddhist, Christian, Moslem, Hitlerite, Bourgeois or Marxist physics.* To think or write the contrary, one must either be very stupid or, what is worse, quite blinded by partisan emotion. If one is attached to a faith, one may set out to show that it is not in contradiction with the laws of science. It is, on the contrary, wholly inadmissible in the middle of the twentieth century that religious or metaphysical beliefs should be used to test a physical law, or as a guide to direct research.

It is no less deplorable that the publication of a physical theory often leads to the classification of its author into this or that metaphysical or political category. This card-indexing is exasperating and leads to the formation of hostile clans. The primordial, vital duty of our period is, on the contrary, to strive towards a better understanding of mankind, setting out from neutral territory; and we must be sufficiently tolerant to begin with to accept that such neutral zones do exist. Every one of us dreams of an ideal society. Tolerance and charity are the two golden rules of my "Earthly paradise".

* * *

The subordination of philosophy to the experimental sciences *where they have common ground* is, in my opinion, the cornerstone of a healthy attitude towards physics and epistemology. Oddly enough, this is now accepted by a good many philosophers but rejected by a whole school of physicists. As a fundamental epistemological attitude is involved, I give a specific example: the reaction of the various schools of thought to Hubble's law (the expansion of the universe).

Although I have studied this question at some length (I have now been working on relativity almost exclusively for some twenty years), I shall not come down either for or against, since I have not thought deeply enough about cosmological problems. In the spirit of this preface, I shall limit myself to considering the expansion of the universe from the epistemological point of view only.

The crude experimental fact is as follows. We know that a substance can be characterized by its spectrum (the iron doublet, for example). When the same doublet is identified in the light from some star, an overall shift is found, towards the red for example. Apart from the gravitational shift which is very small, the phenomenon is regarded *by all physicists* as a Doppler effect, caused by a widening of the relative separation of earth and star; it is regularly used to measure the radial velocities of the stars.

Observation of the galaxies reveals the same basic phenomenon, but with spectral shifts which are much larger and proportional to the distance. The straightforward interpretation of these shifts, irrespective of any cosmological theory, is to attribute them to a Doppler effect, just as in the case of the stars. This is the reaction of the majority of professional astronomers, including in particular those who discovered the phenomenon. Hence Hubble's law: the galaxies are travelling away from us with a velocity proportional to their distance.

The attitude of the theoreticians is very instructive. The first theoretical explorations fall within the framework of gravitational theory. Let us begin with these, therefore.

Einstein's law of gravitation in its original form contained only three terms. Einstein soon tried to apply it to a static universe, and encountered difficulties which he considered insurmountable. To ensure the stability of the universe (before Hubble's law was discovered, an unstable universe seemed unacceptable), he therefore introduced a supplementary term (called the λ term or cosmological term) which mathematically was compatible with his principles but which destroyed the simplicity of his law. Nevertheless, he did introduce it, under experimental pressure so he believed.

Later, Friedmann noticed that Einstein's original equation is admirably suited for designing non-static models; this observation, together with Hubble's experimental discovery, gave unstable universes the freedom of the city. Many authors, Einstein among them, then discarded the term in λ. We should recognize in this an example of perfect submission to experiment, and not to a preconception.

The expansion theory has much to commend it to favour: a Doppler effect interpretation common to both stars and galaxies; integration into the relativistic formalism with or without the term in λ; agreement with certain geological observations concerning an earlier hyperdense state.

The attitude of the adversaries of the theory is, on the contrary, unpalatable to the physicist who wishes to argue purely as a physicist. The only argument which could affect us (disagreement with observation) is scarcely touched upon, and with good reason. Obviously—the authors are quite explicit enough—their revulsion before the theory arises from either ideological or metaphysical reasons. The hypothesis of a hyperdense state (a physical suggestion) is confused with the idea of a Creation (metaphysics). Unfortunately for us, it was given a strong thrust forward by a priest (l'Abbé Lemaître); in consequence, the theory was labelled clerical and reactionary, which shows well enough the unhealthy climate in which the arguments were conducted.

To begin with, such an attitude implies an imperfect understanding of the theory; for, whether we consider the present universe or the "primitive atom", the metaphysical problem of the creation remains intact. Furthermore, even if the idea of a creation were inseparable from the theory, it would be anti-scientific to exclude it *a priori*, just because it appears in certain religious faiths. It is equally inadmissible to claim that, *a priori*, all the physical statements in the Bible or the Koran, taken literally, are all right or all wrong. The eternity of "matter" is acceptable as a working hypothesis, just as is the creation hypothesis, but absurd and detestable as a *ne varietur* statement.

It is a serious fact that we are present at an ideological and even political intrusion, which would like to govern the development of the sciences. This is a mediaeval attitude, in the worst sense of the term.

There was a time when it was prudent to be in good odour with the Inquisition, if one wanted to work on physics or astronomy without risking the stake. Physicists suffered for a long time from the attitude of the Church towards them; the classic example, which is symbolic, is of course that of Galileo. Certainly, towards the end of the eighteenth century, these inquisitional intrusions became far more infrequent and even ceased altogether in the majority of countries. It is none the less true that in the nineteenth century, such practices were still severely enforced in some countries. Everywhere else, they were certainly held in abeyance by the theological authorities, but not clearly and explicitly renounced. Until 1965, the official attitude to Galileo's trial was as follows: Galileo was rightly condemned. He was condemned, not because of his astronomical conclusions but because he tried to justify them by quotations extracted from holy writ. A subtle distinction, such as only theologians and metaphysicians can make! A convenient distinction, above all, which ill conceals the hard facts. Wisdom has finally triumphed; the Church has taken its time, but at last it has rehabilitated Galileo.†

Unfortunately, we now have to defend ourselves on a different front. Nowadays, if certain people are to be believed, no valid physics can be produced without one's being in a materialist (dialectic) state of grace. By acting thus, by lying in the path of the steam roller of the experimental sciences, these physicists render no service at all to their beliefs (which I respect, as I respect all beliefs, but *only in so far as they are tolerant*). I am, moreover, convinced that their ideas do not impose such an attitude in the scientific domain. After this childhood folly, *a subsequent refinement of the concepts will lead them to a wiser attitude* (the evolution of Rome required centuries; we must be indulgent where mere decades are concerned, therefore).

Let us make clear the progress that has been achieved. The *Questions Scientifiques* of 1952, of which I spoke in the 1955 edition of *Kinematics*, contain extremely startling passages, of which the least that I can say is that they sound more like an electoral speech than physics. Consider now, in "Editions de la Nouvelle Critique", the September–October number of 1957 of *Recherches Internationales*, which is devoted to physics. Certainly, we are all considered there to be "bourgeois scholars" which becomes rather comic when we consider the *material* conditions of our life (this does not mean that I find them bad) and those of our Soviet colleagues. This outmoded, inaccurate, ridiculous and exceedingly exasperating vocabulary is to be abandoned.

Remarks like the following may raise a smile: "Lacking the notions of dialectical materialism, which alone allows a true understanding of relativity, the bourgeois physicists [yet again!]..." (p. 189).

I take no offence at being implicitly classified with those who have not understood relativity, for Einstein himself heads the list. Really? Certainly! Such are the enormities which are written in cold blood, by those in possession of Metaphysical "Terewth" with a capital T.

However, I see a sincere desire to discuss (p. 8); I cannot approve too strongly the call for collaboration between philosophers and physicists. I perceive too that another very interesting idea is appearing: philosophy can be enriched, and hence transformed by the development of physics. During the summer of 1961, the press informed us of a

† In some countries, our Muslim colleagues are also in danger of getting into serious difficulties over the earth's rotation. A press release of 9 June 1966 (taken from the Paris *Figaro* for 10 June 1966) sent from Djeddah quotes the following declaration of a scholarly Ulema, Sheik Abdel Aziz ben Abdallah, who is vice-president of the Islamic University of Medina: "Every man who continues to insist that the earth moves round the sun should be called upon to repent and to denounce his error. If not, he will be treated as an apostate and an infidel. He will then have to be executed. All his possessions will be confiscated for the benefit of the Muslim community." Responsibility for this information I leave to the *Figaro*, of course, but it is, alas, all too probably accurate.

new advance in the liberal direction. The political powers recognized openly, it seems, that they are not competent to arbitrate in a conflict of a scientific nature. The Academician Kapitza stated that "dialectical Marxism alone cannot solve all problems.... Dependence upon it ... is an attitude that has held up progress in the scientific domain." A few more efforts and we shall perhaps manage to get on together.

I am amused to read that such authors as Rosenfeld or Vigier are striving to attune their physical interpretations (contradictory, so it seems) to the political ideas common to both. This seems to me wasted effort, and recalls distressingly the period when one had to demonstrate the agreement between one's ideas and the literal meaning of the Bible. In writings on physics, I am constantly finding appeals to authority, with quotations from Lenin, Engels and so on (poor Stalin, much quoted in Marxist physics some twenty years ago, has entirely disappeared); this reminds me of the times when contradictious mouths were closed with a quotation from Aristotle (αυτός ἔφη). I know that this is supposed to be altogether different: the arguments which are advanced do not convince me!

I am perfectly prepared to admit that metaphysical investigations have sometimes caused us to look at scientific problems from a fresh angle. This is true *for all ideologies*, however, and I fear that the occasional progress thus achieved was dearly bought if it led (involuntarily) to entanglement in a dogma.

Let us return to the idea of expansion.† For other authors, it possesses the grave defect of being relativistic, and hence worthless. *A priori*, the theory is undesirable and so anything else will do to replace it. And what is offered? For the moment, a hypothesis which is scarcely satisfactory: light is deemed to age as it travels in such a way as to give the observed result. This is a perfect specimen of a "loaded" theory.

For the issue we are considering, I reiterate, it matters little so far as I am concerned whether the galaxies are receding or their light is getting older; whether matter is eternal, or was created abruptly, or is continually being created. I subscribe to no act of faith on this point; I shall, perhaps, even be led to express doubts about the "reality" of expansion in my book on Cosmology. I can already foresee that these arguments will partially reduce to the choice of a metric, and this choice is logically arbitrary though imposed in practice (*Kinematics*, § [5]). However this may be, when I do choose one interpretation, it will be for physicists' reasons and not to please the shade of St. Thomas or Karl Marx. For the time being, I wish simply to spotlight *two different intellectual attitudes*. One of these, I adopt. I believe, in all frankness, and—I underline this heavily—independently of all political or metaphysical preferences, that the other must be regarded as retrograde; in my opinion, this other attitude is a *sharp reversal* of the spirit of free examination and free research which has been so painfully acquired over the centuries and which is sometimes still contested by the powers that be.

* * *

A healthy attitude to physics also rejects preconceived ideas, such as are found in the schools of Eddington and Milne in particular. These scientists modernize and turn

† Soviet science regarded this as "clerical and reactionary" until 1958, when Ambarzoumian gave his support to the idea of expansion.

to their own account the old illusions, and claim to be able to deduce quantitative physical laws from epistemological considerations, by reason and without recourse to experiment.

* * *

Those who believe that trespassers upon philosophy should be prosecuted will reproach me for penetrating into unknown territory, and for behaving in it like an ignorant novice; but, I am pondering my activity as a physicist, my profession, what I do every day. It would be unlikely if I were constantly wrong: if only non-physicists (professional philosophers) could think coherently about such subjects.

It would be astonishing—although it is all too often implicitly accepted—*if ignorance or superficial knowledge of a subject were the essential qualification for pronouncing correct general ideas about it.*

For a dialogue between physicists and philosophers to be effective, and not to degenerate into a vain polemic, one preliminary is essential: *the method of working must be agreed upon*, the same language spoken. The physicist must reject any interlocutor, in this kind of argument, who is manifestly ignorant of the elements of physics (such as I should call an ostrich-philosopher). The ostrich-philosopher will, of course, reply that there are specifically philosophical problems, a philosophical way of treating them, and so on. Let him be on his guard! For several decades, a scientific philosophy has grown up under pressure of circumstances which stands in the same relation to traditional philosophy as does astronomy to astrology. It ought surely to be possible to avoid this rupture. As an example, let us take the problem of time *in physics*. Anyone who treats this topic without taking into account the relativity of simultaneity and the ideas concerning time-inversion arising from quantum field theory (Stückelberg, Feynmann, de Beauregard) is behaving like an ostrich-philosopher.

The philosopher in turn will expect of the physicist a good knowledge of the present situation of the problem in question; I implore him not to confuse the present state of the problem with its detailed history, however. I do not expect him to have read Newton's *Principia*; I refuse all discussion with him if he does not know the first thing about contemporary theories of gravitation. He in turn may require that I shall have read *Durée et simultaneité* by Bergson, *La Dialectique de la durée* by Bachelard or *Space and Time* by Reichenbach, for example; all the same, if he really seeks collaboration, he will agree that study of the complete works of Hume or Engels is not utterly indispensable.

Some philosopher-readers will, of course, classify me in one of the familiar systems, which will excuse them from any subsequent intellectual exertion. The dialectical materialist will call me an idealist (which is apparently the worst possible insult); the idealist will find me coarsely materialist. *In my role as physicist*, however, all these stances leave me indifferent. *The suggestions advanced in this preface must be considered in context, that is, uniquely in relation to the needs of the physicist.* (I repeat, "needs"; epistemology is not an occupation for armchair pipe-smokers, but a necessity for the physicist today.) The recurrence of the doctrine adopted, in other domains, may be interesting but should have no influence on the adoption or rejection of propositions.

* * *

"Wie lieblich ist es, dass Worte und Töne da sind: sind nicht Worte und Töne Regenbögen und Schein-Brücken zwischen Ewig-Geschiedenem?"† wrote Nietzsche in his *Zarathoustra* (the Convalescent). The ageing Boileau, on the other hand, assures us that "ce qui se conçoit bien s'énonce clairement / Et les mots pour le dire arrivent aisement."‡

Physicists have a leaning for Boileau; they are, however, much less optimistic than many people imagine. To state only what has been clearly conceived would sorely restrict human relationships and even the writings of physicists. Fresnel could not have given his formulae for total reflection and quantum theoreticians would have hesitated to speak of the wave ψ. All the same, this does not mean that modern physicists delight in obscure concepts. Like their predecessors, they progress unceasingly in the direction of clarity and exactitude: they try to arrange that the bridges become less and less imaginary. In the process, however, *the notions of the clear and the obvious themselves undergo modifications*, are seen in a new light. In this sense, we can say that the goal of the physicist's work is not clearly conceived. Even if this goal is the attainment of more clarity, it is in continual development.

I am going to try to classify, in this preface, the ideas which come to my mind when I use the word "truth". Other prefaces deal with the notions of existence, causality and also return to the concept of truth.

Truth in Physical Laws

Truth, that magic and fascinating word, is evocative of the whole human adventure. The Don Quixote asleep in all of us (more and more deeply as the years pass) is ready for every adventure in the search for truth, for every sacrifice in its defence. His inner dream, ideally woven of truth and justice, can at any moment spring into violent action if a pathetic voice cries to him "it is not true" or "it is unjust".

The worthy Sancho Panza (whose influence increases with the years) is, however, diligent in comprehending what is hidden behind the fine words of his master. Before setting his mule into a trot behind Rosinante, he wonders whether this giant is not a simple windmill, or if this army of magicians is not a prosaic caravan of merchants.

Let us imitate him; let us make an effort to scrutinize "the truth", following the guidance of common sense and the patient accumulation of facts, and avoiding as far as possible quixotic impulses. But, you will say, we sometimes tire of this commonplace labour. Well then, read a sonnet by way of invigoration ("J'ai longtemps habité sous de vastes portiques .... Fuir! là-bas, fuir! Je sens que des oiseaux sont ivres"§). Pull on your boots and go off into the mountains. Take your canoe and explore a river. I don't care which, but in scientific research, don't let yourself be carried away by the music of words.

I should repeat willingly on my own account the *Quid est veritas* of Pontius Pilate.

† How delightful it is that there should be words and sounds! Are not words and sounds rainbows and imaginary bridges between ever-separate beings? *Also sprach Zarathoustra:* "Der Genesende".

‡ Well conceived is clearly stated, and the words to express it come easily.

§ Long have I lived 'neath portals vast... Flee! Away, flee! I feel that the birds are drunk...

Perhaps, in his mouth, this was a sign of disillusioned scepticism (and granted the circumstances, who are we to reproach him?). The question remains, however, and in every sphere it requires an answer. To say "this is true" is not enough; what is meant by truth must first be defined.

* * *

The concept of truth is applied to qualitative or quantitative propositions which, in physics, are called laws. Immediately, we meet the difficulty already indicated. What is a proposition? I can elucidate this term by saying that a proposition consists of one or more statements, negations, alternatives, ..., bearing upon certain objects or certain phenomena and their relations. I shall be asked, what is a statement, and above all, what is an object? At the risk of going round in circles, I can only reply *quid potest capere capiat.* I proceed, alerting the reader that, in accordance with the attitude adopted, I shall return to the meaning of these terms later.

Given a proposition, we can by hypothesis think of a different proposition. The propositions to which we shall apply the concept of truth seem therefore to need to be distinguished from obvious propositions (example: this walking-stick is a walking-stick). However, is anything obvious if we take account of the future? The only obvious propositions are no doubt purely formal; they are tautologies.

The concept of truth has several levels, or at least, is regarded differently in the various disciplines. I shall deal essentially with physical truth, for only there am I liable to be at all competent. In another preface (to *Relativité généralisée*, Gauthier-Villars, 1961), I give my opinion as a physicist on certain aspects of the concept in other domains, but with far more reserve. I could certainly suppress these last reflections, but impenetrable barriers between the various intellectual spheres are always bad. Fragmentation is certainly a practical necessity, but it is a sign of our inadequacy; the various specialists should reach out to one another, should contrast their methods of research and their points of view about concepts going under the same word. I risk nothing in saying that they can expect surprises.

* * *

By definition, a proposition is true when it enables us to establish the line of conduct, the behaviour that leads to a goal defined beforehand. This goal might, for example, be to watch a particular star pass the hair-line of a telescope; the line of conduct consists of specifying the orientation of the telescope and the time of observation.

The description of the goal to be achieved and the instructions about the line of conduct require a clear enough vocabulary for action, and this condition is necessary and sufficient. *Practical intelligibility* is required, not logical obviousness, and already this shows us the tolerances within which we may use the "primitive" vocabulary.

For a given term, the limits depend essentially upon the speakers and the period. A word which is now regarded as "primitive" may not have existed a century ago. The phrase "We take an aluminium wire ...", which is immediately clear today, was meaningless 100 years ago.

One of the simplest ways of conferring intelligibility is by naming something, an everyday object which can be displayed, a common phenomenon. Everybody knows what a lens is. There is thus nothing to prevent lenses from being used from the beginning in an optical textbook, before the theory is known, to describe the practical steps taken to study the laws of reflection at a plane mirror.

For the physicist, this transcendental climber as Kant would say, *what succeeds is true*. This empirical, utilitarian attitude is not due to a philosophical preference, *it is the very condition for the existence of physics*.

Thus physical truth equals success of predictions concerning our contacts with the outside world. This assumes that it is possible not to succeed; this 'resistance' of the outside world is what we call the existence of the outside world.

I have already stressed heavily the independence of the experimental sciences with respect to metaphysical systems.

Thus physical truth is amoral and the notions of good and evil are foreign to it.

* * *

Let us analyse the principle features of this truth.

(i) *It is essentially directed towards the future*. True physical laws are based upon past experiments and produce predictions. This may seem paradoxical, for numerous laws relate to past events. To bring out the nuances, let us examine some examples.

Consider the three statements: on 11 February 1958, the ionospheric stations recorded a violent magnetic storm; about four (or six) thousand million years ago, the universe was reduced to the state of a primitive atom; on 2 October 1957, at time t, star A passed the cross-hair of a telescope at some specified observatory.

The first statement is of a historical nature; an accumulation of such facts as this would doubtless allow us to formulate laws, but this isolated fact, *inasmuch as it is isolated*, is of no interest to the physicist. Nevertheless, we notice—and this is rather curious—that even historical statements possess a certain predictive character, and hence in some sense, they too are directed towards the future. If, in fact, the statement is right, this means that every subsequent examination of the archives will confirm it. This "prediction" is not concerned with the fact itself, however.

The second statement has not been experimentally confirmed and falls among the "useful" statements; the concept of truth cannot be applied to it (see below).

The third statement, taken literally, is also of a historical nature, and from this point of view, holds no more interest for us than the first. In fact, however, physics has been able to establish laws in the astronomical field. Implicitly, therefore, the statement must be complemented thus: star A has passed . . . ; therefore it will pass . . .

This is the genuine true physical law; it has been verified experimentally several times and allows forecasts to be made. Such a law has two aspects therefore, one completed and the other potential. Explicitly or not, it contains the time-evolution variable (that is, the date), which must be carefully distinguished from the time-coordinate. This variable is made explicit in astronomy. When it is not, as in most physical laws, it is understood that the phenomena do not depend upon it. Thus the law governing falling bodies involves the time-coordinate but is independent of time-evolution.

We supplement these remarks below, bringing out the notion of a virtually true proposition.

(ii) *We cannot generate the truth by pure reasoning; it necessarily implies some check with the outside world* (in the broad sense, even if this world is regarded as purely phenomenal). The truth requires some sanction other than logic; truth arises from experiment and is concerned with experiment.

Every law is thus subject to this check. We cannot with certainty deduce a law from another which is known to be true because of (iii) and (iv) below.

(iii) The method of deduction (logic) is, in fact, itself under the jurisdiction of the outside world, for it comes from this outside world. In the introduction to his edition of Descartes, Sartre tells us: "Je puis (par le doute) mettre tous les existants entre parenthèses, vide et néant moi-même, je néantis tous ce qui existe.... Se désengluer de l'univers existant et le contempler de haut comme une pure succession de phantasmes."†

The mathematician or the metaphysician perhaps finds this attitude acceptable. The physicist believes that it is impossible for him to "extricate" himself, for if he could, physics would cease to be; furthermore, he suspects that for others such detachment is probably an illusion.

In the proposition "Nihil est in intellectu quod non fuerit prius in sensu, nisi ipse intellectus", which is dragged into all books on the "Guiding Principles of Knowledge", the physicist would delete the last phrase.

It does not seem that many types of logic have been used *in practice*, but in modern physics we meet indications. It is therefore vital that we remain unimpressed by logical taboos; we recall the lot of Euclidian geometry.

Illogical dreams show clearly that our minds can behave in contradiction to the logic of our waking state, but in a way which nevertheless seems perfectly natural during the dream.

(iv) *There is no absolute truth.* A law is true only in a certain approximation; but a true law is definitively established, with its conditions of applicability and of approximation. There are no revolutions in physics. These two features which seem contradictory at first sight must be carefully discriminated.

(v) Truth is operational in nature. To say that a physical law is true is meaningful only if it corresponds to a practicable operation. If the operation can be imagined without contradiction, but not carried out in practice, the law must be merely labelled useful (see below).

(vi) There are no isolated true propositions. The practical operation of verification entails measuring conventions which are so many postulates.

The proposition is also a member of a logical sequence (the theory) but from the point of view of truth, this is less important. In fact, it is the proposition that can be verified, and it can often fit within different sorites. The important thing is the combination of measuring conventions plus proposition to be verified.

* * *

† I can (through doubt) put everything in existence in brackets, myself empty and non-existent, I annihilate everything that exists.... Extricate oneself from the existing universe and contemplate it from above as a pure succession of phantasms.

Let us now return to the meaning of the word proposition, giving it added nuances and enriching it by specifying certain aspects of it. The following distinctions do not seem to have been made clearly; in my opinion, they are vital.

It often happens that a law has never been verified, but is very close to another which is accepted as true and is, above all, part of a very advanced physical theory. We shall say that it is *virtually true*. In the everyday practice of physics, such laws are *placed on the same footing as strictly true laws*.

The theory of diffraction at infinity from an aperture has been confirmed by innumerable experiments, and for apertures of many shapes. The technician requiring an aperture of a particular shape performs his theoretical calculations without any doubts about the correctness of the result, even if he has not yet obtained any experimental results.

We have spoken of *qualitative* and *quantitative propositions*. Are there such things as qualitative propositions in physics, however? Strictly speaking, it is doubtful. Two examples will illustrate this. The proposition, "This ruler lengthens", must be specified with a number, once the standard has been chosen; if it has not yet been chosen, the sentence is meaningless. The proposition, "This body is blue", corresponds to a particular value of wavelength, for the physicist, if the blue is monochromatic. In fact, qualitative propositions are in common use; their employment implies that the subject to which they are applied has not reached a highly developed state.

Some statements are purely formal, and are inadequate to determine an experimental procedure because the words or symbols are not operational. They may be called *empty propositions*. There is no lack of examples: laws of affine geometry before the selection of a metric; tensor equations of general relativity before the reference system has been chosen.

The concept of physical truth is not applicable to such propositions. It was in this sense that I wrote in *Kinematics* (p. 52, § [40]): "How can we speak of properties and laws before defining the measuring conventions?"

A proposition may be regarded as false in the literal sense, and yet be used in the physical sciences. This is the case of Coulomb's law for magnetic poles. It is now agreed that these poles do not exist (I return to this point in another preface in connection with the concept of existence), and this is mentioned when the law is stated. Sometimes the law is not necessarily false, but it cannot be directly verified: it is not operational; most hypotheses and laws fall into this category. We shall call these *useful* (or *convenient*) *propositions*. They are tools, like theories, and the concept of physical truth must not be applied to them. We can only say "everything happens as though . . .".

These propositions are in frequent use, but we must try to reduce the number of them. For in using such tools, our requirements of truth are in danger of becoming blunted, and above all, false problems are created. We shall encounter examples in connection with theories later (magnetic poles, mechanical aether in the theory of light).

* * *

Physical Theories and the Concept of Truth

A physical theory is an ensemble of propositions of various types (true, useful, ...), organized in an order which is determined by the logical process chosen so that each proposition follows from its predecessors.

Physical theories often contain breaks; these are supplementary postulates, explicit or implicit, which are inserted in the process of deduction. Furthermore, they are constantly introducing terms which can only be understood by recourse to observation or experiment.

Thus even in so far as it is a system of propositions, a physical theory is essentially different from mathematical sorites in these two features. It is wrong to say that physical theory is a mathematical sorites subsequently compared with experiment. Such a parallel only leads to often forced analogies; on the contrary, the differences must be firmly underlined.

Mathematical sorites may be compared with a construction which collapses if its foundations are undermined. Here, it would not matter: the upper storeys can remain in place while we alter the cellar. We must therefore look for another metaphor. If mathematical sorites is a vertical construction, physical theory is horizontal; it is a bungalow, in which we can dispose the rooms how we will. Let us now separate its properties.

* * *

A theory may be either good or bad but it is neither true nor false. In fact, theories are tools for classifying known propositions and discovering new ones. A tool may be good or bad: it is neither true nor false. A *tool* must be of *utility*, and thus convenient and productive. This last requirement is vital; what matters is not so much the interlinkage of the results obtained as the power of discovery. A theory never is established for reasons of convenience alone; it must also have proved its fruitfulness by means of its own predictions.

Different tools can sometimes be used for a given sequence of operations; likewise theories. It even happens that theories based upon contradictory hypotheses are useful simultaneously, contrary to the situation of propositions. Some tools are better than others, though, like machine tools. Good physical theories are the machine tools of the brain.

In the history of techniques, the invention of a good tool is more important than the results obtained with it; nevertheless, what the tool has created often survives when the tool is replaced by a better one. The same is true of propositions and theories. Only the former remain while the latter change. The great names in the history of physics are, however, associated with the creation of theories (Kepler, Newton, Fresnel, Einstein, ...).

By way of example—and because one must accept one's responsibilities towards controversial subjects—the law concerning the Langevin traveller possesses, to my mind, all the characteristics of physical truth. It would remain, even if one day the theory of relativity should give way to a more satisfactory theory (and of course I admit the possibility of this).

* * *

One could say that a theory is false if it contains false statements. It is, however, always possible to modify it in such a way as to eliminate the false statements while retaining essential hypotheses; the history of physics proves this. If we accepted this attitude, we should be compelled to say: a theory is false if it contains at least one false statement and true if all its statements are true. The following two consequences would then ensue. Different theories embracing the same true statements would be simultaneously true and, furthermore, what is true today would be false tomorrow. This is certainly the point of view that numerous philosophers of science adopt. I find it preferable, and closer to the intuitive concept of truth, to adopt a different attitude.

* * *

A physical theory is a good one only up to a certain point. In mathematical sorites, the sequence of deductions is true up to infinity, if the starting-point is accepted. In physics, even logically deduced propositions may be physically false beyond a certain proposition, P. Mathematical deduction is a tool of the physicist, but never a definitive criterion of truth.

We have said that a theory contains true propositions and useful propositions. Two theories may be acceptable up to propositions P and Q respectively if all their true propositions are common and if they contain no false propositions. Moreover, they may contain different empty or useful propositions, or even contradictory ones.

* * *

Let us specify *the characteristics that allow us to choose a good theory from several acceptable ones*.

The principal characteristic is fruitfulness, and this is always the definitive criterion. At the moment of discovery, however, physicists are always influenced more or less consciously by other factors.

A good theory must be based upon few postulates; above all, as many as possible of the secondary postulates (rules of thumb) introduced in the course of development must be eliminated. This search for logical unity has played an important role during the decisive periods of physics.

Let us consider the theory of light as an example. The fruitfulness condition came into play when Planck and Einstein advanced a corpuscular theory at a time when the wave theory seemed definitively established. This new theory was manifestly incapable of interpreting interference and diffraction phenomena. For some twenty years, physicists accustomed themselves to using either the wave theory or the corpuscular theory according to the phenomena involved. In building this state of affairs into a system, some considered that the existence of two theories was not an obstacle and that there was no point in attempting to unify them. Yet it is to precisely this attempt that we owe de Broglie wave mechanics, which has proved to be so remarkably fruitful.

There is a third characteristic of good theories, which is made much less of. Ideally, a theory should contain none but true propositions. Perhaps this is impossible, but we should endeavour to reduce the number of empty or convenient propositions.

The magnetic-pole hypothesis enables us to erect a fruitful and logical theory of magnetic phenomena. Even so, it is now better to avoid this hypothesis, and to erect the theory on electric charges and relativistic dynamics. We thus eliminate a superfluous concept; furthermore, only Coulomb's law, which is operationally true, is used—the corresponding magnetic law is only convenient. In this way, the analogies between magnetism and electricity are revealed at their true worth; in particular, we bring out the differences, which were all too often passed over in silence in the old theory when they disturbed the parallelism.

Another important example is the removal of the hypothesis of a mechanical aether, to which many unverifiable properties had to be attributed. I showed in *Kinematics* that in their most advanced stage, the pre-relativistic theories asserted the existence of an absolute system of reference (the aether) and then in principle posited the impossibility of revealing it experimentally.

The eradication of useful propositions and of fictitious elements, so far as this is possible, enables us to detect, and hence eliminate, pseudo-problems (for instance, nugatory questions about the structure and motion of the aether). This is an important step forward.

There are, too, false criteria. The "falseness" of a physical theory about certain facts is not demonstrated by devising another correct theory which interprets the same facts. This would be wasted effort: we know that an infinite number of logically correct theories could be devised.

* * *

Is the correspondence between the theories and the outside world a reason for speaking of the truths of the theories?

When we contemplate a theory in a sufficiently advanced state, we cannot but be struck by the correspondence between our theoretical forms and the outside world, or its reactions at least.

This correspondence must not be overestimated, however. Our attention is drawn to the predictions which are successful, like Le Verrier's famous discovery. The predictions which are unsuccessful are infinitely more numerous; at a given instant in science, however, they are put to one side (this does not mean that they are put out of sight, but merely that we cannot do better for the time being). They really come on to the stage when they fit into a new theory.

Consider the advance of the perihelion of Mercury and the advance of the Moon, for example. These phenomena were treated as ancillary corrections in the Newtonian theory but in later theories they became fundamental.

Our theories are constantly being touched up and sometimes revised from top to bottom; their propositions are valuable only after experimental confirmation. The sorites that the physicist has created is built under the surveillance of the outside world which provides the materials and repeatedly checks the robustness of the construction.

It must therefore be clearly understood that there is no relationship between two ensembles initially independent: one created by the brain (the propositions of a theory) and the other consisting of the phenomena and objects of the outside world. The only reasonable question is as follows: how is it that our brains manage to devise theories, that is, logical structures, which are in agreement with the reactions of the outside world? Or again, why is it that our ordinary logic can be used for our contacts with the outside world?

The physicist thinks that this is not a real problem. Logic (like theories) is created in the human brain under pressure from the outside world, *as one of the conditions of our survival*. We do not ask why a hammer knocks in nails: it was made so that it would. We are not surprised that a river flows along its bed; if we considered river and bed separately beforehand, we might be surprised.

Some schools of philosophy imagine, on the contrary, that the human intellect with its working rules is an entity independent of experiment. The problem of the relationship is then a real one, and very extensive metaphysical systems have been constructed: harmony established beforehand between the two orders (Leibnitz), partial or total suppression of the outside world (idealism) and so on.

The physicist is certainly used to formulating hypotheses of curious appearance; but when he is compelled to do so, he proceeds step by step and he constantly confronts the consequences with experiment. He gapes in amazement at the metaphysicians who devote their lives to the erection of vast theoretical systems upon most unreasonable hypotheses without batting an eyelid.†

As I said in *Kinematics*, in perfect seriousness and not just for effect, he looks on metaphysicians with much the same eye as he looks on poets or surrealist painters. Often he finds the beauty or originality of these designs seductive, but their positive and often aggressive dogmatism provokes in him a mild levity.

I have read somewhere that Buddha already regarded metaphysics as "a footpath of opinions, a thicket of opinions, a jungle of opinions, . . .". We can go further. The natural sciences and the social sciences, a rough school of modesty, have taught us that our intellect is, if left to itself, totally powerless before the simplest problems. Nevertheless, there are still many metaphysicians obstinately trying to find a general interpretation of the Universe by this route alone. Their preoccupations and assertions are worth no more than those of astrologers. The modern profusion of astrologers, healers and metaphysicians (of the kind I am talking about) is a disquieting sign of a collective intellectual infantilism.

To all of you, professors and doctors, political panjandrums and academic pontiffs, proud and smug with your ideological hobby-horses, one single piece of advice: reflect upon chapters XVIII, XIX and XX of Rabelais' *Pantagruel*. The sense of the ridiculous can still save us from tyrannies.

Let us understand one another clearly. *A priori*, I do not deny the utility of research in these topics. *A priori*, I do not even deny that the position of the planet Mars may have

† HORATIO: O day and night, but this is wondrous strange!
HAMLET: And therefore as a stranger, give it welcome.
There are more things in heaven and earth, Horatio,
Than are dreamt of in your philosophy.

One might say, too, that there are more things in the metaphysical systems than in the Universe. The physicist also could certainly develop theories of pure physics, having no relation to reality—for example, he could establish an astronomy with a law of attraction in $1/r^3$. It seems to him more useful to deal with the real universe, however.

some influence on the potato crop, the Stock Exchange or the election of an F.R.S. Improbable it may be, but all *a priori* negation is to be eschewed. I merely say that the methods employed, which were acceptable and even useful 300 years ago (Kepler was a practising astrologer, and according to the historians made the better part of his living out of it), have become *ridiculous*, given our present-day knowledge. They can be blamed upon too strict segregation of the various branches of research, and hence upon ignorance.

* * *

An idea of Duhem is connected with this problem of the relationship between theories and the outside world: the theories are considered as natural classifications. "Plus elle [la théorie physique] se perfectionne, plus nous pressentons que l'ordre dans lequel elle range les lois expérimentales est le reflet d'un ordre ontologique; ... le physicien ... affirme sa foi en un ordre réel dont ses théories sont une image, de jour en jour plus claire et plus fidèle."†

In the issue of *Recherches Internationales* (p. 223) that I have already quoted, I find a similar idea from the pen of Alexandrov (a dialectic materialist): "theory ... *reflects the objective connection of nature* in the logical connection of its concepts and its results" (my italics).

A seductive but debatable point of view. Our theories are certainly not an artificial system, as Duhem says. To base them upon an ontological order is to attribute to them an absolute character, however, and thus to stray well outside the domain of the physicist. In postulating the existence of such an order, a metaphysical choice has already been made. However, Duhem himself writes "en mettant la physique théorique sous la dependance d'une métaphysique, on ne contribue point à lui assurer le bénéfice du consentement universel".‡

I am convinced that this appeal to the absolute should be completely extirpated and that theories should be considered in the same way as standards. We saw in *Kinematics* that we can use any standard we choose to; all the same, some, which are called natural standards, are of overwhelming practical value. Similarly, among all the theories current at a given period of physics, there is one good theory, one *natural* theory of *purely practical* pre-eminence; without straying beyond the proper sphere of the physicist, we cannot go further.

The concepts of natural standard and natural theory form a key idea of relativistic physics; they ought to be extended to the whole of physics.

* * *

† The more it [the physical theory] is perfected, the more we anticipate that the order in which the experimental laws are organized reflects an ontological order; ... the physicist affirms his faith in a real order of which his theories are a representation which daily becomes clearer and more faithful.

‡ In making theoretical physics dependent upon metaphysics, we are doing nothing to confer the benefit of universal consent upon it.

Truth and Beauty

It is undeniable that great theories provoke truly aesthetic feelings in the majority of physicists. Fully to experience them, some training is necessary—sometimes, one must even be a specialist. But is not the same often true of the appreciation of music, for example? Nevertheless, these feelings can be experienced by everyone to a greater or lesser extent, as all readers of the better kinds of *vulgarisation* know full well.

People have wondered whether there is a connection between the beauty of a theory and its truth. We must reframe the question, since we have eliminated the idea of truth from theories. Is there a connection between the beauty of a theory and its usefulness? The reply varies from person to person.

The theoreticians are often guided by the aesthetic appearance of their relationships. This is a powerful stimulus in research, a kind of mystique engendering enthusiasm. Why not employ it? Although the useful aspect of theories is the only solid one, why should we not season them with illusions, if these illusions are agreeable and fruitful and *if we know how to restrict them within the domain of illusions*. There can, of course, be no question of choosing between two theories on aesthetic grounds; sometimes, we are obliged to make the opposite choice.

The Greeks believed that the heavenly bodies travel along circular paths (or rather along epicycloids, which are derived from circles), because in their eyes, the circle was the most perfect curve. This preconceived idea, of an aesthetic nature, was an excellent guide in the earliest research; later, however, it became ossified and an obstacle to progress. When Kepler adopted elliptic trajectories, he did more than replace one curve by another: the whole method was profoundly altered.

We recall the words of Planck, probably irritated by those who regretted the trouble that the quantum had brought upon the wave theory: elegance is the tailor's business. In fact, what at first contact shocks our aesthetic sensibilities sometimes becomes agreeable with use. New canons of beauty are created, in physics as elsewhere.

It is a strange thing that *although what is beautiful may often be useful in physics, beauty and truth never go together*. Theories can be beautiful but are never true. Propositions can be true, but possess no aesthetic qualities. Newton's law, for example, is not beautiful in itself; if the attraction were inversely proportional to distance cubed, our aesthetic sense would not be offended. On the other hand, the theory that is deduced from this proposition moves us by its harmony and breadth.

* * *

Connected with the theories are the meanings of the words "understand", "explain". "Comprehensible" is synonymous with "familiar", "habitual". To explain is to refer back to things already understood.

When, as a schoolboy, I read for the first time some of Rimbaud's poems or Nietzsche's *Zarathoustra*, I was acutely sensitive to the poignant beauty of these writings. My intelligence, however, was not involved at all, and when some of my schoolfriends asked what this or that phrase meant, I was unable to reply. I did not understand these lines at all, and I realized this (What are you reading it for, then? You're batty!). Today, the "Bateau ivre", "Voyelles" or "Das Nachtlied" are no longer shadowy in places, and this illusion of clarity is none other than long familiarity.

Newton's hypothesis concerning action at a distance and, more recently, the hypotheses of relativity have been labelled incomprehensible. Nowadays, however, the first is regarded as a model of clarity, and we are getting accustomed to the second; it is odd that some people who still refuse to accept the latter use Newton's laws as their criterion of lucidity. I can say only this to them: that it is a question of habit. You will come round in time, always provided your mental inertia is not too great.

In short, one does not understand a concept in physics, one gets used to it. One does not explain a new phenomenon, one relates it to known phenomena and one establishes the laws which enable it to be reproduced. One does not state why, but only how.

If we say that the universe is comprehensible and intelligible, we merely mean that we are capable of stating true laws about it; in the final analysis, therefore, this is the same as *asserting* our qualities of adaptation and survival. It is a pseudo-problem to wonder why the universe is comprehensible.

Other meanings have been ascribed to this intelligibility. For example, an intelligible world, for some metaphysicians, equals a world created by an intelligence for a particular end; the harmony and regularity of the laws would be a consequence of this higher intelligence, a proof of its existence. The "regularity" and "harmony" certainly lend themselves to magnificent literary efforts. Behind these words there is, however, only an affirmation that we are able to state laws which enable us to survive. You can regard your existence as a proof of the existence of God if you wish; this is no business of the physicist.

Intelligibility has often been coupled with the possibility of a mechanical model; however, this is simply to reduce complex phenomena to other simpler ones.

* * *

Epistemological Laws and the Concept of Truth

This problem takes me beyond the limits I had set myself; let me merely draw attention to it and *blaze a few trees*.

Some epistemological laws are nearly true in the same way as physical laws. For example: given an ensemble of phenomena, they can be interpreted by means of several theories. The experimental "verification" really consists in setting out these theories.

There are also useful epistemological laws and theories, from which the notion of truth is absent. For example, it seems desirable to slant the physical theories towards the operational point of view, and towards the removal of fictitious entities. We are sometimes influenced in this by practical reasons alone (a better yield of true results). Since the physical theories cannot be called true or false, however, nor *a fortiori* can an epistemological doctrine destined to guide us in our choice of theory.

THE POINT OF VIEW ADOPTED; ADVICE TO THE READER ON HOW BEST TO USE THE BOOK

[1] Mechanics as the Fundamental Science; the Essentially Relativistic Structure of Electromagnetism

Mechanics was long regarded as the fundamental science, to which all other branches of physics should be reducible. This attitude brought signal successes, with the optical theories of the aether, for example, and the development of kinetic theory. Towards the end of the nineteenth century, however, the theories which were proposed in order to interpret the variation of mass with velocity as revealed by cathode rays, undermined this purely mechanistic view of the universe. The laws of electromagnetism gradually acquired more importance, and finally became fundamental. At the climax of this evolution, the theoreticians of relativity insisted that mechanics must be reduced to electricity, and this point of view is still quite widely held. Numerous authors admittedly develop dynamics autonomously, as we shall see at the end of Chapter I, but in the general treatises, electromagnetism is also developed autonomously, with Maxwell's equations postulated as a starting-point.

In this volume, I return to the older standpoint: this was formerly inadequate when everything had to be reduced to *Newtonian* mechanics, but becomes viable again with *relativistic* dynamics. All things considered, and after organizing the material from several points of view, I find it more correct (or more "convenient") epistemologically and more meaningful physically to start from very simple mechanical postulates and deduce electromagnetism from them, than to state Maxwell's equations which are much more involved and contain concepts that can be reduced to simpler ones. *I deduce the whole of electromagnetism by applying relativistic mechanics to the laws of electrostatics*; this brings out the essentially relativistic structure of electromagnetism.

It is often said that high velocities or extremely accurate measurements are necessary for relativistic effects to be observable. Now, on the contrary, because of special circumstances to which we shall return later, relativity will have an appreciable effect even at low velocities (first order in β). The familiar laws of electromagnetism are the best possible confirmation of relativity, as Langevin, I think, has already remarked, but this has not been sufficiently emphasized.

The most severe test of relativity lies in the interaction between moving electric charges, and in the departures from Coulomb's law and Newtonian mechanics that the "magnetic" terms represent.

The enormous forces which act in dynamos and motors are relativistic effects. It is curious that these relativistic (or, as we say, magnetic) terms are very small in comparison with the electrostatic terms (at 1 cm s^{-1}, the ratio is of the order of 10^{-21}) but that the latter are exactly cancelled out in this case: a most remarkable phenomenon. Oersted was therefore the first to make observations on relativistic dynamics. *The whole of electricity thus becomes a chapter of dynamics*: it is the relativistic continuation of the chapter which, in classical mechanics, is devoted to the *Newtonian field*.

These may seem weird, exaggerated statements. Their import will be better understood after Chapter X has been digested. It is not difficult to see from a specific example that the application of Newtonian dynamics to the interaction between charges leads to a self-contradiction, however. For, the basic equation of this dynamics implies that force and acceleration are invariant under any change of Galileian system. Consider two charges at rest in some system; each experiences the Coulomb force but this is not true (owing to the magnetic terms) if we refer the charges to another Galileian system, and this contradicts the invariance properties that we have just mentioned. *It is therefore a logical heresy to continue restricting electromagnetism within the Newtonian framework,*† *and is a historical survival resulting solely from our mental inertia.* Furthermore, something more than a mere tidying up is involved. Some problems, unipolar induction for example, have no satisfactory solution other than in relativistic terms; for the most part, they are passed by in silence, or analysed with the aid of special supplementary hypotheses. How many people, even if highly qualified in electricity, would know how to approach the following very simple problem—let alone perform the calculations? A magnetized sphere (of iron, for example) rotates about its axis with a constant angular velocity; determine the electric and magnetic fields, the polarizations, and the actual charge distribution.

Nevertheless, the phenomena involved are basic. Research workers, both experimentalists and theoreticians, will find that this is a branch of physics in which much is still to be done. *Some questions are almost untouched*; the electrostatics of rotating bodies, for example, hardly exists. From this point of view, magnetic poles have, of course, no part to play: I do not deny the undoubted usefulness of this concept, but it needs to be restricted to its proper place, which is as a tool in calculation.

Some authors arrange the electric and magnetic quantities in a perfectly symmetrical fashion, and introduce magnetic currents and so on. The method that I recommend keeps closer to physical reality. In the macroscopic theory, with our present ideas, magnetic poles and fluids stand in much the same relation to electric charges as does the old caloric to inertial masses in the study of heat.

Nevertheless, I make no predictions about the future of the two concepts (electric charge and magnetic pole) in the microscopic sphere. On this scale, it is difficult (though I do not say impossible) to consider spin as an effect arising from the motion of electric charges. Does this mean a return to magnetic poles? I am not qualified to offer a balanced opinion. I simply *feel* that the fact that a corpuscle is charged or possesses a spin are attributes of it, just as its inertial mass or dimensions are, and that in consequence, electric charges may perhaps also lose their irreducible nature in favour of some higher entity.

However this may be, the advantage of the attitude adopted in the present work is that the magnetic concepts are purged of their parasitic aspects. My point of view, if systematically adopted, involves rethinking the whole theory. This has some unforeseen consequences, which clash with certain currently fashionable ideas. For example, *the Giorgi or mks system of units*, which is widely regarded as a criterion of modernity, *appears artificial*

† The expression "epistemological heresy" would perhaps be more fitting if it were not rather too daring. In pure logic, two apparently contradictory physical theories can always be reconciled: in the present case, these are Newtonian mechanics and the interactions between electric charges. Supplementary concepts or hypotheses have merely to be introduced, and here we should posit a material aether, possessing suitable hydrodynamic properties.

and outmoded and I conclude that it ought to be rejected. These ideas which appeared in the first edition of the present book (1957) are developed at length in another book, in which the electromagnetic theory is wholly rebuilt relativistically (*Electricité macroscopique et relativiste*, Gauthier-Villars, 1963). The publication of this book has enabled me to condense the present English edition, by removing various sections which are better placed in a treatise on electricity. Here, I indicate only the point of departure and the spirit of the method, and show why I have reversed the usual procedure and taken relativistic dynamics as the fundamental science.

The reader had to be forewarned, especially the reader familiar with electromagnetic theory. The development in this book will surprise him. The majority of the questions that he had thought to be purely electromagnetic—the decomposition of a force into a magnetic and an electric force, the transformation of fields, and the idea of vector potential—are in fact all purely mechanical concepts. *Even Maxwell's equations are equations of mechanics*; returned to their proper place, they are seen in quite a new light, and their import is much more exactly determined.

This reconstruction has given me a great deal of pleasure, as I watched the mechanical part grow and the electromagnetic part dwindle. The reader who is prepared to accord me a few days' work will make no complaint. The alterations that he will have to make to some of his habits will be more than compensated by the new features that he will encounter in the landscape, however familiar.

He will experience a sense of mental relief, a spiritual repose akin to that of the archaeologist who, after painful exploration of the various sections of some dead city deep in virgin forest, is finally able to fly over the site and sees the general plan revealed (Angkor is a particularly striking instance of this). A simple guiding idea then coordinates the sparse facts, the fragmentary theories, and provides a welcome alleviation for the brain; furthermore, neglected problems are brought into the open, and avenues for exploration open up before the investigator.

[2] The Use of the Principle of Relativity; the Importance of Changes of System of Reference

No one today can deny the utility of the formulae of relativistic dynamics. Not a few still strive to dissociate them from relativity, however.

It is certainly possible to calculate in a single Galileian system of reference and to regard these formulae as consequences of an impulse postulate which renders the older dynamics more general, without having recourse to relativistic ideas: the time transformation and so on. My first chapter, and the applications to numerous problems of motion, can be read directly without any foreknowledge of relativistic kinematics.

This would only reveal one side of the question, however, and I should even say, a minor aspect. Only by exploiting to the full the relativistic procedures, by systematic use of changes of system, do the more important consequences become apparent. *The whole of electromagnetism rests on the transformation of force.* Fields described by scalar potentials thus lead to the general case of scalar and vector potentials. These ideas ought to find their place in the teaching of electricity, at least at university level, for the origin of the magnetic concepts appears here clearly. It cannot be objected that relativity involves too difficult calculations. Just as the philosopher Diogenes (with whom I feel myself ever more in sympathy) demonstrated motion by walking, I offer to teachers and students two texts, one on relativistic dynamics and the other on relativistic electricity, which can be understood at every level. There is a tendency to put relativity at the end of courses on electromagnetism, as a final crowning peak, but I believe this to be outmoded.

[3] The General Plan of the Work

In this volume, I examine the general laws of the dynamics of a point of constant proper mass (first part) and variable proper mass (second part) and the theory of collisions (third part).

The postulates that are adopted here cease to be valid in some situations, which are discussed in § [6]: very large or very rapidly varying accelerations, for example. This topic is examined in another work: *Rayonnement et dynamique du point chargé*, Gauthier-Villars, 1967.

[4] Advice to the Reader on the Use of the Book

To avoid the reproach made by Montaigne,† I have arranged that the text can be studied or consulted by as wide a range of readers as possible. I have thus included *three different levels*, which can be separated in reading (I might have separated them in fact by adopting a purely pedagogical plan, but this would have been logically inconvenient, and would have separated related topics in an arbitrary fashion). The beginner should omit Chapters III, VI, VII and VIII so as to study the application of relativistic dynamics to concrete cases—interactions between charges—as soon as possible. The calculations are elementary and comprehensible to every student; the physical meaning of the results is made sedulously clear.

On the second level, variational methods will be studied: Chapters VII and VIII. Finally, the reader will be introduced to the use of the tensor calculus in four dimensions, which is indispensable for a thorough training in relativity.

† "La difficulté est une monnoye que les sçavants employent, comme les joueurs de passe-passe, pour ne découvrir la vanité de leur art, . . . " (*Essais*, II, 12). [Difficulty is a currency that scholars employ, like sleight-of-hand, to conceal the vanity of their art . . .] Let me not be accused of want of modesty in classing myself among the "scholars". Today, scholars clutter the streets. Do we not see in the newspapers that hundreds, even thousands of scholars frequently assemble at specialized scientific congresses?

NOTATION

THE numbers indicate the sections in which the corresponding quantities are defined. No ambiguity can arise when the same letter is used for different quantities.

Roman alphabet

A	vector potential	66
$\mathfrak{B}$	component occurring in the relativistic expression for force	43
	magnetic field	119, 136
c	relativistic constant (velocity of light)	
c_0	constant occurring in the system of units	43
curl	vector operator curl	
$\mathfrak{D}$	electric displacement	91, 137
div	divergence operator	
E	component occurring in the relativistic expression for force	43
	electric field	119, 136
E	kinetic energy	14
F	force	5, 157, 174
h	Planck's constant	
$\hbar$	$h/2\pi$	
H	$(=W)$ energy	76
H	magnetic induction	91, 137
$\mathcal{H}$	$(=W+\varepsilon V)$ Hamiltonian	77, 81
i	$\sqrt{-1}$	
	electric current	118
$\mathfrak{J}_m$	magnetic polarization	137
$\mathfrak{J}_e$	electric polarization	137
j	current density	91
K	arbitrary Galileian reference system	
K_0	proper reference system	
	constant occurring in the system of units	117

K^i	$(i = 1, 2, 3, 4)$ four-force	36, 168
L	Lagrangian per unit volume	90
$\mathcal{L}$	Lagrangian or Hamiltonian action	90
$\mathcal{L}_p$	value of $\mathcal{L}$ for a free particle	73
log	decimal logarithm (to the base 10)	
ln	Naperian logarithm (to the base e)	
m	mass	5
m_0	rest mass	5
$\mathcal{M}^{ij}$	four-induction tensor, with components $\mathfrak{D}$ and $\mathbf{H}$	100
$\mathcal{N}^{ij}$	four-field tensor, with components $\mathbf{E}$ and $\mathfrak{B}$	49
$\mathcal{N}^*_{ij}$	the adjoint or dual tensor of $\mathcal{N}^{ij}$	101
0	(zero) this subscript is used to identify proper values	
$\mathbf{p}$	$(= m\mathbf{v})$ momentum	5
p	pressure	
$\mathfrak{P}$	$\left(\mathbf{p}+\frac{\varepsilon}{c_0}\mathbf{A}\right)$ variational momentum	77, 82
P^i	four-current vector, with components $\mathbf{j}$ and $ic\varrho/c_0$	100
Q	electric charge	
	quantity of heat	185
	Q-coefficient	197
S	Jacobi function	79
	entropy	185
t	the time variable	
T	temperature	185
	proper time	
T^{ij}	momentum-energy tensor	188A
$\mathcal{T}$	work	
$\mathbf{u}$	velocity	
U	total energy of a thermodynamic system	185
$\mathbf{v}$	(components v^α, $\alpha = 1, 2, 3$) velocity	34
V^i	$(i = 1, 2, 3, 4)$ four-velocity	34
V	scalar potential	57
$\mathcal{V}$	volume	
$\mathbf{w}$	velocity	
W	$(= mc^2)$ energy	13

Greek alphabet

β	$(= v/c)$ relativistic ratio	
γ	acceleration	
Γ^i	four-acceleration	
ε	constant	43
	electric charge	117
	conversion coefficient	192
λ	wavelength	

μ_0	constant occurring in the system of units	118
	constant	196
ν	frequency	
ϱ	charge density	91
	mass density	185A
σ	surface density	
	surface	
Π^i	four-momentum vector	34
Φ^i	four-potential	66
$\bar{\omega}$	volume	

PART ONE

The Dynamics of Particles with Constant Proper Mass

CHAPTER I

FUNDAMENTAL EQUATIONS

A. THE IMPULSE POSTULATE AND ITS CONSEQUENCES

[5] The Postulate

(a) Remaining faithful to the method advanced in my *Kinematics*, an exact mathematical postulate is stated *ex abrupto*, upon which relativistic point dynamics can be erected. We shall subsequently examine the various arguments which allow us to relate this postulate to other statements, which were formerly held or believed to be clearer, wrongly in my opinion.

The fundamental postulate of pre-relativistic particle dynamics is commonly stated in the form

$$\mathbf{F} = m_0 \boldsymbol{\gamma} \tag{1}$$

The acceleration $\boldsymbol{\gamma}$ of a particle is proportional to the applied force $\mathbf{F}$; the constant of proportionality, m_0, is a positive arithmetic constant, characteristic of the particle and known as its mass. The impulse theorem gives a mathematically equivalent relation:

$$\frac{\mathrm{d}}{\mathrm{d}t}(m_0 \mathbf{v}) = \mathbf{F} \quad \text{or} \quad \mathrm{d}\mathbf{p} = \mathbf{F}\,\mathrm{d}t \tag{2}$$

in which we have written

$$\mathbf{p} = m_0 \mathbf{v}$$

and $\mathbf{v}$ is the velocity of the particle.

The impulse of a force $\mathbf{F}$ during time $\mathrm{d}t$ (that is, $\mathbf{F}\,\mathrm{d}t$) increases the momentum of the particle by $\mathrm{d}\mathbf{p}$.

It is wise to recapitulate the pre-relativistic sense of the words employed: I urge the reader to consult Chapter IX immediately, where a detailed analysis of the notions of force and mass is to be found.

(b) In relativistic dynamics, we take not only the principle of relativity but also the following relation as fundamental postulates:

$$\mathrm{d}\mathbf{p} = \mathbf{F}\,\mathrm{d}t \tag{3}$$

The impulse retains its pre-relativistic definition and the momentum is defined by the relation

$$\mathbf{p} = \frac{m_0 \mathbf{v}}{\sqrt{1-v^2/c^2}} = \frac{m_0 \mathbf{v}}{\sqrt{1-\beta^2}} \tag{4}$$

in which m_0 is the *pre-relativistic or rest mass*, defined in (a); we shall call it the *proper mass*. The *relativistic mass* is by definition the quantity

$$m = \frac{m_0}{\sqrt{1-\beta^2}}, \quad \text{so that} \quad \mathbf{p} = m\mathbf{v} \tag{5}$$

In pre-relativistic dynamics, relations (1) and (2) are equivalent. We shall see later that this is not true in relativity; relation (1) is *not true*, even if m_0 is replaced by m.

The definitions of force and mass for a moving point, and hence the form of the law of motion, are largely arbitrary (Chapter IX). It is only necessary that with respect to a system in which the point is at rest, the pre-relativistic equation is obtained and the force is the same as the force defined in statics (which can be measured with a dynamometer at rest).

There are many physically equivalent ways (identical experimental statements) of fulfilling this condition; I have chosen the one that is usually adopted and which gives the simplest and most convenient theory (in the sense of §§ [5] and [33] of my *Kinematics*). The reader is referred to Chapter IX for a detailed analysis of these questions, but it is as well to be warned from the outset, for the various notations must be compared only with much circumspection.

Remark. In this first Part, m_0 is a constant and can be taken outside the differentiation; when we write

$$\frac{\mathrm{d}}{\mathrm{d}t}\left(\frac{m_0 \mathbf{v}}{\sqrt{1-\beta^2}}\right) = \mathbf{F}$$

the law seems also to be applicable to cases in which the proper mass m_0 varies with time (in rockets, for example). Many authors (including myself, in the first edition) have indeed believed this, but a more thorough-going scrutiny shows that it is an unjustifiable extrapolation. I shall return to this in the following Part.

(c) *The origin of the postulates (3) and (4).* The dynamics which is deduced from the pre-relativistic postulate and is usually known as Newtonian dynamics fails at velocities which are not small in comparison with c.

Originally, this was a result of experimental work on motion at high velocities (the motion of electrons *in vacuo*, for example). There is, moreover, a theoretical reason for this inadequacy: the Newtonian equation is not relativistically invariant. We have already noticed that acceleration is not an invariant quantity (*Kinematics*, § [85]).

The postulate upon which we are basing the new dynamics was of course derived by a sequence of inductions and successive modifications, which cannot be retraced in a didactic work (see the Historical Notes). Nevertheless, I hope that the reader will find it reasonable enough if he bears the following three prerequisites in mind:

— for small values of β, the postulate must lead to Newtonian dynamics;
— it is suggested by experiments on the deflection of electrons by a constant force, normal to the initial velocity; the measurements show that, in this special case, the only correction to be made to Newtonian dynamics is the replacement of m_0 by m;
— it must yield an equation possessing relativistic invariance, and this is indeed quite fundamental.

In Chapter II we shall find a logically more satisfactory point of view: *the relativistic law is a consequence of the pre-relativistic law and the principle of relativity*. If the reader is not rebuffed by four-dimensional calculus, I advise him to read § [38] immediately.

[6] A Preliminary Survey of the Sphere of Validity of the Dynamics Deduced from the Preceding Postulate

The following remarks apply *a fortiori* to Newtonian particle dynamics.

(a) *The point corpuscle*

The present work deals with the point corpuscle. We shall not consider systems of interacting points, the relativistic theory of which is in its infancy.

As an exception, the equations may also be used to study the interaction between two particles.

(b) *The corpuscle free of spin*

To interpret the experimental results, physicists have been obliged to attribute an intrinsic angular momentum or spin to certain particles, the electron for example. We shall not consider this concept at all; the formulae thus cease to be valid when phenomena arising from spin are not negligible. The macroscopic dynamics of particles with spin is postponed to another work.

(c) *The slowly accelerated corpuscle; low energies*

A field of force is given, determined beforehand by static measurements; the postulate of dynamics then allows us to calculate the motion of a particle placed in this field. This is the standpoint of Newtonian dynamics and we have only to generalize the postulate.

For a point charge, for example, the electromagnetic field is given. The forces are calculated from the known formulae, and substituted into the new equation of motion.

This, however, is a special case, as our heading suggests. More generally, the moving particle acts as a source of radiant energy, and to the ordinary forces must be added supplementary terms, known as the radiation reaction. For these terms to be negligible, the acceleration must be small, and so must its variation with time. This is a virtually unexplored domain, the *terra incognita* of macroscopic mechanics. The theories which have been suggested remain within the usual relativistic framework, but we must not be shocked if the facts oblige us to generalize the Lorentz transformation and to modify the meaning of the constant c.

The weak acceleration approximation is adequate for many topics: electron optics and the elementary theory of particle accelerators, for example (the word weak should not be misunderstood: "weak" accelerations may be enormous in comparison with everyday accelerations). The approximation is, on the other hand, not adequate for the full theory of accelerators such as the bevatron. Analogous problems are associated with gravitational radiation.

These problems are treated in my *Rayonnement et dynamique* mentioned above.

(d) *The macroscopic nature of the theory*

The postulate naturally includes as a special case pre-relativistic dynamics, which thus forms one of its spheres of validity.

However, relativistic divergences properly speaking occur mostly for "microscopic" particles (electrons, ...), and we know that the motion of these usually requires the use of quantum mechanics; the classical concepts of trajectory and velocity are abandoned. This is the case in the study of nuclei, atoms and molecules. The postulate that we shall develop employs classical concepts, and yields a purely macroscopic dynamics.

It is therefore imperative to state the range of validity and in particular to examine what we mean by a trajectory. The general criterion is as follows: the macroscopic concepts can be employed when the action (in the Hamiltonian sense, see Chapter VI) during the phenomenon is large in comparison with Planck's constant; or, alternatively, when the dimensions of the region through which the particle moves are large in comparison with the de Broglie wavelength.

(e) *Possible changes of nature of the particles studied*

In most problems of motion, the corpuscles are assumed to undergo no change of nature: an electron remains an electron. The rest energy m_0c^2 (§ [13]) does not intervene, therefore, although it appears in the formulae. This was automatically the case in the old dynamics, but there are today many phenomena of annihilation (two electrons of opposite signs vanish and produce a photon), of creation, or of capture. These phenomena in which the equivalence of matter and energy is involved are studied in Part Three.

[7] The Equations of Motion in Cartesian Coordinates

Let $Oxyz$ be a set of axes, with respect to which we measure the motion of the particle. The Cartesian components of $\mathbf{p}$ are of the form

$$p_x = \frac{m_0}{\sqrt{1-v^2/c^2}} \frac{\mathrm{d}x}{\mathrm{d}t}, \ldots$$

The vector impulse postulate gives

$$X = \frac{\mathrm{d}p_x}{\mathrm{d}t} = \frac{\mathrm{d}}{\mathrm{d}t}\frac{m_0\dfrac{\mathrm{d}x}{\mathrm{d}t}}{\sqrt{1-\dfrac{1}{c^2}\left[\left(\dfrac{\mathrm{d}x}{\mathrm{d}t}\right)^2+\left(\dfrac{\mathrm{d}y}{\mathrm{d}t}\right)^2+\left(\dfrac{\mathrm{d}z}{\mathrm{d}t}\right)^2\right]}}$$

$$= \frac{m_0}{1-v^2/c^2}\left\{\frac{\mathrm{d}^2x}{\mathrm{d}t^2}\sqrt{1-v^2/c^2} + \frac{\mathrm{d}x}{\mathrm{d}t}\frac{1}{c^2\sqrt{1-v^2/c^2}}\left[\frac{\mathrm{d}x}{\mathrm{d}t}\frac{\mathrm{d}^2x}{\mathrm{d}t^2}+\frac{\mathrm{d}y}{\mathrm{d}t}\frac{\mathrm{d}^2y}{\mathrm{d}t^2}+\frac{\mathrm{d}z}{\mathrm{d}t}\frac{\mathrm{d}^2z}{\mathrm{d}t^2}\right]\right\}$$

$$Y = \frac{\mathrm{d}p_y}{\mathrm{d}t} = \frac{m_0}{1-v^2/c^2}\left\{\frac{\mathrm{d}^2y}{\mathrm{d}t^2}\sqrt{1-v^2/c^2} + \frac{\mathrm{d}y}{\mathrm{d}t}\frac{1}{c^2\sqrt{1-v^2/c^2}}\left[\frac{\mathrm{d}x}{\mathrm{d}t}\frac{\mathrm{d}^2x}{\mathrm{d}t^2}+\frac{\mathrm{d}y}{\mathrm{d}t}\frac{\mathrm{d}^2y}{\mathrm{d}t^2}+\frac{\mathrm{d}z}{\mathrm{d}t}\frac{\mathrm{d}^2z}{\mathrm{d}t^2}\right]\right\}$$

with a similar equation for Z. These highly complicated equations are virtually never used to calculate individual problems of motion.

[8] The Relations Between Force and Acceleration; Transverse Mass and Longitudinal Mass

(a) Consider the equations above; let us select the axes so that *at the time t considered*, the velocity lies along the Ox-axis. Thus

$$v = \frac{\mathrm{d}x}{\mathrm{d}t} \qquad \frac{\mathrm{d}y}{\mathrm{d}t} = \frac{\mathrm{d}z}{\mathrm{d}t} = 0$$

and hence

$$\left.\begin{aligned} X &= \frac{\mathrm{d}p_x}{\mathrm{d}t} = \frac{m_0}{(1-v^2/c^2)^{\frac{3}{2}}}\frac{\mathrm{d}^2x}{\mathrm{d}t^2} \\ Y &= \frac{\mathrm{d}p_y}{\mathrm{d}t} = \frac{m_0}{\sqrt{1-v^2/c^2}}\frac{\mathrm{d}^2y}{\mathrm{d}t^2} \end{aligned}\right\} \qquad (1)$$

with an equation for p_z similar to that for p_y.

It is immediately clear that

— the ratio of force to acceleration is not the same for the component parallel to the velocity as for those perpendicular to it;

— this ratio depends upon the velocity and is not constant as in pre-relativistic dynamics.

We denote the components of **F** and **γ** parallel to the velocity by F_t and γ_t respectively, and those normal to it by F_n and γ_n (in general, there are two for each vector).

The equations of motion can thus be written in the ordinary form

$$F_t = m_t\gamma_t \qquad F_n = m_n\,\gamma_n$$

if we set

$$m_t = \frac{m_0}{(1-v^2/c^2)^{\frac{3}{2}}} \qquad m_n = \frac{m_0}{\sqrt{1-v^2/c^2}} \tag{2}$$

so defining a transverse mass m_n and a longitudinal mass m_t.

Equations (1) resemble the pre-relativistic equations, but in fact the difference is much greater than the formulae suggest. Quite apart from the fact that the masses vary with v, it is to be stressed that *the axes Oxyz alter with the time, t*. Equations (2) cannot therefore be employed as they stand to determine the laws of motion; if the motion is to be referred to Cartesian axes, the full equations of § [7] must be used.

Remark. The usual definition of acceleration is used above; another form may also be adopted (§ [25]).

(b) *Vector proof*

Let **t** and **n** be unit vectors, along the tangent and principal normal, and let ϱ be the radius of curvature. Then

$$\frac{d\mathbf{v}}{dt} = \frac{d}{dt}(v\mathbf{t}) = \mathbf{t}\frac{dv}{dt} + \frac{v^2}{\varrho}\mathbf{n}$$

$$\mathbf{F} = \frac{d}{dt}\frac{m_0\mathbf{v}}{\sqrt{1-v^2/c^2}} = \frac{m_0\mathbf{t}}{(1-v^2/c^2)^{\frac{3}{2}}}\frac{dv}{dt} + \frac{m_0\mathbf{n}}{\sqrt{1-v^2/c^2}}\frac{v^2}{\varrho}$$

Scalar multiplication by **n** yields a relation which gives the radius of curvature:

$$\frac{m_0}{\sqrt{1-v^2/c^2}}\frac{v^2}{\varrho} = F_n$$

[9] Properties of the Masses m_t and m_n

It is instructive to plot the curves showing how the masses vary with v.

We notice first that m_t and m_n are equal only for the extreme values of v:

$$v = 0 \qquad m_t = m_n = m_0$$

$$v = c \qquad m_t = m_n = \infty$$

For intermediate values of v, we have

$$\frac{m_n}{m_t} = 1 - \frac{v^2}{c^2} \quad \text{and hence} \quad m_n < m_t$$

When v is small compared with c, it is convenient to expand m_t and m_n as power series:

$$m_t = m_0\left(1 + \frac{3}{2}\frac{v^2}{c^2} + \frac{15}{8}\frac{v^4}{c^4} + \frac{35}{16}\frac{v^6}{c^6} + \dots\right)$$

$$m_n = m_0\left(1 + \frac{1}{2}\frac{v^2}{c^2} + \frac{3}{8}\frac{v^4}{c^4} + \frac{5}{16}\frac{v^6}{c^6} + \dots\right)$$

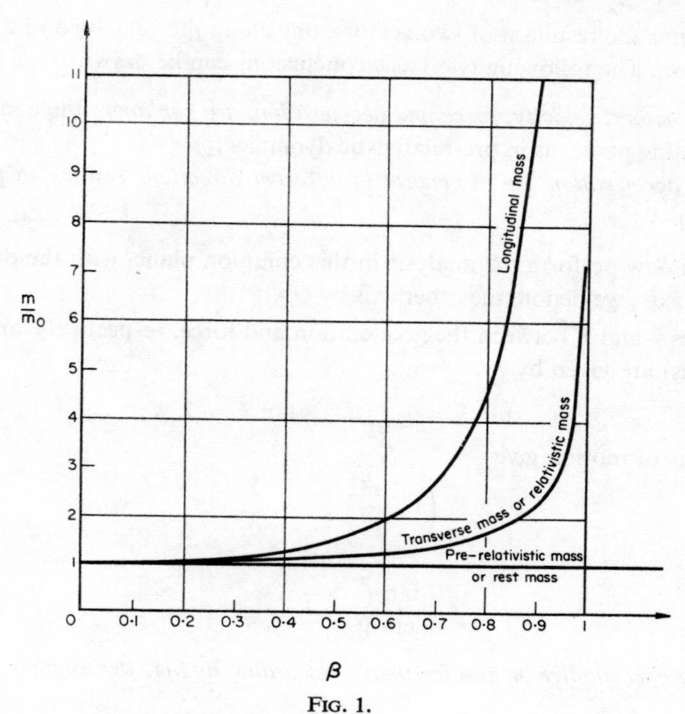

FIG. 1.

The angular coefficients of the tangents are given by

$$\alpha_t = \frac{dm_t}{dv} = \frac{3m_0}{(1-v^2/c^2)^{\frac{5}{2}}}\,\frac{v}{c^2}$$

$$\alpha_n = \frac{dm_n}{dv} = \frac{m_0}{(1-v^2/c^2)^{\frac{3}{2}}}\,\frac{v}{c^2}$$

The two curves are initially tangential to the v-axis; as v tends to c, the tangents become vertical. For arbitrary values of v,

$$\frac{\alpha_n}{\alpha_t} = \frac{1}{3}(1-v^2/c^2) \quad \text{and hence} \quad \alpha_n < \alpha_t$$

These results are plotted in Fig. 1; the horizontal straight line at $+1$ corresponds to pre-relativistic dynamics.

[10] The Relative Positions of the Velocity, Force and Acceleration Vectors

(a) Vectorially, we have

$$\mathbf{F} = \frac{d}{dt}(m\mathbf{v}) = m\frac{d\mathbf{v}}{dt} + \frac{dm}{dt}\mathbf{v} = m\boldsymbol{\gamma} + \frac{dm}{dt}\mathbf{v}$$

The force is thus the resultant of two vectors, one along the velocity and the other along the acceleration. The following two basic conclusions can be drawn:

— *the three vectors, velocity, force and acceleration, are coplanar*; the common plane is the osculating plane, as in pre-relativistic dynamics;
— *force and acceleration are in general in different directions* (unlike in pre-relativistic dynamics).

(b) Let us now perform the analysis in this common plane, with the direction of the velocity as x-axis; we denote the other axis by Oy.

The angles $\hat{\gamma}$ and $\hat{F}$ between the acceleration and force, respectively, and the velocity $\mathbf{v}$ (the Ox-axis) are given by

$$\tan \hat{\gamma} = \gamma_y/\gamma_x; \qquad \tan \hat{F} = Y/X$$

The equations of motion give

$$\left(1-\frac{v^2}{c^2}\right)\frac{\gamma_y}{\gamma_x} = \frac{Y}{X}$$

and hence

$$\frac{\tan \hat{F}}{\tan \hat{\gamma}} = 1-\frac{v^2}{c^2}$$

The angle $\hat{F}$ is smaller or greater than $\hat{\gamma}$ according as they are smaller or greater than 90° (Fig. 2).

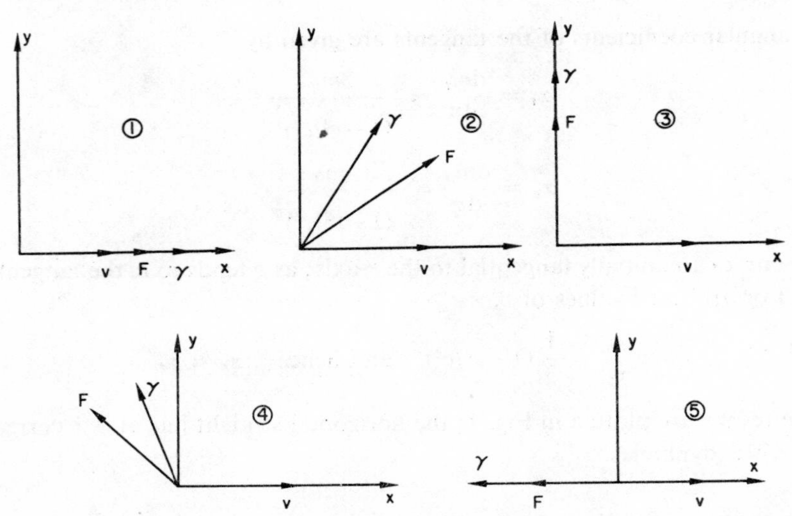

FIG. 2. Relative orientations of the force, velocity and acceleration vectors.

When the force is normal to the velocity, $\tan \hat{F}$ is infinite and hence so is $\tan \hat{\gamma}$; $\mathbf{F}$ and $\boldsymbol{\gamma}$ are then parallel. The same is true if force and velocity are parallel.

The vectors $\mathbf{F}$ and $\boldsymbol{\gamma}$ always lie on the same side of $\mathbf{v}$. When $|\mathbf{v}|$ tends to c, irrespective of $\mathbf{F}$ (of finite magnitude), $\boldsymbol{\gamma}$ tends to zero and becomes perpendicular to $\mathbf{v}$.

(c) *Remark on a difficulty in calculation.* The foregoing analysis allows us to write down the equations of relativistic dynamics by projecting on to the velocity and acceleration, selected as the axes. If γ denotes the magnitude of the acceleration, and F_γ and F_v denote the corresponding components of the force, we obtain

$$F_\gamma = m\frac{\mathrm{d}v}{\mathrm{d}t} = m\gamma = \frac{m_0}{\sqrt{1-\beta^2}}\gamma$$

$$F_v = v\frac{\mathrm{d}m}{\mathrm{d}t} = v\frac{\mathrm{d}}{\mathrm{d}t}\frac{m_0}{\sqrt{1-v^2/c^2}} = \frac{\beta^2 m_0\gamma}{(1-\beta^2)^{\frac{3}{2}}}$$

At first sight, this last result is surprising. We should expect that in the velocity direction, we ought to find the longitudinal mass as a factor whereas here we have a

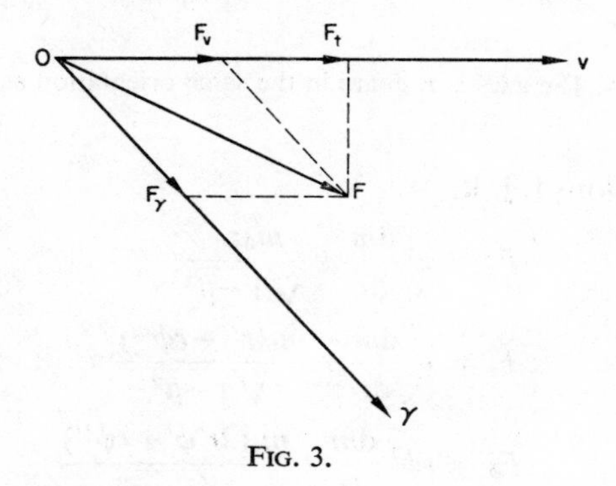

FIG. 3.

supplementary factor β^2. Figure 3 sheds light on this point: in § [8] we have projected F on to v orthogonally, and obtained F_t. Here the axes are oblique and the projections F_v and F_t are in consequence different. The vector analysis above may also be compared with that of § [8].

[11] The Equations of Motion in Cylindrical Polars

The practically minded reader will be grateful to me for giving the formulae used in electron optics immediately, without obliging him first to study the variational methods.

We employ the variables z, r, ϕ of Fig. 4, and project the vectors on to the axes Mz, Mr and $M\phi$. Setting out from the equation of motion

$$\mathbf{F} = \frac{\mathrm{d}}{\mathrm{d}t}(m\mathbf{v})$$

and denoting the unit vectors by **i**, **j**, **k**, we have

$$\mathbf{v} = \mathbf{i}z' + \mathbf{j}r' + \mathbf{k}r\phi'$$

$$\mathbf{F} = \mathbf{v}\frac{\mathrm{d}m}{\mathrm{d}t} + m\frac{\mathrm{d}\mathbf{v}}{\mathrm{d}t} = \mathbf{v}\frac{\mathrm{d}m}{\mathrm{d}t} + m\{\mathbf{i}z'' + \mathbf{j}(r'' - r\phi'^2) + \mathbf{k}(2r'\phi' + r\phi'')\}.$$

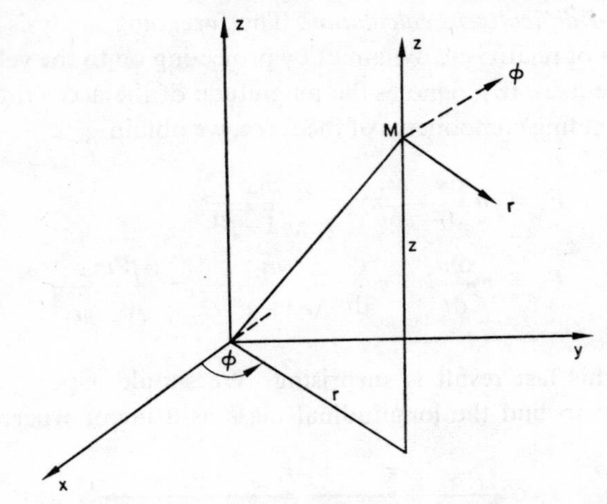

FIG. 4. The axes z, r, ϕ are in the same orientation as x, y, z

Along the three directions **i, j, k**,

$$F_z = z' \frac{dm}{dt} + \frac{m_0 z''}{\sqrt{1-\beta^2}}$$

$$F_r = r' \frac{dm}{dt} + \frac{m_0(r'' - r\phi'^2)}{\sqrt{1-\beta^2}}$$

$$F_\phi = r\phi' \frac{dm}{dt} + \frac{m_0(2r'\phi' + r\phi'')}{\sqrt{1-\beta^2}}$$

in which the component F_ϕ has the same nature as a force, like F_z and F_r. Finally, then,

$$F_z = \frac{d}{dt}\left(\frac{m_0 z'}{\sqrt{1-\beta^2}}\right)$$

$$F_r = \frac{d}{dt}\left(\frac{m_0 r'}{\sqrt{1-\beta^2}}\right) - \frac{m_0 r\varphi'^2}{\sqrt{1-\beta^2}}$$

$$rF_\phi = \frac{d}{dt}\left(\frac{m_0 r^2 \phi'}{\sqrt{1-\beta^2}}\right)$$

[12] The Equations of Motion in Spherical Polars

With the variables r, ϕ, θ of Fig. 5,

$$\mathbf{v} = \mathbf{i}r' + \mathbf{j}r \sin\theta\phi' + \mathbf{k}r\theta'$$

$$m\frac{d\mathbf{v}}{dt} = \mathbf{i}m(r'' - r\theta'^2 - r\phi'^2 \sin^2\theta)$$
$$+\mathbf{j}m(r\phi'' \sin\theta + 2r'\phi' \sin\theta - 2r\theta'\phi' \cos\theta)$$
$$+\mathbf{k}m(r\theta'' + 2r'\theta' + r\phi'^2 \sin\theta\cos\theta)$$

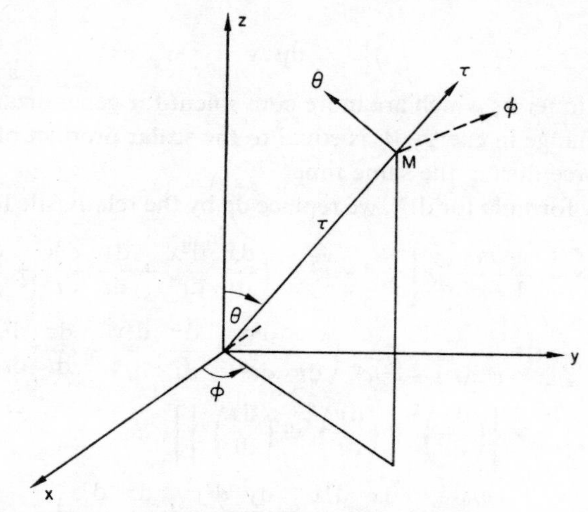

FIG. 5. The axes r, ϕ, θ are in the same orientation as x, y, z.

Hence

$$F_r = r'\frac{dm}{dt} + m(r'' - r\theta'^2 - r\phi'^2 \sin^2\theta)$$

$$F_\phi = r\phi' \cos\theta \frac{dm}{dt} + m(r\phi'' \sin\theta + 2r'\phi' \sin\theta - 2r'\theta'\phi' \cos\theta)$$

$$F_\theta = r\theta' \frac{dm}{dt} + m(r\theta'' + 2r'\theta' + r\phi'^2 \sin\theta \cos\theta)$$

B. THE KINETIC ENERGY THEOREM; THE POSTULATE OF PROPER ENERGY

[13] Relativistic Generalization of the Theorem of Conservation of Energy

In pre-relativistic dynamics, we write for each component

$$X = m\frac{d^2x}{dt^2}, \quad X\,dx = m\frac{d^2x}{dt^2}\,dx = \frac{d}{dt}\left\{\frac{1}{2}m\left(\frac{dx}{dt}\right)^2\right\}dt$$

Adding the respective results for each of the three components leads to the theorem of conservation of energy:

$$d\mathcal{T} = X\,dx + Y\,dy + Z\,dz = d(\tfrac{1}{2}mv^2) = dW$$

in which W denotes the kinetic energy (sometimes called the semi-kinetic energy; in any event, the factor $\frac{1}{2}$ must not be overlooked).

This theorem can be expressed differently:

$$d\mathcal{T} = X\,dx + \ldots = X\,dt\frac{dx}{dt} + \ldots = dp_x v_x + \ldots = d\mathbf{p}\,.\,\mathbf{v}$$

and hence

$$dW = d\mathbf{p}.\mathbf{v}$$

We can state this in terms which are more convenient for generalization to relativity: during time dt, the change in energy W is equal to the scalar product of the velocity and the impulse of the force during the same time.

To obtain the new formula for dW, we replace $d\mathbf{p}$ by the relativistic formula for it:

$$\frac{dW}{dt} = \frac{d\mathbf{p}}{dt}.\mathbf{v} = \frac{m_0}{1-v^2/c^2}\left[\sqrt{1-v^2/c^2}\left(\frac{dx}{dt}\frac{d^2x}{dt^2}+\frac{dy}{dt}\frac{d^2y}{dt^2}+\frac{dz}{dt}\frac{d^2z}{dt^2}\right)\right.$$

$$+\frac{1}{c^2\sqrt{1-v^2/c^2}}\left(\frac{dx}{dt}\frac{d^2x}{dt^2}+\frac{dy}{dt}\frac{d^2y}{dt^2}+\frac{dz}{dt}\frac{d^2z}{dt^2}\right)\times$$

$$\left.\times\left\{\left(\frac{dx}{dt}\right)^2+\left(\frac{dy}{dt}\right)^2+\left(\frac{dz}{dt}\right)^2\right\}\right]$$

$$= \frac{m_0}{(1-\beta^2)^{\frac{3}{2}}}\left(\frac{dx}{dt}\frac{d^2x}{dt^2}+\frac{dy}{dt}\frac{d^2y}{dt^2}+\frac{dz}{dt}\frac{d^2z}{dt^2}\right)$$

The reader can verify that this expression is obtained if, for the energy, we use the expression

$$W = \frac{m_0c^2}{\sqrt{1-v^2/c^2}} = mc^2$$

The calculation could alternatively be performed by replacing the components of force in the formula for work by their relativistic expressions.

A very rapid vector proof can also be given; we have only to form the scalar product of $\mathbf{v}$ and the expression for F given in § [8] (b) to obtain

$$\mathbf{F}.\mathbf{v} = \frac{m_0v}{(1-v^2/c^2)^{\frac{3}{2}}}\frac{dv}{dt} = \frac{d}{dt}\frac{m_0c^2}{\sqrt{1-v^2/c^2}} = \frac{dW}{dt}$$

The preceding arguments would all be equally valid if an arbitrary constant were added to W. *We shall set this constant equal to zero.* This is equivalent to assuming that *a particle at rest has energy* m_0c^2. For the present, this statement may be regarded as a postulate; we shall, however, encounter (§ [149]) arguments which *seem* to lead to the conservation of energy from the impulse postulate.

Fifty years ago, the existence of this rest energy was contested, and it was far more natural to deduce it from other postulates (from electromagnetism, for example). Today, the situation is very different; the simplicity of the expression for the proper energy and its importance in modern physics lead us to treat it as one of the fundamental principles.

Explicitly, then, the relativistic energy theorem is of the form

$$d\mathcal{T} = \mathbf{F}.d\mathbf{s} = d(mc^2)$$

or

$$X\,dx+Y\,dy+Z\,dz = d\frac{m_0c^2}{\sqrt{1-\frac{1}{c^2}\left\{\left(\frac{dx}{dt}\right)^2+\left(\frac{dy}{dt}\right)^2+\left(\frac{dz}{dt}\right)^2\right\}}}$$

Remark 1. The preceding statement can be put into a slightly different form. A mass m travelling at a velocity v is equivalent to an energy mc^2. There is thus an energy current given by

$$mc^2v$$

and we can write

$$\text{Impulse} = \frac{\text{energy current}}{c^2} = \text{mass current}$$

This is a useful expression in the dynamics of continuous media and in electromagnetism.

Remark 2. Suppose that we have a uniformly moving corpuscle in a frame of reference K, with energy W. The coefficient which governs the Lorentz contraction or the time dilatation (*Kinematics*, §§ [63] and [65]) can be written in one of the forms

$$\sqrt{1-\beta^2} = \frac{m_0c^2}{W} = \frac{m_0}{m}$$

[14] Various Kinds of Energy

At low velocities, it is convenient to use the series expansion of W:

$$W = m_0c^2 + \tfrac{1}{2}m_0v^2 + \ldots$$

It is then natural to decompose the energy W into *kinetic energy* E which vanishes when the velocity is zero, given by

$$E = W - m_0c^2 = m_0c^2\left(\frac{1}{\sqrt{1-\beta^2}} - 1\right)$$

and proper energy, m_0c^2.

If the corpuscle is in a field of force which can be derived from a potential, it possesses potential energy. Its total energy thus contains three terms: proper energy, kinetic energy and potential energy.

From the expression for the kinetic energy E, we obtain

$$\beta = \frac{\sqrt{E(E+2m_0c^2)}}{E+m_0c^2};$$

this expression is valid for all systems of units.

The concept of kinetic pseudo-energy

In the theory of relativistic perfect gases, the quantity

$$E^* = \frac{1}{2}\frac{m_0v^2}{\sqrt{1-\beta^2}}$$

occurs and is called the kinetic pseudo-energy. It is a curious mixture of the relativistic and non-relativistic expressions for kinetic energy. The following inequalities may be established:

$$\text{for} \quad \beta > \sqrt{\tfrac{2}{3}} \qquad E > E^*;$$
$$\text{for} \quad \beta < \sqrt{\tfrac{2}{3}} \qquad E < E^*.$$

[15] A Brief Glance at the Consequences of the Relation Between Mass and Energy

This problem dominates modern physics, and it is no exaggeration to say that for better or for worse, it governs the very future of the human race. It is studied at length in Chapter XI, and chiefly in my *Rayonnement et dynamique*, mentioned earlier, where the epistemological aspects are also examined.

For the present, let us merely consider the orders of magnitude involved. The fundamental fact is that mass and energy must be regarded as quantities connected by an equivalence relation. Every exchange of energy between two bodies must entail a variation in mass.

Einstein's formula gives

$$1 \text{ gramme} = 9\times 10^{20} \text{ ergs} = 9\times 10^{13} \text{ joules}$$

The energy contained in 1 gramme of any substance would be enough to lift about 10 million tons through 1 kilometre. Alternatively, we might say that a great deal of energy is required to make very little mass.

For an electron,

$$m_0 = 0{\cdot}90\times 10^{-27} \text{ g}, \quad m_0c^2 = 0{\cdot}81\times 10^{-6} \text{ erg}$$

so that the proper energy is of the order of a microerg.

These orders of magnitude explain why the variation of mass cannot be discerned in everyday thermodynamic or chemical processes. For example,

$$H_2+\tfrac{1}{2}O_2 = H_2O+68{\cdot}400 \text{ cal}$$

The corresponding mass change is

$$\Delta m = 3{\cdot}2\times 10^{-9} \text{ g}$$

for a total mass of 18 g.

[16] Deduction of the Expression for the Impulse from the Energy

We have obtained the energy formula from the impulse postulate, making allowance for the relativistic expression for the mass. It is easy to see that the deduction can be performed in reverse.

We now write the impulse postulate

$$F = \frac{d}{dt}(mv)$$

without making any suppositions about m.

For the work, we have

$$F\, ds = m\frac{dv}{dt}\, ds+v\frac{dm}{dt}\, ds = mv\, dv+v^2\, dm$$

This is equivalent to the energy change, the form of which by hypothesis we know to be

$$mv\ \mathrm{d}v + v^2\ \mathrm{d}m = \mathrm{d}(mc^2) = c^2\ \mathrm{d}m$$

$$mv\ \mathrm{d}v = (c^2 - v^2)\ \mathrm{d}m$$

Finally, therefore, setting m equal to m_0 when v is zero,

$$\Big[\ln m\Big]_{m_0}^{m} = -\tfrac{1}{2}\Big[\ln(c^2 - v^2)\Big]_0^v$$

which does indeed give

$$m = \frac{m_0}{\sqrt{1-\beta^2}}$$

[17] The Energy Corresponding to a Given Momentum; Negative Energies and Positive Energies

(a) From the relations

$$\text{momentum} \qquad \mathbf{p} = m\mathbf{v}$$

$$\text{energy} \qquad W = mc^2$$

we can easily verify the relation

$$p^2 + m_0^2 c^2 = \frac{W^2}{c^2}$$

This enables us to express energy W as a function of p for a given corpuscle (given m_0):

$$W = \pm c\sqrt{p^2 + m_0^2 c^2}$$

It would seem, therefore, that to any given value of the momentum p, we can make a positive energy state and a negative energy state correspond. For a given corpuscle, the positive energy varies from $+m_0c^2$ to $+\infty$ and the negative energy, from $-m_0c^2$ to $-\infty$. The energy can never take values between $-m_0c^2$ and $+m_0c^2$.

(b) *A priori*, two procedures are possible. For the negative energies, we can retain the relations

$$W = mc^2, \qquad \mathbf{p} = m\mathbf{v}$$

The mass m and the proper mass m_0 are then both negative.

Alternatively, we can retain the essentially positive nature of the mass and write

$$W = -mc^2, \qquad \mathbf{p} = -m\mathbf{v}$$

Whichever interpretation is adopted (and the former is usually chosen, see § [239]) the ensuing properties are astonishing. Let us see what are the effects at low velocities: the momentum is in the opposite direction to the velocity, and the acceleration is in the direction opposing the force; reducing the velocity increases the energy.

These are strange but not contradictory properties. We shall study them further in § [239] (the idea of an *antiparticle*).

(c) Let us write the momentum p as a function of the kinetic energy E

$$p = \sqrt{2m_0E + E^2/c^2}$$

In pre-relativistic dynamics, this reduced to

$$p_c = \sqrt{2m_0E}$$

and for given kinetic energy,

$$p > p_c$$

(d) *The relation between the variations of m and of p.* We have

$$W\,\mathrm{d}W = c^2p\,\mathrm{d}p$$

$$\frac{\mathrm{d}m}{m} = \frac{\mathrm{d}W}{W} = \frac{c^2p\,\mathrm{d}p}{W^2} = \frac{c^2p^2}{W^2}\frac{\mathrm{d}p}{p} = \frac{v^2}{c^2}\frac{\mathrm{d}p}{p}$$

Since $v^2/c^2 < 1$, the change of m is negligible if the variation of p is negligible.

[18] Passage to the Pre-relativistic Limit

In kinematics, we obtain the pre-relativistic formula for a mechanical quantity by neglecting velocities v in comparison with c, or by letting c tend to infinity in the relativistic expression. This is no longer true in dynamics, when energy is involved.

In pre-relativistic dynamics, the energy of a particle is defined with respect to an arbitrary origin. The resulting constant is implicitly put equal to zero in purely mechanical calculations, but the possibility of its existence was nevertheless known from thermodynamics (internal energy). To obtain the non-relativistic approximation, the expression for kinetic energy must often be employed

$$E = W - m_0c^2$$

and not the complete expression for the energy. Furthermore, the passage to the limit is not always obtained by letting c tend to infinity. It is clear that such a process gives

$$E = \infty \times 0$$

We have therefore to expand as a power series in β and neglect terms of higher order than β^2.

Remark. Every relativistic formula not containing c explicitly is necessarily identical with its pre-relativistic counterpart.

HISTORICAL AND BIBLIOGRAPHICAL NOTES

The historical study of some important questions must be postponed to later volumes. The evolution of the mass-energy equivalence, for example, can only be fully understood after the theories of electromagnetic field energy have been studied (see *Rayonnement et dynamique*).

[19] Early Ideas about the Dynamics of Variable Masses

(a) Most historians of mechanics point out, as a curiosity, that Laplace [2] already envisaged the possibility of a dynamics more general than Newtonian dynamics. Writing momentum in the form

$$m_0 f(v)$$

he gave the kinetic energy

$$\int m_0 f'(v) v \, dv$$

These were, however, equations that Laplace set down following the natural tendency of mathematicians to generalize. He attached no physical meaning to them.

Abelé and Malvaux [4] mention the less known and much more curious fact that Descartes [1] had envisaged the existence of a limiting velocity for moving particles: "Les puissances naturelles agissent plus ou moins, selon que le sujet est plus ou moins disposé à recevoir leur action: et il est certain qu'une pierre n'est pas également disposée à recevoir un nouveau mouvement, ou une augmentation de vitesse, lorsqu'elle se meut déjà fort vite et lorsqu'elle se meut fort lentement."†

In the same line of thought, though far from physical research, we mention the lectures of Painlevé [3].

(b) The real point of departure lies in the work on the notion of electromagnetic mass (see *Rayonnement et dynamique*). We can perceive three distinct periods: the pre-electromagnetic era (Laplace, Painlevé, etc.), the electromagnetic or just pre-relativistic era (see *Rayonnement et dynamique*) and the relativistic era (which is studied in the following sections).

[1] R. Descartes: *Lettre à Mersenne*, October–November 1631. Quoted by A. Koyré, *Etudes galiléennes*, Hermann, vol. II, 1939, p. 50.

[2] Laplace: *Mécanique céleste*, Première partie, livre I, an VII. Readers to whom Laplace's works are not available may consult Dugas' analysis (ref. [30] of Appendix I, vol. I, p. 338). See also Whittaker (ref. [29] of Appendix I, vol. II, p. 47).

[3] P. Painlevé: *Les Axiomes de la Mécanique*, Gauthier-Villars, 1922, 112 pp. See in particular "Notes sur le principe de l'action et de la réaction", pp. 104–111. Most of these ideas were explained by Painlevé in his lectures at the Faculté des Sciences in Lille, from 1890 onwards.

[4] J. Abelé and P. Malvaux: *Vitesse et univers relativiste*, Paris, Sedes, 1954, p. 37.

[20] The Various Ways of Approaching Relativistic Dynamics

(a) *Deduction, starting from the equations of the electromagnetic field*

This method, the first historically, is one of the fundamental approaches. The laws are deduced from the Maxwell–Lorentz equations in the general treatises of Einstein, von Laue and Costa de Beauregard; I shall study it below (§ [20A]).

(b) *Direct methods*

I have already drawn attention in the first volume of this series to the desire of many authors to liberate kinematics from the postulate about the velocity of light; the same desire was manifested to liberate dynamics from electromagnetism. Let us now consider the various arguments advanced.

We first dispose of a method due to Esclangon. The latter attempted to derive the general laws of dynamics from hypotheses about motion in a constant uniform field of force. Quite apart from the fact that some of his statements are wrong, I find it hardly satisfactory that general equations are extracted by studying a special case of motion. Furthermore, Esclangon's equations are not those usually employed (§ [115]).

† "Natural forces act more or less, according as the subject is more or less ready to receive their action: it is certain that a stone is not equally ready to accept a new motion, or an increase of velocity, when it is already moving very fast and when it is moving very slowly."

The other methods which we shall describe in the following sections are on the whole interesting and suggestive; all the same, their utility resides in the light they shed on certain aspects of the formulae and not in the logical pre-eminence which may be attached to one or other of them. From my own point of view, that of a practical physicist, I find it simpler, clearer and just as logical to postulate the momentum (§ [5]) and proper energy (§ [13]) rules. Nevertheless—and here I am in agreement with the preoccupations of most of these authors—these postulates are also introduced independently of electromagnetism.

(c) *Deduction, starting from the wave-like properties of corpuscles*

This method is described in § [65] (b).

(d) *The method adopted in the present work*

The method can be regarded in two ways, depending upon whether pedagogy (a first course of study) is in question, or a more formal study (reduction of the postulates). In the first case, I state the momentum postulate from the outset, just as in non-relativistic dynamics one sets $\mathbf{F} = m\boldsymbol{\gamma}$.

As I mentioned in § [5] (c), however, this postulate can be deduced from the Newtonian laws and the principle of relativity by applying a general method of attaining relativistic laws. This point of view is expanded in § [38].

I did not feel inclined to adopt this from the outset, to avoid confronting all my readers with the Minkowski formalism. At all events, the *postulate* $d\mathbf{p} = \mathbf{F}\,dt$ must be stated at the beginning of dynamics and if we state the Newtonian law $\mathbf{p} = m_0\mathbf{v}$, application of the principle of relativity yields the relativistic law (§ [38]).

[20A] Relativistic Particle Dynamics, Deduced from Maxwell's Equations

(a) *Pre-relativistic work*

During the closing years of the last century, experimental work on the motion of fast electrons revealed that Newtonian dynamics fails at high velocities. The general idea of the theoreticians was as follows. A moving electron creates an electromagnetic field, to which the theories attribute energy and momentum, following Poynting. A field momentum depending on velocity must thus be added to the particle momentum. Various formulae, in practice very close, were obtained, according to the hypothesis made about the structure of the electron (Abraham, Lorentz, Langevin). I reconsider these at length in *Rayonnement et dynamique* (where a historical study of the concept of electromagnetic mass and some examination of the unitary and dualist theories are to be found).

We mention here only the principal ideas: use of Maxwell's equations, hypotheses on the energy properties of the field and on the structure of the electron. The general conclusion of the authors of that period was: electromagnetism is the fundamental science, and not mechanics.

(b) *Einstein's method*

Einstein [1] retained the general idea that dynamics must be subordinate to electromagnetism, but he dispensed with the special hypothesis about field energy and electron structure. He employed only Maxwell's equations and their consequences for field transformation (§ [71B]). His reasoning (slightly modified for reasons that I shall explain) is as follows.

The equations of dynamics. Let a particle be moving arbitrarily. In general, there is no Galileian system in which it is permanently at rest. Let us refer it to an arbitrary Galileian system, K (axes $Oxyz$). We can of course select this system so that at some instant t, the point is travelling with velocity v parallel to Ox. *At this instant* t, we consider a Galileian system K_0, with axes parallel to those of K, which is travelling with velocity v parallel to Ox. In K_0, *at time* t, the particle is at rest. This system K_0 varies with t. During a very short interval, however, we can treat the particle as though it were at rest in K_0, provided that the acceleration has a negligible effect. We recover the approximation made in *Kinematics* in connection with the rotating disc, and this approximation now governs the whole of dynamics. We can predict that we shall obtain in this way only a first approximation to relativistic dynamics, valid solely for small accelerations.

Pre-relativistic kinematics was the limiting case of relativistic kinematics when the velocities tended to zero, and it is natural to extend the same limit to the equations of classical dynamics.

In K_0, therefore, at the time in question, we write

$$m_0 \frac{d^2x_0}{dt_0^2} = X_0, \quad m_0 \frac{d^2y_0}{dt_0^2} = Y_0, \quad m_0 \frac{d^2z_0}{dt_0^2} = Z_0$$

and we consider how to transform these relations when we introduce the K variables.

Kinematics gives us formulae for transforming the acceleration:

$$\gamma_{0x} = \frac{\gamma_x}{(1-\beta^2)^{\frac{3}{2}}}, \quad \gamma_{0y} = \frac{\gamma_y}{1-\beta^2}, \quad \gamma_{0z} = \frac{\gamma_z}{1-\beta^2}$$

We have now to transform force. Consider a charge ε travelling uniformly, with velocity v with respect to a system K. Suppose that in this system, the charge experiences a field of force $\mathbf{F}(X, Y, Z)$ which we leave unspecified (gravity, for example) and an electric field $\mathbf{E}$. Since the velocity is constant, we must have

$$X = -\varepsilon E_x, \quad Y = -\varepsilon E_y, \quad Z = -\varepsilon E_z$$

In K_0, we have the electric field

$$E_{0x} = E_x, \quad E_{0y} = \frac{E_y}{\sqrt{1-\beta^2}}, \quad E_{0z} = \frac{E_z}{\sqrt{1-\beta^2}}$$

There is also a magnetic field $\mathfrak{B}$ but this does not affect a stationary charge. The force exerted by the electric field must be balanced by the other force (since the charge is at rest in K_0), so that

$$X_0 = -\varepsilon E_{0x} = -\varepsilon E_x, \quad Y_0 = -\varepsilon E_{0y} = -\frac{\varepsilon E_y}{\sqrt{1-\beta^2}}$$

Finally, therefore, we find

$$X_0 = X, \quad Y_0 = \frac{Y}{\sqrt{1-\beta^2}}, \quad Z_0 = \frac{Z}{\sqrt{1-\beta^2}}$$

These formulae are valid for any ε, including the limiting case where the charge is zero.

If we now replace the K_0 variables in the equations of motion by their expressions in terms of the K variables, we obtain

$$\frac{m_0}{(1-\beta^2)^{\frac{3}{2}}} \frac{d^2x}{dt^2} = X, \quad \frac{m_0}{\sqrt{1-\beta^2}} \frac{d^2y}{dt^2} = Y$$

These are the same as the formulae already obtained by Lorentz, but with important differences of interpretation. To begin with, the result is valid for all substances, charged or neutral. Einstein's calculation is very different from those of Lorentz and Abraham. The mass is no longer necessarily electromagnetic: Einstein only states the function $f(v)$ that is valid for all masses.

This is both a simplification and a generalization, perfectly typical of Einstein. The founder of relativity instinctively steers clear of points of detail (the spherical structure of the electron, for example) and unnecessarily specific reasoning (electromagnetic mass) and remains close to essentials. Although the arguments start from Maxwell's equations, the results apply to mechanics in general, and are not necessarily linked to the electromagnetic terminology. The balance is beginning to be redressed in favour of mechanics, though this is naturally in a more general form than Newtonian mechanics. To be exact, we should point out that in the 1905 paper, the equations are written using the force of the system at rest. The difference is purely formal, however; the equations we use today are due to Planck [2].

(c) *The method of von Laue*

Einstein's method was improved, though not substantially altered, by various authors; among these was von Laue, to whom we are indebted for the first general treatise on the theory of relativity [3], and Costa de Beauregard [4]. Starting from electromagnetism, the dynamics of continuous media is first deduced, and appears as a transcription of the formulae with slight changes of wording; from here, we proceed to the dynamics of the particle. This procedure is given in a later volume. If we start from electromagnetism, it is more logical, since the concepts of field and continuous physical medium are very closely related: but, although historically interesting, this method is to be avoided in a work destined above all for physicists. Many years ago I first encountered these topics, in von Laue's work. I still remember my fury as I finally appreciated the detours along which I had been led in order to obtain the elementary formulae of particle dynamics.

(d) *Conclusion*

I have not adopted the course of reducing dynamics to electromagnetism for the following reasons. By electromagnetism, we often understand Maxwell's equations and the postulates concerning the energy of the field. Dynamics, however, is deduced from Maxwell's equations alone, and requires no assumptions whatsoever about the energy postulates. Furthermore, the dynamical relations between energy and momentum are in some cases not satisfied by the field. For experimental reasons, preference is yielded to dynamics. This science has therefore gained, or rather, regained since 1905, the status of an autonomous science. From the epistemological point of view, it seems to me clearer, simpler—in short, more satisfactory—to postulate the equations of dynamics than to postulate Maxwell's equations. I believe that teaching has nothing to lose by this attitude.

Finally, without advocating change for the sake of change, there is undoubtedly much to gain by reconsidering a collection of problems, however familiar, from a new angle.

[1] A. Einstein: Zur Elektrodynamik bewegter Körper. *Ann. Physik* **17** (June 1905) 891–921.
[2] M. Planck: Das Prinzip der Relativität und die Grundgleichungen der Mechanik. *Verh. Dtsch. Ges.* (1906) 136–141.
[3] M. von Laue: ref. [2] of Appendix I.
[4] O. Costa de Beauregard: refs. [7] and [35] of Appendix I.

[21] Langevin's Method

(a) The equations of dynamics are obtained by a combination of relativistic kinematics and postulates about the energy. This method, which was used very early by Langevin in his lectures [1], was reconsidered and developed further by F. Perrin [2] and Allard [3] during the Semaines de Synthèse. Soleillet's lectures [4] may also be consulted.

By means of general arguments about the energy, it is first shown that if a particular energy is measured in a system and has value W, then in any other system moving with velocity u with respect to the former, it will be given by an expression of the form

$$W\phi(u^2)$$

in which ϕ is the universal function to be determined.

In a similar fashion, expressions

$$m_0 f(v^2) \quad \text{and} \quad m_0 g(v^2)\mathbf{v}$$

are obtained for the kinetic energy and the momentum of a particle, where f and g are universal functions to be determined.

I shall not reproduce these arguments here. They are most ingenious, but despite their appearance of complete generality, I remain sceptical (see § [53] of my *Kinematics*), and this lack of faith makes me fear to reproduce them incorrectly.

The reader is referred to the articles listed. I shall, however, reproduce part of the consequences of the argument, for they form a healthy relativistic exercise.

(b) Consider two systems, K_1 and K_2, with velocities u and $-u$ with respect to a third system K_0. In each of the systems K_1 and K_2, we consider two moving particles of unit proper mass, travelling with velocities v and $-v$. The velocities of these four moving bodies in K_0 are obtained with the aid of the formulae for the transformation of velocity.

Suppose first that the velocities u and v are parallel. Because of the symmetry, the two velocities u that are in the same sense as the corresponding velocity v give velocities of the same absolute value in K_0; for the four bodies, therefore, we shall have two velocities w and w'.

Let us calculate in two different ways the energy required to halt all the particles with respect to K_0. Halting them directly in K_0 will require

$$2f(w^2)+2f(w'^2)$$

If we stop them first with respect to K_1 and K_2 respectively, the energy measured in K_1 and K_2 will be

$$4f(v^2)$$

and transforming this to K_0, we find

$$4f(v^2)\,\phi(u^2)$$

We then stop all four bodies in K_0, giving

$$4f(u^2)$$

and this way of stopping thus requires in all the energy

$$4f(v^2)\,\phi(u^2)+4f(u^2)$$

Comparing the two procedures, we see that

$$f(w^2)+f(w'^2) = 2f(v^2)\,\phi(u^2)+2f(u^2)$$

If the velocities v had been perpendicular to the velocities u, the four bodies would all have had the same absolute velocity in K_0, w''. A similar argument leads to

$$f(w''^2) = f(v^2)\,\phi(u^2)+f(u^2)$$

If we replace w, w' and w'' by the expressions for them as functions of u and v, we obtain two functional equations for f and ϕ.

Pre-relativistic kinematics gives

$$f(v^2) = v^2/2, \qquad \phi(v^2) = 1$$

while with relativistic kinematics,

$$f(v^2) = c^2\left(\frac{1}{\sqrt{1-\beta^2}}-1\right), \qquad \phi(v^2) = \frac{1}{\sqrt{1-\beta^2}}$$

Similar arguments would lead to an expression for $g(v^2)$. Penrose and others have recently taken up this method again [5, 6].

[1] P. Langevin: Unpublished lectures given at the Collège de France, see the following references.

[2] F. Perrin: *La Dynamique relativiste et l'inertie de l'énergie*, Paris, Hermann, 1932.

[3] G. Allard: Inertie, matérialisation et dématérialisation. Lecture, published in *L'Energie dans la nature et dans la vie*, Paris, P.U.F., 1949, pp. 103–130.

[4] P. Soleillet, in collaboration with N. Arpiarian: *Eléments de la théorie de la relativité restreinte* (Les Cours de la Sorbonne, C.D.U., Paris, n.d.).

[5] R. Penrose and W. Rindler: Energy conservation as the basis of relativistic mechanics. *Amer. J. Phys.* **33** (1965) 55–69.

[6] J. Ehlers, W. Rindler and R. Penrose: Energy conservation as the basis of relativistic mechanics. *Amer. J. Phys.* **33** (1965) 995–997.

[22] The General Equations of Dynamics, deduced from Hypotheses about Elastic Collisions; Tolman's Method

(a) *The method*

Tolman [1, 2, 3] bases his dynamics upon the following two postulates, together with the principle of relativity:

— the conservation of the total mass of a system of interacting particles

$$\Sigma m = \text{const} \tag{1}$$

— the conservation of the total momentum in each direction

$$\Sigma m u_x = \text{const}, \ldots \tag{2}$$

To avoid all ambiguity, we should mention that we refer to conservation with the passage of time, with respect to a particular system of reference; the quantities in question are not invariant when the reference system is altered.

Consider two identical particles, with velocities u_0 and $-u_0$ along the axis Ox of a Galileian system K_0. During the brief moment of collision, the velocities are zero, and subsequently they become $-u_0$ and u_0; the two principles are obviously satisfied in this case.

We now refer this phenomenon to another Galileian system K, having velocity $-v$ with respect to K_0. Let M denote the total mass of the two particles at the moment of collision, when both have velocity v with respect to K, and let m_1, m_2, u_1 and u_2 be the masses and velocities before the collision. The two principles enable us to write (before and during the collision)

$$m_1+m_2 = M, \qquad m_1u_1+m_2u_2 = Mv$$

From the formulae for the transformation of velocity

$$u_1 = \frac{u_0+v}{1+u_0v/c^2}, \qquad u_2 = \frac{-u_0+v}{1-u_0v/c^2}$$

we obtain

$$\frac{m_1}{m_2} = \frac{1+u_0v/c^2}{1-u_0v/c^2} = \sqrt{\frac{1-u_2^2/c^2}{1-u_1^2/c^2}}$$

(cf. the first formula of § [80] (b)) of *Kinematics*. By hypothesis, the rest masses are equal, so that

$$m = \frac{m_0}{\sqrt{1-\beta^2}} \tag{3}$$

A similar line of reasoning may be followed for particles travelling along different axes.

We then *define* force from the relation

$$\mathbf{F}\,dt = d(m\mathbf{v}) \tag{4}$$

(b) *Critical comments*

This method seems seductive: the postulates and reasoning are simple and the method is independent of electrodynamics. It is frequently adopted in English textbooks [4, 5].

It appears to me, however, that it cannot stand up to a close scrutiny. If the first postulate is to be meaningful, mass must first be defined; the books which use this method are silent on this point, however.

The essence of these postulates may be summarized thus: we call mass the quantity which leads to constant products Σmu_x. In other words, the second postulate is a definition of mass, and the first gives a property of this mass.

To the reader familiar with pre-relativistic dynamics, this first postulate has a deceptively simple appearance. We must not forget that the mass of each corpuscle is variable and it is by no means obvious that the sum must remain constant. Tolman's definition of force is in fact a third postulate. We have, however, already seen that dynamics and in particular the two other postulates (which thus become theorems) can be derived from this postulate alone.

In conclusion, therefore, this method reverts to the one adopted in the present text (§ [5]) once the unnecessary additional statements and deceptively simple appearance have been stripped away.

(c) *Recent texts*

Following the same train of thought, see the articles by Sears [7, 8] and Timotin Alexandru [6].

[1] G. N. Lewis and R. G. Tolman: The principle of relativity and non-Newtonian mechanics. *Phil. Mag.* **18** (1909) 510–523.
[2] R. G. Tolman: Non-Newtonian mechanics, the mass of a moving body. *Phil. Mag.* **23** (1912) 375–380.
[3] R. G. Tolman: Ref. [4] of Appendix I.
[4] G. Møller, Ref. [8] of Appendix I.
[5] P. G. Bergmann: Ref. [6] of Appendix I.
[6] Timotin Alexandru: A supra dinamich relativiste a systemelor de puncte materiale. *Bull. Inst. Polit. Bucaresti* **25** (1963) 105–112.

[7] F. W. SEARS: Some applications of the Brehme diagram. *Amer. J. Phys.* **31** (1963) 269–273.
[8] F. W. SEARS: Another derivation of the expression for the relativistic mass of a particle. *Amer. J. Phys.* **32** (1964) 60.

[23] The Method of H. Malet

The expression for the masses m_t and m_n can be obtained if the formulae for the transformation of velocity, time-interval and force are known (see below, §§ [28] and [30]). The calculation provides a valuable exercise.

Suppose that in a system K we have a uniform field of force, F; a mass m_0 is initially released from rest, and after time $\mathrm{d}t$, is travelling with velocity

$$\mathrm{d}v = \frac{F}{m_0}\,\mathrm{d}t$$

Pre-relativistic dynamics is applicable to this case.

Let K_0 be the system in which the mass is at rest at time t of K (corresponding to t_0 of K_0). In this system, the field is again F.

The body is at rest in K_0 at time t_0, and at time $t_0+\mathrm{d}t_0$ it has velocity

$$\frac{F}{m_0}\,\mathrm{d}t_0$$

In K, at the corresponding time $t+\mathrm{d}t$, the body has velocity $v+\mathrm{d}v$, and the transformation of velocities yields

$$v+\mathrm{d}v = \frac{v+\dfrac{F}{m_0}\,\mathrm{d}t_0}{1+\dfrac{v^2}{c^2}\dfrac{F}{m_0}\,\mathrm{d}t_0}$$

or

$$\frac{F}{m_0}\left(1-\frac{v^2}{c^2}\right)\mathrm{d}t_0 \simeq \mathrm{d}v$$

But

$$\mathrm{d}t = \frac{\mathrm{d}t_0}{\sqrt{1-\beta^2}}$$

and finally

$$F\,\mathrm{d}t = \frac{m_0\,\mathrm{d}v}{(1-\beta^2)^{\frac{3}{2}}}$$

Suppose now that in K the particle is released with velocity v in a field F which is normal to v. In K_0, after time $\mathrm{d}t_0$, the velocity is

$$\frac{F_0\,\mathrm{d}t_0}{m_0}$$

perpendicular to v. In K, after a corresponding time $\mathrm{d}t$, the component normal to v will be

$$\mathrm{d}v = \frac{F_0\,\mathrm{d}t_0}{m_0}\sqrt{1-\beta^2}$$

and in the present case

$$F_0 = \frac{F}{\sqrt{1-\beta^2}}$$

Making allowance for the time transformation, we obtain

$$\frac{m_0\,\mathrm{d}v}{\sqrt{1-\beta^2}} = F\,\mathrm{d}t$$

This reasoning is adapted from a *Note* by H. Malet [1]. This author emphasizes the fact that the transformation of velocities can play the part of a postulate, and in this respect, he paves the way for the work of Abelé and Malvaux [2]. Even though the addition of parallel velocities may reasonably form a point of departure, however, as the latter authors have shown, it can scarcely be maintained that the general velocity transformation is a simple, natural postulate. Furthermore, the force transformation has to be added. It is thus at least as simple to state the impulse postulate directly.

[1] H. MALET: La notion de la variation de la masse, déduite de la seule formule d'addition des vitesses. *C. R. Acad. Sci. Paris* **180** (1925) 425–427.
[2] J. ABELÉ and P. MALVAUX: ref. (4) of § [19], especially pp. 133–143.

[24] The Method of Lalan

This author [2] assumes that relativistic kinematics and the concept of rest mass (taken from classical mechanics) are established. He then proceeds to determine the functions $f(v^2)$ and $g(v^2)$ which appear in the expressions for energy, W, and impulse, p, written in the form

$$W = m_0 f(v^2), \qquad \mathbf{p} = m_0\, g(v^2)\mathbf{v}$$

We retrace his method of showing that there is only one way of selecting these functions such that the following requirement is satisfied (§ [29] (c)): if two systems have the same values of W and $\mathbf{p}$ in one reference system, then they must have the same values of W and $\mathbf{p}$ in every other system of reference.

Consider first two systems, each consisting of two particles of mass m_0; in a reference system K_2, the two particles of the first system have equal and opposite velocities u, parallel to Ox, while the two particles of the second system have equal and opposite velocities u, parallel to Oy. In K_2, they clearly have the same momentum and energy.

We now write down the conditions for this equality to be conserved in another reference system, K_1 (moving at velocity $-v$ with respect to K_2); we find

$$f\left\{\left(\frac{u+v}{1+\dfrac{uv}{c^2}}\right)^2\right\}+f\left\{\left(\frac{u-v}{1-\dfrac{uv}{c^2}}\right)^2\right\} = 2f\left(u^2+v^2-\frac{u^2v^2}{c^2}\right) \tag{1}$$

$$\frac{u+v}{1+\dfrac{uv}{c^2}}\, g\left\{\left(\frac{u+v}{1+\dfrac{uv}{c^2}}\right)^2\right\}+\frac{u-v}{1-\dfrac{uv}{c^2}}\, g\left\{\left(\frac{u-v}{1-\dfrac{uv}{c^2}}\right)^2\right\} = 2ug\left(u^2+v^2-\frac{u^2v^2}{c^2}\right) \tag{2}$$

We write

$$A = \frac{u+v}{1+\dfrac{uv}{c^2}}, \qquad B = \frac{u-v}{1-\dfrac{uv}{c^2}}$$

$$C = \frac{1}{\sqrt{1-A^2/c^2}}, \qquad D = \frac{1}{\sqrt{1-B^2/c^2}}$$

$$f(A^2) = F(C) = f\{c^2(1-1/C^2)\}$$

$$f(B^2) = F(D) = f\{c^2(1-1/D^2)\}$$

and we seek a function $P(C, D)$, such that

$$f\left(u^2+v^2-\frac{u^2v^2}{c^2}\right) = F(P(C, D)) = f\{c^2(1-1/P^2)\}$$

This yields

$$P^2 = \frac{c^4}{c^4+u^2v^2-c^2(u^2+v^2)}$$

Furthermore

$$C^2 = \frac{(c^2+uv)^2}{c^4+u^2v^2-c^2(u^2+v^2)}, \qquad D^2 = \frac{(c^2-uv)^2}{c^4+u^2v^2-c^2(u^2+v^2)}$$

and hence

$$2P = C+D$$

so that relation (1) becomes

$$F(C)+F(D) = 2F\left(\frac{C+D}{2}\right) \tag{3}$$

Similarly,

$$g(A^2) = G(C) = g\{c^2(1-1/C^2)\}$$

giving a relation, (4) say, in C and D, analogous to (3), for (2).

From (3) and (4), we deduce

$$F(C) = HC+H_1, \quad G(C) = K$$

or

$$f(v^2) = \frac{H}{\sqrt{1-v^2/c^2}}+H_1, \quad g(v^2) = \frac{K}{\sqrt{1-v^2/c^2}}$$

in which H, H_1 and K are constants.

We now consider the four numbers, which represent the momentum and energy of an arbitrary system with respect to K_2:

$$L_2 = \Sigma\frac{m_0K}{\sqrt{1-v^2/c^2}}\frac{dx_2}{dt_2}, \qquad M_2 = \Sigma\frac{m_0K}{\sqrt{1-v^2/c^2}}\frac{dy_2}{dt_2}$$

$$N_2 = \Sigma\frac{m_0K}{\sqrt{1-v^2/c^2}}\frac{dz_2}{dt_2}, \qquad O_2 = \Sigma m_0\left(\frac{H}{\sqrt{1-v^2/c^2}}+H_1\right)$$

It must be possible to express the numbers L_1, M_1, N_1 and O_1 for K_1 as functions of L_2, M_2, N_2 and O_2, independently of Σm_0 (so that these quantities can remain equal to those of another system having different masses, if necessary). Whence,

$$H_1 = 0$$

and finally,

$$W = \frac{m_0H}{\sqrt{1-v^2/c^2}}, \qquad \mathbf{p} = \frac{m_0K\mathbf{v}}{\sqrt{1-v^2/c^2}}$$

H and K depend upon the units employed.

Remark. A similar calculation can be made in classical mechanics; it leads to the expressions

$$m_0(Hv^2+H_1) \quad \text{and} \quad m_0K\mathbf{v}$$

but there is then no invariance relationship which allows the constant H_1 to be determined.

This method may be compared with a most interesting paper by Frank [1].

[1] P. FRANK: Energetische Ableitung der Formeln für die longitudinale und transversale Masse des Massenpunktes. *Ann. Physik* **39** (1912) 693–703.

[2] V. LALAN: Sur une définition axiomatique de l'impulsion et de l'énergie, *C. R. Acad. Sci. Paris* **198** (1934) 1211–1213.

[25] A Variant, Proposed by Abelé and Malvaux, upon the Basic Equations

These authors [1] set out from the fact that "si l'on exprime l'accélération dans le mouvement rectiligne par la simple dérivée dv/dt, on applique dans le temps une loi de composition *additive* des vitesses alors que, dans l'espace, la relativité a introduit une loi de composition non-additive".† This interesting remark led them to replace the ordinary acceleration (which they designate "*formal*") by the expression

$$\frac{dv/dt}{1-v^2/c^2}$$

In the general case, they then obtain as components of the "real" acceleration $\mathbf{\Gamma}$, using the axes of § [8],

$$\Gamma_t = \frac{\gamma_t}{1-\beta^2}, \qquad \Gamma_n = \gamma_n$$

† "If the acceleration in rectilinear motion is expressed as the simple derivative dv/dt, an *additive* law of composition of velocities in time is applied, whereas in space, relativity has introduced a non-additive law of composition."

and for the dynamical equations

$$F_t = \frac{m_0}{\sqrt{1-\beta^2}} \Gamma_t, \quad F_n = \frac{m_0}{\sqrt{1-\beta^2}} \Gamma_n$$

The longitudinal and transverse masses are equal.

[1] J. ABELÉ and P. MALVAUX: Ref. [4] of § [19].

[26] Further References

[1] A. O'LEARY: Conservation of energy and mass; arguments in favour of discarding the quantity "relativistic mass". *Amer. J. Phys.* **1** (1947) 280–282.
[2] O. ONICESCU: Sur la mécanique du point matériel. *Bull. Soc. Sci. Mat. Phys. Roumanie* **1** (1957) 461–465.
[3] D. PARK: Relativistic mechanics of a particle. *Amer. J. Phys.* **27** (1959) 311–313.
[4] L. REFF: Rest mass of a free particle. *Amer. J. Phys.* **28** (1960) 745–746.
[5] A. KRYALA: A new derivation of relativistic dynamics. *Amer. Mat. Soc. Not.* **8** No. 3, p. 274.
[6] H. OHNO: On the concept "momentum" in relativistic dynamics. *Science reports Soc. Res. Phys. Chem.* **8** (1962) 35–41.
[7] N. J. IONESCU-PALLAS: Conséquences dynamiques de la relativité. *Revue roumaine de physique* **8** (1963) 309–337.

[27] On Some Changes of Variables

(a) *The variable* ψ, *defined by* $\text{th}\,\psi = \beta$

The principal quantities have the following forms:

$$\frac{1}{\sqrt{1-\beta^2}} = \text{ch}\,\psi, \quad \frac{\beta}{\sqrt{1-\beta^2}} = \text{sh}\,\psi$$

Energy: $$mc^2 = m_0 c^2 \,\text{ch}\,\psi$$

Kinetic energy: $$(m-m_0)c^2 = 2m_0c^2\,\text{sh}^2\frac{\psi}{2}$$

Impulse: $$m\beta c = \frac{1}{c}\, m_0 c^2\,\text{sh}\,\psi$$

The Lorentz transformation for the variables $x, y, z, l\,(= ict)$ is thus given by the matrix

$$\begin{pmatrix} \text{ch}\,\psi & 0 & 0 & i\,\text{sh}\,\psi \\ 0 & 1 & 0 & 0 \\ 0 & 0 & 1 & 0 \\ -i\,\text{sh}\,\psi & 0 & 0 & \text{ch}\,\psi \end{pmatrix}$$

Cosmic-ray specialists frequently use the variable ψ.

(b) *The variable* ϕ *defined by* $\sin\phi = \beta$

The principal quantities now become

$$m = \frac{m_0}{\sqrt{1-\beta^2}} = \frac{m_0}{\cos\phi} = m_0 \sec\phi$$

$$p = mv = m_0 c \tan\phi$$

$$W = mc^2 = m_0c^2 \sec\phi$$

$$E = (m-m_0)c^2 = m_0(\sec\phi - 1)c^2$$

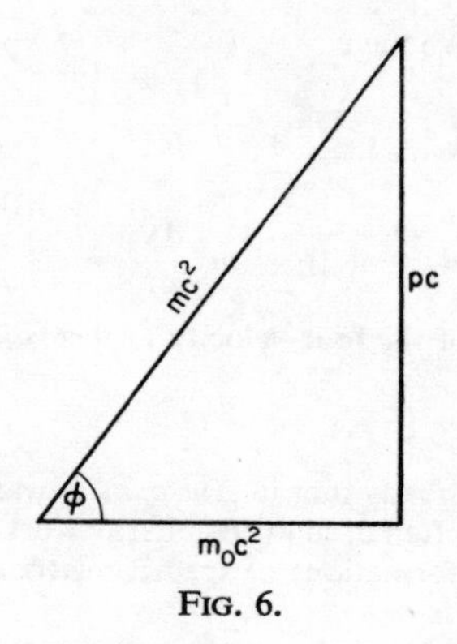

FIG. 6.

This is illustrated in Fig. 6. The transformation is useful because trigonometric tables can be used for intermediate calculations.

Let us calculate the expression for the kinetic energy, using this variable, starting from the impulse postulate

$$F\,dt = d(mv)$$

or

$$F\,dt = d(m_0c\tan\phi) = \frac{m_0c\,d\phi}{\cos^2\phi}$$

The kinetic energy E is defined by

$$dE = F\,ds = Fv\,dt = \frac{m_0c^2\sin\phi}{\cos^2\phi}\,d\phi$$

$$E = m_0c^2\int_0^\phi \frac{\sin\phi\,d\phi}{\cos^2\phi} = \left[\frac{m_0c^2}{\cos\phi}\right]_0^\phi$$

(c) *The variable u, defined by* $u = E/m_0c^2$

We define the parameter u as the ratio of the kinetic energy E to the rest energy, m_0c^2. The squared amplitude of the momentum is given by

$$p^2 = -m_0^2c^2 + \frac{(E+m_0c^2)^2}{c^2} = \frac{E^2}{c^2} + 2m_0E$$

$$= m_0^2c^2\left(\frac{E^2}{m_0^2c^4} + \frac{2E}{m_0c^2}\right)$$

$$\therefore\ p = m_0c\sqrt{u^2+2u}$$

Similarly

$$m = m_n = m_0(1+u)$$

$$m_t = m_0(1+u)^3$$

$$W = mc^2 = m_0c^2(1+u)$$

$$\lambda = h/p = \frac{h}{m_0c\sqrt{u^2+2u}}$$

(d) *The idea of reduced velocity*

This is the name often given to the vector

$$\mathbf{V} = \frac{\mathbf{v}}{\sqrt{1-v^2/c^2}}$$

in which $\mathbf{v}$ is the ordinary velocity. We have

$$V^2 = \frac{v^2}{1-v^2/c^2}, \quad v^2 = \frac{V^2}{1+V^2/c^2}, \quad 1-v^2/c^2 = \frac{1}{1+V^2/c^2}$$

The dynamical equation becomes

$$\mathbf{F} = m_0 \frac{d\mathbf{V}}{dt}$$

We know that $\mathbf{V}$ is the spatial part of the four-velocity (*Kinematics*, § [138]).

Bibliographical Notes

The variable φ, which we have already met in *Kinematics*, was suggested by the earliest relativistic authors, Weyl, for example, [1]. It is often used in cosmic-ray work. Becker [5] develops a highly original geometrical formulation of the transformations of special relativity, with detailed applications in collision theory.

Bryant [3] has systematically employed the variable ϕ; see also Arzeliès [4] and Sears [8].

[1] H. WEYL: *Space, Time, Matter*. Translated by H. Brose from the fourth German edition, Dover, 1st ed. 1922; p. 182 of the 1950 printing.
[2] H. ARZELIÈS: ref. [31] of Appendix I; p. 205 of the French edition of 1955, p. 253 of the English edition of 1966.
[3] L. H. BRYANT: Simplification of relativistic mass equations. *Amer. J. Phys.* **25** (1957) 484–485.
[4] H. ARZELIÈS: ref. [32] of Appendix I, Vol. II, pp. 294 and 444.
[5] J. BECKER: Divertissements astronautico-relativistes. *Perspectives* X (1961) 57–116.
[6] J. W. DEWDNEY: Relativistic relations among mass, velocity, momentum and kinetic energy. *Amer. J. Phys.* **28** (1960) 562.
[7] C. W. NELSON: Relativity nomograph. *Amer. J. Phys.* **29** (1961) 278–279.
[8] F. W. SEARS: Geometrical representation of relations between changes in mass, energy and momentum in special relativity. *Amer. J. Phys.* **32** (1964) 59–60.

CHAPTER II

TRANSFORMATION OF THE MAIN QUANTITIES; FOUR-VECTORS

THE equations of relativistic dynamics must conform to the principle of relativity, that is, they must retain their form in any change of Galileian system of reference (we say that they must be invariant). This enables us to establish the transformation formulae for the various quantities.

A. THE TRANSFORMATION FORMULAE IN ORDINARY TERMS

[28] Transformation from a System at Rest, K_0, to an Arbitrary Galileian System, K

Let K be the Galileian system to which we refer the particle. Consider some arbitrary given instant, and select the Ox-axis along the tangent to the trajectory. Denoting the components of force by X, Y, the equations of motion become

$$X = \frac{m_0}{(1-\beta^2)^{\frac{3}{2}}}\gamma_x, \qquad Y = \frac{m_0}{\sqrt{1-\beta^2}}\gamma_y \tag{1}$$

or

$$X = m_t\gamma_x, \qquad Y = m_n\gamma_y \tag{1'}$$

To obtain the transformation formulae, we first employ the proper system K_0 in which the particle is stationary at the time in question. In K_0

$$X_0 = m_0\gamma_{0x}, \qquad Y_0 = m_0\gamma_{0y} \tag{2}$$

We have already shown in kinematics that the transformation formulae for the components of acceleration are

$$\gamma_{0x} = \frac{\gamma_x}{(1-\beta^2)^{\frac{3}{2}}}, \qquad \gamma_{0y} = \frac{\gamma_y}{1-\beta^2}$$

and these formulae give the acceleration in any system in terms of the acceleration in the proper system. The resultants are related thus:

$$\gamma_0^2 = \frac{\gamma_y^2}{(1-\beta^2)^2} + \frac{\gamma_x^2}{(1-\beta^2)^3} = \frac{\gamma^2}{(1-\beta^2)^2} + \frac{\beta^2\gamma_x^2}{(1-\beta^2)^3}$$

Comparison of equations (1) and (2) yields the formulae for the transformation of force:

$$X_0 = X, \qquad Y_0 = \frac{Y}{\sqrt{1-\beta^2}}$$

For mass and energy,

$$m = m_n = \frac{m_0}{\sqrt{1-\beta^2}}, \qquad m_t = \frac{m_0}{(1-\beta^2)^{\frac{3}{2}}}$$

$$W = mc^2 = \frac{m_0c^2}{\sqrt{1-\beta^2}} \quad \text{so that} \quad W = \frac{W_0}{\sqrt{1-\beta^2}}$$

In K_0, the momentum is zero. In K, we write

$$p = p_x = \frac{m_0 v}{\sqrt{1-\beta^2}} = \frac{m_0c^2}{\sqrt{1-\beta^2}}\frac{v}{c^2}$$

and hence

$$p = p_x = \frac{v}{c^2}\frac{W_0}{\sqrt{1-\beta^2}}$$

[29] The General Case, in which the Particle is at Rest in Neither System; Transformation of the Energy (or Mass) and Momentum

The axes of the two systems are arranged parallel, so that the Lorentz transformation can be employed, but they are arbitrary in the sense that the velocity of the particle is parallel to none of the axes.

Let u denote the velocity of the particle (u_{1x}, u_{1y}, u_{1z} in K_1; u_{2x}, u_{2y}, u_{2z} in K_2) and let v denote that of K_2 with respect to K_1 ($\beta c = v$).

(a) *Transformation of mass (or energy)*

In K_1 and K_2, the mass is given by

$$m_1 = \frac{m_0}{\sqrt{1-u_1^2/c^2}} \quad \text{and} \quad m_2 = \frac{m_0}{\sqrt{1-u_2^2/c^2}}$$

respectively, so that

$$m_1 = m_2\sqrt{\frac{1-u_2^2/c^2}{1-u_1^2/c^2}}$$

and, with the aid of the formulae of § [80] of *Kinematics*,

$$m_1 = m_2\frac{1+vu_{2x}/c^2}{\sqrt{1-\beta^2}} \tag{1}$$

and similarly

$$W_1 = \frac{W_2 + vp_{2x}}{\sqrt{1-\beta^2}}$$

(b) *Transformation of momentum*

We wish to express the components of the momentum $\mathbf{p}$ in K_1 as functions of the values in K_2.

By definition,

$$p_{1x} = m_1 u_{1x}$$

in K_1. Replacing m_1 and u_{1x} by their values, we find

$$p_{1x} = m_2 \frac{1 + vu_{2x}/c^2}{\sqrt{1-\beta^2}} \frac{u_{2x}+v}{1+vu_{2x}/c^2} = \frac{p_{2x}+m_2 v}{\sqrt{1-\beta^2}}$$

Similarly

$$p_{1y} = m_1 u_{1y} = m_2 u_{2y} = p_{2y}, \quad p_{1z} = p_{2z}$$

Thus

$$p_{1x} = \frac{p_{2x}+m_2 v}{\sqrt{1-\beta^2}}, \quad p_{1y} = p_{2y}, \quad p_{1z} = p_{2z} \tag{2}$$

or

$$p_{1x} = \frac{p_{2x}+vW_2/c^2}{\sqrt{1-\beta^2}}$$

For the magnitudes of the momenta, it is clear that

$$p_1^2 - p_{1x}^2 = p_2^2 - p_{2x}^2$$

or, if we denote the angles between $\mathbf{p}_1$ and Ox, and $\mathbf{p}_2$ and Ox by δ_1 and δ_2,

$$p_1^2 \sin^2 \delta_1 = p_2^2 \sin^2 \delta_2$$

Collecting up on each side the quantities which refer to a single system, we obtain

$$p_1^2 = \frac{p_2^2}{1-\beta^2}\left\{1+\frac{v^2}{u_2^2}+\frac{2p_{2x}m_2 v - \beta^2(p_{2y}^2+p_{2z}^2)}{p_2^2}\right\}$$

$$= \frac{p_2^2}{1-\beta^2}\left(1+\frac{v^2}{u_2^2}+\frac{2v}{u_2}\cos\delta_2 - \beta^2 \sin^2\delta_2\right)$$

(c) *If two corpuscles have the same energy and momentum in a reference system, this equality is conserved in every other reference system*

This can be shown immediately. It is one of the reasons why momentum and energy can be grouped as a single mathematical entity (§ [34]).

(d) *Exercises; Panthria's problems*

A particle at rest in K_2 possesses momentum p_1 in K_1. What value, p_1', does p_1 take (in K_1) if in K_2 the particle has momentum p_2 parallel to p_1?

To perform the calculation, we first determine the velocity v of K_2 with respect to K_1:

$$v = \frac{cp_1}{\sqrt{p_1^2+m_0^2c^2}}$$

Furthermore, the energy in K_2 corresponding to p_2 is

$$W_2 = c\sqrt{p_2^2+m_0^2c^2}$$

The change of system then gives

$$p_1' = p_1\sqrt{1+\frac{p_2^2}{m_0^2c^2}}+p_2\sqrt{1+\frac{p_1^2}{m_0^2c^2}}$$

whereas pre-relativistically,

$$p_1' = p_1+p_2$$

This problem is exactly analogous to the "addition of velocities". Here, we "add" two momenta, measured in different systems of reference.

The same question can be applied to the kinetic energies. Let E_1 be the energy in K_1 of a particle at rest in K_2. Find E_1' in K_1 when the particle has energy E_2 in K_2, the velocities involved being parallel. We find

$$E_1' = \left\{\sqrt{E_1\left(1+\frac{E_2}{2m_0c^2}\right)}+\sqrt{E_2\left(1+\frac{E_1}{2m_0c^2}\right)}\right\}^2$$

Pre-relativistically,

$$E_1' = (\sqrt{E_1}+\sqrt{E_2})^2$$

Remark. These calculations can of course be made more general by answering the same questions for any quantity whatsoever.

[30] Force, Impulse, Work

(a) *The components of force*

A force **F** of arbitrary origin, the components of which are X_1, Y_1, Z_1 in K_1 and X_2, Y_2, Z_2 in K_2, acts on a particle travelling with arbitrary velocity u. We are required to calculate X_1, Y_1, Z_1 as functions of the data for K_2.

The component X_1

In K_1, the projection on to O_1x_1 of the dynamical formula gives

$$X_1 = \frac{dp_{1x}}{dt_1}$$

p_{1x} is given by formula (2) of § [29]; hence

$$X_1 = \frac{\mathrm{d}}{\mathrm{d}t_2}\left(\frac{p_{2x}+m_2 v}{\sqrt{1-\beta^2}}\right)\frac{\mathrm{d}t_2}{\mathrm{d}t_1} = \left(\frac{\mathrm{d}p_{2x}}{\mathrm{d}t_2}+v\frac{\mathrm{d}m_2}{\mathrm{d}t_2}\right)\frac{1}{1+vu_{2x}/c^2}$$

$$= \frac{1}{1+vu_{2x}/c^2}\left(X_2+v\frac{\mathrm{d}m_2}{\mathrm{d}t_2}\right)$$

We have still to evaluate $\mathrm{d}m_2/\mathrm{d}t_2$. To do this, we apply the energy theorem in K_2:

$$\mathrm{d}W_2 = c^2\,\mathrm{d}m_2 = X_2\,\mathrm{d}x_2+Y_2\,\mathrm{d}y_2+Z_2\,\mathrm{d}z_2$$

and X_1 becomes

$$X_1 = X_2+\frac{vu_{2y}/c^2}{1+vu_{2x}/c^2}Y_2+\frac{vu_{2z}/c^2}{1+vu_{2x}/c^2}Z_2$$

This formula can alternatively be written

$$X_1 = \frac{1}{1+vu_{2x}/c^2}\left\{X_2+\frac{v}{c^2}(\mathbf{u}_2.\mathbf{F}_2)\right\}$$

The component Y_1

We have

$$Y_1 = \frac{\mathrm{d}p_{1y}}{\mathrm{d}t_1} = \frac{\mathrm{d}p_{2y}}{\mathrm{d}t_2}\frac{\mathrm{d}t_2}{\mathrm{d}t_1}$$

$$\therefore\quad Y_1 = Y_2\frac{\sqrt{1-\beta^2}}{1+vu_{2x}/c^2}$$

The component Z_1

The calculation is identical with that for Y_1:

$$Z_1 = Z_2\frac{\sqrt{1-\beta^2}}{1+vu_{2x}/c^2}$$

Expressions required in Chapter X

In the expression for X_1, we replace u_{2y} and u_{2z} by their expressions (*Kinematics*, § [80]):

$$u_{2y} = u_{1y}\frac{1+vu_{2x}/c^2}{\sqrt{1-\beta^2}}$$

so that

$$X_1 = X_2+\frac{vu_{1y}/c^2}{\sqrt{1-\beta^2}}Y_2+\frac{vu_{1z}/c^2}{\sqrt{1-\beta^2}}Z_2$$

In the expressions for Y_1 and Z_1, we make the substitution (*Kinematics*, § [80])

$$1+\frac{vu_{2x}}{c^2} = \frac{1-\beta^2}{1-vu_{1x}/c^2}$$

whence

$$Y_1 = Y_2 \frac{1 - vu_{1x}/c^2}{\sqrt{1-\beta^2}}$$

$$Z_1 = Z_2 \frac{1 - vu_{1x}/c^2}{\sqrt{1-\beta^2}}$$

The formulae that we have just proved lead to the following two observations, long since made by various authors. *If the force depends only on the position variables* (x_2, y_2, z_2, t_2) *in* K_2, *it will depend on position and velocity in* K_1. If, in an arbitrary system, force depends on position and velocity, then it will depend only on these quantities in any other system. As we shall learn, these remarks are fundamental to the relativistic concept of the electromagnetic field.

We may also say that, if two moving particles A and B with different velocities are acted upon by equal forces in a system K, they will be acted upon by unequal forces in every other system; if under the same conditions, the forces in K are equal and opposite, this will not be true in any other system. In general, therefore, *the idea that action and reaction are equal has no place in relativistic dynamics.*

(b) *Total force*

Suppose to begin with that one system is the rest system, K_0. Let α, α_0 denote the angles between $\mathbf{F}$, $\mathbf{F}_0$ and the relative velocity of the systems:

$$\tan \alpha = \sqrt{1-\beta^2} \tan \alpha_0$$

The angle α *does not transform like the angle between two straight lines* (*Kinematics*, § [68]); we shall return to this point in §§ [40], [132] (c) and [186].

We have

$$F_0 = F \sqrt{\frac{1-\beta^2 \cos^2 \alpha}{1-\beta^2}}$$

and inversely

$$F = F_0 \sqrt{1-\beta^2 \sin^2 \alpha_0}$$

In the case of two arbitrary systems, we denote the angle between $\mathbf{F}_2$ and $\mathbf{u}_2$ by θ_2 (Fig. 7), and find

$$\cot \alpha_1 = \frac{\cos \alpha_2 + \dfrac{\beta u_2}{c} \cos \theta_2}{\sqrt{1-\beta^2} \sin \alpha_2}$$

$$F_1^2 = \frac{F_2^2}{\left(1+\dfrac{vu_{2x}}{c^2}\right)^2} \left(1 - \beta^2 \sin^2 \alpha_2 + \frac{2vu_2}{c^2} \cos \alpha_2 \cos \theta_2 + \frac{\beta^2 u_2^2}{c^2} \cos^2 \theta_2\right)$$

(c) *Transformation of impulse*

In K_1, the impulse $\mathrm{d}\mathbf{I}_1 = \mathbf{F}_1\, \mathrm{d}t_1$ has components

$$\mathrm{d}I_{1x} = X_1\, \mathrm{d}t_1, \qquad \mathrm{d}I_{1y} = Y_1\, \mathrm{d}t_1, \qquad \mathrm{d}I_{1z} = Z_1\, \mathrm{d}t_1$$

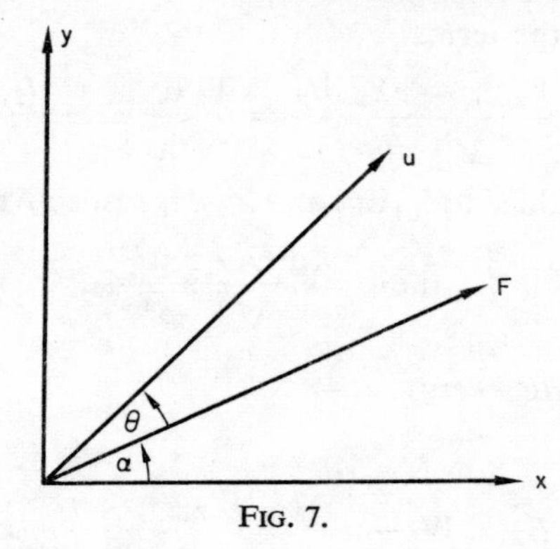

FIG. 7.

and using the force and time transformations,

$$dI_{1x} = \left\{X_2 + \frac{v/c^2}{1+vu_{2x}/c^2}(u_{2y}Y_2+u_{2z}Z_2)\right\}\frac{dt_2+\frac{v}{c^2}\,dx_2}{\sqrt{1-\beta^2}}$$

$$= \frac{1}{\sqrt{1-\beta^2}}\left(X_2\,dt_2+\frac{v}{c^2}\,\mathbf{F}_2.\mathbf{u}_2\,dt_2\right)$$

or

$$dI_{1x} = \frac{dI_{2x}+\frac{v}{c^2}\,d\mathcal{T}_2}{\sqrt{1-\beta^2}}$$

For the transverse components,

$$dI_{1y} = Y_2\frac{\sqrt{1-\beta^2}}{1+vu_{2x}/c^2}\,\frac{dt_2+\frac{v}{c^2}\,dx_2}{\sqrt{1-\beta^2}} = Y_2\,dt_2 = dI_{2y}$$

We have demonstrated the connection between energy and momentum earlier; here, impulse and work are connected by a change of system of reference. The above expression for the transformation of dI_x is the most suggestive, for it leads to the Minkowski formalism (§ [36]). We can, however, also write

$$dI_{1x} = \frac{1}{\sqrt{1-\beta^2}}\left(dI_{2x}+\frac{v}{c^2}\,d\mathbf{I}_2\,.\,\mathbf{u}_2\right)$$

(d) *Transformation of work*

Work transforms like mass or energy, in consequence of the equivalence relation $d\mathcal{T} = c^2\,dm$. It is nevertheless interesting to recover the formulae from the definition of work. We have

$$d\mathcal{T}_1 = \mathbf{F}_1.\mathbf{u}_1\,dt_1 = X_1\,dx_1+Y_1\,dy_1+Z_1\,dz_1$$

$$= \left\{X_2+\frac{v/c^2}{1+vu_{2x}/c^2}(u_{2y}Y_2+u_{2z}Z_2)\right\}\frac{dx_2+v\,dt_2}{\sqrt{1-\beta^2}}$$

$$+(Y_2\,dy_2+Z_2\,dz_2)\frac{\sqrt{1-\beta^2}}{1+vu_{2x}/c^2}$$

The component X_2 gives the terms

$$\frac{X_2\,\mathrm{d}x_2+vX_2\,\mathrm{d}t_2}{\sqrt{1-\beta^2}}=\frac{X_2\,\mathrm{d}x_2+v\,\mathrm{d}I_{2x}}{\sqrt{1-\beta^2}}$$

The terms in Y_2 and Z_2 reduce to $Y_2\,\mathrm{d}y_2$ and $Z_2\,\mathrm{d}z_2$ respectively. Hence

$$\mathrm{d}\mathcal{T}_1=\frac{\mathrm{d}\mathcal{T}_2+v\,\mathrm{d}I_{2x}}{\sqrt{1-\beta^2}}$$

(e) *Transformation of kinetic energy*

We have

$$E_1=W_1-m_0c^2=\frac{W_2+vp_{2x}}{\sqrt{1-\beta^2}}-m_0c^2$$

$$=\frac{E_2+vp_{2x}}{\sqrt{1-\beta^2}}+m_0c^2\left(\frac{1}{\sqrt{1-\beta^2}}-1\right)$$

This transformation lacks the symmetry of the preceding formulae. We shall learn that kinetic energy does not figure in the Minkowski formalism; this is an argument in favour of the total energy W, which is the true relativistic energy.

[31] The Angular Momentum about the Origin of Coordinates

Consider a particle in arbitrary motion, and its angular momentum about the origin, O. This origin is of course arbitrary and hence has no particular physical significance.

In a system K_2, the components are

$$m_{2x}=y_2p_{2z}-z_2p_{2y}$$
$$m_{2y}=z_2p_{2x}-x_2p_{2z}$$
$$m_{2z}=x_2p_{2y}-y_2p_{2x}$$

We perform the Lorentz transformation, to go over into a system K_1, with respect to which K_2 has velocity v. We introduce the angular momentum about the origin of K_1. We have (§ [29])

$$m_{2x}=y_1p_{1z}-z_1p_{1y}=m_{1x}$$

$$m_{2y}=z_1\frac{p_{1x}-vW_1/c^2}{\sqrt{1-\beta^2}}-\frac{x_1-vt_1}{\sqrt{1-\beta^2}}p_{1z}$$

$$=\frac{1}{\sqrt{1-\beta^2}}\left\{m_{1y}-\beta\left(\frac{z_1W_1}{c}-p_{1z}ct_1\right)\right\}$$

$$=\frac{1}{\sqrt{1-\beta^2}}\{m_{1y}-m_1v(z_1-v_{1z}t_1)\}$$

$$m_{2z}=\frac{x_1-vt_1}{\sqrt{1-\beta^2}}p_{1y}-y_1\frac{p_{1x}-vW_1/c^2}{\sqrt{1-\beta^2}}$$

$$=\frac{1}{\sqrt{1-\beta^2}}\{m_{1z}-m_1v(y_1-v_{1y}t_1)\}$$

Since the motion is in general arbitrary, no further simplification is possible.

[32] Transformation of the Proper Angular Momenta

It is conceivable that the above transformations might be required in some problem. It is to be stressed, however, that the angular momentum about the origin has no particular physical properties, and furthermore, it leads us to consider *different* momenta in each system (since the origins O_1 and O_2 are not only different but in relative motion). It is therefore not surprising that $\mathbf{m}_1$ and $\mathbf{m}_2$ cannot be expressed as functions of one another. Although the preceding exercise is logically sound it does not fit into the category of relativistic calculations, where the Lorentz transformation is usually applied to the expression for the *same* quantity in two different systems.

It is more interesting to consider angular momenta when there is some point C, about which to take moments, which is the same for all systems of reference. Central forces are a case in point: the same quantity is being transformed. In a concrete example, the nucleus of a hydrogen atom might be the point C, and the planetary electron the particle whose motion we analyse.

Henceforward, we assume that a Galileian system exists in which C is stationary.

The moment about C in the system K_0 in which C is at rest and in which the particle has velocity u_0 will be called the proper angular momentum. It has components

$$m_{0x} = \frac{m_0}{\sqrt{1-u_0^2/c^2}}(y_0u_{0z}-z_0u_{0y})$$

$$m_{0y} = \frac{m_0}{\sqrt{1-u_0^2/c^2}}(z_0u_{0x}-x_0u_{0z})$$

$$m_{0z} = \frac{m_0}{\sqrt{1-u_0^2/c^2}}(x_0u_{0y}-y_0u_{0x})$$

We now derive the expression for this angular momentum in a system K, with respect to which C has velocity v and the particle the velocity u. This problem does not arise in pre-relativistic dynamics.

In the system K, it *seems* natural to define the angular momentum about C by the components

$$m_x = \frac{m_0}{\sqrt{1-u^2/c^2}}(yu_z-zu_y)$$

$$m_y = \frac{m_0}{\sqrt{1-u^2/c^2}}\{z(u_x-v)-(x-vt)u_z\}$$

$$m_z = \frac{m_0}{\sqrt{1-u^2/c^2}}\{(x-vt)u_y-y(u_x-v)\}$$

The velocity with respect to the axes K_0 associated with C, (u_x-v, u_y, u_z) in terms of K, appears in these expressions. This means that we must use the vector addition law for velocities (*Kinematics*, § [78]). If we express the K_0 variables in terms of those of K in the expressions for $m_{0x}, \ldots$, for example, we obtain

$$m_{0x} = m_x, \quad m_{0y} = \frac{m_y}{\sqrt{1-\beta^2}}, \quad m_{0z} = \frac{m_z}{\sqrt{1-\beta^2}}$$

Remark. We shall see in § [39] that the above definition cannot be retained. Even now we can see that it contains a major defect: it is abnormal for the radical to contain the total velocity **u** and not the relativistic velocity.

B. THE MINKOWSKI FORMALISM

[33] Recapitulation of the Ideas of Kinematics

In a given system, we characterize an event by its three spatial coordinates (x, y, z in Cartesian coordinates) and by its time coordinate t. We saw in *Kinematics* that it is advantageous to replace the variable t by one of the quantities

$$u = ct \quad \text{or} \quad l = iu = ict$$

Below, we shall use the variables x, y, z, l exclusively, and we shall write them in the contravariant form

$$x^1 = x, \quad x^2 = y, \quad x^3 = z, \quad x^4 = ict$$

or

$$x^i \ (i = 1, 2, 3, 4)$$

Latin indices always take the values 1, 2, 3, 4 and Greek indices 1, 2, 3. In these variables, the world-element of length takes the form (*Kinematics*, p. 254)

$$\begin{aligned} ds^2 &= dx^2+dy^2+dz^2+dl^2 \\ &= (dx^1)^2+(dx^2)^2+(dx^3)^2+(dx^4)^2 \end{aligned}$$

which gives a Euclidean metric. In this case, there is no need to distinguish between co-variance and contravariance. Nevertheless, we shall employ the correct indicial notation x^i (a contravariant quantity) and x_i (a covariant quantity) for two reasons. This notation allows us to simplify the expressions with the aid of *Einstein's summation convention*: if the same index occupies the two positions, above and below, summation is implied. For example, we write

$$ds^2 = \sum_i (dx^i)^2 = dx_i \, dx^i$$

Furthermore, the correct notation prepares us for generalizations. Nevertheless, if the reader is not familiar with tensors, he can disregard the position of the indices, upper or lower ($x^i = x_i$). He need only remember the summation convention. These indices only indicate contra- or covariance in the general notation, x^i or x_i. When the variables x, y, z, l are employed explicitly, they will be used as indices without any idea of variance being attached to them. P'_x, for example, is quite simply the component of P along Ox in K'.

We now state the transformation formulae for vectors and tensors in matrix form (the special Lorentz transformation) where K' has velocity v with respect to K:

$$\begin{pmatrix} P'_x \\ P'_y \\ P'_z \\ P'_l \end{pmatrix} = \begin{pmatrix} \frac{1}{\sqrt{1-\beta^2}} & 0 & 0 & \frac{i\beta}{\sqrt{1-\beta^2}} \\ 0 & 1 & 0 & 0 \\ 0 & 0 & 1 & 0 \\ \frac{-i\beta}{\sqrt{1-\beta^2}} & 0 & 0 & \frac{1}{\sqrt{1-\beta^2}} \end{pmatrix} \begin{pmatrix} P_x \\ P_y \\ P_z \\ P_l \end{pmatrix} \tag{1}$$

and, for a general Lorentz transformation characterized by the matrix $\boldsymbol{\alpha}$,

$$\{P'\} = \boldsymbol{\alpha}.\{P\}$$

For second-order tensors

$$\mathfrak{T}' = \boldsymbol{\alpha}\mathfrak{T}\boldsymbol{\alpha}_t \tag{2}$$

in which t indicates that the second matrix is transposed.

[34] The Four-momentum

The quantities dx^i form the contravariant components of a vector. Since dT is an invariant, the components of world-velocity, defined by (*Kinematics*, § [138])

$$V^i = \frac{dx^i}{dT}$$

are thus also contravariant. The same is true of the components of world-momentum

$$\Pi^i = m_0 V^i = m_0 \frac{dx^i}{dT}$$

The Cartesian components of this vector are

$$\Pi_x = \frac{m_0}{\sqrt{1-\beta^2}} \frac{dx}{dt}, \qquad \Pi_y = \frac{m_0}{\sqrt{1-\beta^2}} \frac{dy}{dt}$$

$$\Pi_z = \frac{m_0}{\sqrt{1-\beta^2}} \frac{dz}{dt}, \qquad \Pi_l = \frac{i m_0 c}{\sqrt{1-\beta^2}} = i\frac{W}{c}$$

Its spatial component is therefore momentum, and its time component is proportional to the energy of the particle. The squared modulus has the (invariant) value

$$\Pi_i \Pi^i = -m_0^2 c^2$$

Remark. Some authors define world-velocity as the ratio dx/dt, which leads to formal differences. It seems more natural to use the proper time dT, the more general form of dt.

[35] The Four-acceleration: Recapitulation

Using the indicial notation, we write (in terms of the contravariant components of velocity)

$$\Gamma^i = \frac{\mathrm{d}V^i}{\mathrm{d}T}$$

The Cartesian components are, explicitly,

$$\Gamma_x = \frac{\mathrm{d}V_x}{\mathrm{d}T} = \frac{1}{\sqrt{1-\beta^2}}\frac{\mathrm{d}}{\mathrm{d}t}\left(\frac{1}{\sqrt{1-\beta^2}}\frac{\mathrm{d}x}{\mathrm{d}t}\right), \ldots$$

$$\Gamma_l = \frac{\mathrm{d}V_l}{\mathrm{d}T} = \frac{ic}{\sqrt{1-\beta^2}}\frac{\mathrm{d}}{\mathrm{d}t}\left(\frac{1}{\sqrt{1-\beta^2}}\right)$$

The relation

$$V_i\Gamma^i = V_x\Gamma^x + V_y\Gamma^y + V_z\Gamma^z + V_l\Gamma^l = 0$$

shows that $\mathbf{\Gamma}$ *is normal to the world-velocity, and hence, to the world-line. It is a space-like vector.* Since the modulus is an invariant, we can evaluate it in any system whatsoever. In the rest-system, we find simply

$$\Gamma_x = \frac{\mathrm{d}^2x}{\mathrm{d}t^2}, \ldots, \quad \Gamma_l = 0$$

so that *the modulus of* Γ *is equal to the ordinary acceleration, calculated in the rest-system.*

[36] The Four-vector Force and the Four-vector Impulse (or World-impulse)

We have just seen that the vector Π^i gathers into a single concept the energy and momentum of a corpuscle. The world-force in its turn connects force and power. If **v** is the particle velocity, the components of the world-force are

$$K_x = \frac{F_x}{\sqrt{1-\beta^2}}, \quad K_y = \frac{F_y}{\sqrt{1-\beta^2}}, \quad K_z = \frac{F_z}{\sqrt{1-\beta^2}}$$

$$K_l = \frac{i}{c}\frac{\mathbf{v}\,.\,\mathbf{F}}{\sqrt{1-\beta^2}}$$

This four-vector is perpendicular to the four-velocity and we can show that

$$K_iV^i = 0$$

or, in Cartesians,

$$K_xV_x + K_yV_y + K_zV_z + K_lV_l = 0$$

The world-impulse is thus, by definition

$$\mathbf{I} = \mathbf{K}\,\mathrm{d}T$$

Remark 1. If formula (1) of § [33] is applied to the four-vectors K^i and Π^i, we recover the transformation formulae established by direct methods in the first part of this chapter.

Remark 2. I believe that it is worth while to distinguish between momentum and impulse; we must not be too strict, however, for many authors confuse them. Likewise, the product $K^i\,\mathrm{d}T$ may be called either four-work of four-energy.

[37] The Dynamical Law for a Particle, in Vector Form

If we employ vectors with four components, the three equations for impulse and the energy equation can be summarized in a single world-vector relation, *independent of the system of reference:*

$$K^i = \frac{\mathrm{d}\Pi^i}{\mathrm{d}T}$$

which expresses the equality between the world-impulse and the increase of world-momentum. We may also write

$$K^i = \frac{\mathrm{d}}{\mathrm{d}T}\left(m_0 \frac{\mathrm{d}x^i}{\mathrm{d}T}\right) = m_0 \frac{\mathrm{d}V^i}{\mathrm{d}T} = m_0 \Gamma^i$$

These statements are proved as follows. For the spatial components, we have explicitly

$$\frac{F_x}{\sqrt{1-\beta^2}} = \frac{\mathrm{d}}{\mathrm{d}t}\left(\frac{m_0 v_x}{\sqrt{1-\beta^2}}\right)\frac{\mathrm{d}t}{\mathrm{d}T} = \frac{1}{\sqrt{1-\beta^2}}\,\frac{\mathrm{d}}{\mathrm{d}t}\left(\frac{m_0 v_x}{\sqrt{1-\beta^2}}\right)$$

and we do indeed find

$$F_x = \frac{\mathrm{d}p_x}{\mathrm{d}t}$$

For the time component

$$\frac{i}{c}\,\frac{\mathrm{d}W}{\mathrm{d}t}\,\frac{1}{\sqrt{1-\beta^2}} = \frac{i}{c}\,\frac{\mathbf{v}\cdot\mathbf{F}}{\sqrt{1-\beta^2}} = \frac{i}{c}\,\frac{\mathrm{d}W}{\mathrm{d}t}\,\frac{1}{\sqrt{1-\beta^2}}$$

which is an identity.

Remark. We shall meet in Chapter III (§ [49]) another four-dimensional form of the equations of dynamics.

[38] Relativistic Dynamics Deduced from Newtonian Dynamics and the Principle of Relativity; the General Method of Obtaining Relativistic Laws

(a) Let us now consider the question from another point of view. We disregard §§ [34], [36] and [37], and assume that only the principle of relativity (the Lorentz transformation) and the *kinematic* quantities, world-velocity and acceleration, are given.

A general method of seeking a relativistic law is to take the pre-relativistic law (if the latter is applicable in a particular reference system) and apply the Lorentz transformation.

In the present case, the Newtonian law

$$\mathbf{F}_0 = m_0 \boldsymbol{\gamma}_0 \tag{1}$$

is very satisfactory at low velocities; it is therefore rigorously true in the reference system K_0 in which the moving body is at rest at the time in question. We therefore take (1) as a postulate. The relativistic law we are seeking is a vector relation in four dimensions, so that our first task is to cast (1) into world-vector form. We are already familiar with the four-vector acceleration $\mathbf{\Gamma}_0$, which in K_0 has spatial component γ_0 and time-component zero.

We *define* a four-vector force by the following condition: in K_0, the spatial component is $\mathbf{F}_0$ and the time-component must vanish. Relation (1) may also be written

$$\mathbf{K}_0 = m_0 \mathbf{\Gamma}_0 \tag{2}$$

In another system K, related to K_0 through the simple Lorentz transformation, the components of $\mathbf{K}$ are given by the matrix

$$\begin{pmatrix} K_x \\ K_y \\ K_z \\ K_l \end{pmatrix} = \begin{pmatrix} \frac{1}{\sqrt{1-\beta^2}} & 0 & 0 & -\frac{i\beta}{\sqrt{1-\beta^2}} \\ 0 & 1 & 0 & 0 \\ 0 & 0 & 1 & 0 \\ \frac{i\beta}{\sqrt{1-\beta^2}} & 0 & 0 & \frac{1}{\sqrt{1-\beta^2}} \end{pmatrix} \begin{pmatrix} F_{0x} \\ F_{0y} \\ F_{0z} \\ 0 \end{pmatrix}$$

whence

$$\begin{aligned} K_x &= \frac{F_{0x}}{\sqrt{1-\beta^2}}, \qquad & K_y &= F_{0y} \\ K_z &= F_{0z}, \qquad & K_l &= \frac{i\beta F_{0x}}{\sqrt{1-\beta^2}} \end{aligned} \tag{3}$$

Relation (2) is a world-expression; it is therefore independent of the reference system, and we may also write

$$\mathbf{K} = m_0 \mathbf{\Gamma} \quad \text{or} \quad K^i = m_0 \Gamma^i \tag{4}$$

Substituting the components (3) and the components of $\mathbf{\Gamma}$ (§ [35]) into (4), we find

$$K_x = \frac{F_{0x}}{\sqrt{1-\beta^2}} = m_0 \Gamma_x = \frac{1}{\sqrt{1-\beta^2}} \frac{\mathrm{d}}{\mathrm{d}t} \left(\frac{m_0}{\sqrt{1-\beta^2}} \frac{\mathrm{d}x}{\mathrm{d}t} \right)$$

or

$$F_{0x} = \frac{\mathrm{d}}{\mathrm{d}t} \left(\frac{m_0}{\sqrt{1-\beta^2}} \frac{\mathrm{d}x}{\mathrm{d}t} \right) \tag{5}$$

Likewise

$$\begin{aligned} K_y &= F_{0y} = m_0 \Gamma_y = \frac{1}{\sqrt{1-\beta^2}} \frac{\mathrm{d}}{\mathrm{d}t} \left(\frac{m_0}{\sqrt{1-\beta^2}} \frac{\mathrm{d}y}{\mathrm{d}t} \right) \\ K_z &= F_{0z} = m_0 \Gamma_z = \frac{1}{\sqrt{1-\beta^2}} \frac{\mathrm{d}}{\mathrm{d}t} \left(\frac{m_0}{\sqrt{1-\beta^2}} \frac{\mathrm{d}z}{\mathrm{d}t} \right) \\ K_l &= \frac{i\beta F_{0x}}{\sqrt{1-\beta^2}} = m_0 \Gamma_l = \frac{1}{\sqrt{1-\beta^2}} \frac{\mathrm{d}}{\mathrm{d}t} \left(\frac{i m_0 c}{\sqrt{1-\beta^2}} \right) \end{aligned} \tag{6}$$

Relations (5) and (6) lead us quite naturally to define a new quantity, with components

$$\frac{m_0}{\sqrt{1-\beta^2}}\frac{\mathrm{d}x}{\mathrm{d}t} = m_0\frac{\mathrm{d}x}{\mathrm{d}T}, \ldots, \frac{im_0c}{\sqrt{1-\beta^2}}$$

or $m_0\mathbf{V}$, in which $\mathbf{V}$ is the four-velocity. This is the world-momentum, which we denote by $\boldsymbol{\Pi}$.

Relation (4) can thus also be written

$$K^i = \frac{\mathrm{d}\Pi^i}{\mathrm{d}T} \tag{7}$$

Relations (5) and (6) form the law of dynamics, in the system K, stated explicitly in terms of the usual variables. We notice that the notions of mass and force in K have not yet been defined (that is, their meaning for a particle in motion). These definitions are therefore not at all necessary to describe the phenomena.

(b) *Definition of the three-dimensional force and the mass in K.* Consider the first three equations of (5) and (6). It is quite natural to *define* momentum $\mathbf{p}$ in K by the three components

$$p_x = \frac{m_0}{\sqrt{1-\beta^2}}\frac{\mathrm{d}x}{\mathrm{d}t}, \ldots$$

and this leads us to *define* mass, by writing

$$\mathbf{p} = m\mathbf{v}, \quad m = \frac{m_0}{\sqrt{1-\beta^2}}$$

The corresponding relations of (5) and (6) become

$$F_{0x} = \frac{\mathrm{d}p_x}{\mathrm{d}t}, \quad F_{0y}\sqrt{1-\beta^2} = \frac{\mathrm{d}p_y}{\mathrm{d}t}$$

To simplify the notation, we *define* a new quantity $\mathbf{F}$, which we call force in K, with the components

$$F_x = F_{0x}, \quad F_y = F_{0y}\sqrt{1-\beta^2}, \quad F_z = F_{0z}\sqrt{1-\beta^2}$$

The fourth relation (6) gives

$$vF_{0x} = \frac{\mathrm{d}}{\mathrm{d}t}\left(\frac{m_0c^2}{\sqrt{1-\beta^2}}\right)$$

This is the energy theorem.

We notice that the power is given by vF_{0x} or vF_x, or in general, $\mathbf{v}.\mathbf{F}$. The definitions of $\mathbf{F}$ and work are thus compatible.

Important remark. These concepts are common, but one could manage without them. Only m_0 and F_0 need be given. We shall return to this in Chapter IX.

[39] Angular Momentum

(a) *Angular momentum about the origin*

This quantity is defined by

$$m^{ik} = x^i \Pi^k - x^k \Pi^i \tag{1}$$

where Π^i is the world-angular momentum and x^i are the coordinates of the moving point M. We thus obtain the spatial components

$$m^{23} = -m^{32} = yp_z - zp_y = m_x, \ldots \tag{2}$$

This is the angular momentum defined in § [31]. With $x^4 = ict$, the time components have the form

$$m^{14} = -m^{41} = xicm - ictp_x, \ldots \tag{3}$$

If we consider the origin and the moving point M simultaneously ($t = 0$), we bring out the components of the baryocentric momentum, $m_x, \ldots$.

(b) *Proper angular momentum*

We define this quantity by

$$m^{ik} = m_0 \{x^i(U^k - V^k) - x^k(U^i - V^i)\} \tag{4}$$

in which U^i is the four-velocity of M and V^i is that of the centre C about which the angular momenta are taken. For the spatial components, for example, we have

$$m^{13} = -m^{31} = m_y = \frac{m_0}{\sqrt{1 - u^2/c^2}}(zu_x - xu_z) - \frac{m_0}{\sqrt{1 - v^2/c^2}}(zv_x - xv_z)$$

while for the simple Lorentz transformation, we have

$$m_y = \frac{m_0}{\sqrt{1 - u^2/c^2}}(zu_x - xu_z) - \frac{m_0}{\sqrt{1 - v^2/c^2}}zv$$

We compare this with the definition of § [32], which gave

$$m_y = \frac{m_0}{\sqrt{1 - u^2/c^2}}(zu_x - xu_z) - \frac{m_0}{\sqrt{1 - u^2/c^2}}zv$$

Remark: For the components of angular momentum density, see § [41].

HISTORICAL AND BIBLIOGRAPHICAL NOTES

[40] On the Transformation of Force, Energy and Impulse

The transformation formulae which first appeared in the literature dealt with the fields **E** and $\mathfrak{B}$, and were given by H. A. Lorentz and Poincaré in particular (§ [71B]). Einstein [1] set out the equations of dynamics in a system K, by using the force in K_0, expressed as a function of **E** and $\mathfrak{B}$ (§ [43]). The force transformation can be deduced from these, but it appears explicitly only in later works, von Laue [2], for example, where the formulae for impulse and energy are also given. Interesting complementary comments have been made by Pathria [3, 4]. Very recently some authors have questioned the correctness of the transformation formulae for the force, quite unjustifiably in my opinion (§ [186]).

[1] A. EINSTEIN: ref. [1] of § [20](b).
[2] M. VON LAUE: ref. [2] of Appendix I.
[3] R. K. PATHRIA: Resultant momentum and kinetic energy in relativistic mechanics. *Amer. J. Phys.* **24** (1956) 47–48.
[4] R. K. PATHRIA: On transformations in relativistic mechanics. *Amer. J. Phys.* **24** (1956) 411–412

[41] On the Concept of Angular Momentum

(a) The definition of angular momentum about the origin is given by various authors [1 – 6] in the form I have adopted in § [39](a). On the other hand, the definition that I have given in § [39](b) for the proper angular momentum *does not seem to have been suggested hitherto*. In the first French edition of the present book (1957), I wrongly presented the definition of § [32] as a perfectly acceptable one, following L. de Broglie in this; in fact, it must be replaced by the definition given in § [39](b).

(b) The Dirac theory attributes a magnetic moment (and hence angular momentum) to the electron; this quantity is represented by a four-vector. At first, this was considered to be paradoxical, since the physical idea of angular momentum seemed to imply tensor variance. The explanation lies in the fact that the Dirac theory involves densities. Let us denote the angular momentum of an element of volume by dm_{ij}. If we represent the element of volume by the tensor

$$d\varpi_{ijk} = dx_i \times dx_j \times dx_k$$

then we shall define the momentum density by a four-vector σ^i:

$$\sigma^k d\varpi_{ijk} = dm_{ij}$$

If we represent the volume element by the invariant

$$d\varpi_0 = dx_1 \times dx_2 \times dx_3 \times dx_4$$

then we shall define the momentum density by a tensor:

$$\sigma_{ij} d\varpi_0 = dm_{ij}$$

Each definition has its advantages. The writings of Costa de Beauregard [3], who has made a special study of these problems, should be consulted; he adopts the vector σ^k.

[1] L. DE BROGLIE: La variance relativiste du moment cinétique d'un corps en rotation. *J. Math. pures et appl.* **15** (1936) 89–95.
[2] L. DE BROGLIE: *Théorie générale des particules à spin.* Paris, Gauthier-Villars, 1943, pp. 44–60.
[3] O. COSTA DE BEAUREGARD: refs. [7] and [35] of Appendix I.
[4] L. DE BROGLIE: *La Théorie des particules de spin* $\frac{1}{2}$. Paris, Gauthier-Villars, 1952, pp. 40–51.
[5] C. MØLLER: ref. [8] of Appendix I, p. 110.
[6] J. L. SYNGE: ref. [13] of Appendix I, p. 216.

[42] On the World Continuum of Minkowski and its Applications in Dynamics

The idea of regarding the time variable as the fourth dimension, to be added to the three spatial dimensions, is quite old. It is already to be found in the *Encyclopédie* over the signature of d'Alembert [01], and also in Lagrange's work [02]. Only with the discovery of the Lorentz transformation, and above all, of the time transformation, did this idea become important and attract the attention of the physicists.

Poincaré [1], in investigating the invariants of the Lorentz group, envisaged

$$x, y, z, \sqrt{-1}t$$

"comme les coordonnés dans l'espace à quatre dimensions".† He states that in this space, the distance

$$x^2+y^2+z^2-t^2$$

between two points is invariant, and that "la transformation de Lorentz n'est qu'une rotation de cet espace autour de l'origine, regardée comme fixe".‡

This point of view was considerably extended in 1908 by Hargreaves [3] and Minkowski [2] simultaneously, in two fundamental papers. Hargreaves' work is virtually forgotten, but Bateman [8] and Whittaker (in his historical work, see Appendix I, ref. [29]) have rightly stressed its merits: it contains a four-dimensional analysis of the field equations.

Nevertheless, credit for systematically studying the possibilities of four-dimensional calculus must indisputably go to Minkowski; his work immediately attracted the attention of physicists and mathematicians. The object of his first article [2] was to deduce the equations of electromagnetism for moving media, starting from the familiar equations for media at rest, with the aid of the principle of relativity; for discussion of this topic, the reader is referred to another volume (*Milieux conducteurs et polarisables en mouvement*, Gauthier-Villars, 1959).

In the course of analysing this problem, Minkowski established the basic properties of world-vectors and tensors (world or "space-time", "espace-temps", "Raum-Zeit"), and discovered the most important of these. He gives the world-equations for the electromagnetic field *in vacuo*, for example (see my Chapter VIII, and the tensor there denoted by $\mathcal{N}$: § [101]). In the second part of his memoir, the equations of particle dynamics are obtained in much the same way as I have derived them above. The ideas of world-line and proper time are introduced in the course of his discussion.

The geometrical interpretation of the Lorentz transformation and its consequences are studied in an article of 1909 [4], which is particularly devoted to questions of kinematics.

We should also mention two articles by Minkowski which were published posthumously, one edited by M. Born [5] and the other by Lorentz and Einstein [10]. This tool of Minkowski's was developed systematically by several authors. We have to thank Sommerfeld [6], Bateman [8] and Kafka [9] for important contributions. Von Laue [7] produced the first general treatise on relativistic physics, using world-vectors and tensors systematically.

The fullest and most elegant presentation of Minkowski's methods is incontestably to be found in the works of Costa de Beauregard [11] and Synge [12].

The amalgamation of the space and time variables into a single mathematical entity possesses important epistemological consequences, which have been much studied. Meyerson [13] gives a general survey of the question, with references to pre-1924 work. For more recent discussion, see § [43A] of my *Kinematics*.

Remark. The reader who wishes to know what the non-relativists think about the concept of space-time may read E. Sevin [14] or du Bourg de Bozas [15], for example.

[01] J. D'ALEMBERT: *Encyclopédie*, Vol. 4, 1754, p. 1010, the word "Dimension".
[02] L. DE LAGRANGE: *Oeuvres*, Paris, 1867–1892, Vol. 9, p. 337.
[1] H. POINCARÉ: Sur la dynamique de l'électron. *Rend. Circ. Mat. Palermo* **21** (1906) 129-176; *Oeuvres complètes*, Gauthier-Villars, Vol. IX, pp. 494–586 (see especially p. 542).

† "As the coordinates in four-dimensional space."
‡ "The Lorentz transformation is no more than a rotation of this space about the origin, which is regarded as fixed."

[2] H. MINKOWSKI: Die Grundgleichungen für die elektromagnetischen Vorgänge in bewegten Körpern. *Nachr, Göttingen*, 1908, pp. 53–111.

[3] R. HARGREAVES: Integral forms and their connexion with physical equations. *Trans. Camb. Phil. Soc.* (1908) 107–122.

[4] H. MINKOWSKI: Raum und Zeit. *Phys. Z.* (1909) 104–111.

[5] H. MINKOWSKI: Eine Ableitung der Grundgleichungen für die elektromagnetischen Vorgänge in bewegten Körpern vom Standpunktes der Elektronentheorie. *Math. Ann.* **68** (1910) 526–551. Edited by M. Born (Aus dem Nachlass von Herman Minkowski). This paper is preceded by a new printing of the 1908 article (Ref. [2]), pp. 472–525.

[6] A. SOMMERFELD: Zur Relativitätstheorie. Vierdimensionale Vektoranalysis. *Ann. Physik* **32** (1910) 749–776; *ibid.* **33** (1910) 650–689.

[7] VON LAUE: ref. [2] of Appendix I.

[8] H. BATEMAN: The electromagnetic vectors. *Phys. Rev.* **12** (1918) 450–481.

[9] H. KAFKA: Zur vierdimensionalen Analysis. *Ann. Physik* **58** (1919) 1–54.

[10] LORENTZ, EINSTEIN and MINKOWSKI: *Das Relativitätsprinzip*, Teubner, 1922. (This gives the text of a lecture, delivered at Cologne on 21 September 1908.)

[11] O. COSTA DE BEAUREGARD: refs. [7] and [35] of Appendix I.

[12] J. L. SYNGE: ref. [13] of Appendix I.

[13] E. MEYERSON: *La Déduction relativiste*, Paris, Payot, 1924; see especially pp. 97–110.

[14] E. SEVIN: ref. [10] of § [115]. This author introduces a four-dimensional space (plus time); his ideas are very odd.

[15] DU BOURG DE BOZAS: *Les Espaces-temps du Docteur Martiny*, Imprimerie de La Chapelle, Montligeon (Orne), 1952.

CHAPTER III

RELATIVISTIC DECOMPOSITION OF FORCE (HEAVISIDE, THOMSON, LORENTZ)

This chapter is fundamental to a proper understanding of electromagnetism

[43] The two Heaviside–Thomson–Lorentz Components of the Total Force

In § [30], we saw that in general force depends upon the velocity, **u**; for some particular frame of reference, it may not depend upon **u**, but for any other frame, it will. It is therefore of interest to express it as the sum of two terms, only one of which depends upon **u**.

(a) Consider a corpuscle, acted upon by a force $\mathbf{F}_2$, with components X_2, Y_2, Z_2, which may be independent of velocity, in the reference system K_2.

In another system K, with respect to which K_2 has velocity v parallel to Ox, the particle has velocity **u** and experiences a force **F** which has components (§ [30])

$$\begin{aligned} X &= X_2 + \frac{vu_y/c^2}{\sqrt{1-v^2/c^2}}\,Y_2 + \frac{vu_z/c^2}{\sqrt{1-v^2/c^2}}\,Z_2 \\ Y &= \frac{1-vu_x/c^2}{\sqrt{1-v^2/c^2}}\,Y_2, \quad Z = \frac{1-vu_x/c^2}{\sqrt{1-v^2/c^2}}\,Z_2 \end{aligned} \tag{1}$$

These formulae suggest that we introduce the vectors **P** and **Q**, with components

$$\mathbf{P}\begin{cases} X_2 \\ \dfrac{Y_2}{\sqrt{1-v^2/c^2}} \\ \dfrac{Z_2}{\sqrt{1-v^2/c^2}} \end{cases}; \qquad \mathbf{Q}\begin{cases} 0 \\ \dfrac{-(v/c^2)\,Z_2}{\sqrt{1-v^2/c^2}} \\ \dfrac{(v/c^2)\,Y_2}{\sqrt{1-v^2/c^2}} \end{cases} \tag{2}$$

Formulae (1) can then be written in vector form

$$\mathbf{F} = \mathbf{P} + \mathbf{u}\times\mathbf{Q}$$

in which **P** is a polar vector and **Q** is an axial vector.

The total force **F** reduces to the component **P** when the velocity **u** vanishes, that is, when we use the Lorentz transformation to change into the rest system K_0.

This decomposition is obviously suggested by the expression for the electromagnetic force. Whereas this expression was used only in electricity in pre-relativistic physics, however, it occurs in relativistic dynamics as a mechanical idea of general application.

It is well known that the consequences of applying Newtonian dynamics to the motion of charged particles in an electromagnetic field often seem strange to our pre-relativistic attitudes. The reason for this is that *even at low velocities, relativistic dynamics governs this motion.*

"Par ma foi! il y a plus de quarante ans que je dis de la prose sans que j'en susse rien, et je vous suis le plus obligé du monde de m'avoir appris cela"† said Monsieur Jourdain to the Philosopher.

In just the same way, physicists have been using relativistic dynamics without being aware of it. Much becomes clear if we resolutely adopt this point of view.

Henceforward, *the reader should avoid confusing relativistic dynamics and the dynamics of high velocities; the first is the more general title.*

(b) In terms of total force, the impulse and energy equations are of the form

$$\mathbf{F}\,\mathrm{d}t = \mathrm{d}(m\mathbf{v}), \qquad \mathbf{F}.\mathrm{d}\mathbf{s} = \mathrm{d}W = c^2\,\mathrm{d}m$$

Using the new components, they become

$$(\mathbf{P}+\mathbf{u}\times\mathbf{Q})\,\mathrm{d}t = \mathrm{d}(m\mathbf{v})$$
$$\mathbf{P}.\mathrm{d}\mathbf{s} = \mathrm{d}W = c^2\,\mathrm{d}m$$

since the component $\mathbf{u}\times\mathbf{Q}$ does no work.

To obtain formulae which we can use in electricity without alteration, we introduce the constants of proportionality ε and c_0 and the two vectors $\mathbf{E}$ and $\mathfrak{B}$, and in future we shall write

$$\mathbf{F} = \varepsilon\left(\mathbf{E}+\frac{\mathbf{u}}{c_0}\times\mathfrak{B}\right)$$

Later we shall encounter the vectors $\mathbf{E}$ and $\mathfrak{B}$ in the form of electric field and magnetic field, whence the notation. I should like to insist yet again, however, that for the present, the reader must regard their definitions as purely mechanical. There is of course no reason not to bear these fields in mind, if some familiar mental picture is desired. Similarly, ε is here just a constant, which we are perfectly free to put equal to unity.

Remark. For $\mathfrak{B}$, two notations are possible. Most of the time, it will be regarded as a vector, defined by three components, which in Cartesians we write $\mathcal{B}_x$, $\mathcal{B}_y$, $\mathcal{B}_z$. Physically, it is preferable to regard it as a tensor, the components of which are written in terms of those of the vector as follows:

$$\begin{pmatrix} \mathcal{B}_{xx} & \mathcal{B}_{xy} & \mathcal{B}_{xz} \\ \mathcal{B}_{yx} & \mathcal{B}_{yy} & \mathcal{B}_{yz} \\ \mathcal{B}_{zx} & \mathcal{B}_{zy} & \mathcal{B}_{zz} \end{pmatrix} = \begin{pmatrix} 0 & \mathcal{B}_z & -\mathcal{B}_y \\ -\mathcal{B}_z & 0 & \mathcal{B}_x \\ \mathcal{B}_y & -\mathcal{B}_x & 0 \end{pmatrix}$$

The vector $\mathfrak{B}$ is the dual or adjoint vector of the tensor $\mathcal{B}_{ij}$.

† "By the Lord! Forty years and more I have been speaking in prose without knowing a thing about it and I am very much obliged to you for telling me" (Molière: *Le Bourgeois Gentilhomme*).

The force formula is thus

$$F_x = \varepsilon E_x + \frac{\varepsilon}{c_0}(u_x \mathcal{B}_{xx} + u_y \mathcal{B}_{xy} + u_z \mathcal{B}_{xz})$$

or in matrix and tensor notation

$$\{F\} = \varepsilon\{E\} + \frac{\varepsilon}{c_0}\{u\}\{\mathcal{B}\}$$

$$F_i = \varepsilon E_i + \frac{\varepsilon}{c_0} u^j \mathcal{B}_{ij}$$

The scalar product

$$\mathbf{u} \cdot (\mathbf{u} \times \mathfrak{B}) = u^i (\mathbf{u} \times \mathfrak{B})_i$$

is zero:

$$\mathcal{B}_{ij} u^i u^j = 0$$

[44] Transformation of the Components (or Fields) E and $\mathfrak{B}$

(a) Consider a field $\mathbf{E}_1$ and a field $\mathfrak{B}_1$ in K_1. A particle travelling with velocity $\mathbf{u}_1$ in K_1 thus experiences a force

$$\mathbf{F}_1 = \varepsilon\left(\mathbf{E}_1 + \frac{\mathbf{u}_1}{c_0} \times \mathfrak{B}_1\right)$$

or in rectangular Cartesians,

$$X = \varepsilon\left\{E_x + \frac{1}{c_0}(u_y \mathcal{B}_z - u_z \mathcal{B}_y)\right\}, \ \ldots$$

In another reference system K_2, travelling with velocity $\mathbf{v}$ with respect to K_1, the particle will have velocity $\mathbf{u}_2$; the force and fields become $\mathbf{F}_2$, $\mathbf{E}_2$ and $\mathfrak{B}_2$ respectively, with

$$\mathbf{F}_2 = \varepsilon\left(\mathbf{E}_2 + \frac{\mathbf{u}_2}{c_0} \times \mathfrak{B}_2\right)$$

The constant ε is treated as an invariant *by convention.*

As usual, we select the x-axes to be parallel to the relative velocity $\mathbf{v}$.

(b) The force transformation (§ [30]) gives, for the Y component,

$$Y_1 = Y_2 \frac{1 - v u_{1x}/c^2}{\sqrt{1-\beta^2}}, \qquad \beta = v/c$$

Introducing the expression for Y in terms of $\mathbf{E}$ and $\mathfrak{B}$, we obtain

$$E_{1y} + \frac{1}{c_0}(u_{1z}\mathcal{B}_{1x} - u_{1x}\mathcal{B}_{1z})$$
$$= \frac{1 - v u_{1x}/c^2}{\sqrt{1-\beta^2}}\left\{E_{2y} + \frac{1}{c_0}(u_{2z}\mathcal{B}_{2x} - u_{2x}\mathcal{B}_{2z})\right\}$$

In addition, the velocity transformation yields

$$u_{2x} = \frac{u_{1x}-v}{1-vu_{1x}/c^2}, \qquad u_{2z} = u_{1z}\frac{\sqrt{1-\beta^2}}{1-vu_{1x}/c^2}$$

Hence

$$E_{1y}+\frac{1}{c_0}(u_{1z}\mathcal{B}_{1x}-u_{1x}\mathcal{B}_{1z})$$

$$= \frac{1-vu_{1x}/c^2}{\sqrt{1-\beta^2}}\left\{E_{2y}+\frac{1}{c_0}\left(\frac{u_{1z}\sqrt{1-\beta^2}}{1-vu_{1x}/c^2}\mathcal{B}_{2x}-\frac{u_{1x}-v}{1-vu_{1x}/c^2}\mathcal{B}_{2z}\right)\right\}$$

$$= \frac{1-vu_{1x}/c^2}{\sqrt{1-\beta^2}}E_{2y}+\frac{1}{c_0}\left(u_{1z}\mathcal{B}_{2x}-\frac{u_{1x}-v}{\sqrt{1-\beta^2}}\mathcal{B}_{2z}\right)$$

Rearranging,

$$\left(E_{1y}-\frac{E_{2y}}{\sqrt{1-\beta^2}}-\frac{v\mathcal{B}_{2z}}{c_0\sqrt{1-\beta^2}}\right)+\left(-\frac{\mathcal{B}_{1z}}{c_0}+\frac{vE_{2y}}{c^2\sqrt{1-\beta^2}}+\frac{\mathcal{B}_{2z}}{c_0\sqrt{1-\beta^2}}\right)u_{1x}$$
$$+\frac{u_{1z}}{c_0}(\mathcal{B}_{1x}-\mathcal{B}_{2x}) = 0$$

The relation above must be true for any values of u_{1x} and u_{1z}; the three quantities in brackets must thus be separately equal to zero, giving

$$E_{1y} = \frac{E_{2y}+(v/c_0)\mathcal{B}_{2z}}{\sqrt{1-\beta^2}}$$

$$\mathcal{B}_{1z} = \frac{\mathcal{B}_{2z}+(c_0\beta/c)E_{2y}}{\sqrt{1-\beta^2}}$$

$$\mathcal{B}_{1x} = \mathcal{B}_{2x}.$$

(c) If we perform the same calculations on

$$Z_1 = Z_2\frac{1-vu_{1x}/c^2}{\sqrt{1-\beta^2}}$$

we find

$$\left(E_{1z}-\frac{E_{2z}}{\sqrt{1-\beta^2}}+\frac{v}{c_0}\frac{\mathcal{B}_{2y}}{\sqrt{1-\beta^2}}\right)+\left(\frac{\mathcal{B}_{1y}}{c_0}-\frac{\mathcal{B}_{2y}}{c_0\sqrt{1-\beta^2}}+\frac{vE_{2z}}{c^2\sqrt{1-\beta^2}}\right)u_{1x}$$
$$+\frac{u_{1y}}{c_0}(\mathcal{B}_{2x}-\mathcal{B}_{1x}) = 0$$

so that

$$E_{1z} = \frac{E_{2z}-(v/c_0)\mathcal{B}_{2y}}{\sqrt{1-\beta^2}}; \qquad \mathcal{B}_{1y} = \frac{\mathcal{B}_{2y}-(c_0\beta/c)E_{2z}}{\sqrt{1-\beta^2}}$$

(d) Finally, consider the transformation of the remaining component

$$X_1 = \frac{1}{1+vu_{2x}/c^2}\left(X_2+\frac{v}{c^2}\mathbf{F}_2.\mathbf{u}_2\right)$$

$$= \frac{1}{1+vu_{2x}/c^2}\left(X_2+\frac{v}{c^2}\varepsilon\mathbf{E}.\mathbf{u}_2\right)$$

Substituting for X_1 and X_2 we obtain

$$\left(1+\frac{vu_{2x}}{c^2}\right)\left\{E_{1x}+\frac{1}{c_0}(u_{1y}\mathcal{B}_{1z}-u_{1z}\mathcal{B}_{1y})\right\}$$

$$-\left\{E_{2x}+\frac{1}{c_0}(u_{2y}\mathcal{B}_{2z}-u_{2z}\mathcal{B}_{2y})\right\}=\frac{\beta}{c}\,\mathbf{E}_2\,.\,\mathbf{u}_2$$

Using the transformations for $\mathcal{B}_{1y}$ and $\mathcal{B}_{1z}$, this relation reduces to

$$E_{1x}=E_{2x}$$

(e) In conclusion, for the simple Lorentz transformation, we obtain the formulae

$$E_{1x}=E_{2x} \qquad \mathcal{B}_{1x}=\mathcal{B}_{2x}$$

$$E_{1y}=\frac{E_{2y}+(v/c_0)\mathcal{B}_{2z}}{\sqrt{1-\beta^2}} \qquad \mathcal{B}_{1y}=\frac{\mathcal{B}_{2y}-(c_0\beta/c)E_{2z}}{\sqrt{1-\beta^2}}$$

$$E_{1z}=\frac{E_{2z}-(v/c_0)\mathcal{B}_{2y}}{\sqrt{1-\beta^2}} \qquad \mathcal{B}_{1z}=\frac{\mathcal{B}_{2z}+(c_0\beta/c)E_{2y}}{\sqrt{1-\beta^2}}$$

In vector notation, we denote the components of each vector normal and parallel to the velocity by the suffices n and t, and the transformation formulae adopt the form

$$\mathbf{E}_{2n}=\frac{\mathbf{E}_{1n}+(\mathbf{v}/c_0)\times\mathfrak{B}_{1n}}{\sqrt{1-\beta^2}} \qquad \mathbf{E}_{2t}=\mathbf{E}_{1t}$$

$$\mathfrak{B}_{2n}=\frac{\mathfrak{B}_{1n}-(c_0\mathbf{v}/c^2)\times\mathbf{E}_{1n}}{\sqrt{1-\beta^2}} \qquad \mathfrak{B}_{2t}=\mathfrak{B}_{1t}$$

Figure 8a shows the transformations for $\mathbf{E}_{2n}$ and $\mathfrak{B}_{2n}$.

The vector formulae for the transformation of the resultants $\mathbf{E}$ and $\mathfrak{B}$ can also be written

$$\mathbf{E}_2=\frac{\mathbf{E}_1}{\sqrt{1-\beta^2}}+\frac{\mathbf{v}}{v^2}\,\mathbf{v}\,.\,\mathbf{E}_1\left(1-\frac{1}{\sqrt{1-\beta^2}}\right)+\frac{1}{\sqrt{1-\beta^2}}\,\frac{\mathbf{v}}{c_0}\times\mathfrak{B}_1$$

$$\mathfrak{B}_2=\frac{\mathfrak{B}_1}{\sqrt{1-\beta^2}}+\frac{\mathbf{v}}{v^2}\,\mathbf{v}\,.\,\mathfrak{B}_1\left(1-\frac{1}{\sqrt{1-\beta^2}}\right)-\frac{1}{\sqrt{1-\beta^2}}\,\frac{c_0\mathbf{v}}{c^2}\times\mathbf{E}_1$$

These relations are valid for axes which are inclined to the relative velocity at the same angle, and this relative velocity v may be in any direction (*Kinematics*, § [52]). They are of little interest; if general formulae are required, it is better to employ four-dimensional formalism (Chapter VIII).

Remark (1). *To first order*, the resultants may be written

$$\mathbf{E}_2=\mathbf{E}_1+\frac{\mathbf{v}}{c_0}\times\mathfrak{B}_1, \qquad \mathfrak{B}_2=\mathfrak{B}_1-\frac{c_0\mathbf{v}}{c^2}\times\mathbf{E}_1$$

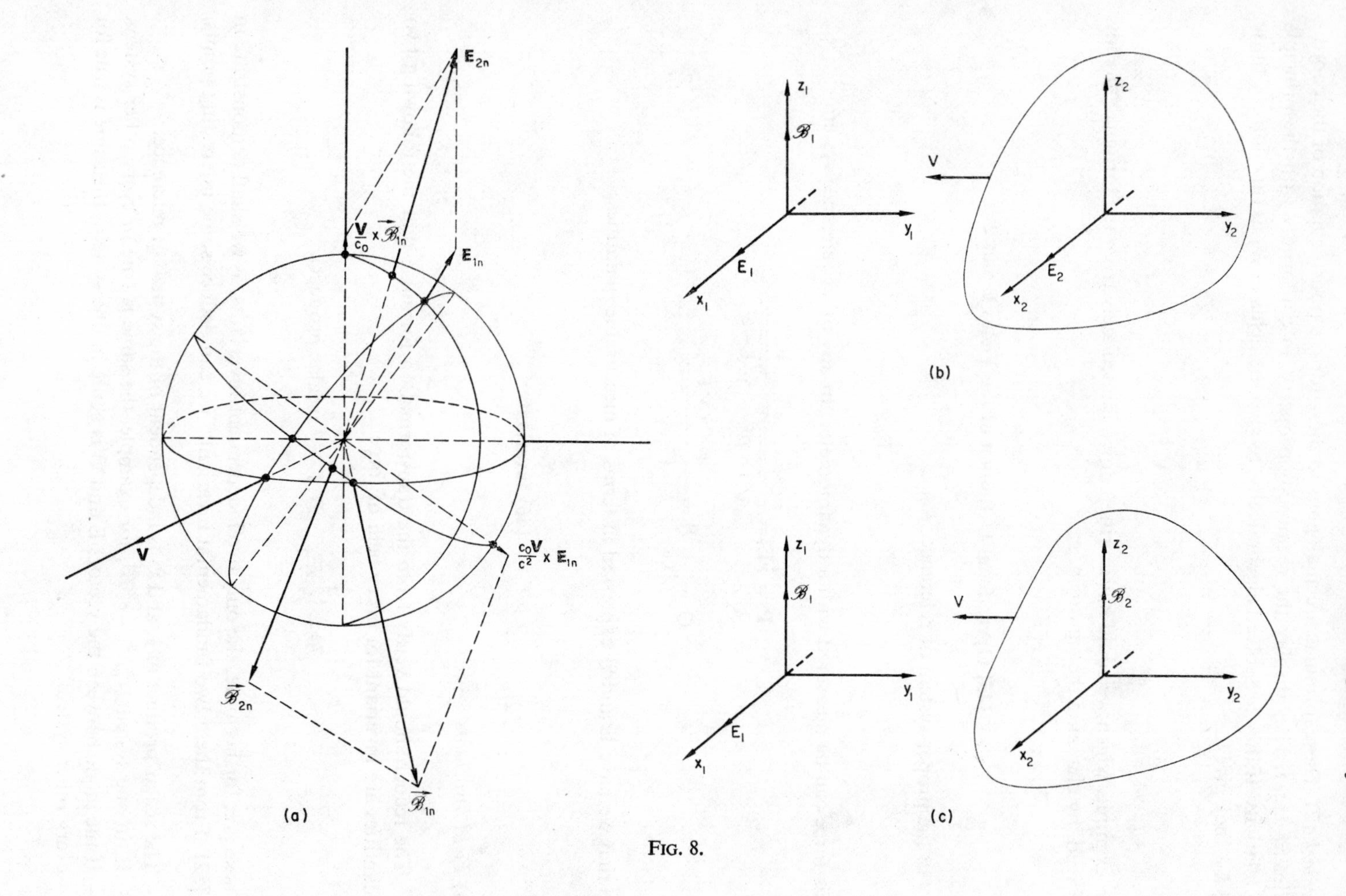

E_{2n}
$\frac{V}{c_0} \times \vec{\mathscr{B}}_{1n}$
E_{1n}
V
$\frac{c_0 V}{c^2} \times E_{1n}$
$\vec{\mathscr{B}}_{2n}$
$\vec{\mathscr{B}}_{1n}$
(a)
z_1
$\mathscr{B}_1$
y_1
E_1
x_1
V
z_2
y_2
E_2
x_2
(b)
z_1
$\mathscr{B}_1$
y_1
E_1
x_1
V
z_2
$\mathscr{B}_2$
y_2
x_2
(c)

FIG. 8.

Remark (2). These formulae are analogous to those for the transformation of the coordinates x, y, z, t. They display the reciprocity property, in conformity with the principle of relativity. If they are solved algebraically, to give the fields of K_1 as functions of those of K_2, they give

$$E_{1y} = \frac{E_{2y}+(v/c_0)\mathcal{B}_{2z}}{\sqrt{1-\beta^2}}$$

for example, which we could have written down immediately from the formulae already derived, by changing the sign of v.

[45] Operational Definition of the Fields E and $\mathfrak{B}$

In the proper system of reference, K_0,

$$\mathbf{F}_0 = \varepsilon\mathbf{E}_0;$$

this force can be measured with a dynamometer. In any other reference system,

$$\mathbf{P} = \varepsilon\mathbf{E} = \frac{\varepsilon E_0}{\sqrt{1-\beta^2}} = \frac{\mathbf{F}_0}{\sqrt{1-\beta^2}}$$

$$\mathbf{Q} = \frac{\varepsilon}{c_0}\mathfrak{B} = -\frac{\varepsilon}{c^2}\frac{\mathbf{v}\times\mathbf{E}_0}{\sqrt{1-\beta^2}}$$

so that we have **E** and $\mathfrak{B}$ expressed in terms of measurable quantities.

[46] Invariants

(a) *Field invariants*

The reader should verify from the transformation formulae that the following two quantities are invariant for every field of force:

$$\mathcal{B}^2 - \frac{c_0^2}{c^2}E^2, \quad \mathfrak{B}.\mathbf{E} \qquad \text{(scalar product)}$$

These are, furthermore, the only two independent invariants, as we shall demonstrate in § [52]. From these two fundamental invariants, we can deduce some interesting points.

— The scalar product of **E** and $\mathfrak{B}$ is independent of the system of reference.
— If, in one system, $c_0^2E^2 < c^2\mathcal{B}^2$, for example, the same is true in every other system.
— If the angle between the vectors **E** and $\mathfrak{B}$ is acute in one system, the same is true for every other system.

(b) *On the possibility of making one of the fields vanish*

If **E** and $\mathfrak{B}$ are orthogonal in one system, they remain orthogonal in every other system, since

$$\mathfrak{B}.\mathbf{E} = 0$$

If the invariant $\mathbf{E}.\mathfrak{B}$ is zero, we can always find a system in which one of the fields vanishes (**E** or $\mathfrak{B}$, depending on the sign of the other invariant). If, conversely, there exists a system in which one of the fields is zero, the two fields must be orthogonal.

Given fields $\mathbf{E}_1$, $\mathfrak{B}_1$ in K_1, let us determine the system K_2 in which $\mathfrak{B}_2$ is zero. We see immediately that the relative velocity v of K_2 with respect to K_1 must be normal to $\mathfrak{B}_1$; only thus can $\mathcal{B}_{2x}$ vanish when $\mathcal{B}_{1x}$ is zero. The further conditions are

$$\mathcal{B}_{1y} + \frac{c_0\beta}{c} E_{1z} = \mathcal{B}_{1z} - \frac{c_0\beta}{c} E_{1y} = 0$$

or

$$\frac{c_0\beta}{c} = -\frac{\mathcal{B}_{1y}}{E_{1z}} = \frac{\mathcal{B}_{1z}}{E_{1y}}$$

This equation tells us v, and also gives

$$E_{1y}\mathcal{B}_{1y} + E_{1z}\mathcal{B}_{1z} = 0$$

so that only if $\mathbf{E}_1$ and $\mathfrak{B}_1$ are perpendicular can the problem be solved. Let us assume that the second invariant is also zero:

$$\mathcal{B}^2 - \frac{c_0^2}{c^2} E^2 = 0$$

The above relation gives

$$\left(\frac{c_0\beta}{c}\right)^2 = \left(\frac{\mathcal{B}_{1y}}{E_{1z}}\right)^2 = \left(\frac{\mathcal{B}_{1z}}{E_{1y}}\right)^2 = \frac{\mathcal{B}_1^2}{E_1^2 - E_{1x}^2} = \frac{(c_0^2/c^2)\,E_1^2}{E_1^2 - E_{1x}^2} = \frac{c_0^2}{c^2}\,\frac{1}{1 - E_{1x}^2/E_1^2}$$

whence

$$\beta = \frac{1}{\sqrt{1 - E_{1x}^2/E_1^2}}$$

There is no solution for which $\beta < 1$. In the special case $E_{1x} = 0$, we have $\beta = 1$, $v = c$. One of the fields can be removed only by considering a system K_2 travelling at velocity c with respect to K_1. This is the limiting case where our formulae cease to be valid, however, for their derivation has involved multiplication by $\sqrt{1-\beta^2}$. For $\beta < 1$, therefore, one field cannot be cancelled unless only the invariant $\mathbf{E}.\mathfrak{B}$ is zero.

Let us select axes Ox_1 and Oz_1 parallel to $\mathbf{E}_1$ and $\mathfrak{B}_1$ respectively. One of the fields can be cancelled by transforming to a system K_2 travelling in the direction Oy_1. Two cases are possible. If $c\mathcal{B}_1 < c_0E_1$, $\mathcal{B}_2$ can be removed by giving K_2 the velocity

$$v = -c^2\mathcal{B}_1/c_0E_1$$

In this case (Fig. 8b),

$$\mathcal{B}_2 = 0, \quad E_2 = E_1\sqrt{1-\beta^2} = \sqrt{E_1^2-(c^2/c_0^2)\,\mathcal{B}_1^2}$$

If $c_1\mathcal{B}_1 > c_0 E_1$, E_2 can be cancelled by giving K_2 the velocity

$$v = -c_0 E_1/\mathcal{B}_1$$

and now (Fig. 8c)

$$E_2 = 0, \quad \mathcal{B}_2 = \mathcal{B}_1\sqrt{1-\beta^2} = \sqrt{\mathcal{B}_1^2-(c_0^2/c^2)\,E_1^2}$$

These two cases can be reduced to a single calculation if changes of reference system with $v > c$ are employed (see Vol. II of the French edition, § [70]).

(c) *On the possibility of making the two fields parallel*

Suppose that $\mathbf{E}.\mathfrak{B} \neq 0$; we can then find a system in which the two fields are parallel. We set out from a system K_1, and attempt to determine a system K_2 having the required property. We take the axis Ox_2 in K_2 perpendicular to the direction of $\mathbf{E}_2$ and $\mathfrak{B}_2$, which is common by hypothesis. From this, the special transformation gives

$$E_{2x} = E_{1x} = 0, \quad \mathcal{B}_{2x} = \mathcal{B}_{1x} = 0$$

Since the two fields are parallel, the vector product $\mathbf{E}_2\times\mathfrak{B}_2$ is zero and using the above equations, this condition reduces to

$$E_{2y}\mathcal{B}_{2z}-E_{2z}\mathcal{B}_{2y} = 0$$

Hence, with the aid of the field transformation, we find

$$\left(E_{1y}-\frac{v}{c_0}\mathcal{B}_{1z}\right)\left(\mathcal{B}_{1z}-\frac{c_0\beta}{c}E_{1y}\right)-\left(E_{1z}+\frac{v}{c_0}\mathcal{B}_{1y}\right)\left(\mathcal{B}_{1y}+\frac{c_0\beta}{c}E_{1z}\right)$$

$$= (E_{1y}\mathcal{B}_{1z}-E_{1z}\mathcal{B}_{1y})(1+\beta^2)-\frac{v}{c_0}\left\{\mathcal{B}_{1y}^2+\mathcal{B}_{1z}^2+\frac{c_0^2}{c^2}(E_{1y}^2+E_{1z}^2)\right\}$$

$$= (E_{1y}\mathcal{B}_{1z}-E_{1z}\mathcal{B}_{1y})(1+\beta^2)-\frac{v}{c_0}\left(\mathcal{B}^2+\frac{c_0^2}{c^2}E^2\right) = 0$$

in which v is the velocity of K_2 with respect to K_1. Given an arbitrary system K_1, therefore, the velocity of the system K_2 in which the fields become parallel will be determined by

$$\frac{v}{1+v^2/c^2} = c_0\frac{E_{1y}\mathcal{B}_{1z}-E_{1z}\mathcal{B}_{1y}}{\mathcal{B}^2+\dfrac{c_0^2}{c^2}E^2}$$

or in vector form

$$\frac{\mathbf{v}}{1+v^2/c^2} = c_0\frac{\mathbf{E}\times\mathfrak{B}}{\mathcal{B}^2+\dfrac{c_0^2}{c^2}E^2}$$

(d) *Application to the calculation of trajectories*

Suppose that we are given fields $\mathbf{E}_1$, $\mathfrak{B}_1$ in a system K_1, and some problem involving motion. It may be advantageous to consider the problem (after transforming the initial conditions) in another system K_2 in which the field is simpler, and perhaps reduces to a single component. The return to the system K_1 is then a purely kinematic problem. One has only to be on one's guard: the trajectory is not an instantaneous geometrical curve, since its different points correspond to different values of time t.

[47] The Formulae in Cylindrical Coordinates

In terms of the cylindrical components of $\mathbf{E}$ and $\mathfrak{B}$, the components of force F_z, F_r, F_ϕ (in the notation of § [11]) are given by

$$F_z = \varepsilon\left\{E_z+\frac{1}{c_0}(v_r\mathcal{B}_\phi - v_\phi\mathcal{B}_r)\right\} = \varepsilon\left\{E_z+\frac{1}{c_0}\left(\mathcal{B}_\phi\frac{dr}{dt} - \mathcal{B}_r r\frac{d\phi}{dt}\right)\right\}$$

$$F_r = \varepsilon\left\{E_r+\frac{1}{c_0}(v_\phi\mathcal{B}_z - v_z\mathcal{B}_\phi)\right\} = \varepsilon\left\{E_r+\frac{1}{c_0}\left(\mathcal{B}_z r\frac{d\phi}{dt} - \mathcal{B}_\phi\frac{dz}{dt}\right)\right\}$$

$$F_\phi = \varepsilon\left\{E_\phi+\frac{1}{c_0}(v_z\mathcal{B}_r - v_r\mathcal{B}_z)\right\} = \varepsilon\left\{E_\phi+\frac{1}{c_0}\left(\mathcal{B}_r\frac{dz}{dt} - \mathcal{B}_z\frac{dr}{dt}\right)\right\}$$

In practice, it frequently so happens that the component $\mathfrak{B}_\phi$ vanishes (electron optics). To obtain the equations of motion, the above expressions are substituted into the formulae of § [11].

[48] The Formulae in Spherical Coordinates

These become

$$F_r = \varepsilon\left\{E_r+\frac{1}{c_0}(v_\phi\mathcal{B}_\theta - v_\theta\mathcal{B}_\phi)\right\} = \varepsilon\left\{E_r+\frac{1}{c_0}\left(\mathcal{B}_\theta r\sin\theta\frac{d\phi}{dt} - \mathcal{B}_\phi r\frac{d\theta}{dt}\right)\right\}$$

$$F_\phi = \varepsilon\left\{E_\phi+\frac{1}{c_0}(v_\theta\mathcal{B}_r - v_r\mathcal{B}_\theta)\right\} = \varepsilon\left\{E_\phi+\frac{1}{c_0}\left(\mathcal{B}_r r\frac{d\theta}{dt} - \mathcal{B}_\theta\frac{dr}{dt}\right)\right\}$$

$$F_\theta = \varepsilon\left\{E_\theta+\frac{1}{c_0}(v_r\mathcal{B}_\phi - v_\phi\mathcal{B}_r)\right\} = \varepsilon\left\{E_\theta+\frac{1}{c_0}\left(\mathcal{B}_\phi\frac{dr}{dt} - \mathcal{B}_r r\sin\theta\frac{d\phi}{dt}\right)\right\}$$

where θ denotes the co-latitude (Fig. 5).

[49] The Force Tensor $\mathcal{N}^{ij}$ (Minkowski–de Beauregard)

The tensor $\mathfrak{B}$ and the vector **E** in *xyz*-space may be united to form an antisymmetric tensor in *xyzl*-space, with matrix

$$\mathcal{N} = \begin{pmatrix} 0 & \mathcal{B}_z & -\mathcal{B}_y & -\dfrac{ic_0}{c}E_x \\ -\mathcal{B}_z & 0 & \mathcal{B}_x & -\dfrac{ic_0}{c}E_y \\ \mathcal{B}_y & -\mathcal{B}_x & 0 & -\dfrac{ic_0}{c}E_z \\ \dfrac{ic_0}{c}E_x & \dfrac{ic_0}{c}E_y & \dfrac{ic_0}{c}E_z & 0 \end{pmatrix}.$$

Using the formulae of § [44], it can be verified that the components of $\mathcal{N}$ do indeed transform like those of a tensor.

With these components, the four-vector force is of the form

$$K_i = \frac{\varepsilon}{c_0}\mathcal{N}_{ij}V^j = \frac{\varepsilon}{c_0}\mathcal{N}_{ij}\frac{\mathrm{d}x^j}{\mathrm{d}T} \tag{1}$$

which leads to the dynamical equation

$$m_0\Gamma^i = m_0\frac{\mathrm{d}V^i}{\mathrm{d}T} = \frac{\varepsilon}{c_0}\mathcal{N}^{ij}\frac{\mathrm{d}x_j}{\mathrm{d}T}$$

or alternatively

$$m_0\,\mathrm{d}V_i = \frac{\varepsilon}{c_0}\mathcal{N}_{ij}\,\mathrm{d}x^j$$

or

$$\mathrm{d}\Pi_i = \frac{\varepsilon}{c_0}\mathcal{N}_{ij}\,\mathrm{d}x^j$$

[50] Operational Definition of the Tensor $\mathcal{N}_{ij}$

We have deduced the matrix $\mathcal{N}_{ij}$ from the properties of the components of **E** and $\mathfrak{B}$, established in §§ [43] and [44]. The rigorous definition is as follows. The four-vector force K_i acting on a body moving with four-velocity V_i is given, and the second-order tensor $\mathcal{N}_{ij}$ is *defined* by relation (1) of § [49].

This does *not* define $\mathcal{N}_{ij}$ uniquely, for by saying that K_i and V_i are given, we mean that their components in a particular system of reference are given. Relation (1) provides only four equations, whereas the tensor $\mathcal{N}_{ij}$ has sixteen components.

To make the definition unequivocal, certain supplementary hypotheses are necessary. K_i and V_i may be given in any reference system whatsoever; let us select the system K_0 in which the particle is at rest. We then have

$$K_{0x} = F_{0x}, \qquad K_{0y} = F_{0y}, \qquad K_{0z} = F_{0z}, \qquad K_{0l} = 0$$
$$V_{0x} = V_{0y} = V_{0z} = 0, \qquad V_{0l} = ic$$

Applying (1),

$$(\mathcal{N}_{14})_0 = -\frac{ic_0}{c}\frac{F_{0x}}{\varepsilon}, \qquad (\mathcal{N}_{24})_0 = -\frac{ic_0}{c}\frac{F_{0y}}{\varepsilon}$$

$$(\mathcal{N}_{34})_0 = -\frac{ic_0}{c}\frac{F_{0z}}{\varepsilon}, \qquad (\mathcal{N}_{44})_0 = 0$$

Thus only the fourth column of the matrix $\mathcal{N}_{ij}$ is fixed. If we make *the additional hypothesis* that the matrix is *antisymmetric*, we fix the fourth row.

The $\mathcal{N}_{\alpha\beta}$ ($\alpha, \beta = 1, 2, 3$) remain to be determined; we are therefore at liberty to put them equal to *zero*, and finally, in K_0, we find

$$\mathcal{N}_0^{ij} = \begin{pmatrix} 0 & 0 & 0 & -\frac{ic_0}{c}\frac{F_{0x}}{\varepsilon} \\ 0 & 0 & 0 & -\frac{ic_0}{c}\frac{F_{0y}}{\varepsilon} \\ 0 & 0 & 0 & -\frac{ic_0}{c}\frac{F_{0z}}{\varepsilon} \\ \frac{ic_0}{c}\frac{F_{0x}}{\varepsilon} & \frac{ic_0}{c}\frac{F_{0y}}{\varepsilon} & \frac{ic_0}{c}\frac{F_{0z}}{\varepsilon} & 0 \end{pmatrix}$$

To transform to another system, K, we use the appropriate matrix. The antisymmetry of $\mathcal{N}_{ij}$ is conserved.

For the simple Lorentz transformation,

$$x^1 = \frac{x_0^1 - i\beta x_0^4}{\sqrt{1-\beta^2}}, \qquad x^2 = x_0^2$$

$$x^3 = x_0^3, \qquad x^4 = \frac{x_0^4 + i\beta x_0^1}{\sqrt{1-\beta^2}}$$

we have

$$\mathcal{N}_{ij} = \{\alpha\}\{\mathcal{N}_0^{ij}\}\{\alpha\}_t$$

with

$$\alpha = \begin{pmatrix} \frac{1}{\sqrt{1-\beta^2}} & 0 & 0 & \frac{-i\beta}{\sqrt{1-\beta^2}} \\ 0 & 1 & 0 & 0 \\ 0 & 0 & 1 & 0 \\ \frac{i\beta}{\sqrt{1-\beta^2}} & 0 & 0 & \frac{1}{\sqrt{1-\beta^2}} \end{pmatrix}$$

and $\{\alpha\}_t$ is the transpose of $\{\alpha\}$.

Thus

$$\mathcal{N}^{ij} = \begin{pmatrix} 0 & \frac{c_0\beta}{c}\frac{F_{0y}}{\varepsilon\sqrt{1-\beta^2}} & \frac{c_0\beta}{c}\frac{F_{0z}}{\varepsilon\sqrt{1-\beta^2}} & -\frac{ic_0}{c}\frac{F_{0x}}{\varepsilon} \\ -\frac{c_0\beta}{c}\frac{F_{0y}}{\varepsilon\sqrt{1-\beta^2}} & 0 & 0 & -\frac{ic_0}{c}\frac{F_{0y}}{\varepsilon\sqrt{1-\beta^2}} \\ -\frac{c_0\beta}{c}\frac{F_{0z}}{\varepsilon\sqrt{1-\beta^2}} & 0 & 0 & -\frac{ic_0}{c}\frac{F_{0z}}{\varepsilon\sqrt{1-\beta^2}} \\ \frac{ic_0}{c}\frac{F_{0x}}{\varepsilon} & \frac{ic_0}{c}\frac{F_{0y}}{\varepsilon\sqrt{1-\beta^2}} & \frac{ic_0}{c}\frac{F_{0z}}{\varepsilon\sqrt{1-\beta^2}} & 0 \end{pmatrix}$$

[51] Other Hypotheses

(a) *Symmetric tensors*

There is nothing to prevent our making $\mathcal{N}_0^{ij}$ symmetric:

$$\mathcal{N}_0^{ij} = \begin{pmatrix} 0 & 0 & 0 & -\dfrac{ic_0}{c}\dfrac{F_{0x}}{\varepsilon} \\ 0 & 0 & 0 & -\dfrac{ic_0}{c}\dfrac{F_{0y}}{\varepsilon} \\ 0 & 0 & 0 & -\dfrac{ic_0}{c}\dfrac{F_{0z}}{\varepsilon} \\ -\dfrac{ic_0}{c}\dfrac{F_{0x}}{\varepsilon} & -\dfrac{ic_0}{c}\dfrac{F_{0y}}{\varepsilon} & -\dfrac{ic_0}{c}\dfrac{F_{0z}}{\varepsilon} & 0 \end{pmatrix}$$

and so

$$\mathcal{N}^{ij} = \begin{pmatrix} -\dfrac{2c_0\beta}{c}\dfrac{F_{0x}}{\varepsilon(1-\beta^2)} & -\dfrac{c_0\beta}{c}\dfrac{F_{0y}}{\varepsilon\sqrt{1-\beta^2}} & -\dfrac{c_0\beta}{c}\dfrac{F_{0z}}{\varepsilon\sqrt{1-\beta^2}} & -\dfrac{ic_0}{c}\dfrac{1+\beta^2}{1-\beta^2}\dfrac{F_{0x}}{\varepsilon} \\ -\dfrac{c_0\beta}{c}\dfrac{F_{0y}}{\varepsilon\sqrt{1-\beta^2}} & 0 & 0 & -\dfrac{ic_0}{c}\dfrac{F_{0y}}{\varepsilon\sqrt{1-\beta^2}} \\ -\dfrac{c_0\beta}{c}\dfrac{F_{0z}}{\varepsilon\sqrt{1-\beta^2}} & 0 & 0 & -\dfrac{ic_0}{c}\dfrac{F_{0z}}{\varepsilon\sqrt{1-\beta^2}} \\ -\dfrac{ic_0}{c}\dfrac{1+\beta^2}{1-\beta^2}\dfrac{F_{0x}}{\varepsilon} & -\dfrac{ic_0}{c}\dfrac{F_{0y}}{\varepsilon\sqrt{1-\beta^2}} & -\dfrac{ic_0}{c}\dfrac{F_{0z}}{\varepsilon\sqrt{1-\beta^2}} & \dfrac{2c_0\beta}{c}\dfrac{F_{0x}}{\varepsilon(1-\beta^2)} \end{pmatrix}$$

(b) *Tensors without symmetry*

We write

$$\mathcal{N}_0^{ij} = \begin{pmatrix} 0 & 0 & 0 & -\dfrac{ic_0}{c}\dfrac{F_{0x}}{\varepsilon} \\ 0 & 0 & 0 & -\dfrac{ic_0}{c}\dfrac{F_{0y}}{\varepsilon} \\ 0 & 0 & 0 & -\dfrac{ic_0}{c}\dfrac{F_{0z}}{\varepsilon} \\ 0 & 0 & 0 & 0 \end{pmatrix}$$

whence

$$\mathcal{N}^{ij} = \begin{pmatrix} -\dfrac{c_0\beta}{c}\dfrac{F_{0x}}{\varepsilon(1-\beta^2)} & 0 & 0 & -\dfrac{ic_0}{c}\dfrac{F_{0x}}{\varepsilon(1-\beta)^2} \\ -\dfrac{c_0\beta}{c}\dfrac{F_{0y}}{\varepsilon\sqrt{1-\beta^2}} & 0 & 0 & -\dfrac{ic_0}{c}\dfrac{F_{0y}}{\varepsilon\sqrt{1-\beta^2}} \\ -\dfrac{c_0\beta}{c}\dfrac{F_{0z}}{\varepsilon\sqrt{1-\beta^2}} & 0 & 0 & -\dfrac{ic_0}{c}\dfrac{F_{0z}}{\varepsilon\sqrt{1-\beta^2}} \\ -\dfrac{ic_0\beta^2}{c}\dfrac{F_{0x}}{\varepsilon(1-\beta^2)} & 0 & 0 & \dfrac{c_0\beta}{c}\dfrac{F_{0x}}{\varepsilon(1-\beta^2)} \end{pmatrix}$$

If we form the product $(\varepsilon/c_0)\mathcal{H}_{ij}V^j$, we do of course obtain the same result for each of the three preceding matrices:

$$K^i = \{\mathcal{H}^{ij}\} \begin{pmatrix} \dfrac{V_x}{\sqrt{1-\beta^2}} \\ 0 \\ 0 \\ \dfrac{ic}{\sqrt{1-\beta^2}} \end{pmatrix} = \begin{pmatrix} \dfrac{F_{0x}}{\sqrt{1-\beta^2}} \\ F_{0y} \\ F_{0z} \\ \dfrac{i\beta F_{0x}}{\sqrt{1-\beta^2}} \end{pmatrix} = \begin{pmatrix} \dfrac{F_x}{\sqrt{1-\beta^2}} \\ \dfrac{F_y}{\sqrt{1-\beta^2}} \\ \dfrac{F_z}{\sqrt{1-\beta^2}} \\ \dfrac{i}{c}\dfrac{\mathbf{v}\cdot\mathbf{F}}{\sqrt{1-\beta^2}} \end{pmatrix}$$

For the components of $\mathbf{F}$ expressed in terms of those of $\mathbf{F}_0$, we thus recover the expressions already obtained in § [28]; they would be obtained immediately by writing

$$\{K^i\} = \{\alpha\}\{K_0^i\}$$

[52] Investigation of the Invariants

Let λ be one of the characteristic values or eigenvalues of the matrix $\mathcal{H}$. The characteristic equation of the matrix is then written

$$\begin{vmatrix} -\lambda & \mathcal{B}_z & -\mathcal{B}_y & -\dfrac{ic_0}{c}E_x \\ -\mathcal{B}_z & -\lambda & \mathcal{B}_x & -\dfrac{ic_0}{c}E_y \\ \mathcal{B}_y & -\mathcal{B}_x & -\lambda & -\dfrac{ic_0}{c}E_z \\ \dfrac{ic_0}{c}E_x & \dfrac{ic_0}{c}E_y & \dfrac{ic_0}{c}E_z & -\lambda \end{vmatrix} = \lambda^4 + \left(\mathcal{B}^2 - \frac{c_0^2}{c^2}E^2\right)\lambda^2 - \frac{c_0^2}{c^2}(\mathbf{E}.\mathfrak{B})^2 = 0$$

The two independent invariants—the sum of the characteristic values and the product of the characteristic values—are deduced from the invariance of this eigenvalue equation.

Remark. In the French edition, I adopted rather too hasty a pace, liable to lead the reader into error. There, to speed the calculations, I selected one of the vectors to be parallel to one of the axes ($\mathbf{E}$ parallel to Oz), and took the plane containing the two vectors as the plane xOz. This gave

$$\begin{vmatrix} -\lambda & \mathcal{B}_z & 0 & 0 \\ -\mathcal{B}_z & -\lambda & 0 & 0 \\ 0 & -\mathcal{B}_x & -\lambda & -\dfrac{ic_0}{c}E \\ 0 & 0 & \dfrac{ic_0}{c}E & -\lambda \end{vmatrix} = \lambda^4 + \left(\mathcal{B}_x^2 + \mathcal{B}_z^2 - \frac{c_0^2}{c^2}E^2\right)\lambda^2 + \frac{c_0^2}{c^2}E^2\mathcal{B}^2 = 0$$

Thus $(\mathbf{E}.\mathfrak{B})^2$ appeared as $E^2\mathcal{B}^2$, which is not true in the general case.

[53] The Angular Momentum Theorem

Setting out from (§ [49])

$$d\Pi^i = \frac{\varepsilon}{c_0} \mathcal{H}^{ij} \, dx_j \tag{1}$$

we write

$$x^i \, d\Pi^j - x^j \, d\Pi^i = \frac{\varepsilon}{c_0} (x^i \mathcal{H}^{jk} - x^j \mathcal{H}^{ik}) \, dx_k \tag{2}$$

The vectors dx^i and Π^i are collinear, so that we may write

$$d(x^i \Pi^j - x^j \Pi^i) = \frac{\varepsilon}{c_0} (x^i \mathcal{H}^{jk} - x^j \mathcal{H}^{ik}) \, dx_k \tag{3}$$

In fact, the tensor $\Pi^j \, dx^i - \Pi^i \, dx^j$ is zero. The antisymmetric tensor $m^{ij} = x^i \Pi^j - x^j \Pi^i$ is by definition the world-angular momentum about the origin $x^i = 0$. The spatial components of m^{ij} are the component of the ordinary angular momentum

$$m^{23} = -m^{32} = yp_z - zp_y = m_x$$
$$m^{31} = -m^{13} = zp_x - xp_z = m_y$$
$$m^{12} = -m^{21} = xp_y - yp_x = m_z$$

The components $m^{\alpha 4}$ are

$$m^{\alpha 4} = x^\alpha \Pi^4 - x^4 \Pi^\alpha = x^\alpha imc - ictmv^\alpha = imc(x^\alpha - v^\alpha t)$$

If $t = 0$, the $m^{\alpha 4}$ are the components of moment of the mass, or barycentric moment, about the origin of the spatial coordinates.

The world-moment of the forces about the point $x^i = 0$ is obtained by contracting the third-order tensor

$$N^{ijk} = \frac{\varepsilon}{c_0} (x^i \mathcal{H}^{jk} - x^j \mathcal{H}^{ik})$$

which is antisymmetric in i and j, with dx^k. In four-dimensional form, therefore, the angular momentum theorem is written

$$dm^{ij} = N^{ijk} \, dx_k$$

HISTORICAL AND BIBLIOGRAPHICAL NOTES

[54] On the Heaviside–Thomson–Lorentz Force

The formula of § [43] is first encountered in physics as the expression for the force acting on a charge ε travelling with velocity $\mathbf{v}$ in an electric field $\mathbf{E}$ and a magnetic field $\mathfrak{B}$.

Heaviside [2] seems to have been the first to give the component involving $\mathfrak{B}$ correctly; Thomson [1] had already derived an expression involving a vector product, but he had wrongly included a coefficient

$\frac{1}{2}$. The complete formula including both **E** and $\mathfrak{B}$ was used by H. A. Lorentz [3] in a general theory of electromagnetic phenomena. Finally, Schwarzschild [4] set out a formula taking advantage of the existence of potentials for the electric and magnetic fields, and introduced variational methods; we shall meet his work again in Chapter V. For the history of the transformation formulae, the reader is referred to § [40].

C. de Beauregard [5, 6] has shown that the Heaviside–Lorentz expression and the corresponding Minkowski formalism have a general mechanical import, seen from the relativistic point of view. It is his remarks on this topic that have led me to "mechanize" the equations of electromagnetism in a systematic fashion. I do not believe that this standpoint has so far been sufficiently exploited.

[1] J. J. Thomson: On the electric and magnetic effects produced by the motion of electrified bodies. *Phil. Mag.* **11** (1881) 227.
[2] O. Heaviside: On the electromagnetic effects due to motion of electrification through a dielectric. *Phil. Mag.* **27** (1889) 324.
[3] H. A. Lorentz: La théorie électromagnétique de Maxwell, et son application aux corps mouvants. *Arch. Néerl.* **25** (1892) 363–552.
[4] K. Schwarzschild: Zur Elektrodynamik. *Nachr. Göttingen* (1903) 126–141.
[5] O. Costa de Beauregard: Définition covariante de la force. *C. R. Acad. Sci. Paris* **221** (1945) 743–745.
[6] O. Costa de Beauregard: ref. [7] of Appendix I, pp. 77 and 104; see also ref. [35].

Notice that §§ [55] *and* [56] *have been removed.*

CHAPTER IV

FIELDS OF FORCE DERIVABLE FROM SCALAR POTENTIALS

A. THE MOTION OF A POINT IN A FIELD OF FORCE WHICH POSSESSES A SCALAR POTENTIAL; THE ANALOGY WITH GEOMETRICAL OPTICS

[57] The Trajectory Equation for Plane Parallel Equipotential Surfaces

(a) We shall employ the intrinsic equations, which are written in terms of the principal normal and the tangent to the trajectory at the point being considered. The first will give us the trajectory, and the second the equation of motion along it.

Let **F** be a field of force with scalar potential; we write

$$\mathbf{F} = -\varepsilon \operatorname{grad} V$$

in which ε is the same constant as in § [43]. To give a concrete example, this corresponds to the case of a charge ε (algebraic value) moving in an electric field; nevertheless, the problem is, of course, quite general, and purely mechanical. The gradient vector is given in various coordinate systems in § [66].

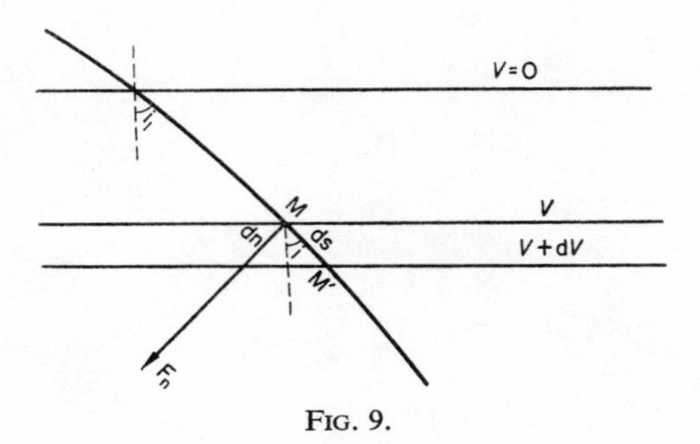

FIG. 9.

Let V and $V+\mathrm{d}V$ be two equipotential surfaces, infinitesimally close to one another, which a trajectory intersects at points M, M' at an angle of incidence i (Fig. 9). The component of force normal to the trajectory is

$$F_n = -\varepsilon \frac{\partial V}{\partial n} = -\varepsilon \frac{\mathrm{d}V \sin i}{\mathrm{d}s \cos i}$$

This component may also be written

$$F_n = mv^2/\varrho = -mv^2\, \mathrm{d}i/\mathrm{d}s$$

in which ϱ is the radius of curvature of the trajectory at M. If i increases, ϱ becomes smaller, hence the minus sign.

Using the relativistic expression for the mass, we can calculate mv^2:

$$mv^2 = mc^2 - m_0^2c^4/mc^2$$

Let m_1 be the mass of the corpuscle *as it crosses the zero equipotential surface*. The energy theorem gives

$$mc^2 - m_1c^2 = -\varepsilon V$$

$$mc^2 = \varepsilon(V_1 - V), \quad \text{with} \quad V_1 = \frac{m_1c^2}{\varepsilon}$$

Substituting into the expression for mv^2,

$$mv^2 = \varepsilon(V_1 - V) - \frac{\varepsilon^2 V_0^2}{\varepsilon(V_1 - V)} = \frac{\varepsilon}{V_1 - V}\{(V_1 - V)^2 - V_0^2\}$$

with

$$V_0 = \frac{m_0c^2}{\varepsilon}$$

and we may confirm that

$$(mv^2)_{V=0} = \varepsilon(V_1 - V_0^2/V_1) = m_1c^2 - m_0^2c^4/m_1c^2$$

Setting the two expressions for F_n equal,

$$\varepsilon \frac{\mathrm{d}V}{\mathrm{d}s} \frac{\sin i}{\cos i} = \frac{\varepsilon}{V_1 - V}\{(V_1 - V)^2 - V_0^2\} \frac{\mathrm{d}i}{\mathrm{d}s}$$

or

$$\frac{1}{2} \frac{\mathrm{d}\{(V_1 - V)^2 - V_0^2\}}{(V_1 - V)^2 - V_0^2} + \frac{\mathrm{d}(\sin i)}{\sin i} = 0$$

Integrating,

$$\sqrt{(V_1 - V)^2 - V_0^2}\, \sin i = \text{const}$$

Denoting the value of i where V is zero by i_1, we write

$$\boxed{\sqrt{(V_1 - V)^2 - V_0^2}\, \sin i = \sqrt{V_1^2 - V_0^2}\, \sin i_1}$$

If, in particular, the particle is released from rest,

$$V_1 = V_0, \quad \sqrt{V^2 - 2VV_0}\, \sin i = 0$$

The angle i remains equal to zero and the corpuscle describes a trajectory perpendicular to the equipotential surfaces; it is therefore a straight line.

To the Newtonian approximation, we have

$$F_n = -mv^2 \frac{di}{ds} = -2\varepsilon(E_1 - V) \frac{di}{ds} = -\varepsilon \frac{dV \sin i}{\cos i}$$

in which E_1 is the kinetic energy of the particle where it enters the field. Integrating,

$$\sqrt{E_1 - V} \sin i = \sqrt{E_1} \sin i_1$$

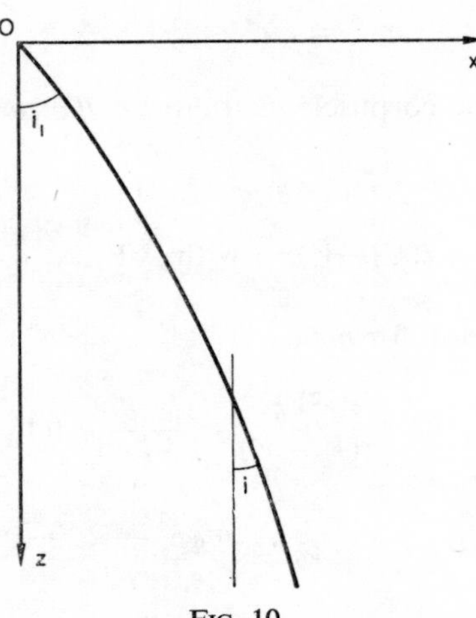

FIG. 10.

(b) Let us take the plane xOy parallel to the equipotential surfaces, measuring z from the zero surface (Fig. 10). The relation becomes

$$\frac{dx}{dz} = \frac{\sin i_1}{\sqrt{n^2 - \sin^2 i_1}} \qquad (\text{since } \tan i = dx/dz)$$

in which

$$n^2 = \frac{\{V(z) - V_1\}^2 - V_0^2}{V_1^2 - V_0^2}$$

By analogy with optics, n is the refractive index.

[58] The Trajectory Equation in the General Case

(a) *The general case*

We shall see in § [86] that the above expression for the refractive index remains valid for equipotential surfaces of arbitrary shape. The product $n \sin i$ is in general no longer constant, and to obtain the trajectory equation each special case must be analysed separately, using Fermat's principle or Maupertuis' theorem. We mention that in the general case a particle which starts from rest begins by following a line of force, but subsequently strays from it. *Trajectories and lines of force are different curves; only if the lines of force are straight and the particle is released from rest do they coincide.*

(b) *Spherical equipotential surfaces*

We assume the refractive index to be of the form

$$n = f(r)$$

in which r is the radius vector from the centre of the spheres. In the optics of inhomogeneous media, it is shown that the trajectory is plane, and satisfies the equation

$$nr \sin i = C = \text{const}$$

where i is the angle between the radius vector and the trajectory at each point.

In polar coordinates,

$$\tan i = r\, d\theta/dr$$

whence

$$\frac{dr}{d\theta} = \frac{r}{\tan i} = \frac{r\sqrt{n^2r^2 - C^2}}{C}$$

(c) *Rotationally symmetric equipotential surfaces; the plane trajectory case*

For plane or spherical equipotential surfaces, the trajectories are planar irrespective of the direction of the initial velocity. In the present case, this is only so if the axis of revolution and the initial velocity are coplanar, and this is what we shall assume.

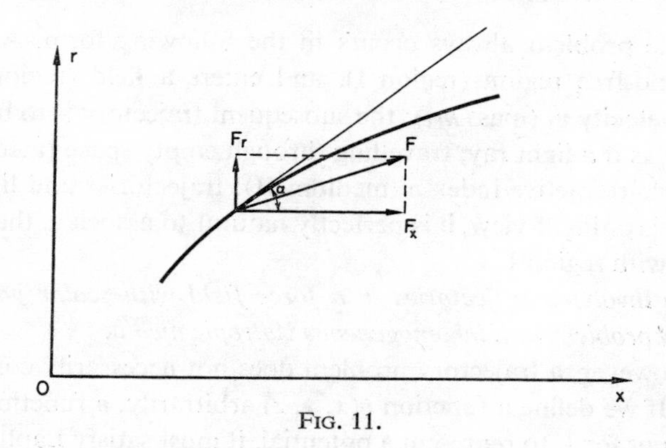

FIG. 11.

Let Ox be the axis of revolution, and r the distance of an arbitrary point M from the axis; $V(x, r)$ denotes the potential. We characterize each trajectory by its equation $r(x)$, and we denote the angle between tangent and axis by α (Fig. 11). To obtain the differential equation, we apply the reasoning of § [57] to the present case.

We write

$$F_n = mv^2/\varrho$$

and now

$$\varrho = \frac{(1+r'^2)^{\frac{3}{2}}}{r''}$$

$$F_n = F_x \sin\alpha - F_r \cos\alpha = \frac{\varepsilon(V'_x r' - V'_r)}{\sqrt{1+r'^2}}$$

taking F_n positive in the sense which gives a positive projection on Or. Using the expression already found for mv^2

$$\varepsilon \frac{V'_x r' - V'_r}{\sqrt{1+r'^2}} = \frac{\varepsilon}{V_1 - V} \frac{r''}{(1+r'^2)^{\frac{3}{2}}} \{(V_1-V)^2 - V_0^2\}$$

or

$$(1+r'^2)(V'_x r' - V'_r)(V_1 - V) - r''\{(V_1-V)^2 - V_0^2\} = 0$$

If the velocity of the corpuscle is zero at the zero equipotential ($V_1 = V_0$), the equation simplifies to

$$(1+r'^2)(V'_x r' - V'_r)(V - V_0) + r''(V^2 - 2VV_0) = 0$$

This is the relation commonly employed in electron optics. The origin of potential is then the cathode voltage, and the electrons are assumed to be emitted with negligible velocity.

To obtain the Newtonian approximation (low velocities) we write

$$V_1 = V_0, \quad V \ll V_0$$

$$(V - V_1)^2 - V_0^2 \simeq V - 2VV_1 \simeq -2VV_1$$

$$(1+r'^2)(V'_x r' - V'_r) + 2Vr'' = 0$$

[59] Analogy with the Optics of Inhomogeneous Media

In practice, the problem always occurs in the following form. A body is initially travelling in a field-free region (region I), and enters a field (region II) at angle of incidence i_1 with velocity v_1 (mass m_1); the subsequent trajectory is to be determined.

This is exactly as if a light ray, travelling through empty space (medium I), falls on a medium of variable refractive index n (medium II); trajectories and light rays in II are identical. From this point of view, it is perfectly natural to associate the reference refractive index (unity) with region I.

Every problem involving trajectories in a force field with scalar potential can be reduced to an optical problem with inhomogeneous isotropic media.

Conversely, however, a trajectory problem does not necessarily correspond to every optical problem. If we define a function $n(x, y, z)$ arbitrarily, a function V does indeed correspond to it, but for V to represent a potential, it must satisfy Laplace's equation.

B. THE WAVE-CORPUSCLE DUALITY; ELEMENTARY CONSIDERATIONS

[60] The de Broglie Waves Associated with a Corpuscle

(a) *General Remarks*

The analogy between trajectories and rays has long been recognized in pre-relativistic physics (Hamilton). It was made more exact and generalized by Louis de Broglie, who above all attached a deep-seated physical meaning to it, and it thus furnished a springboard for quantum mechanics. We know that in the geometrical approximation to the latter, the concept of trajectories retains its meaning, and it is therefore not surprising that in this case, quantum concepts can be employed in macroscopic mechanics. With every moving particle, we associate a wavelength, given by the de Broglie formula (the origin of which we study in § [61])

$$\lambda = h/mv$$

It is interesting to recalculate the refractive index along these lines.

(b) *The general expression for the refractive index*

Let W denote the constant total energy of the particle. Where the latter enters the field

$$W = \frac{m_0c^2}{\sqrt{1-v_1^2/c^2}}, \qquad v_1^2 = c^2\left(1-\frac{m_0^2c^4}{W^2}\right)$$

At an arbitrary point in the field where the potential is V

$$W = \frac{m_0c^2}{\sqrt{1-v^2/c^2}}+P$$

$$v^2 = c^2\left\{1-\frac{m_0^2c^4}{(W-P)^2}\right\}$$

where P denotes the potential energy acquired by the particle in the field.

The associated de Broglie wavelengths are given by

$$\lambda_1 = \frac{h\sqrt{1-\beta_1^2}}{m_0v_1}, \qquad \lambda = \frac{h\sqrt{1-\beta^2}}{m_0v}$$

so that the refractive index is given by

$$n = \frac{\lambda_1}{\lambda} = \frac{v}{v_1}\sqrt{\frac{1-\beta_1^2}{1-\beta^2}}$$

$$= \sqrt{\frac{1-\dfrac{m_0^2c^4}{(W-P)^2}}{1-m_0^2c^4/W^2}}\,\frac{W-P}{W} = \sqrt{\frac{(W-P)^2-m_0^2c^4}{W^2-m_0^2c^4}}$$

We do indeed recover the result that in the general relativistic case, the refractive index is not equal to the ratio of the velocities.

(c) *The Newtonian approximation*

We must now introduce the initial kinetic energy, by writing

$$W = m_0c^2+E$$

After simplification, the quantity under the square root becomes

$$\frac{E^2+2m_0c^2E+P^2-2P(m_0c^2+E)}{E^2+2m_0c^2E}$$

which reduces, as c tends to infinity, to

$$\frac{E-P}{E}$$

Finally,

$$n = \sqrt{1-\frac{P}{E}} = \sqrt{1-\frac{P}{W-m_0c^2}}$$

With the notation employed above

$$W = m_1c^2 = \varepsilon V_1$$
$$P = m_1c^2 - mc^2 = \varepsilon V$$
$$W-P = \varepsilon(V_1 - V)$$

so that

$$\text{relativistic refractive index:} \quad n = \sqrt{\frac{(V_1-V)^2-V_0^2}{V_1^2-V_0^2}}$$

$$\text{Newtonian refractive index:} \quad n = \sqrt{1-\frac{V}{V_1-V_0}}$$

If the index is to be used only for calculating trajectories, this expression may be simplified; since any arbitrary factor may be introduced (we write merely $n \sin i$ = constant)

$$n = \sqrt{V_1-V_0-V}$$

Taking the origin of potential at the point where the particle velocity is zero

$$n = \sqrt{-V}$$

[61] Relativistic Properties of the Wave-parameters of a Corpuscle, to the Geometrical Approximation

(a) The formula given above for the de Broglie wavelength today forms part of the vast edifice of quantum mechanics. It is nevertheless valuable to examine de Broglie's early arguments, because of their intrinsic interest for the relativistic physicist. Furthermore, and this is most important, although the wave-like picture of the corpuscle upon which they are based was for a long time abandoned in the development of quantum mechanics, it now seems once again to offer fresh possibilities.

At the beginning of the century, physicists were confronted with two attitudes towards the structure of light: light is a wave-phenomenon, or light consists of particles (photons, see below, § [62]). Which attitude was to be adopted depended upon the phenomenon to be studied. In seeking a theory which would encompass both situations, Louis de Broglie was led to suggest that in general some wave process must be associated with every corpuscle (and not with photons alone).

Let us consider, as he did, a particle of mass m_0, *at rest* in a Galileian system K_0. It is quite natural to associate with it, as "wave process", a stationary wave, which in complex notation is written

$$\psi_0 = ae^{2\pi i\nu_0 t_0} \qquad (a = \text{const})$$

In a system K, with respect to which K_0 has velocity v, we must use the variable t, such that

$$t_0 = \frac{t - \frac{v}{c^2} x}{\sqrt{1-\beta^2}}$$

The wave then becomes

$$\psi = a e^{2\pi i \nu (t - x/u)}$$

with

$$\nu = \frac{\nu_0}{\sqrt{1-\beta^2}} \quad \text{and} \quad u = c^2/v$$

We obtain a wave of frequency ν, propagating along Ox with phase velocity u. This phase velocity is always greater than c. It corresponds to a group velocity U, given by

$$Uu = c^2$$

Thus the particle velocity v is equal to the group velocity of the associated wave.

(b) To relate the mechanical quantities m_0 and v to the wave quantities ν and u, we use the Planck–Einstein relation which gives the photon energy in wave form (§ [62]), and which we apply to an arbitrary corpuscle:

$$W = h\nu$$

We are thus led to write

$$m_0 c^2 = h\nu$$

whence

$$\nu_0 = \frac{m_0 c^2}{h}, \qquad \lambda = \frac{u}{\nu} = \frac{h\sqrt{1-\beta^2}}{m_0 v} = \frac{h}{mv}$$

(c) The relativistic physicist is immediately struck by an important point: *the de Broglie frequency does not transform like the frequency of a clock* (*Kinematics*, §[63]). We come to the same conclusion if we transform the energy given by Planck's formula, retaining the same constant h for all systems (see below, in connection with the photon). This frequency may be perhaps connected with some genuine vibratory process inside the corpuscle, but with the usual formulae such a process can never act as time variable, for it does not advance uniformly with respect to relativistic time (*Kinematics*, § [62]). Its advance is such that the process and the wave ψ along which the particle is travelling remain in phase. We can thus say that if the wave ψ is given, the motion of the particle is governed by this phase relationship. If we pursue this line of reasoning, we find ourselves considering the wave ψ as the only reality; the corpuscle becomes a localized oscillatory phenomenon, a singular zone of the wave, with a well-defined motion. This attitude is far from the present day ideas which dominate ordinary quantum mechanics where the corpuscles are the physical reality and the wave ψ is a purely mathematical function, which characterizes their behaviour. The de Broglie approach, on the contrary, is related to the unitary theory in the sense of Mie and Born, which we study in *Rayonnement et Dynamique* and, more generally, to modern attempts to create a theory of fundamental particles.

(d) Without attempting to predict the future of this point of view in any way, we might make the following observation. If the frequency is attributed to a real process, the former must be defined completely. Basing their reasoning on the fact that the only measurable quantities are λ and v (or U), some authors have believed that ν and u are to some extent indeterminate. Louis de Broglie has, however, shown in his *Optique électronique* that the *relativistic variance condition* leads to the opposite conclusion.

[62] The Photon *in vacuo*

Many phenomena have very simple explanations if we accept that light of frequency ν consists of particles (called photons), each of energy

$$W = h\nu$$

h is Planck's constant ($h = 6{\cdot}62\times 10^{-27}$ erg/s). These corpuscles have finite energy at velocity c and their rest mass must thus be zero; it should be realized, however, that this concept here has no meaning, for there is no reference system in which the photon is at rest: its velocity is c in every system. The momentum vector $\mathbf{p}$ has the indeterminate form

$$\mathbf{p} = \frac{0\times c}{0}$$

which we avoid by writing

$$\mathbf{p} = \frac{W}{c^2}\mathbf{v}$$

We thus find that *the photon may be regarded as a corpuscle of zero proper mass, the energy and momentum of which are given by*

$$W = h\nu \qquad p = \frac{h\nu}{c}$$

We might say that for the photon, all the energy is kinetic; the rest energy is, so to say, zero, but in fact this has no meaning.

[63] The Doppler Effect (Transformation of the Frequency of a Photon)

Consider two systems, K_1 and K_2; K_2 has velocity v with respect to K_1. Combining the above relations with those of § [29], we obtain the transformation formulae for the frequency.

If the photon is travelling parallel to Ox (and hence to v), we have

$$\nu_1 = \sqrt{\frac{1+\beta}{1-\beta}}\,\nu_2$$

In the general case, the photon velocity makes an arbitrary angle δ_2 with the axis Ox_2 (or v)

$$p_{2x} = \frac{h\nu_2}{c} \cos \delta_2, \qquad p_{2y} = \frac{h\nu_2}{c} \sin \delta_2$$

From the energy transformation

$$\nu_1 = \frac{1+\beta \cos \delta_2}{\sqrt{1-\beta^2}} \nu_2$$

and by transforming the momentum parallel to v

$$\cos \delta_2 = \frac{\cos \delta_1 - \beta}{1-\beta \cos \delta_1}$$

This enables us to write down the relation between ν_1 and ν_2, using the angle δ_1:

$$\nu_1 = \frac{\sqrt{1-\beta^2}}{1-\beta \cos \delta_1} \nu_2$$

We could of course have obtained this formula immediately, by applying the relativity principle: we had only to exchange suffices and replace β by $-\beta$. Setting δ_1 equal to zero, we recover the special case studied above.

If we observe normal to the velocity in K_1, we have

$$\cos \delta_1 = 0, \qquad \cos \delta_2 = -\beta, \qquad \nu_1 = \nu_2 \sqrt{1-\beta^2}$$

This formula is not reciprocal in form, since the observation is not made in the same way in K_1 and K_2.

These formulae form the relativistic theory of the Doppler effect *in vacuo*, and we shall study it at greater length in the volume on relativistic optics.

We perceive that the corpuscular attitude to light in its most elementary form leads straightforwardly to the correct theory of this phenomenon, provided we apply relativistic dynamics. The reader should notice in passing three important points:

—the existence of a transverse Doppler effect ($\delta_1 = \pi/2$), already pointed out in *Kinematics* (§ [64]);

—the fact that the source (and its motion) has no effect;

—the use of relativistic dynamics allows us to consider light either from the corpuscular or from the wave point of view at will (in ordinary optics, the Doppler effect is considered only from the latter point of view).

Remark. Two particles can be distinguished absolutely by their rest masses. On the contrary, the comparison of two photons has only a relative meaning and depends on the system of reference. Any frequency whatsoever can be attributed to a given photon by selecting the system of reference suitably.

As an exercise, the reader might verify the following statements. If two photons with frequencies ν_1 and ν_2 are travelling in the same direction in a system K, the ratio ν_1/ν_2 is invariant. If, in K, two photons with frequencies ν_1 and ν_2 are travelling in different directions, there exists a reference system in which the two photons have the same frequency and are travelling in opposite directions.

[64] Transformation of the Wavelength

Since the velocity is always c, the transformation for λ can be derived from that for frequency, ν, immediately. In the case of propagation parallel to v,

$$\lambda_1 = \lambda_2 \sqrt{\frac{1-\beta}{1+\beta}} = \lambda_2 \frac{\sqrt{1-\beta^2}}{1+\beta}$$

and this expression does possess the reciprocity property.

The attention of the reader is drawn to this formula, which provokes an interesting question of principle. In deriving the wavelength transformation, there are two mistakes which might be made: we might think that

—either the wavelength should transform like a length, with the Lorentz contraction;
—or, recalling that lengths are measured with the aid of an optical standard, that the wavelength of a given light beam is the same in every system.

The above expression shows that both beliefs are wrong. For the Lorentz transformation applies to the distance between two given points A and B, which are the same irrespective of the system of measurement. In the present case, λ is the distance between two points M and M' at which the phase is the same at a given instant. The last phrase contains the explanation of the paradox. Two simultaneous events at M at M' (same phase) for one system are not simultaneous in another system. When we change systems, one of the points is altered, and we measure the distance between M and M''. There is thus a double effect, involving non-simultaneity and the Lorentz contraction.

HISTORICAL AND BIBLIOGRAPHICAL NOTES

[65] On Fields with Scalar Potentials and the Corresponding Optical Analogues

(a) The formulae of the present chapter are used above all in electron optics, where they are basic. I shall therefore give the bibliography in the corresponding volume, since most of the publications deal with applications. The references given below to L. de Broglie [4, 5, 6] and P. Chanson [7] are to be regarded simply as a preliminary study. Nor can I consider even briefly the question of the wave–corpuscle duality without going beyond the scope of the present text. I merely mention, as a mark of respect, that the concept of the photon is due to Einstein [1], and that the wave-like properties of corpuscles in general were discovered by L. de Broglie [4].

(b) *Dynamics deduced from the wave-properties of corpuscles.* This method was used for the first time by Haas [2, 3] and later, independently, by Schlegel [9]. As postulates, we take relativistic kinematics and the two formulae

$$W = h\nu \tag{1}$$

$$\lambda = h/mv \tag{2}$$

in which m is regarded simply as a parameter of proportionality. Applying relativistic kinematics to (2) yields

$$\frac{c^2}{v} = u = \lambda\nu \tag{3}$$

Substituting in (3) for ν and λ the values taken from (1) and (2), we find

$$W = mc^2$$

From the de Broglie formula for transformation of frequency,

$$h\nu = \frac{h\nu_0}{\sqrt{1-\beta^2}} \quad \text{and} \quad m = \frac{m_0}{\sqrt{1-\beta^2}}$$

Hence

$$W = \frac{W_0}{\sqrt{1-\beta^2}}$$

It is then natural to define the kinetic energy to be

$$E = W - W_0 = W_0\left(\frac{1}{\sqrt{1-\beta^2}} - 1\right)$$

This increase of kinetic energy is caused by work done by the forces. For rectilinear motion and force parallel to the velocity, we write

$$\int_0^x X\,\mathrm{d}x = m_0c^2\left(\frac{1}{\sqrt{1-\frac{1}{c^2}\left(\frac{\mathrm{d}x}{\mathrm{d}t}\right)^2}} - 1\right)$$

so that

$$\frac{\mathrm{d}}{\mathrm{d}t}\int_0^x X\,\mathrm{d}x = \frac{\mathrm{d}}{\mathrm{d}x}\left(\int_0^x X\,\mathrm{d}x\right)\frac{\mathrm{d}x}{\mathrm{d}t}$$

$$= m_0c^2\frac{\mathrm{d}}{\mathrm{d}t}\left(\frac{1}{\sqrt{1-\frac{1}{c^2}\left(\frac{\mathrm{d}x}{\mathrm{d}t}\right)^2}} - 1\right)$$

or

$$X\frac{\mathrm{d}x}{\mathrm{d}t} = m_0c^2\frac{\mathrm{d}x}{\mathrm{d}t}\frac{\mathrm{d}^2x}{\mathrm{d}t^2}\left\{1-\frac{1}{c^2}\left(\frac{\mathrm{d}x}{\mathrm{d}t}\right)^2\right\}^{-\frac{3}{2}}$$

which is the equation involving longitudinal mass. A similar calculation would yield transverse mass. In this connection, we mention the articles of Stiegler [8, 11, 12, 14] and Jankovic [13].

[1] A. Einstein: Ueber einen die Erzeugung und Verwandlung des Lichtes betreffenden heuristischen Gesichtspunkt. *Ann. Physik* **17** (1905) 132–148.
[2] A. Haas: Ueber die Ableitung der fundamentalen Relativitäts-theoretischen Sätze aus der Broglieschen Hypothese der Phasenwellen. *Phys. Z.* **28** (1927) 632–634.
[3] A. Haas: *La Mécanique ondulatoire et les nouvelles théories quantiques.* Gauthier-Villars, 1937, pp. 32–37 (2nd French ed., translated from the 4th and 5th editions of the original; 1st ed. 1929).
[4] L. de Broglie: Recherches sur la théorie des quanta. Thèse, Paris, 1924; *Ann. Physique* **3** (1925) 22–128.
[5] L. de Broglie: *Optique électronique et corpusculaire.* Hermann, 1950, pp. 38–41 and 267–269.
[6] L. de Broglie: Une interprétation nouvelle de la mécanique ondulatoire est-elle possible? *Conférences Palais de la Découverte*, série A, No. 201.
[7] P. Chanson: Les éléments optiques des lentilles électroniques et le microscope à protons. *Ann. Physique* **2** (1947) 333–413.
[8] K. D. Stiegler: Sur les rapports entre le principe de Maupertuis–Lagrange et celui de Fermat d'une part et la théorie de la relativité restreinte et la mécanique ondulatoire d'autre part. *Bull. Acad. Roy. Belgique* **39** (1953) 1052–1063.
[9] R. Schlegel: Wave and inertial properties of matter. *Amer. J. Phys.* **22** (1954) 77–82.
[10] L. de Broglie: *Une tentative d'interprétation causale et non linéaire de la mécanique ondulatoire.* Gauthier-Villars, 1956.
[11] K. Stiegler: L'hypothèse d'ondes corpuscules et la théorie de la relativité restreinte. *Nuovo Cimento* **8** (1958) 922–926.
[12] K. Stiegler: On the mechanical foundation of the theory of special relativity. *Nuovo Cimento* **8** (1959) 873–879.
[13] Z. Jankovic: Remarks on the paper by K. Stiegler: "On the mechanical foundation of the theory of special relativity". *Nuovo Cimento* **16** (1960) 569.
[14] K. Stiegler: Antwort auf die Bemerkungen von Z. Jankovic zu meiner Abhandlung: "On the mechanical foundation of the theory of special relativity". *Nuovo Cimento* **16** (1960) 579–581.

CHAPTER V

FIELDS OF FORCE DERIVABLE FROM A SCALAR POTENTIAL AND A VECTOR POTENTIAL

[66] General Remarks; Definition of the Potentials

(a) *Preliminary remarks, by way of introduction*

The idea of scalar potential, the only type commonly considered in classical mechanics, is only a special case. We may use such a potential only when the component $\mathfrak{B}$ is zero. As a matter of fact, *the vector* **E** *may certainly be defined as the gradient of a scalar, but the tensor* $\mathfrak{B}$ *may not.*

We know that such a tensor can be obtained from a vector **A**, using the curl operator:

$$\mathfrak{B} = \text{curl } \mathbf{A}$$

Just as the operator grad yields a vector from a scalar, so curl gives an antisymmetric second rank tensor from a vector. In generalizing the mechanical concept of potential, we are thus led to consider two functions, the scalar V and the vector **A**.

An example will show that this point of view is perfectly compatible with the relativistic transformation. Suppose that in a reference system K_2, a force field reduces to the single component $\mathbf{E}_2$ that can be derived from a scalar potential V_2

$$E_{2x} = -\frac{\partial V_2}{\partial x_2}, \ldots$$

We now go over to a system K_1. The formulae of kinematics, we recall, are of the form

$$\frac{\partial}{\partial x_2} = \frac{1}{\sqrt{1-\beta^2}}\left(\frac{\partial}{\partial x_1}+\frac{v}{c^2}\frac{\partial}{\partial t_1}\right)$$

$$\frac{\partial}{\partial y_2} = \frac{\partial}{\partial y_1} \qquad \frac{\partial}{\partial z_2} = \frac{\partial}{\partial z_1}$$

$$\frac{\partial}{\partial t_2} = \frac{1}{\sqrt{1-\beta^2}}\left(\frac{\partial}{\partial t_1}+v\frac{\partial}{\partial x_1}\right)$$

In K_1, we have the fields

$$E_{1x} = E_{2x} = -\frac{1}{\sqrt{1-\beta^2}}\left(\frac{\partial}{\partial x_1}+\frac{v}{c^2}\,\frac{\partial}{\partial t_1}\right)V_2$$

$$E_{1y} = \frac{E_{2y}}{\sqrt{1-\beta^2}} = -\frac{1}{\sqrt{1-\beta^2}}\,\frac{\partial V_2}{\partial y_1}$$

$$E_{1z} = \frac{E_{2z}}{\sqrt{1-\beta^2}} = -\frac{1}{\sqrt{1-\beta^2}}\,\frac{\partial V_2}{\partial z_1}$$

$$\mathcal{B}_{1x} = 0$$

$$\mathcal{B}_{1y} = -\frac{c_0\beta}{c\sqrt{1-\beta^2}}E_{1z} = \frac{c_0\beta}{c\sqrt{1-\beta^2}}\,\frac{\partial V_2}{\partial z_1}$$

$$\mathcal{B}_{1z} = \frac{c_0\beta}{c\sqrt{1-\beta^2}}E_{1y} = -\frac{c_0\beta}{c\sqrt{1-\beta^2}}\,\frac{\partial V_2}{\partial y_1}$$

We see that $\mathfrak{B}_1$ is the curl of a vector $\mathbf{A}_1$, having components

$$A_{1x} = \frac{c_0\beta}{c\sqrt{1-\beta^2}}V_2, \qquad A_{1y} = A_{1z} = 0$$

In addition, we define in K_1 the scalar potential

$$V_1 = \frac{V_2}{\sqrt{1-\beta^2}}$$

and the field $\mathbf{E}_1$ is thus related to *two* potentials—the scalar V_1 and the vector $\mathbf{A}_1$—by the expressions

$$\mathbf{E}_1 = -\text{grad } V_1 - \frac{1}{c_0}\,\frac{\partial \mathbf{A}_1}{\partial t}$$

(b) *Definition of fields with potentials*

The foregoing remarks make the following definition obvious. We shall say that the field $(\mathbf{E}, \mathfrak{B})$ is derivable from a potential if there exist a vector $\mathbf{A}$ and a scalar V such that

$$\mathfrak{B} = \text{curl } \mathbf{A}, \quad \mathbf{E} = -\text{grad } V - \frac{1}{c_0}\,\frac{\partial \mathbf{A}}{\partial t} \tag{1}$$

We shall show in § [69] that this definition is invariant.

(c) *The four-dimensional definition*

Let us represent the total force, apart from a constant factor, by the covariant tensor $\mathcal{H}_{ij}$ (§ [49]). Such a tensor may be derived from a covariant four-vector Φ_i by the defining relation

$$\mathcal{H}_{ij} = \frac{\partial \Phi_j}{\partial x^i} - \frac{\partial \Phi_i}{\partial x^j}$$

This is a general expression, valid for all coordinate systems; it provides a generalization of the elementary notion of potential. In $xyzl$ space, we denote the components of Φ_i by

$$\Phi_x = A_x, \quad \Phi_y = A_y, \quad \Phi_z = A_z, \quad \Phi_l = \frac{ic_0}{c} V$$

Using the definition of $\mathcal{M}$ in terms of Φ, we recover (1); Φ is called the four-potential.

(d) *The components of* **E** *and* $\mathfrak{B}$ *in the commoner orthogonal coordinate systems*

Cartesian coordinates, xyz

$$\mathcal{B}_x = \frac{\partial A_z}{\partial y} - \frac{\partial A_y}{\partial z}, \ldots$$

Orthogonal curvilinear coordinates

We select coordinates u, v, w, such that the element of length is of the form

$$d\sigma^2 = e_u^2\, du^2 + e_v^2\, dv^2 + e_w^2\, dw^2$$

For the gradient of a function V, we have components

$$(\text{grad } V)_u = \frac{1}{e_u} \frac{\partial V}{\partial u}, \ldots$$

and for the curl of a vector **A**

$$(\text{curl } \mathbf{A})_u = \frac{1}{e_v e_w} \left\{ \frac{\partial (e_w A_w)}{\partial v} - \frac{\partial (e_v A_v)}{\partial w} \right\}, \ldots$$

We now apply these relations to two important special cases.

Cylindrical coordinates z, r, φ (Fig. 4):

$$\mathcal{B}_z = \frac{1}{r} \left\{ \frac{\partial (r A_\phi)}{\partial r} - \frac{\partial A_r}{\partial \phi} \right\}$$

$$\mathcal{B}_r = \frac{1}{r} \left\{ \frac{\partial A_z}{\partial \phi} - \frac{\partial (r A_\phi)}{\partial z} \right\}$$

$$\mathcal{B}_\phi = \frac{\partial A_r}{\partial z} - \frac{\partial A_z}{\partial r}$$

$$E_z = -\frac{\partial V}{\partial z} - \frac{1}{c_0} \frac{\partial A_z}{\partial t}$$

$$E_r = -\frac{\partial V}{\partial r} - \frac{1}{c_0} \frac{\partial A_r}{\partial t}$$

$$E_\phi = -\frac{1}{r} \frac{\partial V}{\partial \phi} - \frac{1}{c_0} \frac{\partial A_\phi}{\partial t}$$

Spherical polar coordinates r, θ, ϕ (Fig. 5):

$$\mathcal{B}_r = \frac{1}{r^2 \sin\theta}\left\{\frac{\partial(rA_\theta)}{\partial\phi} - \frac{\partial(r\sin\theta A_\phi)}{\partial\theta}\right\}$$

$$\mathcal{B}_\theta = \frac{1}{r\sin\theta}\left\{\frac{\partial(r\sin\theta A_\phi)}{\partial r} - \frac{\partial A_r}{\partial\phi}\right\}$$

$$\mathcal{B}_\phi = \frac{1}{r}\left\{\frac{\partial A_r}{\partial\theta} - \frac{\partial(rA_\theta)}{\partial r}\right\}$$

$$E_r = -\frac{\partial V}{\partial r} - \frac{1}{c_0}\frac{\partial A_r}{\partial t}$$

$$E_\theta = -\frac{1}{r}\frac{\partial V}{\partial\theta} - \frac{1}{c_0}\frac{\partial A_\theta}{\partial t}$$

$$E_\phi = -\frac{1}{r\sin\theta}\frac{\partial V}{\partial\phi} - \frac{1}{c_0}\frac{\partial A_\phi}{\partial t}$$

The axes are correctly orientated only if the coordinates are taken in the order r, θ, ϕ.

(e) *The equations of dynamics using the four-potential*

In the relation (§ [49])

$$\mathrm{d}\Pi^i = \frac{\varepsilon}{c_0}\mathcal{H}^{ij}\,\mathrm{d}x_j$$

we replace $\mathcal{H}^{ij}$ by its expression as a function of Φ^i:

$$\mathrm{d}\Pi^i = \frac{\varepsilon}{c_0}\left(\frac{\partial\Phi^j}{\partial x_i} - \frac{\partial\Phi^i}{\partial x_j}\right)\mathrm{d}x_j$$

so that

$$\mathrm{d}\Pi^i + \frac{\varepsilon}{c_0}\frac{\partial\Phi^i}{\partial x_j}\,\mathrm{d}x_j = \frac{\varepsilon}{c_0}\frac{\partial\Phi^j}{\partial x_i}\,\mathrm{d}x_j$$

or

$$\mathrm{d}\mathcal{P}^i = \frac{\varepsilon}{c_0}\frac{\partial\Phi^j}{\partial x_i}\,\mathrm{d}x_j \quad \text{where} \quad \mathcal{P}^i = \Pi^i + \frac{\varepsilon}{c_0}\Phi^i$$

(f) *The Herz vector*

We can define a vector, $\boldsymbol{\pi}$, by the relations

$$\mathbf{A} = \mathrm{curl}\,\boldsymbol{\pi}, \qquad V = -\mathrm{div}\,\boldsymbol{\pi}$$

This is the Herz vector commonly used in electromagnetism, which here has a purely mechanical interpretation. We shall not use it.

[67] The Expression for the Lorentz Force for Fields with Potentials; the Schwarzschild Function

We have

$$F_x = \varepsilon\left\{E_x + \frac{1}{c_0}(v_y B_z - v_z B_y)\right\}$$

$$= \varepsilon\left\{-\frac{\partial V}{\partial x} - \frac{1}{c_0}\frac{\partial A_x}{\partial t} + \frac{1}{c_0}\left[v_y\left(\frac{\partial A_y}{\partial x} - \frac{\partial A_x}{\partial y}\right) - v_z\left(\frac{\partial A_x}{\partial z} - \frac{\partial A_z}{\partial x}\right)\right]\right\}$$

$$= \varepsilon\left\{-\frac{\partial V}{\partial x} + \frac{1}{c_0}\left[v_x\frac{\partial A_x}{\partial x} + v_y\frac{\partial A_y}{\partial x} + v_z\frac{\partial A_z}{\partial x} - \frac{\partial A_x}{\partial t} - v_x\frac{\partial A_x}{\partial x} - v_y\frac{\partial A_x}{\partial y}\right.\right.$$

$$\left.\left. - v_z\frac{\partial A_x}{\partial z}\right]\right\}$$

Writing

$$\mathscr{L} = \varepsilon\left\{V - \frac{1}{c_0}(v_x A_x + v_y A_y + v_z A_z)\right\} = \varepsilon\left(V - \frac{\mathbf{v}}{c_0}\cdot\mathbf{A}\right)$$

we obtain Schwarzschild's expression:

$$F_x = \frac{\mathrm{d}}{\mathrm{d}t}\frac{\partial \mathscr{L}}{\partial v_x} - \frac{\partial \mathscr{L}}{\partial x}$$

We recall that

$$\frac{\mathrm{d}}{\mathrm{d}t} = \frac{\partial}{\partial t} + v_x\frac{\partial}{\partial x} + v_y\frac{\partial}{\partial y} + v_z\frac{\partial}{\partial z}$$

[68] Gauge Invariance

The quantities **A** and V are not completely defined by the fields. If $F(x, y, z, t)$ is an arbitrary (differentiable) function, the two expressions

$$\mathbf{A} \quad \text{and} \quad \mathfrak{A} = \mathbf{A} + c_0 \,\mathrm{grad}\, F$$

yield the same value for $\mathfrak{B}$, since

$$\mathrm{curl}\,\mathrm{grad}\, F = 0$$

Let us therefore replace **A** by $\mathfrak{A}$ in the right-hand side of the formula

$$\mathbf{E} = -\mathrm{grad}\, V - \frac{1}{c_0}\frac{\partial \mathbf{A}}{\partial t}$$

We obtain

$$-\mathrm{grad}\, V - \frac{1}{c_0}\left(\frac{\partial \mathbf{A}}{\partial t} + c_0\,\mathrm{grad}\,\frac{\partial F}{\partial t}\right) = -\mathrm{grad}\left(V + \frac{\partial F}{\partial t}\right) - \frac{1}{c_0}\frac{\partial \mathbf{A}}{\partial t}$$

To obtain the same value of **E**, we must replace

$$V \quad \text{by} \quad V - \frac{\partial F}{\partial t}$$

Thus we shall have the same fields $\mathbf{E}$ and $\mathfrak{B}$ if we replace possible values of the potentials, $\mathbf{A}$ and V, by

$$\mathfrak{A} = \mathbf{A} + c_0 \text{ grad } F \quad \text{and} \quad \mathcal{V} = V - \frac{\partial F}{\partial t}$$

The invariance of the fields under this transformation is known as *gauge invariance* (invariance de jauge, Eichinvarianz). In some calculations, this property enables us to simplify the expressions for V and $\mathbf{A}$.

[69] Transformation of the Potentials

(a) We derive the transformations from those for the fields $\mathbf{E}$ and $\mathfrak{B}$ and from the expressions for these fields in terms of their potentials.

The transformation

$$E_{2x} = E_{1x}$$

can be written

$$-\frac{\partial V_2}{\partial x_2} - \frac{1}{c_0}\frac{\partial A_{2x}}{\partial t_2} = -\frac{\partial V_1}{\partial x_1} - \frac{1}{c_0}\frac{\partial A_{1x}}{\partial t_1}$$

or

$$\frac{1}{\sqrt{1-\beta^2}}\left\{\left(\frac{\partial}{\partial x_1} + \frac{v}{c^2}\frac{\partial}{\partial t_1}\right)V_2 + \frac{1}{c_0}\left(\frac{\partial}{\partial t_1} + v\frac{\partial}{\partial x_1}\right)A_{2x}\right\}$$

$$= \frac{\partial V_1}{\partial x_1} + \frac{1}{c_0}\frac{\partial A_{1x}}{\partial t_1}$$

The first two relations (below) are obtained by comparison.

The transformation

$$\mathcal{B}_{2x} = \mathcal{B}_{1x}$$

becomes

$$\frac{\partial A_{2z}}{\partial y_2} - \frac{\partial A_{2y}}{\partial z_2} = \frac{\partial A_{1z}}{\partial y_1} - \frac{\partial A_{1y}}{\partial z_1}$$

or

$$\frac{\partial A_{2z}}{\partial y_1} - \frac{\partial A_{2y}}{\partial z_1} = \frac{\partial A_{1z}}{\partial y_1} - \frac{\partial A_{1y}}{\partial z_1}$$

The last pair of relations (1) are obtained by comparison.

Finally, therefore, we obtain the transformation formulae

$$V_1 = \frac{V_2 + \frac{v}{c_0}A_{2x}}{\sqrt{1-\beta^2}}, \qquad A_{1x} = \frac{A_{2x} + \frac{c_0 v}{c^2}V_2}{\sqrt{1-\beta^2}} \tag{1}$$

$$A_{1y} = A_{2y}, \qquad A_{1z} = A_{2z}$$

(b) In four-dimensional notation, we have the following relation directly with the special Lorentz transformation *(Kinematics*, p. 273):

$$\begin{pmatrix} A_{2x} \\ A_{2y} \\ A_{2z} \\ \frac{ic_0}{c} V_2 \end{pmatrix} = \begin{pmatrix} \frac{1}{\sqrt{1-\beta^2}} & 0 & 0 & \frac{i\beta}{\sqrt{1-\beta^2}} \\ 0 & 1 & 0 & 0 \\ 0 & 0 & 1 & 0 \\ -\frac{i\beta}{\sqrt{1-\beta^2}} & 0 & 0 & \frac{1}{\sqrt{1-\beta^2}} \end{pmatrix} \begin{pmatrix} A_{1x} \\ A_{1y} \\ A_{1z} \\ \frac{ic_0}{c} V_1 \end{pmatrix}$$

which yields (1), on multiplying out the matrices.

(c) *A potential invariant*. The following quantity is invariant:

$$A_x^2+A_y^2+A_z^2-\frac{c_0^2}{c^2} V^2 = A^2-\frac{c_0^2}{c^2} V^2;$$

it is the square of the modulus of the four-potential, Φ.

[70] The Purely Mechanical Significance of the Faraday–Maxwell Equations

(a) *The local form*

These equations are a consequence of the very definitions of fields with potentials. We set out from the expressions

$$\mathbf{E} = -\text{grad } V-\frac{1}{c_0}\frac{\partial \mathbf{A}}{\partial t} \tag{1}$$

$$\mathfrak{B} = \text{curl } \mathbf{A} \tag{2}$$

from which we eliminate the potentials. Taking the curl of **E**,

$$\text{curl } \mathbf{E} = -\text{curl grad } V-\frac{1}{c_0}\frac{\partial}{\partial t}\text{ curl } \mathbf{A}$$

The first term on the right-hand side vanishes since curl grad is always zero; we replace curl **A** by $\mathfrak{B}$, and obtain

$$\text{curl } \mathbf{E} = -\frac{1}{c_0}\frac{\partial \mathfrak{B}}{\partial t} \tag{3}$$

From (2),

$$\text{div } \mathfrak{B} = 0 \tag{4}$$

since div curl is always zero.

Conversely, if (3) and (4) are satisfied, potential functions exist. In fact, relation (4) shows that $\mathfrak{B}$ may be regarded as the curl of some vector **A**. Substituting into (3), we obtain

$$\text{curl } \mathbf{E}+\frac{1}{c_0}\frac{\partial}{\partial t}(\text{curl } \mathbf{A}) = \text{curl}\left(\mathbf{E}+\frac{1}{c_0}\frac{\partial \mathbf{A}}{\partial t}\right) = 0$$

The vector in parentheses, whose curl vanishes, can be regarded as the gradient of a scalar, V, which proves the statement above (the sign is selected by convention).

Although these formulae were originally obtained in electricity and are still used uniquely in electrical calculations, they have a *general mechanical sphere of application* in special relativity.

That Maxwell's equations have been adaptable, without modification, to the various theories which have been developed during a full century has often been remarked upon with admiration ("Ist es ein Gott, der diese Zeichen schrieb?",† asked Boltzmann). Their mechanical interpretation explains their permanence, despite transformations which go far beyond the limits of electromagnetism. This generalization entails a limitation: Maxwell's equations may only be employed in Galileian reference systems. They must be freshly examined if they are to be applied to accelerating systems (see *Milieux conducteurs et polarisables en mouvement)*. We have already made a similar remark in *Kinematics* (§ [47], p. 66).

(b) *Expressions in the commoner systems of orthogonal coordinates*

For curl, see § [66]. The divergence takes the following forms:

Cartesian coordinates

$$\operatorname{div} \mathfrak{B} = \frac{\partial \mathfrak{B}_x}{\partial x} + \frac{\partial \mathfrak{B}_y}{\partial y} + \frac{\partial \mathfrak{B}_z}{\partial z}$$

Orthogonal curvilinear coordinates

$$\operatorname{div} \mathfrak{B} = \frac{1}{e_u e_v e_w} \left\{ \frac{\partial}{\partial u}(e_v e_w \mathcal{B}_u) + \frac{\partial}{\partial v}(e_w e_u \mathcal{B}_v) + \frac{\partial}{\partial w}(e_u e_v \mathcal{B}_w) \right\}$$

Cylindrical polar coordinates

$$\operatorname{div} \mathfrak{B} = \frac{\partial \mathcal{B}_z}{\partial z} + \frac{1}{r}\left(\frac{\partial (r\mathcal{B}_r)}{\partial r} + \frac{\partial \mathcal{B}_\phi}{\partial \phi}\right)$$

Spherical polar coordinates

$$\operatorname{div} \mathfrak{B} = \frac{1}{r^2}\frac{\partial}{\partial r}(r^2 \mathcal{B}_r) + \frac{1}{r \sin\theta}\left\{\frac{\partial \mathcal{B}_\phi}{\partial \phi} + \frac{\partial(\sin\theta \mathcal{B}_\theta)}{\partial \theta}\right\}$$

(c) *The integral form*

We consider a fixed closed loop C in the field $\mathbf{E}$, $\mathfrak{B}$. We now calculate the loop integral of the vector $\mathbf{E}$ along the curve C:

$$\oint_C \mathbf{E} \, . \, d\mathbf{l} = -\oint_C \operatorname{grad} V \, . \, d\mathbf{l} - \frac{1}{c_0}\frac{\partial}{\partial t}\oint_C \mathbf{A} \, . \, d\mathbf{l}$$

† "Were these symbols the writing of a God?"

The first integral vanishes, unless there are points A, B, ... on C at which the function V has discontinuities, V_A, V_B, ...

The second integral gives

$$\frac{\partial}{\partial t}\oint_C \mathbf{A}\,.\,\mathrm{d}\mathbf{l} = \frac{\partial}{\partial t}\iint_S \text{curl}\,\mathbf{A}\,.\,\mathbf{n}\,\mathrm{d}S$$

$$= \frac{\partial}{\partial t}\iint \mathfrak{B}\,.\,\mathbf{n}\,\mathrm{d}S = \frac{\partial \mathcal{F}}{\partial t}$$

S is an arbitrary surface containing the contour C; we have used the conservative property of the vectors curl $\mathbf{A}$ or $\mathfrak{B}$. $\mathcal{F}$ denotes the flux of $\mathfrak{B}$ through C.

Finally,

$$\oint_C \mathbf{E}\,.\,\mathrm{d}\mathbf{l} = \Sigma V_A - \frac{1}{c_0}\frac{\partial \mathcal{F}}{\partial t}$$

We may replace the ∂ by d, for by hypothesis the curve C is fixed and fully defined, so that $\mathcal{F}$ depends only upon t.

In electrical applications of this formula, the V_A correspond to localized contact electromotive forces and the other term is the induced electromotive force.

If there are no terms V, the integral vanishes if $\mathcal{F}$ is independent of time.

Important remark. Let us consider a field obtained from $\mathbf{A}$ alone. The transformation of the line integral into a surface integral implies that $\mathbf{A}$ is continuous in space (with respect to x, y, z); the same is true of $\partial\mathbf{A}/\partial t$ and hence of $\mathbf{E}$ also. However, curl $\mathbf{A}$ and hence $\mathfrak{B}$ may be discontinuous. We could say the same about the continuity of $\mathbf{E}$ if we wrote

$$\iint_S \text{curl}\,\mathbf{E}\,.\,\mathbf{n}\,\mathrm{d}S = \oint_C \mathbf{E}\,.\,\mathrm{d}\mathbf{l}$$

Let us assume that $\mathfrak{B}$ is not zero and is uniform within some cylinder (Fig. 12) and zero outside. The loop integral is different from zero only if C encloses the cylinder, and the form of C only appears in the number of times the cylinder is circled. At each point of C

$$\mathfrak{B} = 0, \qquad \frac{\partial \mathfrak{B}}{\partial t} = 0, \qquad \text{curl}\,\mathbf{E} = 0, \qquad \mathbf{E} \neq 0$$

Although $\mathbf{E}$ is not zero, its curl vanishes. This case is basic to a good understanding of the phenomena of electromagnetic induction.

(d) *The tensor expression*

Introducing the tensor $\mathcal{B}_{ij}$ and denoting the coordinates x_i, we have

$$\frac{\partial E_i}{\partial x_j} - \frac{\partial E_j}{\partial x_i} = \frac{1}{c_0}\frac{\partial \mathcal{B}_{ij}}{\partial t}$$

$$\frac{\partial \mathcal{B}_{ij}}{\partial x_k} + \frac{\partial \mathcal{B}_{jk}}{\partial x_i} + \frac{\partial \mathcal{B}_{ki}}{\partial x_j} = 0$$

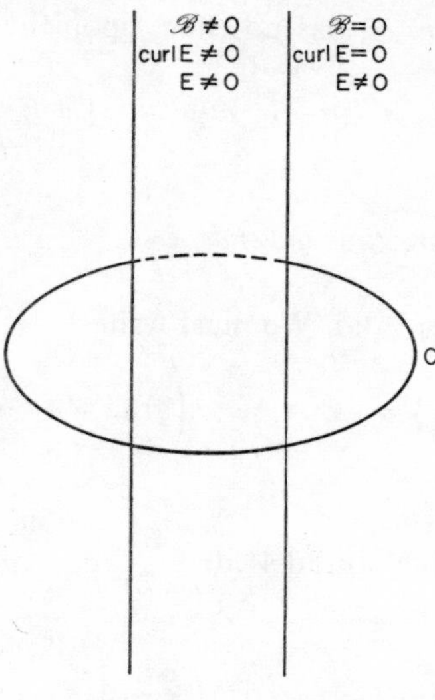

FIG. 12.

Despite the three suffices, the second expression yields only one equation in the present case of three dimensions; in Cartesians,

$$\frac{\partial \mathcal{B}_{xy}}{\partial z}+\frac{\partial \mathcal{B}_{yz}}{\partial x}+\frac{\partial \mathcal{B}_{zx}}{\partial y}=0$$

which we can, moreover, extract directly from the usual formalism with $\mathcal{B}_x$, ..., with the aid of the table in § [43] (remark). The formulae above will prepare the reader for the four-dimensional generalizations.

[70A] General Dynamical Theorems

(a) *Conservation of total energy for* V, **A** *independent of time*

This conservation law is valid here, just as in pre-relativistic dynamics; using the fact that *the component* $\mathfrak{B}$ *does no work*, we have

$$\varepsilon\,\mathrm{d}V = -\mathbf{F}\,.\,\mathrm{d}\mathbf{s} = -\mathrm{d}(mc^2)$$

$$\varepsilon(V-V_a) = \left(\frac{m_0}{\sqrt{1-v_0^2/c^2}}-m\right)c^2 \tag{1a}$$

$$\varepsilon V+mc^2 = \varepsilon V_a+\frac{m_0c^2}{\sqrt{1-v_0^2/c^2}} = \mathcal{H} \tag{1b}$$

in which $\mathcal{H}$ *denotes the total energy*, the sum of the proper energy, the kinetic energy and the potential energy. V_a denotes the potential for velocity v_0 (in § [57], we assumed that V_a was zero).

Since the proper energy, m_0c^2, has no effect upon the motion, it is sometimes convenient to employ

$$\mathcal{H}' = \mathcal{H} - m_0c^2 = \varepsilon V + E \tag{2}$$

instead of $\mathcal{H}$.

(b) *The case where V and* **A** *are time-dependent*

Equation (1a) is no longer valid. We must write

$$\frac{\mathrm{d}}{\mathrm{d}t}\left(\frac{m_0c^2}{\sqrt{1-\beta^2}}\right) = \varepsilon \mathbf{E} \,.\, \mathbf{v} = -\varepsilon\left(\operatorname{grad} V + \frac{1}{c_0}\frac{\partial \mathbf{A}}{\partial t}\right) . \frac{\mathrm{d}\mathbf{r}}{\mathrm{d}t} \tag{3}$$

Using the fact that

$$\mathrm{d}V = \operatorname{grad} V . \mathrm{d}\mathbf{r} + \frac{\partial V}{\partial t}\mathrm{d}t \tag{4}$$

we find

$$\mathrm{d}\left(\frac{m_0c^2}{\sqrt{1-\beta^2}}\right) = -\varepsilon\left(\mathrm{d}V - \frac{\partial V}{\partial t}\mathrm{d}t + \frac{1}{c_0}\frac{\partial \mathbf{A}}{\partial t}\cdot \mathrm{d}\mathbf{r}\right)$$

$$\frac{\mathrm{d}\mathcal{H}}{\mathrm{d}t} = \mathrm{d}\left(\frac{m_0c^2}{\sqrt{1-\beta^2}} + \varepsilon V\right) = \varepsilon\left(\frac{\partial V}{\partial t}\mathrm{d}t - \frac{1}{c_0}\frac{\partial \mathbf{A}}{\partial t}\cdot \mathrm{d}\mathbf{r}\right) \tag{5}$$

The right-hand side characterizes the time variation of $\mathcal{H}$.

(c) *The expression for the velocity*

The particle starts from rest at a point where the potential is zero (choice of origin); let us calculate the velocity v after the particle has travelled through potential V. The energy theorem gives

$$\frac{m_0c^2}{\sqrt{1-v^2/c^2}} - m_0c^2 = -\varepsilon V$$

$$v = c\sqrt{1 - \frac{1}{(1-V/V_0)^2}} \quad \text{with} \quad V_0 = \frac{m_0c^2}{\varepsilon} \tag{6}$$

In pre-relativistic dynamics, we should have found

$$\frac{1}{2}m_0v^2 = -\varepsilon V, \quad v = \sqrt{-\frac{2\varepsilon V}{m_0}}$$

The two curves are shown in Fig. 13.

Equation (6) may be written

$$v = \sqrt{-\frac{2\varepsilon V}{m_0}}\,\frac{\sqrt{1-V/2V_0}}{1-V/V_0} \tag{7}$$

thus displaying the effect of the relativistic factor. The momentum is hence

$$p = \frac{m_0 v}{\sqrt{1-v^2/c^2}} = \sqrt{-2\varepsilon m_0 V(1-V/2V_0)} \tag{8}$$

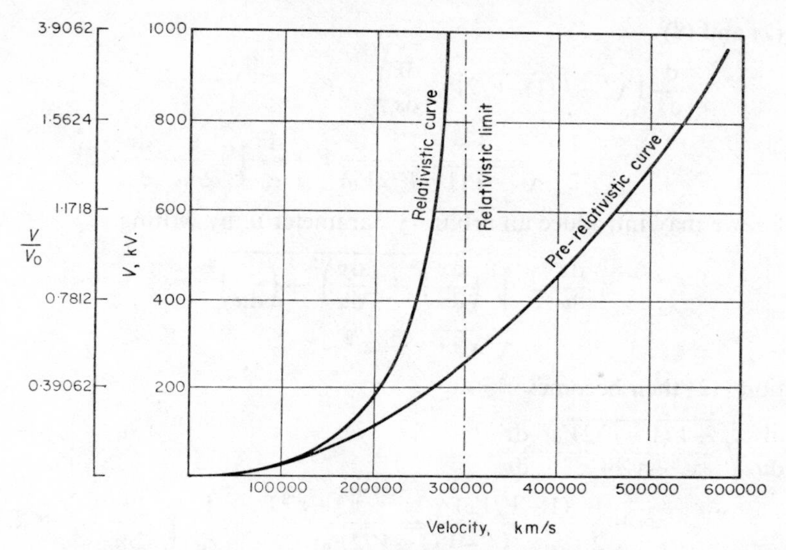

FIG. 13. The velocity as a function of the potential difference (the scale on the right corresponds to electrons).

(d) *Calculation of the radius of curvature*

In the expression for ϱ given in § [8] (b), we replace v^2 by the value given above; for F_n, we have explicitly

$$F_n = \varepsilon E_n + \frac{\varepsilon v}{c_0} \mathcal{B}_b$$

in which $\mathcal{B}_b$ is the component of $\mathfrak{B}$ along the binormal to the trajectory. Thus

$$\frac{1}{\varrho} = \frac{(1-V/V_0)F_n}{-2\varepsilon V(1-V/2V_0)}$$

$$= \frac{1-V/V_0}{2V(1-V/V_0)} \frac{\partial V}{\partial n} + \sqrt{\frac{\varepsilon}{2m_0}} \frac{B_b}{\sqrt{-V(1-V/2V_0)}} \tag{9}$$

(e) *The differential equation of the trajectory*

We define this curve by the expression for the radius vector $\mathbf{r}$ as a function of arc length s, $\mathbf{r}(s)$. We have

$$\mathbf{v} = \frac{\mathrm{d}\mathbf{r}}{\mathrm{d}t} = \frac{\mathrm{d}\mathbf{r}}{\mathrm{d}s}\frac{\mathrm{d}s}{\mathrm{d}t} = v\frac{\mathrm{d}\mathbf{r}}{\mathrm{d}s} \tag{10}$$

The general dynamical equation then becomes

$$v\frac{\mathrm{d}}{\mathrm{d}s}\left(\frac{m_0 v}{\sqrt{1-v^2/c^2}}\frac{\mathrm{d}\mathbf{r}}{\mathrm{d}s}\right) = \varepsilon\left(\mathbf{E} + \frac{v}{c_0}\frac{\mathrm{d}\mathbf{r}}{\mathrm{d}s}\times\mathfrak{B}\right) \tag{11}$$

or, using (7) and (8)

$$\frac{d}{ds}\left(\sqrt{-V(1-V/2V_0)}\,\frac{d\mathbf{r}}{ds}\right)$$

$$= \frac{1}{2}\cdot\frac{1-V/V_0}{\sqrt{-V(1-V/2V_0)}}\,\mathbf{E}+\frac{1}{c_0}\sqrt{\frac{\varepsilon}{2m_0}}\,\frac{d\mathbf{r}}{ds}\times\mathfrak{B} \tag{12}$$

Instead of s, we may introduce an arbitrary parameter u, by writing

$$\frac{ds}{du} = \sqrt{\left(\frac{dx}{du}\right)^2+\left(\frac{dy}{du}\right)^2+\left(\frac{dz}{du}\right)^2}$$

$$= \sqrt{x'^2+y'^2+z'^2} \tag{13}$$

and equation (12) then becomes

$$\frac{d}{du}\sqrt{\frac{-V(1-V/2V_0)}{x'^2+y'^2+z'^2}}\,\frac{d\mathbf{r}}{du}$$

$$= -\frac{1}{2}\,\frac{(1-V/V_0)\sqrt{x'^2+y'^2+z'^2}\,\mathbf{E}}{\sqrt{-V(1-V/2V_0)}}+\frac{1}{c_0}\sqrt{\frac{\varepsilon}{2m_0}}\,\frac{d\mathbf{r}}{du}\times\mathfrak{B} \tag{14}$$

One of the coordinates, x say, can be selected as the parameter u, and two differential equations for $y(x)$ and $z(x)$ are then obtained. The time parameter may also be chosen.

In pre-relativistic dynamics, (14) reduces to

$$\frac{d}{du}\sqrt{\frac{-V}{x'^2+y'^2+z'^2}}\,\frac{d\mathbf{r}}{du} = -\frac{1}{2}\sqrt{\frac{x'^2+y'^2+z'^2}{-V}}\,\mathbf{E}$$

$$+\frac{1}{c_0}\sqrt{\frac{\varepsilon}{2m_0}}\,\frac{d\mathbf{r}}{du}\times\mathfrak{B} \tag{15}$$

(f) *Optical analogues*

The reader will find some indications in § [86].

HISTORICAL AND BIBLIOGRAPHICAL NOTES

[71] On the origin of the potentials V and $\mathbf{A}$

(a) *The scalar potential V*

The concept of potential for a force-field is employed by Laplace [1], in the general theory of the Newtonian field. He does not use the term "potential", but describes the quantity clearly ("... la somme des molécules divisées par leurs distances respectives au point attiré"†) and gives the equation for this quantity *in vacuo*. The concept is later applied in electricity and magnetism by Poisson [2] and Green [3]; the latter seems to be the first to use the word potential.

† "The sum of the molecules divided by their distances from the point attracted."

In studying magnetism, Gauss [4] perfected the general theory of potential, introducing equipotential surfaces, for example.

(b) *The vector potential* **A**

This concept appears in the work of Neumann [5], on the interactions between currents, in the formula

$$\mathbf{A} = \oint_C \frac{i\,\mathrm{d}\mathbf{s}}{r}$$

He showed that the electromotive force exerted by C on a circuit element is obtained by differentiating **A** with respect to time.

The formula which is used today, giving the total field **E** as a function of V and **A** (§ [66]) was already given in Kirchhoff [6].

[1] Laplace: *Mém. Acad.* 1785, p. 113.
[2] Poisson: ref. [5] of § [143].
[3] G. Green: *An Essay on the Application of Mathematical Analysis to the Theories of Electricity and Magnetism*, Nottingham, 1828. Reprinted in *The Mathematical Papers of the late G. Green*, p. 1.
[4] C. F. Gauss: *Théorèmes généraux sur les forces d'attraction et de répulsion agissant en raison inverse du carré des distances.*
[5] F. Neumann: Recherches sur la théorie mathématique de l'induction. *J. Math.* **13** (1848) 113 (French translation of the German text).
[6] G. Kirchhoff: Ueber die Bewegung der Elektricität in Leitern. *Ann. Physik* **102** (1857) 529–544.

[71A] On the Maxwell–Lorentz Equations

These have been deduced here from mechanical postulates; they are the consequences of applying relativistic dynamics to a Newtonian field of force.

Historically, the development was in the opposite direction. Maxwell's equations were obtained by inductive reasoning or by arguments based on analogy, with the experimental observations on electromagnetic phenomena as starting point. From these equations, the theoreticians were led to relativistic dynamics.

It is most interesting for the relativistic physicist to follow the evolution of ideas which led first Maxwell and then Lorentz to the form of their equations. This question is also of the utmost interest to the philosophers: we recognize in fact that the mathematical form remains inviolate while very different physical meanings are successively attached to it. An examination of this would be out of place in the present volume. Here, we have merely indicated (§ [20] (b)) the final stage: the passage from the field equations—assumed to be known—to relativistic dynamics, with the aid of the transformation formulae.

[71B] On the Transformation of Fields, Potentials and Densities

(a) In the preceding pages, the vectors **E** and $\mathfrak{B}$ have been defined in purely mechanical terms; their transformation formulae follow from those for the transformation of force and Maxwell's equations come afterwards.

Historically, the sequence was reversed. The vector **E** was the electric field, defined as the force exerted upon unit charge *at rest;* $\mathfrak{B}$ or **H** (with $\mathfrak{B} = \mu_0\mathbf{H}$, see § [137]) denoted the magnetic field, defined, in the course of studying the properties of magnets, to be the couple exerted on a magnetic dipole of unit magnetic moment. With these definitions, nothing is said about the action of the fields on moving charges.

In his important paper "Versuche einer Theorie ..." [1], H. A. Lorentz sought to establish the influence of the earth's motion on electromagnetic phenomena. He accepted the existence of aether, virtually at rest in the absolute system of pre-relativistic mechanics (that of the sun or the stars), and in this privileged system, he wrote the field equations (ϱ denoting the charge density) in the form

$$\operatorname{div} \mathbf{E}_1 = \frac{4\pi\varrho_1}{K_0}, \qquad \operatorname{div} \mathbf{H}_1 = 0$$

$$\operatorname{curl} \mathbf{E}_1 = -\frac{\mu_0}{c_0}\frac{\partial \mathbf{H}_1}{\partial t}, \qquad \operatorname{curl} \mathbf{H}_1 = \frac{K_0}{c_0}\frac{\partial \mathbf{E}_1}{\partial t}+\frac{4\pi}{c_0}\varrho_1\mathbf{u}_1$$

He then showed that on neglecting second-order quantities, these formulae are unaffected in form if they are transformed to the earth's system, using the following formulae:

$$x_1 = x_2+vt, \qquad y_1 = y_2, \qquad z_1 = z_2$$

$$t_1 = t_2+\frac{vx_2}{c^2}, \qquad u_1 = u_2+v$$

provided we put

$$\mathbf{E}_1 = \mathbf{E}_2+\frac{\mu_0\mathbf{v}}{c_0}\times\mathbf{H}, \qquad \mathbf{H}_1 = \mathbf{H}_2+\frac{K_0\mathbf{v}}{c_0}\times\mathbf{E}$$

and ϱ is unchanged. These formulae are correct to first order. Larmor [2] performed the analysis retaining terms of second order. Lorentz [3] gave general formulae for the fields, but did not discard the idea that the aether provides a privileged system of reference. Moreover, his formulae for density and velocity are wrong: he wrote

$$\varrho_1 = \frac{\varrho_2}{\sqrt{1-\beta^2}} \quad \text{instead of} \quad \varrho_1 = \varrho_2\frac{1+vu_{2x}/c^2}{\sqrt{1-\beta^2}}$$

and

$$u_1 = \frac{u_2+v}{1+v^2/c^2} \quad \text{instead of} \quad u_1 = \frac{u_2+v}{1+vu_2/c^2}$$

He himself recognized that the goal he sought (equations in the same form in both systems) had not been fully attained: the equation in div **E** is of the form

$$\operatorname{div} \mathbf{E}_2 = \left(1-\frac{vu_{2x}}{c^2}\right)\varrho_2$$

Einstein [4] and Poincaré [5] made the necessary corrections. The potential transformation is to be found first in the works of H. A. Lorentz, then Einstein and then Poincaré.

(b) The expression for the Heaviside–Thomson force can be derived from the Maxwell–Lorentz formulae and the transformation relations; there is no need to state the former as a supplementary postulate.

Consider first the action of an electric field on a moving charge: let a charge ε move with velocity v with respect to a system K, in which the electric field **E** acts. At the time in question, we select rectangular axes such that Ox is parallel to v. In the system K_0 of the moving charge, there is an electromagnetic field, with electric components

$$E_{0x} = E_x, \qquad E_{0y} = \frac{E_y}{\sqrt{1-\beta^2}}, \qquad E_{0z} = \frac{E_z}{\sqrt{1-\beta^2}}$$

In K_0, the components of force are

$$F_{0x} = \varepsilon E_{0x} = E_x, \qquad F_{0y} = \varepsilon E_{0y} = \frac{\varepsilon E_y}{\sqrt{1-\beta^2}}, \ldots$$

so that in K,

$$F_x = F_{0x} = \varepsilon E_x, \qquad F_y = F_{0y}\sqrt{1-\beta^2} = \varepsilon E_y$$

For any velocity, therefore, we have

$$\mathbf{F} = \varepsilon\mathbf{E}$$

Consider now the action of a static magnetic field on a charge in uniform motion. In the system K, with respect to which the charge has velocity v parallel to Ox, we have by hypothesis a purely magnetic field, $\mathfrak{B}$. In K_0, in which the charge is stationary, only the electric field acts on the charge, exerting a force $\mathbf{F}_0$:

$$F_{0x} = \varepsilon E_{0x} = 0$$

$$F_{0y} = \varepsilon E_{0y} = -\frac{1}{c_0}\frac{\varepsilon v}{\sqrt{1-\beta^2}}\mathcal{B}_z$$

$$F_{0z} = \varepsilon E_{0z} = \frac{1}{c_0}\frac{\varepsilon v}{\sqrt{1-\beta^2}}\mathcal{B}_y$$

so that in the system K

$$F_x = 0, \quad F_y = -\frac{\varepsilon v}{c_0}\mathcal{B}_z, \quad F_z = \frac{\varepsilon v}{c_0}\mathcal{B}_y$$

This argument seems to be due to Tolman [6].

[1] H. A. Lorentz: *Versuch einer Theorie der elektrischen und optischen Erscheinungen in bewegten Körpern.* E. J. Brill, Leiden, 1895; reprinted by Teubner, Leipzig, 1906.

[2] Larmor: *Aether and Matter*, 1900, p. 173.

[3] H. A. Lorentz: Electromagnetic phenomena in a system moving with any velocity smaller than that of light. *Proc. Konig. Acad. Amsterdam* **6** (1904) 809–831.

[4] A. Einstein: ref. [1] of Appendix I.

[5] H. Poincaré: Sur la dynamique de l'électron. *Rend. Circ. Mat. Palermo* **21** (1906) 129–176; *Oeuvres Complètes*, Gauthier-Villars, Vol. IX, pp. 494–586.

[6] R. C. Tolman: Note on the derivation from the principle of relativity of the fifth fundamental equation of the Maxwell-Lorentz theory. *Phil. Mag.* **21** (1911) 296 – 301.

[7] B. Podolsky: On the Lorentz transformation of charge and current densities. *Phys. Rev.* **72** (1947) 624–626.

CHAPTER VI

VARIATIONAL METHODS: EQUATIONS OF MOTION

[72] Introduction

The methods that are usually classified under the heading "analytical mechanics" lead to three types of equations: the principle of least action, Hamilton's canonical equations and the Hamilton–Jacobi equation. These may be regarded as theorems derived from the impulse postulate, and this is normally the point of view adopted in classical mechanics (for the relativistic working, see § [88A]). Alternatively, any one may be taken as a fundamental postulate, and we shall adopt this point of view below, with the principle of least action as starting point. The reader will thus become acquainted with an aspect of the formalism of dynamics which is of frequent use in modern physics. It is even rather intriguing to realize that this formalism is no less useful as a research procedure than in classifying systematically results obtained by other means.

Another reason for the relativistic importance of these methods lies in the invariance of the conclusions drawn. Their use ensures that the requirements of the principle of relativity are automatically satisfied.

The method is only appropriate in a higher course on dynamics. In essence, we shall strive to provide a variational definition of all the mechanical quantities; as we shall see, however, these definitions remain empty of meaning until we compare them with the quantities defined in the preceding chapters.

A. THE ARBITRARY FIELD OF FORCE

[73] Hamilton's Principle

(a) *The pre-relativistic formalism*

We first recapitulate the analysis to be found in all the treatises on classical mechanics, in order to connect the discussion below with topics with which the reader will be familiar.

Consider a particle acted upon by frictionless connections and a field of force. Let us suppose that at each instant t, a virtual displacement compatible with the connections at time t is imposed upon the particle, *away from its real position.* As the components of this virtual displacement (there is no need to specify the coordinates), we choose infinitesimally small functions, continuous in time, which vanish at the two ends of a given time-interval, $t_1 - t_0$. A virtual variation of kinetic energy δE and a virtual variation of the work done by the applied force $\delta\mathcal{T}$ are associated with the virtual displacement at time t.

Hamilton's principle is written

$$\int_{t_0}^{t_1} (\delta E + \delta\mathcal{T})\, \mathrm{d}t = 0$$

or

$$\delta \int_{t_0}^{t_1} \mathcal{L}\, \mathrm{d}t = 0 \quad \text{where} \quad \mathcal{L} = E + \mathcal{T}$$

The function $\mathcal{L}$ is the Hamiltonian action.

The term E is characteristic of the free particle, and the term $\mathcal{T}$ of the existence of a field of force, that is to say, of the interaction between the particle and the other physical phenomena. If we regard matter as the only physical reality, we may say that $\mathcal{T}$ is characteristic of the interaction between the particle in question and other particles. To make this distinction clearly visible and to simplify subsequent generalization, we shall write

$$\mathcal{L}_p = E, \quad \mathcal{L}_i = \mathcal{T}$$

where the suffices p and i signify particle and interaction. Thus we write

$$\delta \int_{t_0}^{t_1} \mathcal{L}\, \mathrm{d}t = \delta \int_{t_0}^{t_1} (\mathcal{L}_p + \mathcal{L}_i)\, \mathrm{d}t = 0 \tag{1}$$

This integral, taken over an arbitrary time-interval, is the Hamiltonian integral or the action integral.

The meaning of this equation is that in the real motion of a particle between two arbitrary times t_0 and t_1, the action integral is stationary.

In other words, for a path of finite length, the integral must take either a maximum or minimum value. If the path is infinitesimal, the integral must pass through a minimum, hence the name: the principle of least action.

This principle may be taken as starting-point for pre-relativistic dynamics; as we shall show, the other equations of mechanics—the equations of Lagrange, for example—are deduced from it.

In the special case of a scalar potential, V,

$$\mathrm{d}\mathcal{T} = -\varepsilon\, \mathrm{d}V$$

and Hamilton's equation takes the form

$$\delta \int_{t_0}^{t_1} (E - \varepsilon V)\, \mathrm{d}t = 0$$

The function $\mathcal{L}$ is then the difference between the kinetic and potential energies.

(b) *The relativistic formalism*

We shall demonstrate that the general expression (1) for the variational formula may be used as the basis for relativistic dynamics.

Let us disregard, from the point of view of logic, all the foregoing work. Any mechanical problem will be characterized uniquely by the function $\mathcal{L}$; in terms of $\mathcal{L}$, all the common quantities must be defined (field of force, ...) and the laws of motion derived, and we must prove that the results obtained in the preceding chapters are recovered.

Together with formula (1), we *postulate*

$$\mathcal{L}_p = -m_0c^2\sqrt{1-\beta^2};$$

for the present, the function $\mathcal{L}_i$ remains indeterminate (in general, it is *not* the work, and indeed this quantity has not yet been defined). We shall see in § [102] how the choice of $\mathcal{L}_p$ can be made to seem quite natural, by the use of four-dimensional reasoning.

From its very expression, the function $\mathcal{L}_p$ is characteristic of a free particle, since no extraneous quantity (which could be interpreted as a field of force) figures in it. The force field will then be characterized by $\mathcal{L}_i$. The coefficient m_0 is the variational definition of the proper mass.

Important remark. Before proceeding further, we should notice that Hamilton's principle is *not* generalized by simply substituting the relativistic expression for the kinetic energy or the total energy into the ordinary formula; the quantity $\mathcal{L}_p$ is different from both E and W.

If we expand these quantities, we find

$$\begin{aligned} W &= m_0c^2+\tfrac{1}{2}\,m_0v^2+\tfrac{3}{8}\,m_0c^2\beta^4+\tfrac{5}{16}\,m_0c^2\beta^6+\ \dots \\ E &= \tfrac{1}{2}\,m_0v^2+\tfrac{3}{3}\,m_0c^2\beta^4+\tfrac{5}{16}\,m_0c^2\beta^6+\ \dots \\ \mathcal{L}_p &= -\,m_0c^2+\tfrac{1}{2}\,m_0v^2+\tfrac{1}{8}\,m_0c^2\beta^4+\tfrac{1}{16}\,m_0c^2\beta^6+\ \dots \end{aligned}$$

The fact that the kinetic energy is used in the pre-relativistic form of Hamilton's principle arises solely because the three quantities have the same limit, when only terms in v^2 are retained and the rest energy is neglected. We notice, furthermore, that the constant term m_0c^2 does not affect the equations of motion which are obtained by differentiating $\mathcal{L}$. It is for this reason that at low velocities, we can write $\mathcal{L}_p$ in the form

$$\tfrac{1}{2}m_0v^2 \quad \text{instead of} \quad -m_0c^2+\tfrac{1}{2}m_0v^2$$

Terminology

$\mathcal{L}$ is the *Hamiltonian action*, or succinctly, the *action; $\mathcal{L}_p$* is the *Lagrange function* or *Lagrangian*, and is in general different from the kinetic energy. These two names are frequently confused: $\mathcal{L}$ is often called the Lagrangian.

[74] The Equations of Lagrange

Using the generalized coordinates q_i,

$$\delta \mathscr{L} = \sum \left(\frac{\partial \mathscr{L}}{\partial q_i} \delta q_i + \frac{\partial \mathscr{L}}{\partial q_i'} \delta q_i' \right), \quad \text{with} \quad q_i' = \frac{\mathrm{d}q_i}{\mathrm{d}t}$$

This expression does not contain the term

$$\frac{\partial \mathscr{L}}{\partial t} \delta t$$

since we are considering a virtual variation, calculated without varying the time.

The quantity $\delta q_i'$ is the difference between the real and altered values of q_i'; it is hence the time derivative of the variation of q_i:

$$\delta q_i' \, \mathrm{d}t = \mathrm{d}(\delta q_i)$$

and we can therefore write

$$\int_{t_0}^{t_1} \frac{\partial \mathscr{L}}{\partial q_i'} \delta q_i' \, \mathrm{d}t = \int_{t_0}^{t_1} \frac{\partial \mathscr{L}}{\partial q_i'} \mathrm{d}(\delta q_i) = \left[\frac{\partial \mathscr{L}}{\partial q_i'} \delta q_i \right]_{t_0}^{t_1} - \int_{t_0}^{t_i} \delta q_i \frac{\mathrm{d}}{\mathrm{d}t} \frac{\partial \mathscr{L}}{\partial q_i'} \mathrm{d}t$$

By hypothesis, the variations vanish at the limits so that the integrated term is zero.

Hamilton's relation then gives

$$\int_{t_0}^{t_1} \sum \left(\frac{\partial \mathscr{L}}{\partial q_i} - \frac{\mathrm{d}}{\mathrm{d}t} \frac{\partial \mathscr{L}}{\partial q_i'} \right) \delta q_i \, \mathrm{d}t = 0$$

This equation must be satisfied for all δq_i, and irrespective of the range of integration. We thus obtain the *equations of Lagrange, which are applicable in both relativistic and pre-relativistic mechanics, provided that the correct expression for $\mathscr{L}$ is employed:*

$$\frac{\mathrm{d}}{\mathrm{d}t} \frac{\partial \mathscr{L}}{\partial q_i'} - \frac{\partial \mathscr{L}}{\partial q_i} = 0$$

Separating the terms in $\mathscr{L}_p$ and $\mathscr{L}_i$, we have

$$\frac{\mathrm{d}}{\mathrm{d}t} \frac{\partial \mathscr{L}_p}{\partial q_i'} - \frac{\partial \mathscr{L}_p}{\partial q_i} = \frac{\partial \mathscr{L}_i}{\partial q_i} - \frac{\mathrm{d}}{\mathrm{d}t} \frac{\partial \mathscr{L}_i}{\partial q_i'}$$

The left-hand side is characteristic of the corpuscle alone and its motion, while the right-hand side contains the forces acting upon it.

We can thus write

$$\frac{\mathrm{d}}{\mathrm{d}t} \frac{\partial \mathscr{L}_p}{\partial q_i'} - \frac{\partial \mathscr{L}_p}{\partial q_i} = Q_i$$

where we define a new quantity

$$Q_i = \frac{\partial \mathscr{L}_i}{\partial q_i} - \frac{\mathrm{d}}{\mathrm{d}t} \frac{\partial \mathscr{L}_i}{\partial q_i'}$$

and this is defined, in variational terms, as the "force in the direction of the variable q_i". In the special case for which $\mathcal{L}_i$ is independent of the q_i', this reduces to

$$Q_i = \frac{\partial \mathcal{L}_i}{\partial q_i}$$

We can go no further until the function $\mathcal{L}_i$ is given.

[75] Explicit Equations of Motion; Variational Definition of the Generalized Momentum

(a) Let us call

$$p_i = \frac{\partial \mathcal{L}_p}{\partial q_i'}$$

the generalized momentum.

(b) *Cartesian coordinates*

We show that, in this case, the p_i are the components of the momentum (§ [5]). With the pre-relativistic definition of $\mathcal{L}_p$, we should have

$$p_x = \frac{\partial}{\partial x'}\left\{\frac{1}{2}m_0(x'^2+y'^2+z'^2)\right\} = m_0 x' = m_0\frac{\mathrm{d}x}{\mathrm{d}t}$$

Application of the equations of Lagrange leads to the Cartesian equations of motion; we notice that $\partial\mathcal{L}_p/\partial x = 0$.

In relativistic mechanics,

$$p_x = \frac{\partial}{\partial x'}\left(-m_0c^2\sqrt{1-\frac{x'^2+y'^2+z'^2}{c^2}}\right) = \frac{m_0x'}{\sqrt{1-\beta^2}}$$

Since $\mathcal{L}_p$ does not depend upon the coordinates here, the Lagrange equations take the form

$$\frac{\mathrm{d}p_x}{\mathrm{d}t} = \frac{\partial\mathcal{L}_i}{\partial x} - \frac{\mathrm{d}}{\mathrm{d}t}\frac{\partial\mathcal{L}_i}{\partial x'}$$

To recover the equations of motion of Chapter I, therefore, the function $\mathcal{L}_i$ must be chosen in each case in such a way that the force components are given by

$$X = \frac{\partial\mathcal{L}_i}{\partial x} - \frac{\mathrm{d}}{\mathrm{d}t}\frac{\partial\mathcal{L}_i}{\partial x'}$$

This is the variational definition of force in Cartesian coordinates.

(c) *Cylindrical coordinates*

Denoting the coordinates by z, r, ϕ, we write $\mathcal{L}_p$ in the form

$$\mathcal{L}_p = -m_0c^2\sqrt{1-\frac{z'^2+r'^2+r^2\phi'^2}{c^2}}$$

whence

$$p_z = \frac{\partial \mathcal{L}_p}{\partial z'} = \frac{m_0 z'}{\sqrt{1-\beta^2}}; \qquad p_r = \frac{\partial \mathcal{L}_p}{\partial r'} = \frac{m_0 r'}{\sqrt{1-\beta^2}}$$

$$p_\phi = \frac{\partial \mathcal{L}_p}{\partial \phi'} = \frac{m_0 r^2 \phi'}{\sqrt{1-\beta^2}}; \qquad \frac{\partial \mathcal{L}_p}{\partial z} = 0$$

$$\frac{\partial \mathcal{L}_p}{\partial r} = \frac{m_0 r \phi'^2}{\sqrt{1-\beta^2}}; \qquad \frac{\partial \mathcal{L}_p}{\partial \phi} = 0$$

It is clear that p_z and p_r are the components of the momentum in the directions z and r; p_ϕ is the moment of the component normal to the r–z plane about Oz.

Explicitly, the Lagrange equations are now

$$\frac{\mathrm{d}}{\mathrm{d}t}\frac{m_0 z'}{\sqrt{1-\beta^2}} = \frac{\partial \mathcal{L}_i}{\partial z} - \frac{\mathrm{d}}{\mathrm{d}t}\frac{\partial \mathcal{L}_i}{\partial z'}$$

$$\frac{\mathrm{d}}{\mathrm{d}t}\frac{m_0 r'}{\sqrt{1-\beta^2}} - \frac{m_0 r \phi'^2}{\sqrt{1-\beta^2}} = \frac{\partial \mathcal{L}_i}{\partial r} - \frac{\mathrm{d}}{\mathrm{d}t}\frac{\partial \mathcal{L}}{\partial r'}$$

$$\frac{\mathrm{d}}{\mathrm{d}t}\frac{m_0 r^2 \phi'}{\sqrt{1-\beta^2}} = \frac{\partial \mathcal{L}_i}{\partial \phi} - \frac{\mathrm{d}}{\mathrm{d}t}\frac{\partial \mathcal{L}_i}{\partial \phi'}$$

Comparison with the expressions given in § [11] yields the variational definition of the components of force along z, r and ϕ:

$$F_z = \frac{\partial \mathcal{L}_i}{\partial z} - \frac{\mathrm{d}}{\mathrm{d}t}\frac{\partial \mathcal{L}_i}{\partial z'}$$

$$F_r = \frac{\partial \mathcal{L}_i}{\partial r} - \frac{\mathrm{d}}{\mathrm{d}t}\frac{\partial \mathcal{L}_i}{\partial r'}$$

$$F_\phi = \frac{1}{r}\left(\frac{\partial \mathcal{L}_i}{\partial \phi} - \frac{\mathrm{d}}{\mathrm{d}t}\frac{\partial \mathcal{L}_i}{\partial \phi'}\right)$$

Remark. For a single particle, the Lagrange formalism is no more convenient than direct calculation, using the equations of Chapter I.

[76] Variational Expression for the Energy of a Particle

Consider the quantity

$$H = \sum p_i q_i' - \mathcal{L}_p$$

In Cartesians,

$$H = x'p_x + y'p_y + z'p_z - \mathcal{L}_p = \frac{m_0 v^2}{\sqrt{1-\beta^2}} + m_0 c^2 \sqrt{1-\beta^2} = \frac{m_0 c^2}{\sqrt{1-\beta^2}}$$

It is therefore the energy of the particle: the sum of the proper energy and the kinetic energy.

[77] Hamilton's Canonical Equations

(a) We define two new quantities

$$\mathcal{P}_i = \frac{\partial \mathcal{L}}{\partial q_i'}, \qquad \mathcal{H} = \sum \mathcal{P}_i q_i' - \mathcal{L}$$

$\mathcal{P}_i$ is the momentum conjugate to q_i and $\mathcal{H}$ is Hamilton's function or the Hamiltonian of the motion. In Cartesian coordinates, Hamilton's principle becomes

$$\delta \int_{M_0}^{M_1} \mathcal{H}\,dt - \mathcal{P}_x\,dx - \mathcal{P}_y\,dy - \mathcal{P}_z\,dz = 0$$

If, in $\mathcal{H}$ we replace

$$\mathcal{P}_i \quad \text{by} \quad \partial\mathcal{L}/\partial q_i'$$

the function $\mathcal{H}$ will depend upon the $2r+1$ variables (r not greater than 3)

$$q_1, \ldots, q_r, \quad q_1', \ldots, q_r', \quad t$$

We calculate the partial derivatives of $\mathcal{H}$ with respect to $\mathcal{P}_i$ and q_i:

$$\frac{\partial \mathcal{H}}{\partial \mathcal{P}_i} = q_i', \qquad \frac{\partial \mathcal{H}}{\partial q_i} = -\frac{\partial \mathcal{L}}{\partial q_i}$$

Notice that

$$\frac{\partial}{\partial q_i} \sum \mathcal{P}_j q_j' = \sum \left(\frac{\partial \mathcal{P}_j}{\partial q_i} q_j' + \mathcal{P}_j \frac{\partial q_j'}{\partial q_i} \right) = 0$$

If we regard the q_i' as auxiliary variables, defined by the set of equations

$$q_i' = \frac{dq_i}{dt}$$

we have a system of $2r$ first-order differential equations, including those of Lagrange, which give the q_i, q_i' as functions of t.

The $2r$ equations, in the notation introduced above, are of the form

$$\frac{d\mathcal{P}_i}{dt} = -\frac{\partial \mathcal{H}}{\partial q_i}, \qquad \frac{dq_i}{dt} = \frac{\partial \mathcal{H}}{\partial \mathcal{P}_i}$$

They are valid in classical and relativistic mechanics.

Remark. These equations could be taken as the fundamental postulate; in § [83], we shall give an example of the reasoning for a special case.

[78] Poisson Brackets

The time derivative of an arbitrary function

$$F(q_i, \mathcal{P}_i, t)$$

is written

$$\frac{dF}{dt} = \frac{\partial F}{\partial t} + \sum \left(\frac{\partial F}{\partial q_i} q_i' + \frac{\partial F}{\partial \mathcal{P}_i} \mathcal{P}_i' \right)$$

$$= \frac{\partial F}{\partial t} + \sum \left(\frac{\partial F}{\partial q_i} \frac{\partial \mathcal{H}}{\partial \mathcal{P}_i} - \frac{\partial F}{\partial \mathcal{P}_i} \frac{\partial \mathcal{H}}{\partial q_i} \right)$$

The Poisson brackets of two arbitrary functions A, B, of q_i and $\mathcal{P}_i$ is the name given to the expression

$$[A, B] = \sum \left(\frac{\partial A}{\partial q_i} \frac{\partial B}{\partial \mathcal{P}_i} - \frac{\partial A}{\partial \mathcal{P}_i} \frac{\partial B}{\partial q_i} \right)$$

With this notation,

$$\frac{dF}{dt} = \frac{\partial F}{\partial t} + [F, \mathcal{H}]$$

This equation is useful in developing quantum mechanics in matrix notation.

[79] The Jacobi Function and the Hamilton–Jacobi Equation

(a) *Definition of the Jacobi function; the general Hamilton–Jacobi equation*

The Hamilton integral, from a fixed point M_0 to a current point M is a function of the four variables q_i and t which define M. This function, with the sign reversed† is known as the Jacobi function:

$$S(q_i, t) = -\int_{M_0}^{M} \mathcal{L}\, dt = \int_{M_0}^{M} \mathcal{H}\, dt - \sum \mathcal{P}_i dq_i$$

whence

$$\mathcal{H} = \frac{\partial S}{\partial t}, \quad \mathcal{P}_i = -\frac{\partial S}{\partial q_i}$$

If, in the expression for $\mathcal{H}$, we replace the $\mathcal{P}_i$ by the value above, we obtain

$$\mathcal{H} = -\left(\sum \frac{\partial S}{\partial q_i} q_i' + \mathcal{L} \right)$$

† Many authors write $S = \int \mathcal{L}\, dt$; the choice of sign is purely a question of convention.

The function S is thus the solution of the equation known as the Hamilton–Jacobi equation:

$$\frac{\partial S}{\partial t} = -\left(\sum \frac{\partial S}{\partial q_i} q_i' + \mathcal{L}\right)$$

(b) *Procedure for Forming the Hamilton–Jacobi Equation*

Consider this equation and the definition of $\mathcal{H}$,

$$\mathcal{H} = \sum \mathcal{P}_i q_i' - \mathcal{L}$$

We see that the Hamilton–Jacobi equation is obtained formally from the Hamiltonian, $\mathcal{H}$, by replacing the $\mathcal{P}_i$ by $-\partial S/\partial q_i$ and setting the result equal to $\partial S/\partial t$.

This is equivalent to making the substitutions

$$\mathcal{P}_i \to -\frac{\partial S}{\partial q_i}, \qquad \mathcal{H} \to \frac{\partial S}{\partial t}$$

This rule will be familiar to theoreticians of quantum mechanics.

Remark. The substitution is correct, irrespective of the coordinates employed. We note that the rules of quantum mechanics that are extracted from it are valid only in Cartesian coordinates.

(c) *Finite equations of motion*

The Hamilton–Jacobi equation is a first order partial differential equation. Let us suppose that a complete integral of this equation is known, that is, a solution which depends on three non-additive arbitrary constants α, β, γ:

$$S(q_i, t, \alpha, \beta, \gamma)$$

It can be shown (the pre-relativistic demonstration remains valid here) that the finite equations of motion are given by the relations

$$\frac{\partial S}{\partial \alpha} = a, \qquad \frac{\partial S}{\partial \beta} = b, \qquad \frac{\partial S}{\partial \gamma} = c, \qquad -\frac{\partial S}{\partial q} = \mathcal{P}_i$$

in which a, b, c are three new arbitrary constants, defining one of the possible motions of the particle.

B. THE FIELD OF FORCE WITH POTENTIALS V AND A

[80] The Expression for $\mathcal{L}$

(a) Hitherto we have given only the expression for the function $\mathcal{L}_p$ corresponding to a free particle; we have said nothing of $\mathcal{L}_i$, the form of which depends upon the field in question.

In the present case, we shall show that to recover the equation of motion of § [5], we must write

$$\mathscr{L} = -m_0c^2\sqrt{1-\beta^2} - \varepsilon\left(V - \frac{\mathbf{v}}{c_0}\cdot\mathbf{A}\right)$$

in which V is a scalar function and $\mathbf{A}$ is a vector function, the physical meanings of which have still to be specified; $\mathbf{v}\,.\,\mathbf{A}$ denotes the scalar product, given in Cartesians, for example, by

$$v_xA_x + v_yA_y + v_zA_z = x'A_x + y'A_y + z'A_z$$

(b) Let us write down the equations of Lagrange in Cartesian coordinates; we find

$$\frac{\partial\mathscr{L}}{\partial x'} = \frac{m_0x'}{\sqrt{1-\beta^2}} + \frac{\varepsilon}{c_0}A_x$$

$$\frac{\mathrm{d}}{\mathrm{d}t}\left(\frac{\partial\mathscr{L}}{\partial x'}\right) = \frac{\mathrm{d}}{\mathrm{d}t}\frac{m_0x'}{\sqrt{1-\beta^2}} + \frac{\varepsilon}{c_0}\left(\frac{\partial A_x}{\partial t} + \frac{\partial A_x}{\partial x}x' + \frac{\partial A_x}{\partial y}y' + \frac{\partial A_x}{\partial z}z'\right)$$

$$\frac{\partial\mathscr{L}}{\partial x} = -\varepsilon\frac{\partial V}{\partial x} + \frac{\varepsilon}{c_0}\left(x'\frac{\partial A_x}{\partial x} + y'\frac{\partial A_y}{\partial x} + z'\frac{\partial A_z}{\partial x}\right)$$

so that

$$\frac{\mathrm{d}}{\mathrm{d}t}\frac{m_0x'}{\sqrt{1-\beta^2}} = -\frac{\partial V}{\partial x} + \frac{\varepsilon}{c_0}\left\{y'\left(\frac{\partial A_y}{\partial x} - \frac{\partial A_x}{\partial y}\right) - z'\left(\frac{\partial A_x}{\partial z} - \frac{\partial A_z}{\partial x}\right)\right\} - \frac{\varepsilon}{c_0}\frac{\partial A_x}{\partial t}$$

$$= \varepsilon\left\{E_x + \frac{1}{c_0}(v_yB_z - v_zB_y)\right\} = X$$

We thus recover the equations of motion if V and $\mathbf{A}$ are identified with a scalar and a vector potential, respectively.

(c) Some authors add the constant m_0c^2 to $\mathscr{L}_p$, and hence to $\mathscr{L}$; the physical conclusions concerning the motion are unaffected, but the Hamiltonian no longer contains the rest energy.

In the expression for $\mathscr{L}$, let us replace V and $\mathbf{A}$ by

$$V - \frac{\partial F}{\partial t} \quad \text{and} \quad \mathbf{A} + c_0\,\mathrm{grad}\,F$$

respectively (gauge transformation, § [68]). The sum of the supplementary terms of $\mathscr{L}$,

$$-\varepsilon\left(-\frac{\partial F}{\partial t} - \mathbf{v}\,.\,\mathrm{grad}\,F\right) = \varepsilon\frac{\mathrm{d}F}{\mathrm{d}t},$$

is equal to a total derivative. Integrated, the result will depend only on the limits and cannot therefore affect a variational calculation, since the limits are assumed to be fixed.

[81] The Expression for the Hamiltonian $\mathcal{H}$

In Cartesian coordinates, the first terms are written explicitly:

$$\mathcal{P}_i q_i' = \frac{m_0}{\sqrt{1-\beta^2}}(v_x^2+v_y^2+v_z^2)+\frac{\varepsilon}{c_0}(v_x A_x+v_y A_y+v_z A_z)$$

$$= \frac{m_0 v^2}{\sqrt{1-\beta^2}}+\frac{\varepsilon}{c_0}\mathbf{v}.\mathbf{A}$$

whence

$$\mathcal{H} = \frac{m_0 v^2}{\sqrt{1-\beta^2}}+m_0 c^2\sqrt{1-\beta^2}+\varepsilon V$$

$$= \frac{m_0 c^2}{\sqrt{1-\beta^2}}+\varepsilon V$$

Thus the function $\mathcal{H}$ is expressed in terms of a single potential, V, even in the presence of a vector potential, $\mathbf{A}$.

Taking into account the fact that the component $\mathfrak{B}$ does no work, we see that the Hamiltonian $\mathcal{H}$ is equal to the sum of the rest energy, the kinetic energy and the potential energy (§ [70A]).

If we eliminate $v/\sqrt{1-\beta^2}$ between the two relations

$$\mathcal{H} = \frac{m_0 c^2}{\sqrt{1-\beta^2}}+\varepsilon V, \quad \mathbf{p} = \frac{m_0 \mathbf{v}}{\sqrt{1-\beta^2}}$$

we obtain

$$\mathcal{H}(x, y, z, p_x, p_y, p_z, t) = \pm c\sqrt{m_0^2 c^2+p_x^2+p_y^2+p_z^2}+V(x, y, z, t)$$

The function $\mathcal{H}$ is transcendental. Only in pre-relativistic dynamics, where the proper energy term is suppressed, does the square root disappear. *This fact plays an important role in particle wave mechanics.* Introducing $\mathcal{P}$, this formula becomes

$$\mathcal{H} = \pm c\sqrt{m_0^2 c^2+\sum_{x, y, z}\left(\mathcal{P}_x-\frac{\varepsilon}{c_0}A_x\right)^2}+\varepsilon V$$

We have

$$\mathcal{P}_x = \frac{\partial \mathcal{L}}{\partial x'} = p_x+\frac{\partial \mathcal{L}_i}{\partial x'} = p_x+\frac{\varepsilon}{c_0}A_x$$

Low velocities. Taking into account the remark made in § [73](b) about the proper energy, we find that to this approximation

$$\mathcal{L} = \frac{m_0 v^2}{2}-\varepsilon\left(V-\frac{\mathbf{v}}{c_0}\cdot\mathbf{A}\right)$$

$$\mathcal{H} = \frac{1}{2m_0}\sum_{x, y, z}\left(\mathcal{P}_x-\frac{\varepsilon}{c_0}A_x\right)^2+\varepsilon V$$

$$\mathcal{P}_x = m_0 v_x+\frac{\varepsilon}{c_0}A_x$$

[82] Remark Concerning the Concepts of Momentum and Energy

Many authors use the term momentum, or even impulse, to designate the quantity

$$\mathfrak{P} = \mathbf{p} + \frac{\varepsilon}{c_0}\mathbf{A}$$

which is the resultant of the genuine momentum $\mathbf{p}$ (in the sense of Chapter I) and the vector $\varepsilon\mathbf{A}/c_0$.

In connection with electrical phenomena, $\mathfrak{P}$ is often called the electromagnetic momentum. The quantity $\mathbf{p}$ is described as the kinetic momentum, and the other term as the potential momentum. *All these names I find regrettable. In particle dynamics, only the vector* $\mathbf{p}$ *deserves to be called momentum.*

Likewise, I have thought it right to distinguish between the energy H and the Hamiltonian $\mathcal{H}$ (§§ [76] and [77]). For, in the case of an arbitrary force field, $\mathcal{H}$ is not connected with the concept of energy. Only with conservative fields is the concept of total energy related to the Hamiltonian, $\mathcal{H}$, which becomes the sum of the energy H (the actual energy) and the potential energy.

[83] The Usual Equations, Deduced from Hamilton's Canonical Equations

To familiarize the reader with the manipulation of $\mathfrak{P}$ and $\mathcal{H}$, let us take the canonical equations as postulate, given the above expressions for $\mathfrak{P}$ and $\mathcal{H}$, and derive the Cartesian equations from them.

Low velocities

We start from

$$\frac{\mathrm{d}\mathfrak{P}}{\mathrm{d}t} = -\operatorname{grad}\mathcal{H} \tag{1}$$

From the definition of $\mathfrak{P}$,

$$\frac{\mathrm{d}\mathcal{P}_x}{\mathrm{d}t} = \frac{\mathrm{d}}{\mathrm{d}t}\left(m_0 v_x + \frac{\varepsilon}{c_0}A_x\right) = m_0\frac{\mathrm{d}v_x}{\mathrm{d}t} + \frac{\varepsilon}{c_0}\left(\frac{\partial A_x}{\partial t} + \frac{\partial A_x}{\partial x}v_x + \frac{\partial A_x}{\partial y}v_y + \frac{\partial A_x}{\partial z}v_z\right)$$

From the definition of $\mathcal{H}$

$$(-\operatorname{grad}\mathcal{H})_x = \frac{\varepsilon}{m_0 c_0}\left\{\left(\mathcal{P}_x - \frac{\varepsilon}{c_0}A_x\right)\frac{\partial A_x}{\partial x} + \left(\mathcal{P}_y - \frac{\varepsilon}{c_0}A_y\right)\frac{\partial A_y}{\partial x}\right.$$
$$\left. + \left(\mathcal{P}_z - \frac{\varepsilon}{c_0}A_z\right)\frac{\partial A_z}{\partial x}\right\} - \varepsilon\frac{\partial V}{\partial x}$$
$$= \frac{\varepsilon}{c_0}\left(v_x\frac{\partial A_x}{\partial x} + v_y\frac{\partial A_y}{\partial x} + v_z\frac{\partial A_z}{\partial x}\right) - \varepsilon\frac{\partial V}{\partial x}$$

Using the identity

$$v_x \frac{\partial A_x}{\partial x}+v_y \frac{\partial A_y}{\partial x}+v_z \frac{\partial A_z}{\partial x}$$
$$= v_x \frac{\partial A_x}{\partial x}+v_y \frac{\partial A_x}{\partial y}+v_z \frac{\partial A_x}{\partial z}+v_y \left(\frac{\partial A_y}{\partial x}-\frac{\partial A_x}{\partial y}\right)-v_z \left(\frac{\partial A_x}{\partial z}-\frac{\partial A_z}{\partial x}\right)$$

and substituting in (1), we do indeed recover the equation

$$m_0 \frac{\mathrm{d}v_x}{\mathrm{d}t} = -\varepsilon \frac{\partial V}{\partial x}-\frac{\varepsilon}{c_0} \frac{\partial A_x}{\partial t}+\frac{\varepsilon}{c_0}(v_y\mathcal{B}_z-v_z\mathcal{B}_y)$$

[84] An Important Application of the Canonical Equations; the Lagrange Invariant for a Beam of Particles (Gabor)

(a) At time t, we describe a closed curve C in a beam of particles *free of mutual interactions.* At a subsequent time t', the particles which were on C will describe a different curve, C'. We now demonstrate that, if $\mathcal{H}$ is conservative (V and **A** independent of time), the loop integral of the vector $\mathfrak{P}$ around the curve C is conserved in time, if we calculate it along successive curves $C, C', \ldots$ If, in particular,

$$C = \oint_C \mathfrak{P}.\mathrm{dl} = 0$$

the equality is conserved. In this special case, we say that the beam is irrotational.

In the following calculations, we denote the elements of length in question at a given time by dx, and we denote differentiations carried out in time by the symbol D. The change in C in time dt (which we have to show to be equal to zero) is given by

$$\frac{\mathrm{D}C}{\mathrm{D}t} = \frac{\mathrm{D}}{\mathrm{D}t}\int \mathcal{P}_x\,\mathrm{d}x+\mathcal{P}_y\,\mathrm{d}y+\mathcal{P}_z\,\mathrm{d}z$$
$$= \int \frac{\mathrm{D}\mathcal{P}_x}{\mathrm{D}t}\,\mathrm{d}x+\frac{\mathrm{D}\mathcal{P}_y}{\mathrm{D}t}\,\mathrm{d}y+\frac{\mathrm{D}\mathcal{P}_z}{\mathrm{D}t}\,\mathrm{d}z+\mathcal{P}_x\frac{\mathrm{D}\mathrm{d}x}{\mathrm{D}t}+\mathcal{P}_z\frac{\mathrm{D}\mathrm{d}y}{\mathrm{D}t}+\mathcal{P}_z\frac{\mathrm{D}\mathrm{d}z}{\mathrm{D}t}$$
$$= \int \frac{\mathrm{D}\mathcal{P}_x}{\mathrm{D}t}\,\mathrm{d}x+\frac{\mathrm{D}\mathcal{P}_y}{\mathrm{D}t}\,\mathrm{d}y+\frac{\mathrm{D}\mathcal{P}_z}{\mathrm{D}t}\,\mathrm{d}z+\mathcal{P}_x\,\mathrm{d}\frac{\mathrm{D}x}{\mathrm{D}t}+\mathcal{P}_y\,\mathrm{d}\frac{\mathrm{D}y}{\mathrm{D}t}+\mathcal{P}_z\,\mathrm{d}\frac{\mathrm{D}z}{\mathrm{D}t}$$

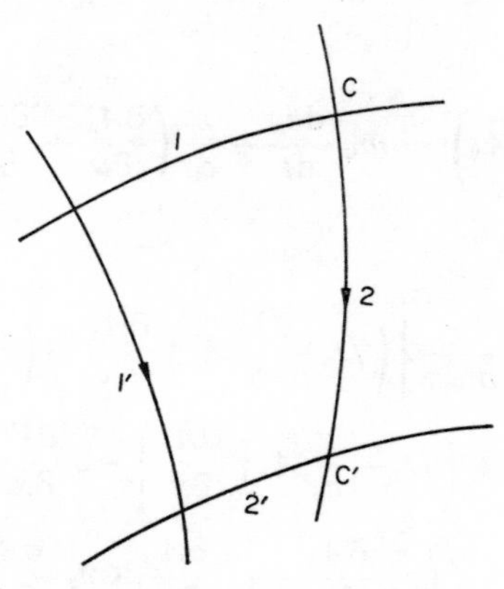

FIG. 14.

The same result is obtained (Fig. 14), whether we first carry out the simultaneous displacement dx, and then follow the trajectory (path 1, 2) or first follow the trajectory, then carry out the simultaneous displacement (path 1', 2').

Thus

$$\begin{aligned}\frac{\mathrm{D}C}{\mathrm{D}t} &= \int \frac{\mathrm{D}\mathcal{P}_x}{\mathrm{D}t}\,\mathrm{d}x+\frac{\mathrm{D}\mathcal{P}_y}{\mathrm{D}t}\,\mathrm{d}y+\frac{\mathrm{D}\mathcal{P}_z}{\mathrm{D}t}\,\mathrm{d}z \\ &\quad +\mathrm{d}(\mathcal{P}_x v_x+\mathcal{P}_y v_y+\mathcal{P}_z v_z)-(v_x\,\mathrm{d}\mathcal{P}_x+v_y\,\mathrm{d}\mathcal{P}_y+v_z\,\mathrm{d}\mathcal{P}_z) \\ &= -\int \frac{\partial\mathcal{H}}{\partial x}\,\mathrm{d}x+\frac{\partial\mathcal{H}}{\partial y}\,\mathrm{d}y+\frac{\partial\mathcal{H}}{\partial z}\,\mathrm{d}z \\ &\quad +\frac{\partial\mathcal{H}}{\partial\mathcal{P}_x}\,\mathrm{d}\mathcal{P}_x+\frac{\partial\mathcal{H}}{\partial\mathcal{P}_y}\,\mathrm{d}\mathcal{P}_y+\frac{\partial\mathcal{H}}{\partial\mathcal{P}_z}\,\mathrm{d}\mathcal{P}_z \\ &= -\int \mathrm{d}\mathcal{H} = 0\end{aligned}$$

which proves the theorem.

Using Stokes' theorem, this circuit integral can be cast into the form of the flux of the vector curl $\mathfrak{P}$ through any surface bounded by C. This flux is thus a constant of the motion also. The same is true for any element of surface, and the quantity

$$\text{curl}\ \mathfrak{P}\ \mathrm{d}S$$

is thus a constant.

(b) *The case of axially symmetrical fields; Busch's theorem.* The vector potential at a point M is then tangent to the circle through M, centred on the symmetry axis. If a group of particles lies on a circle C_1, centred on the axis, at time t_1, then at every later instant t the group will lie on a circle C_2, also centred on the axis. Lagrange's theorem then gives

$$r_1\left(v_1+\frac{\varepsilon}{m_0 c_0}A_1\right) = r_2\left(v_2+\frac{\varepsilon}{m_0 c_0}A_2\right)$$

by virtue of the symmetry, whence

$$\begin{aligned}r_1 v_1 - r_2 v_2 &= \frac{\varepsilon}{m_0 c_0}(r_2 A_2 - r_1 A_1) \\ &= \frac{\varepsilon}{m_0 c_0}(\mathcal{F}_2-\mathcal{F}_1)\end{aligned}$$

in which $\mathcal{F}$ is the flux of $\mathfrak{B}$. This last relation is obtained by applying Stokes' theorem to the vector $\mathbf{A}$.

We conclude, therefore, that between two positions M_1 and M_2, the change in angular momentum of a corpuscle about the symmetry axis is proportional to the difference between the flux of $\mathfrak{B}$ through the coaxial circles passing through M_1 and M_2 respectively.

Remark. These relations are primarily of interest in electron optics.

[85] The Hamilton–Jacobi Equation for a Time-dependent Potential

(a) *The pre-relativistic equation in Cartesians*

Into the expression for $\mathcal{H}$ at low velocities, we make the substitution given in § [79]

$$\frac{\partial S}{\partial t} = \frac{1}{2m_0} \sum_{x, y, z} \left(\frac{\partial S}{\partial x} + \frac{\varepsilon}{c_0} A_x \right)^2 + \varepsilon V$$

or, in the case when only a scalar potential is present,

$$\frac{\partial S}{\partial t} = \frac{1}{2m_0} \left\{ \left(\frac{\partial S}{\partial x} \right)^2 + \left(\frac{\partial S}{\partial y} \right)^2 + \left(\frac{\partial S}{\partial z} \right)^2 \right\} + \varepsilon V$$

(b) *The relativistic equation*

Squaring to remove the square root in the relativistic expression for $\mathcal{H}$, we have

$$\frac{1}{c^2} (\mathcal{H} - \varepsilon V)^2 - \sum_{x, y, z} \left(\mathcal{P}_x - \frac{\varepsilon}{c_0} A_x \right)^2 = m_0^2 c^2$$

Substituting

$$\sum_{x, y, z} \left(\frac{\partial S}{\partial x} + \frac{\varepsilon}{c_0} A_x \right)^2 - \frac{1}{c^2} \left(\frac{\partial S}{\partial t} - \varepsilon V \right)^2 + m_0^2 c^2 = 0$$

or in vector form

$$\left(\nabla S + \frac{\varepsilon}{c_0} \mathbf{A} \right)^2 - \frac{1}{c^2} \left(\frac{\partial S}{\partial t} - \varepsilon V \right)^2 + m_0^2 c^2 = 0$$

The function S', defined by

$$S' = S - \int_{M_0}^{M_1} m_0 c^2 \, dt = S - m_0 c^2 (t - t_0)$$

is sometimes employed instead of S. The proper energy term is thus eliminated, without affecting the motion.

[86] The Time-independent Potential; the General Concept of Refractive Index

(a) *Maupertuis' formulation*

Hamilton's principle, for given total energy $\mathcal{H}_0$, is now written

$$\delta \int_{M_0}^{M_1} \sum \mathcal{P}_i \, dq_i = 0$$

The real trajectory joining two points M_0 and M_1, for given $\mathcal{H}_0$, is such that this integral is stationary; the integral is known as the Maupertuis action.

In pre-relativistic mechanics, for a scalar potential, the Maupertuis integral takes the familiar form

$$\int_{M_0}^{M_1} \sum \mathcal{P}_i \, dq_i = \int_{M_0}^{M_1} m_0 \mathbf{v} . d\mathbf{s}$$

In relativistic mechanics, we retain the same expression, except that m_0 is replaced by the relativistic mass:

$$\int_{M_0}^{M_1} \sum \mathcal{P}_i \, dq_i = \int_{M_0}^{M_1} \frac{m_0}{\sqrt{1-\beta^2}} \left(\frac{dx}{dt} dx + \frac{dy}{dt} dy + \frac{dz}{dt} dz \right)$$

$$= \int_{M_0}^{M} m\mathbf{v} . d\mathbf{s}$$

In the general case (both V and $\mathbf{A}$), the integral becomes

$$\int \left(\mathbf{p} + \frac{\varepsilon}{c_0} \mathbf{A} \right) . \mathbf{t} \, ds$$

in which $\mathbf{t}$ is a unit vector along the tangent.

Comparing with Fermat's integral

$$\int n \, ds$$

we see that the refractive index corresponding to the field (V, $\mathbf{A}$) is proportional to

$$\sqrt{-2\varepsilon m_0 V(1 - V/2V_0)} + \frac{\varepsilon}{c_0} \mathbf{A} . \mathbf{t}$$

This is the generalization of the formulae of § [57]. The refractive index contains an isotropic term (p) and an anisotropic term ($\mathbf{A} . \mathbf{t}$). *In the case of an arbitrary potential* (V *and* $\mathbf{A}$), *the trajectories are calculated like light rays in an anisotropic medium.*

We can confirm that, as in § [80], a gauge transformation has no effect on the trajectories, although the refractive index then contains a supplementary term.

(b) *The Hamilton–Jacobi equation*

The constant total energy $\mathcal{H}_0$ can now be taken outside the integration:

$$S = \mathcal{H}_0(t - t_0) - \int_{M_0}^{M_1} \sum \mathcal{P}_i \, dq_i = \mathcal{H}_0(t - t_0) - S_1$$

so that

$$\frac{\partial S}{\partial t} = \mathcal{H}_0, \qquad -\frac{\partial S}{\partial q_i} = \frac{\partial S_1}{\partial q_i}$$

The Hamilton–Jacobi equation then becomes

$$\mathcal{H}_0 = \sum \frac{\partial S_1}{\partial q_i} q_i' - \mathcal{L}$$

In Cartesian coordinates, the pre-relativistic equation is

$$\mathscr{H}_0 = \frac{1}{2m_0} \sum_{x,y,z} \left(\frac{\partial S_1}{\partial x} - \frac{\varepsilon}{c_0} A_x \right)^2 + \varepsilon V$$

while the relativistic equation is

$$\sum_{x,y,z} \left(\frac{\partial S_1}{\partial x} - \frac{\varepsilon}{c_0} A_x \right)^2 - \frac{1}{c^2} (\mathscr{H}_0 - \varepsilon V)^2 + m_0^2 c^2 = 0$$

A complete integral will thus be of the form

$$S(q_i, \alpha, \beta, \mathscr{H}_0) = \mathscr{H}_0(t - t_0) - S_1(q_i, \alpha, \beta, \mathscr{H}_0)$$

and the finite equations of motion are now

$$\frac{\partial S_1}{\partial \alpha} = a, \quad \frac{\partial S_1}{\partial \beta} = b, \quad \frac{\partial S}{\partial \mathscr{H}_0} = t - \frac{\partial S_1}{\partial \mathscr{H}_0} = c$$

The first two equations, which do not contain time, define the trajectory, and the third governs the motion along it.

HISTORICAL AND BIBLIOGRAPHICAL NOTES

[87] The Principle of Least Action in Optics and in Newtonian Dynamics

A historical study of the birth and evolution of this principle would require a whole volume, or several chapters at the very least; I cannot presume to summarize the subject in a few lines. Nevertheless, I feel that I must provide the reader with a few general indications, for I am certain that they will help him to grasp the spirit of the variational methods, the abstract nature of which often repels the non-mathematician. For further details, the reader is referred to the various histories of mechanics, and in particular to the works of R. Dugas [1] and P. Brunet [10] which make very pleasant reading and which can guide the reader who desires to study the topic further. In the references, I give only a few milestones, merely mentioning the principal authors.

The principle of least action seems to have been employed correctly in physics for the first time by Fermat [2]. This is one of the classic stages in optics, obtained by Fermat from the "principe si commun et si établi que la Nature agit toujours par les voies les plus courtes"†. The novelty of Fermat's work does not, of course, lie in this vague statement but in his mathematical interpretation of "the shortest paths" or "the least action": the fundamental concept of optical path-length. Leibnitz and the Cartesians refused to accept Fermat's arguments, and interminable controversies, of more metaphysical than physical a nature, ensued.

The principle was introduced into mechanics by Maupertuis [3], who gave the correct definition of "action", but his formulation is based upon simple analogies. A fresh quarrel over the question sprang up, in which many scholars took part (d'Alembert, Koenig, Euler), and even Voltaire. It is clear that an effort was being made to liberate the principle from its metaphysical trappings, and especially from its connection with the idea of the final cause; the important result is Euler's mathematical proof [4] of Maupertuis' statement.

With Lagrange [5], Maupertuis' work took its place in the ensemble of theories of dynamics and was cast definitively into its modern form, freed of any taint of metaphysics.

---

† "Principle, so common and well-established, that Nature always follows the shortest paths."

Hamilton [6] systematically developed geometrical optics and later, dynamics, with the variational principle which is now named after him as his starting-point; he gave the canonical equations, the function $\mathcal{L}$ and in certain cases he used the function $\mathcal{H}$. Furthermore, his parallel development of the optical and mechanical theories drew attention to the formal analogy between these two branches; it was this analogy that led Louis de Broglie to the discovery of wave mechanics.

Jacobi [7] generalized Hamilton's theory, and gave it its definitive form.

Among the modern treatises, we mention the works of Whittaker [8], Mercier [9] and Pars [11].

[1] R. Dugas: ref. [30] of Appendix I.

[2] Fermat: Letter to G. de la Chambre, 1 January 1662; Memoir "Synthesis ad refractiones", probably dating from February 1662.

[3] Maupertuis: Accord de différentes lois de la Nature qui avaient jusqu'ici paru incompatibles. Memoir read before the Académie des Sciences, 15 April 1744.

[4] Euler: *Dissertation sur le principe de moindre action, avec l'examen des objections que M. le Professeur Koenig a faites contre ce principe*, Berlin, 1753.

[5] Louis de Lagrange: *Mécanique analytique*, 1st ed. 1788, last edition published during Lagrange's lifetime, 1811.

[6] Sir W. R. Hamilton: The fundamental papers are, for optics, "Theory of systems of rays", and for dynamics, "Essays on a general method in dynamics" (1834 and 1835). The works of this author have been collected in *The Mathematical Papers of Sir William Rowan Hamilton*, Cambridge U.P., Vol. I, 1931; Vol. II, 1940.

[7] Jacobi: *Vorlesungen über Dynamik*, delivered at Koenigsberg in 1842–1843; published by Reimer, Berlin, 1866.

[8] E. Whittaker: *Analytical Dynamics*, Cambridge U.P., 1st ed. 1904, 4th ed. 1952.

[9] A. Mercier: *Principes de Mécanique analytique*, Gauthier-Villars, 1955 (Collection de physique théorique ..., general editor, J. L. Destouches).

[10] P. Brunet: *Etude historique sur le principe de moindre action*, Paris, Hermann, 1938.

[11] L. Pars: *A Treatise on Analytical Dynamics*, Cambridge U.P., 1965.

Remark. I strongly urge the reader to glance through and study the works of Lagrange, Hamilton and Jacobi; they are astonishingly modern.

[88] The Principle of Least Action in Relativistic Dynamics

(a) In the immediately pre-Einstein period, this principle was employed using the terminology of electromagnetism, like the majority of the formulae of relativistic dynamics.

Schwarzschild [3] gives the Lagrangian $\mathcal{L}_i$ of § [80], which he calls the electrokinetic potential; it is calculated directly as I have indicated in § [67]. He sets out the variational principle for the equations of motion in the case of low velocities. He was led to the expression that I have given in § [67] by introducing Lorentz retarded potentials into his electrokinetic potential; at this point, he perceived that his function could be obtained from Clausius' electrodynamic potential, provided that the electric potential was added and that the quantities present were replaced by their retarded values. This remark connects the evolution of the principle of least action with the concept of electrodynamic potential, briefly mentioned in § [145]; a thorough historical study ought to stress this point.

Schwarzschild then extended the method to the derivation of the field equations, starting with the Lagrangian of § [94]. The idea of such a derivation was not a new one, however: it was from this standpoint that Helmholtz [2] built up the electrodynamics of Herz, and that H. A. Lorentz [1] obtained the equations of his theory of electrons, although these two authors performed their calculations in extremely different ways. Schwarzschild employed a variational principle similar to that of Helmholtz, but which led finally to the equations of Lorentz.

I also draw attention to an article by Herglotz [5] and an important paper by Abraham [4], in which variational methods frequently occur.

The full expression for $\mathcal{L}$ (that given in § [80]) is to be found in Planck's work [7, 8, 9].

The refractive index of an anisotropic medium, to which every field given by potentials V, $\mathbf{A}$ can be reduced, is given by W. Glaser [15, 16].

(b) *The use of proper time as independent variable in the case of static potentials*

By this means, von Laue [10] removes the square roots and gives the relativistic equations a Newtonian appearance.

If v is the velocity of the particle (v is a function of t), the proper time is defined by

$$dT = dt\sqrt{1-v^2/c^2}$$

To obtain the inverse relation, we must replace v by the corresponding quantity which depends upon T, that is, by the spatial component of the four-velocity.

From

$$V_x = \frac{dx}{dT} = \frac{v_x}{\sqrt{1-v^2/c^2}}$$

we derive

$$1-\frac{v^2}{c^2} = \frac{1}{1+V^2/c^2} \quad \text{with} \quad V^2 = V_x^2+V_y^2+V_z^2$$

Thus

$$dt = dT\sqrt{1+V^2/c^2}$$

It can be shown that the energy equation (in which the potential is *provisionally* denoted by U)

$$\frac{m_0c^2}{\sqrt{1-v^2/c^2}}+\varepsilon U = C \qquad (C = \text{constant})$$

takes the Newtonian form, when T is introduced,

$$\frac{1}{2}m_0V^2+\frac{C-\varepsilon U}{2m_0c^2} = -\frac{m_0c^2}{2}$$

We then find:

— The equations of motion in Cartesians

$$\frac{d}{dT}\left(m_0\frac{dx}{dT}+\frac{\varepsilon}{c_0}A_x\right) = \frac{\partial}{\partial x}\left\{\frac{(C-\varepsilon U)^2}{2m_0c^2}+\frac{\varepsilon}{c_0}\mathbf{V}.\mathbf{A}\right\};$$

— Hamilton's equation

$$\mathcal{L} = \frac{1}{2}m_0V^2+\frac{(C-\varepsilon U)^2}{2m_0c^2}+\frac{\varepsilon}{c_0}\mathbf{V}.\mathbf{A}$$

where

$$\delta\int_{T_1}^{T_2}\mathcal{L}\,dT = 0.$$

— The Hamiltonian

$$\mathcal{H} = \frac{1}{2m_0}V^2-\frac{(C-\varepsilon U)^2}{2m_0c^2} = \frac{1}{2m_0}\left\{\left(\mathcal{P}_x-\frac{\varepsilon}{c_0}A_x\right)^2+\dots\right\}-\frac{(C-\varepsilon U)^2}{2m_0c^2}$$

The method entails first seeking the finite equations of motion in the form $x(T), \dots$; only later does one return to the ordinary time-variable t.

(c) *The notation in Minkowski space; generalizations*

The transcription into Minkowski terminology of the equations of this chapter is to be found in Chapter VIII.

Two reproaches have been made to the usual formulation of Hamilton's principle. A regular canonical formalism in Minkowski space is not possible, and furthermore, in quantum mechanics, the time and energy operators appear as canonical conjugates. This suggests that time should be considered not as a parameter but as a dynamical variable, to which a "moment" is attributed. Much research has been carried out along these lines. The reader is referred to the articles listed in the bibliography from [12] onwards, and in particular, to the articles by Schay [42] and Linder [48].

Remark. The profusion of these articles should not mislead the reader into believing that the equations given in §§ [73]–[86] are but a provisional makeshift. These equations are in fact in continualuse in electron optics and accelerator theory.

[1] H. A. Lorentz: La théorie électromagnétique de Maxwell et son application aux corps mouvants. *Arch. Néerl.* **25** (1892) 363–552.
[2] H. von Helmholtz: Das Princip der kleinsten Wirkung in der Elektrodynamik. *Wissenschaft. Abhandl.*, Leipzig **3** (1895) 476–504. Contains references to earlier work.
[3] K. Schwarzschild: Zur Elektrodynamik. I. Zwei Formen des Prinzips kleinsten Action in der elektronen Theorie. *Nachr. Göttingen* **3** (1903) 126–131.
[4] M. Abraham: Prinzipien der Dynamik des Elektrons. *Ann. Physik* **10** (1903) 105–179.
[5] G. Herglotz: Zur Elektronentheorie. *Nachr. Göttingen* (1903) 357–382.
[6] H. Poincaré: Ref. [1] of § [42].
[7] M. Planck: Das Prinzip der Relativität und die Grundgleichungen der Mechanik. *Deutsch. Phys. Ges.* (1906) 136–141.
[8] M. Planck: Zur Dynamik bewegter Systeme. *Sitz. Berichte Akad. Wiss.* (1907) 542–570.
[9] M. Planck: Zur Dynamik bewegter Systeme. *Ann. Physik* **26** (1908) 1–34.
[10] von Laue: Ref. [2] of Appendix I.
[11] R. Dugas: Sur l'interprétation géométrique de la méthode de Jacobi dans le cas d'un point de masse variable. *C. R. Acad. Sci., Paris*, **182** (1926) 566–568.
[12] A. Sommerfeld: *Vorlesungen über theoretische Physik*, Bd. III, pp. 267–275. Wiesbaden, 1948.
[13] Becker: Reference suppressed.
[14] L. Nordheim and E. Fues: *Handbuch der Physik*, Berlin 1927, Vol. 5.
[15] W. Glaser: Ueber geometrisch optische Abbildung durch Elektronenstrahlen. *Z. Phys.* **80** (1933) 451–464.
[16] W. Glaser: Ueber optische Abbildung durch mechanische Systeme und die Optik allgemeiner Medien. *Ann. Physik* **18** (1933) 557–585.
[17] Th. de Donder: *Théorie invariantive du calcul des variations*, Paris, 1935.
[18] P. A. M. Dirac: Forms of relativistic dynamics. *Rev. Mod. Physics* **21** (1939) 392–399.
[19] P. A. M. Dirac: La théorie de l'électron et du champ électromagnétique. *Ann. Inst. H. Poincaré* **9** (1939) 13–49.
[20] O. Costa de Beauregard: Sur la dynamique analytique du point électriquement chargé. *C. R. Acad. Sci., Paris* **214** (1942) 58–60.
[21] D. Gabor: Application of Hamiltonian dynamics to electronic problems. *Proc. I. R. E.* **33** (1945) 792–805.
[22] O. Costa de Beauregard: Ref. [7] of Appendix I, p. 151.
[23] L. de Broglie: ref. [4] of § [65], pp. 28–38.
[24] W. Glaser: *Handbuch der Physik* **33**, 123–395. Springer, Berlin, 1956.
[25] J. Abelé: Introduction à la notion d'action et au principe de l'action stationnaire. *Rev. Quest. Sci.*, 20th January 1948, pp. 25–42.
[26] J. Abelé: Quelques réflexions sur la notion de point materiel. *Rev. Gén. Sci.* **49** (1952) 70–80.
[27] G. Falk: Eine kanonische Formulierung der Relativitätsmechanik und ihr quantentheoretisches Analogon. *Z. Phys.* **132** (1952) 44–53.
[28] W. Macke: Begrundung der speziellen Relativitätstheorie aus der Hamiltonschen Mechanik. *Z. Naturforsch.* **7a** (1952) 76–78.
[29] G. Szamosi: Variational principle and potential in relativistic dynamics. *Acta Phys. Hung.* **6** (1956) 207–215.
[30] L. Infeld: *Max Planck Festschrift 1958*. Berlin, 1958.
[31] F. Sauter: Zur Lorentz-invarianten Formulierung der kanonischen Bewegungsgleichungen in der Punktmechanik. *Z. Phys.* **156** (1959) 275–286.
[32] P. Caldirola: Formulazione lagrangiana e hamiltoniana del moto classico dell' elettrone irragiante. *Rend. Ist. Lombardo* **A 93** (1959) 439–445.
[33] M. Phan Mau Quan: Sur la dynamique analytique du point en relativité restreinte. *C. R. Acad. Sci., Paris* **251** (1960) 639–641.
[34] M. Phan Mau Quan: Sur la dynamique analytique du point en relativité restreinte. *Oesterreich. Mat. Ges., Congress*, 12th–17th September 1960, pp. 66–67.
[35] A. Pignedoli: Ueber die Lagrangeschen Gleichungen der Mechanik eines raschen Teilschens. *Ibid.*, pp. 60–61.
[36] F. Gascon: Un principio de acción para la mecánica relativista. *An. Soc. Esp. Fis. Quim.* **57** (1961) 253–256.
[37] G. Kalman: Lagrangian formalism in relativistic dynamics. *Phys. Rev.* **123** (1961) 384–390.
[38] M. Pham Mau Quan: Sur la dynamique analytique du point en relativité. *Acta Phys. Austriaca* **14** (1961) 232–238.

[39] H. RUND: Dynamics of particles with internal "spin". *Ann. Physik* **7** (1961) 17–27.
[40] H. RUND: Note on the Lagrangian Formalism in relativistic mechanics. *Nuovo Cimento* **23** (1962) 227–232.
[41] Z. BORELOWSKI: Homogeneous variational principle with third order derivatives in the special theory of relativity. *Acta Phys. Pol.* **21** (1962) 609–635.
[42] G. SCHAY: On relativistic one-particle dynamics. *Suppl. Nuovo Cimento* (10) **26** (1962) 291–304.
[43] This reference in the French edition omitted here.
[44] H. STEPHANI: Zur formulierung der vierdimensionalen kanonischen Mechanik eines Massenpunktes bei Verwendung der Eigenzeit als Parameter. *Wiss. Z.* **11** (1962) 131–134.
[45] M. SCHWARTZ: Lagrangian and Hamiltonian formalisms with supplementary conditions. *J. Math. Phys.* **5** (1964) 903–907.
[46] S. HJALMARS: Some remarks on time and energy as conjugate variables. *Nuovo Cimento* **25** (1962) 355–364.
[47] S. HJALMARS and A. LINDER: Further studies on time and energy as conjugate variables. *Phys. Lett.* **4** (1963) 122–123.
[48] A. LINDER: Generalized variational principles in classical mechanics. *Suppl. Nuovo Cimento* **3** (1965) 118–146.
[49] P. M. PEARLE: Relativistic classical mechanics with time as a dynamic variable. *Phys. Rev.* **168** (1968) 1429–1444.

[88A] Direct Demonstration of the Variational Equations, Starting from the Impulse Postulate

We start from

$$\frac{\mathrm{d}}{\mathrm{d}t}\,\frac{m_0\mathbf{v}}{\sqrt{1-v^2/c^2}} = \mathbf{F} = \varepsilon\mathbf{E}+\frac{\varepsilon}{c_0}\,\mathbf{v}\times\mathfrak{B} \tag{1}$$

or

$$\frac{\mathrm{d}}{\mathrm{d}t}\,\frac{m_0x'}{\sqrt{1-v^2/c^2}} = \varepsilon E_x+\frac{\varepsilon}{c_0}\,(y'\mathcal{B}_z-z'\mathcal{B}_y)\,\ldots \tag{2}$$

In addition

$$\mathfrak{B} = \text{curl}\,\mathbf{A} \qquad \mathbf{E} = -\,\text{grad}\,V-\frac{1}{c_0}\,\frac{\partial\mathbf{A}}{\partial t} \tag{3}$$

and hence, for (2)

$$\begin{aligned}\frac{\mathrm{d}}{\mathrm{d}t}\,\frac{m_0x'}{\sqrt{1-v^2/c^2}}+\frac{\varepsilon}{c_0}\left(\frac{\partial A_x}{\partial t}+y'\,\frac{\partial A_x}{\partial y}+z'\,\frac{\partial A_x}{\partial z}+x'\frac{\partial A_x}{\partial x}\right)\\ = -\,\varepsilon\,\frac{\partial V}{\partial x}+\frac{\varepsilon}{c_0}\left(\frac{\partial A_x}{\partial x}\,x'+\frac{\partial A_y}{\partial x}\,y'+\frac{\partial A_z}{\partial x}\,z'\right)\\ = \frac{\partial}{\partial x}\left(-\,\varepsilon V+\frac{\varepsilon}{c_0}\,\mathbf{A}\,.\,\mathbf{v}\right)\end{aligned}$$

or

$$\frac{\mathrm{d}}{\mathrm{d}t}\left(\frac{m_0x'}{\sqrt{1-v^2/c^2}}+\frac{\varepsilon}{c_0}\,A_x\right) = \frac{\partial}{\partial x}\left(-\,\varepsilon V+\frac{\varepsilon}{c_0}\,\mathbf{A}\,.\,\mathbf{v}\right) \tag{4}$$

But

$$\frac{m_0x'}{\sqrt{1-\dfrac{x'^2+y'^2+z'^2}{c^2}}} = \frac{\partial}{\partial x'}\left(-\,m_0c^2\sqrt{1-\frac{1}{c^2}\,(x'^2+y'^2+z'^2)}\right)$$

$$A_x = \frac{\partial}{\partial x'}\,(A_x.x') = \frac{\partial}{\partial x'}\,\mathbf{v}\,.\,\mathbf{A}$$

and so

$$\frac{\mathrm{d}}{\mathrm{d}t}\,\frac{\partial}{\partial x'}\left(-\,m_0c^2\sqrt{1-v^2/c^2}+\frac{\varepsilon}{c_0}\,\mathbf{v}\,.\,\mathbf{A}\right) = \frac{\partial}{\partial x}\left(-\,\varepsilon V+\frac{\varepsilon}{c_0}\,\mathbf{A}\,.\,\mathbf{v}\right) \tag{5}$$

Using the fact that

$$\frac{\partial}{\partial x}(-m_0c^2\sqrt{1-v^2/c^2}) = 0 \qquad \frac{\partial V}{\partial x'} = 0$$

expression (5) may be written

$$\frac{d}{dt}\frac{\partial}{\partial x'}\left\{-m_0c^2\sqrt{1-v^2/c^2}-\varepsilon\left(V-\frac{1}{c_0}\mathbf{v}\cdot\mathbf{A}\right)\right\} = \frac{\partial}{\partial x}\left\{-m_0c^2\sqrt{1-v^2/c^2}-\varepsilon\left(V-\frac{1}{c_0}\mathbf{v}\cdot\mathbf{A}\right)\right\}$$

Finally

$$\frac{d}{dt}\frac{\partial \mathcal{L}}{\partial x'} = \frac{\partial \mathcal{L}}{\partial x}$$

if we write

$$\mathcal{L} = -m_0c^2\sqrt{1-v^2/c^2}-\varepsilon\left(V-\frac{1}{c_0}\mathbf{v}\cdot\mathbf{A}\right).$$

CHAPTER VII

VARIATIONAL METHODS: EQUATIONS FOR FIELDS

[89] General Remarks

We now attempt to exploit the variational method, to investigate the general properties of fields of force. In this chapter we consider only the case in which the field, otherwise arbitrary, can be derived from a potential (V scalar and $\mathbf{A}$ vector). We shall thus characterize the field by a Lagrange function depending upon V, $\mathbf{A}$ and their derivatives. *Here we consider only first derivatives, namely the fields* $\mathfrak{B}$ *and* $\mathbf{E}$.

As we shall see, the theory can be developed in two ways, identical at heart but leading to very different formalisms. From one point of view, Maxwell's equations are obtained, and these are the basis of the macroscopic theory of fields. This we shall analyse first, and it finds continual use in electromagnetism. Alternatively, the calculations can be performed in such a way that the results appear in the form of Lagrange's and Hamilton's equations, with the same appearance as the corresponding equations for particles. This point of view too is fundamental: it constitutes one of the starting-points of quantum field theory.

In short, the topics studied in the present chapter provide the key to all the field theories, macroscopic or quantum.

A. MAXWELL'S EQUATIONS

[90] The Form of the Lagrangian

As we are studying the properties of the field at every point, we consider a Lagrangian per unit volume, L, and we write Hamilton's principle in the form

$$\delta \int_{t_0}^{t_1} \mathcal{L}(V, \mathbf{A}, \mathbf{E}, \mathfrak{B})\, \mathrm{d}t = \delta \int_{t_0}^{t_1} \left\{ \iiint_D L(V, \mathbf{A}, \mathbf{E}, \mathfrak{B})\, \mathrm{d}\varpi \right\} \mathrm{d}t = 0$$

The expression for L will depend upon the interactions between the particles; if these are left unspecified, we shall obtain formulae applicable to very general fields of force.

[91] The Equations of Ampère and Maxwell

Variation of the terms in V and $\mathbf{A}$ yields

$$\iiint_D \left(\frac{\partial L}{\partial V} \delta V + \frac{\partial L}{\partial A_x} \delta A_x + \frac{\partial L}{\partial A_y} \delta A_y + \frac{\partial L}{\partial A_z} \delta A_z \right) \mathrm{d}\varpi$$

$$= \iiint_D (-\varrho \, \delta V + j_x \, \delta A_x + j_y \, \delta A_y + j_z \, \delta A_z) \, \mathrm{d}\varpi$$

where we have written

$$\varrho = -\partial L / \partial V, \qquad \mathbf{j} = \partial L / \partial \mathbf{A}$$

We employ the letters ϱ and $\mathbf{j}$ because the quantities thus defined do, as we shall see, correspond to the densities which figure in the ordinary Maxwell–Lorentz theory. Here, however, they have a purely mechanical and more general meaning.

For the term in $\mathbf{E}$,

$$\iiint_D \left(\frac{\partial L}{\partial E_x} \delta E_x + \frac{\partial L}{\partial E_y} \delta E_y + \frac{\partial L}{\partial E_z} \delta E_z \right) \mathrm{d}\varpi$$

$$= -\iiint_D \left\{ \frac{\partial L}{\partial E_x} \delta \left(\frac{\partial V}{\partial x} + \frac{1}{c_0} \frac{\partial A_x}{\partial t} \right) + \ldots \right\} \mathrm{d}\varpi$$

$$= -\iiint_D \left\{ \frac{\partial L}{\partial E_x} \left(\frac{\partial}{\partial x} \delta V + \frac{1}{c_0} \frac{\partial}{\partial t} \delta A_x \right) + \ldots \right\} \mathrm{d}\varpi$$

changing the order of the operators δ and ∂.

Likewise, variation of the terms in $\mathfrak{B}$ gives

$$\iiint_D \left(\frac{\partial L}{\partial \mathcal{B}_x} \delta \mathcal{B}_x + \frac{\partial L}{\partial \mathcal{B}_y} \delta \mathcal{B}_y + \frac{\partial L}{\partial \mathcal{B}_z} \delta \mathcal{B}_z \right) \mathrm{d}\varpi$$

$$= \iiint_D \left\{ \frac{\partial L}{\partial \mathcal{B}_x} \delta \left(\frac{\partial A_z}{\partial y} - \frac{\partial A_y}{\partial z} \right) + \frac{\partial L}{\partial \mathcal{B}_y} \delta \left(\frac{\partial A_x}{\partial z} - \frac{\partial A_z}{\partial x} \right) \right.$$

$$\left. + \frac{\partial L}{\partial \mathcal{B}_z} \delta \left(\frac{\partial A_y}{\partial x} - \frac{\partial A_x}{\partial y} \right) \right\} \mathrm{d}\varpi$$

We collect up the terms involving a single component of $\mathbf{A}$; thus for the terms in δA_x, for example,

$$\iiint_D \left(\frac{\partial L}{\partial \mathcal{B}_y} \delta \frac{\partial A_x}{\partial z} - \frac{\partial L}{\partial \mathcal{B}_x} \delta \frac{\partial A_x}{\partial y} \right) \mathrm{d}\varpi$$

$$= \iiint_D \left(\frac{\partial L}{\partial \mathcal{B}_y} \frac{\partial}{\partial z} \delta A_x - \frac{\partial L}{\partial \mathcal{B}_z} \frac{\partial}{\partial y} \delta A_x \right) \mathrm{d}\varpi$$

To simplify the expressions, we write

$$\mathfrak{D} = 4\pi \frac{\partial L}{\partial \mathbf{E}}, \qquad \mathbf{H} = -4\pi \frac{\partial L}{\partial \mathfrak{B}}$$

in which the coefficients $\pm 4\pi$ have been introduced so that we finally recover the familiar formulae. The terms in **E** and $\mathfrak{D}$ are of the form

$$-\frac{1}{4\pi} \iiint_D \left\{ \mathcal{D}_x \left(\frac{\partial}{\partial x} \delta V + \ldots \right) + \ldots \right\} d\varpi$$

Integrating by parts,

$$\iiint_D \mathcal{D}_x \frac{\partial}{\partial x} \delta V \, d\varpi = - \iiint_D \delta V \frac{\partial \mathcal{D}_x}{\partial x} \, d\varpi$$

The integrated term vanishes, since the field variations vanish at the limits of integration (infinity).

Finally, variation of these terms gives

$$\frac{1}{4\pi} \iiint_D \left\{ \delta V \left(\frac{\partial \mathcal{D}_x}{\partial x} + \frac{\partial \mathcal{D}_y}{\partial y} + \frac{\partial \mathcal{D}_z}{\partial z} \right) + \frac{1}{c_0} \left(\frac{\partial \mathcal{D}_x}{\partial t} \delta A_x + \frac{\partial \mathcal{D}_y}{\partial t} \delta A_y + \frac{\partial \mathcal{D}_z}{\partial t} \delta A_z \right) \right\} d\varpi$$

the second expression in brackets is obtained after integrating by parts with respect to time.

We may apply a similar reasoning to the terms in $\mathfrak{B}$ and **H**. Integrating by parts and dropping the integrated term,

$$-\frac{1}{4\pi} \iiint_D \left\{ \left(\frac{\partial H_z}{\partial y} - \frac{\partial H_y}{\partial z} \right) \delta A_x + \left(\frac{\partial H_x}{\partial z} - \frac{\partial H_z}{\partial x} \right) \delta A_y + \left(\frac{\partial H_y}{\partial x} - \frac{\partial H_x}{\partial y} \right) \delta A_z \right\} d\varpi.$$

If the sum of these variations is to vanish, each separate factor of the terms δV, δA_x, δA_y and δA_z must be equal to zero:

$$\frac{1}{4\pi} \left(\frac{\partial \mathcal{D}_x}{\partial x} + \frac{\partial \mathcal{D}_y}{\partial y} + \frac{\partial \mathcal{D}_z}{\partial z} \right) - \varrho = 0$$

$$j_x + \frac{1}{4\pi c_0} \frac{\partial \mathcal{D}_x}{\partial t} - \frac{1}{4\pi} \left(\frac{\partial H_z}{\partial y} - \frac{\partial H_y}{\partial z} \right) = 0 \ldots$$

which yields the group of equations

$$\operatorname{div} \mathfrak{D} = 4\pi\varrho, \qquad \operatorname{curl} \mathbf{H} = \frac{1}{c_0} \frac{\partial \mathfrak{D}}{\partial t} + 4\pi \mathbf{j}$$

We realize that *these expressions are valid for all fields with potentials V and* **A**. Nevertheless, the reader should not lose sight of the fact that, for the present, *this is but an empty form.* Only afier the expression for L has been specified will it acquire a meaning.

Tensor notation. Using the summation convention, the relations above may be written

$$\frac{\partial H_{ij}}{\partial x_j} = \partial^j H_{ij} = \frac{1}{c_0}\frac{\partial \mathcal{D}_i}{\partial t} + 4\pi j_i$$

$$\frac{\partial \mathcal{D}_i}{\partial x_i} = \partial^i \mathcal{D}_i = 4\pi\varrho$$

The current density j_i is a vector density (the vector **v** multiplied by the scalar density ϱ); for the equations to be tensorially homogeneous, the quantity H_{ij} must be regarded as an antisymmetric second-order tensor density. This tensor notation brings out clearly the difference between the equations above and those of Chapter V (§ [70](d)); this difference is usually masked by the vector notation.

Remark. Integral form of the relation involving curl **H**. Following the reasoning of § [70](c), we write, for any curve C,

$$\oint_C \mathbf{H}.\mathbf{dl} = \frac{1}{c_0}\frac{\partial}{\partial t}\iint_S \mathfrak{D}\,\mathbf{dS} + 4\pi \iint_S \mathbf{j}.\mathbf{dS}$$

In the steady state, we have simply

$$\oint_C \mathbf{H}.\mathbf{dl} = 4\pi \iint_S \mathbf{j}\,\mathbf{dS}$$

[92] General Reminder of Maxwell's Equations for Fields with Potentials

In conclusion, we have obtained the following results for all such fields.

The field of force is defined by

$$\mathbf{E} = -\text{grad } V - \frac{1}{c_0}\frac{\partial \mathbf{A}}{\partial t}, \quad \mathfrak{B} = \text{curl } \mathbf{A}.$$

For any V and **A**, the field satisfies the Faraday–Maxwell relations

$$\text{curl } \mathbf{E} = -\frac{1}{c_0}\frac{\partial \mathfrak{B}}{\partial t}, \quad \text{div } \mathfrak{B} = 0.$$

A particular potential is characterized by its Lagrange function $L(V, \mathbf{A}, \mathbf{E}, \mathfrak{B})$. Application of Hamilton's principle leads us to define two other vectors

$$\mathfrak{D} = 4\pi\frac{\partial L}{\partial \mathbf{E}}, \quad \mathbf{H} = -4\pi\frac{\partial L}{\partial \mathfrak{B}}$$

which satisfy the Ampère–Maxwell relations

$$\text{div } \mathfrak{D} = 4\pi\varrho, \quad \text{curl } \mathbf{H} = \frac{1}{c_0}\frac{\partial \mathfrak{D}}{\partial t} + 4\pi\mathbf{j}$$

with

$$\varrho = -\partial L/\partial V, \quad \mathbf{j} = \partial L/\partial \mathbf{A}$$

In applications to electrical phenomena, **H** and $\mathfrak{D}$ correspond to the magnetic induction and the electric induction (or electric displacement) respectively; once again, however, these vectors here have a purely *mechanical* meaning.

[93] The Relativistic Invariance of these Equations; Transformation of the Vectors H and $\mathfrak{D}$

The above equations were written down for an arbitrary Galileian system; by definition, they are invariant. We have already seen, in § [44], how the quantities **E** and $\mathfrak{B}$ transform so that this invariance is conserved. We consider now the formulae governing **H** and $\mathfrak{D}$.

For a system K_2, using Cartesian coordinates, the equations are of the form

$$\frac{\partial E_{2z}}{\partial y_2}-\frac{\partial E_{2y}}{\partial z_2}=-\frac{1}{c_0}\frac{\partial \mathcal{B}_{2x}}{\partial t_2}, \dots$$

$$\frac{\partial H_{2z}}{\partial y_2}-\frac{\partial H_{2y}}{\partial z_2}=\frac{1}{c_0}\frac{\partial \mathcal{D}_{2x}}{\partial t_2}+4\pi j_{2x}, \dots$$

$$\frac{\partial \mathcal{B}_{2x}}{\partial x_2}+\frac{\partial \mathcal{B}_{2y}}{\partial y_2}+\frac{\partial \mathcal{B}_{2z}}{\partial z_2}=0$$

$$\frac{\partial \mathcal{D}_{2x}}{\partial x_2}+\frac{\partial \mathcal{D}_{2y}}{\partial y_2}+\frac{\partial \mathcal{D}_{2z}}{\partial z_2}=4\pi\varrho_2$$

Let us now consider another system K_1, using the conventions of the special Lorentz transformation: corresponding axes parallel to those of K_2 and similarly orientated, the velocity v of K_2 with respect to K_1 in the positive O_1x_1 direction.

We find

$$\frac{\partial}{\partial x_2}=\frac{1}{\sqrt{1-\beta^2}}\left(\frac{\partial}{\partial x_1}+\frac{v}{c^2}\frac{\partial}{\partial t_1}\right)$$

$$\frac{\partial}{\partial y_2}=\frac{\partial}{\partial y_1}, \qquad \frac{\partial}{\partial z_2}=\frac{\partial}{\partial z_1}$$

$$\frac{\partial}{\partial t_2}=\frac{1}{\sqrt{1-\beta^2}}\left(\frac{\partial}{\partial t_1}+v\frac{\partial}{\partial x_1}\right)$$

The second Ampère–Maxwell group then becomes

$$\frac{\partial H_{2z}}{\partial y_1}-\frac{\partial H_{2y}}{\partial z_1}=\frac{1}{c_0\sqrt{1-\beta^2}}\left(\frac{\partial \mathcal{D}_{2x}}{\partial t_1}+v\frac{\partial \mathcal{D}_{2x}}{\partial x_1}\right)+4\pi j_{2x}$$

$$\frac{\partial H_{2x}}{\partial z_1}-\frac{1}{\sqrt{1-\beta^2}}\left(\frac{\partial H_{2z}}{\partial x_1}+\frac{v}{c^2}\frac{\partial H_{2z}}{\partial t_1}\right)$$
$$=\frac{1}{c_0\sqrt{1-\beta^2}}\left(\frac{\partial \mathcal{D}_{2y}}{\partial t_1}+v\frac{\partial \mathcal{D}_{2y}}{\partial x_1}\right)+4\pi j_{2y}$$

$$\frac{1}{\sqrt{1-\beta^2}}\left(\frac{\partial H_{2y}}{\partial x_1}+\frac{v}{c^2}\frac{\partial H_{2y}}{\partial t_1}\right)-\frac{\partial H_{2x}}{\partial y_1}$$
$$=\frac{1}{c_0\sqrt{1-\beta^2}}\left(\frac{\partial \mathcal{D}_{2z}}{\partial t_1}+v\frac{\partial \mathcal{D}_{2z}}{\partial x_1}\right)+4\pi j_{2z}$$

These three relations can be written in K_1 in the same form as in K_2 if we employ the following transformations for the components of **H** and $\mathfrak{D}$; they are analogous to those given in § [44] for **E** and $\mathfrak{B}$.

$$\mathcal{D}_{2x} = \mathcal{D}_{1x} \qquad H_{2x} = H_{1x}$$

$$\mathcal{D}_{2y} = \frac{\mathcal{D}_{1y} - \frac{c_0\beta}{c} H_{1z}}{\sqrt{1-\beta^2}} \qquad H_{2y} = \frac{H_{1y} + \frac{v}{c_0}\mathcal{D}_{1z}}{\sqrt{1-\beta^2}}$$

$$\mathcal{D}_{2z} = \frac{\mathcal{D}_{1z} + \frac{c_0\beta}{c} H_{1y}}{\sqrt{1-\beta^2}} \qquad H_{2z} = \frac{H_{1z} - \frac{v}{c_0}\mathcal{D}_{1y}}{\sqrt{1-\beta^2}}$$

With these formulae, we in fact have

$$\frac{1}{\sqrt{1-\beta^2}}\left\{\left(\frac{\partial H_{1z}}{\partial y_1} - \frac{v}{c_0}\frac{\partial \mathcal{D}_{1y}}{\partial y_1}\right) - \left(\frac{\partial H_{1y}}{\partial z_1} + \frac{v}{c_0}\frac{\partial \mathcal{D}_{1z}}{\partial z_1}\right)\right\}$$

$$= \frac{1}{\sqrt{1-\beta^2}}\frac{1}{c_0}\left(\frac{\partial \mathcal{D}_{1x}}{\partial t_1} + v\frac{\partial \mathcal{D}_{1x}}{\partial x_1}\right) + 4\pi j_{2x}$$

In order to obtain the same expression as in K_2, ϱ and **j** must also be transformed; leaving A and B to be determined, we write

$$j_{2x} = Aj_{1x} + B\varrho_1$$

The preceding equation gives

$$\frac{\partial H_{1z}}{\partial y_1} - \frac{\partial H_{1y}}{\partial z_1} = \frac{1}{c_0}\cdot\frac{\partial \mathcal{D}_{1x}}{\partial t_1} + 4\pi j_{1x}$$

$$\frac{\partial \mathcal{D}_{1x}}{\partial x_1} + \frac{\partial \mathcal{D}_{1y}}{\partial y_1} + \frac{\partial \mathcal{D}_{1z}}{\partial z_1} = 4\pi\varrho_1$$

if we write

$$A = \frac{1}{\sqrt{1-\beta^2}}, \qquad B = \frac{-v/c_0}{\sqrt{1-\beta^2}}$$

Following the same argument for the other components, we obtain the following transformations for ϱ and the components of **j**:

$$\varrho_2 = \frac{\varrho_1 - \frac{c_0 v}{c^2} j_{1x}}{\sqrt{1-\beta^2}} \qquad j_{2x} = \frac{j_{1x} - \frac{v}{c_0}\varrho_1}{\sqrt{1-\beta^2}}$$

$$j_{2y} = j_{1y} \qquad j_{2z} = j_{1z}$$

Remark. In applying these formulae to electromagnetism, we recognize ϱ and **j** as charge density and current density respectively.

[94] Examples of Lagrangians

The fields of force which lead to the electrodynamics of Maxwell and Lorentz (Chapter X) correspond to the Lagrangian

$$L = -(\varrho V - \mathbf{j}\,.\,\mathbf{A}) + \frac{1}{8\pi}\left(K_0 E^2 - \frac{\mathcal{B}^2}{\mu_0}\right)$$

in which ϱ and $\mathbf{j}$ are the densities, at the point in question, of the distribution of particles acting, and K_0, μ_0 are two positive constants characterizing the units.

From this example, we can see the origin of the general definitions of ϱ and $\mathbf{j}$ in terms of L, given in § [91]. With this Lagrangian,

$$\mathfrak{D} = K_0\mathbf{E} \quad \text{and} \quad \mathbf{H} = \mathfrak{B}/\mu_0$$

The two groups of vectors thus differ only in a constant of proportionality.

Born and Infeld have created an electrodynamics based upon the following Lagrangian:

$$L(\mathbf{E}, \mathfrak{B}) = \frac{K_0 E_m^2}{4\pi}\left(1 - \sqrt{1 + F - G^2}\right)$$

in which

$$F = -\frac{1}{K_0 E_m^2}\left(K_0 E^2 - \frac{\mathcal{B}^2}{\mu_0}\right), \qquad G = \frac{1}{\sqrt{K_0 u_0}}\,\frac{\mathfrak{B}\,.\,\mathbf{E}}{E_m^2}$$

E_m is a constant which arises in the theory (maximum field).

The point of view of these authors is very different from that adopted in Chapter X of the present work; for them, the only physical reality consists of the field, and the Lagrangian contains no terms in ϱ and $\mathbf{j}$ since there are no terms in V and $\mathbf{A}$. The calculations given above remain valid none the less, and the terms in ϱ and $\mathbf{j}$ in the final equations are deleted.

Bopp and Podolsky use a Lagrangian which contains derivatives of the field.

I shall not go further into these questions here; the rich possibilities of the method are apparent, and familiarity with it is essential if generalized electrodynamics is to be studied.

This topic is discussed at length in my *Rayonnement et Dynamique*.

B. THE LAGRANGE AND HAMILTON EQUATIONS FOR A FIELD

[95] The Concept of the Coordinates of a Field

A field of force is characterized at every point by the vectors $\mathbf{E}_i(t)$ and $\mathfrak{B}_i(t)$, just as an ensemble of particles is characterized by coordinates $q_i(t)$ and their derivatives $q_i'(t)$. Since the field is continuous, it possesses an infinite number of degrees of freedom: it is comparable only with an infinite ensemble of particles. From this point of view, the $\mathbf{E}_i(t)$ and $\mathfrak{B}_i(t)$ may be regarded as the generalized coordinates of the field.

We now exploit this analogy, so as to obtain expressions for the field in the same form as the Lagrange and Hamilton equations.

If the field has potentials, the six functions **E** and $\mathfrak{B}$ may be replaced by the four components of the potential. Let us carry out the calculations for the simple case in which only a scalar potential exists at each point,

$$V(x, y, z, t)$$

If we regard the value at each point as the field coordinate, we write

$$V_{xyz}(t) \quad \text{by analogy with} \quad q_i(t)$$

[96] The Equations of Lagrange for a Field with Scalar Potential

For particles, the variational principle is written

$$\delta \int_{t_1}^{t_2} \mathscr{L}\, \mathrm{d}t = 0 \quad \text{with} \quad \delta q_i(t_1) = \delta q_i(t_2) = 0$$

The Lagrangian

$$\mathscr{L}(q_i, q_i', t)$$

is a function of time and a functional of the possible paths, $q_i(t)$.

For the field, we write

$$\delta \int_{t_1}^{t_2} \left\{ \iiint L\left(V, \frac{\partial V}{\partial x}, \ldots, \frac{\partial V}{\partial t}, t\right) \mathrm{d}\varpi \right\} \mathrm{d}t = 0 \tag{1}$$

The Lagrangian density must obviously contain the derivatives of V since by hypothesis, the field is continuous throughout space. Higher-order derivatives could also be present, but although we shall assume that this is not the case, the reader must not be surprised if he meets advanced theories depending upon more general Lagrangians; I only attempt to give some idea of this type of argument. Hamilton's principle is now of the form

$$\int_{t_1}^{t_2} \left(\iiint \delta L\, \mathrm{d}\varpi \right) \mathrm{d}t = 0$$

with

$$\delta V_{xyz}(t_1) = \delta V_{xyz}(t_2) = 0 \tag{2}$$

so that, denoting the time derivative of V by V',

$$\int_{t_1}^{t_2} \left[\iiint \left\{ \frac{\partial L}{\partial V} \delta V + \sum_{xyz} \frac{\partial L}{\partial \left(\dfrac{\partial V}{\partial x}\right)} \delta\left(\frac{\partial V}{\partial x}\right) + \frac{\partial L}{\partial V'} \delta V' \right\} \mathrm{d}\varpi \right] \mathrm{d}t = 0$$

Changing the order of δ and ∂, we find

$$\int_{t_1}^{t_2} \left[\iiint \left\{ \frac{\partial L}{\partial V} \delta V + \sum_{xyz} \frac{\partial L}{\partial \left(\dfrac{\partial V}{\partial x}\right)} \frac{\partial}{\partial x}(\delta V) + \frac{\partial L}{\partial V'} \frac{\partial}{\partial t}(\delta V) \right\} \mathrm{d}\varpi \right] \mathrm{d}t = 0$$

The term under the summation sign can be integrated by parts with respect to the spatial coordinates, and the surface integral may be assumed to vanish. The second term can be integrated by parts with respect to time, and using relations (2) we may set the double integral equal to zero. Finally,

$$\int_{t_1}^{t_2}\left[\iiint\left\{\frac{\partial L}{\partial V}-\sum_{xyz}\frac{\partial}{\partial x}\frac{\partial L}{\partial\left(\frac{\partial V}{\partial x}\right)}-\frac{\partial}{\partial t}\frac{\partial L}{\partial V'}\right\}\delta V\,\mathrm{d}\varpi\right]\mathrm{d}t = 0$$

This equation must be satisfied for any value of δV, which is the case only if

$$\frac{\partial L}{\partial V}-\sum_{xyz}\frac{\partial}{\partial x}\frac{\partial L}{\partial\left(\frac{\partial V}{\partial x}\right)}-\frac{\partial}{\partial t}\frac{\partial L}{\partial V'} = 0 \tag{3}$$

This expression may be cast into the Lagrange form by introducing the Lagrange integral, $\mathcal{L}$. By analogy with the quantities

$$\frac{\partial \mathcal{L}}{\partial q_i} \quad \text{and} \quad \frac{\partial \mathcal{L}}{\partial q_i'}$$

of particle theory, we must now define the derivatives at every point of $\mathcal{L}$ with respect to V and V'. For this, we divide space into elements of volume $\delta\varpi_i$, and we treat $\mathcal{L}$ as the limit of the sum

$$\sum_i L\left\{V_i, \left(\frac{\partial V}{\partial x}\right)_i, \ldots\right\}\delta\varpi_i$$

as $\delta\varpi_i$ tends to zero; V_i is the average value of V within the element of volume.

We follow a similar argument for the time; we replace the integrals given above by the limit of the sum

$$\sum_i\left\{\frac{\partial L}{\partial V}-\sum_{xyz}\frac{\partial}{\partial x}\frac{\partial L}{\partial\left(\frac{\partial V}{\partial x}\right)}\right\}_i\delta V_i\,\delta\varpi_i-\sum_i\left(\frac{\partial L}{\partial V'}\right)_i\delta V_i'\,\delta\varpi_i$$

where δV_i and $\delta V_i'$ are independent variations.

Let us now suppose that all the δV_i and $\delta V_i'$ are all zero save one, δV_j. We shall call the quantity

$$\frac{\partial \mathcal{L}}{\partial V} = \lim_{\delta\varpi_j\to 0}\frac{\delta\mathcal{L}}{\delta V_j\delta\varpi_j}$$

the *functional derivative of* $\mathcal{L}$ with respect to V at the point in question. We therefore have

$$\frac{\partial \mathcal{L}}{\partial V} = \frac{\partial V}{\partial x}-\sum_{xyz}\frac{\partial}{\partial x}\frac{\partial L}{\partial\left(\frac{\partial L}{\partial V}\right)}$$

and likewise,

$$\frac{\partial \mathcal{L}}{\partial V'} = \lim_{\delta\varpi_j\to 0}\frac{\delta L}{\delta V'\delta\varpi_j} = \frac{\partial L}{\partial V'}$$

With this notation, relations (3) become

$$\frac{\partial}{\partial t}\frac{\partial \mathcal{L}}{\partial V'}-\frac{\partial \mathcal{L}}{\partial V}=0 \tag{4}$$

[97] Hamilton's Equations

Just as in the particle case, we define a quantity

$$\mathcal{P}_j=\frac{\delta \mathcal{L}}{\delta V_j'}=\left(\frac{\partial \mathcal{L}}{\partial V'}\right)_j \delta\varpi_i$$

which we call the momentum canonically conjugate to V_j.

From this definition and the Lagrange equation, we find

$$\mathcal{P}_j'=\left(\frac{\partial \mathcal{L}}{\partial V}\right)_j \delta\varpi_j$$

We then define, again by analogy, the Hamilton function

$$\mathcal{H}=\sum_i \mathcal{P}_i V_i'-\mathcal{L}=\sum_i\left(\frac{\partial \mathcal{L}}{\partial V'}\right)_i V_i'\delta\varpi_i-\mathcal{L}$$

We can then define the Hamiltonian density H, to be

$$\mathcal{H}=\iiint H\,\mathrm{d}\varpi$$

with

$$H=lV'-\mathcal{L},\quad l=\frac{\partial \mathcal{L}}{\partial V'}=\frac{\partial L}{\partial V'} \tag{5}$$

Taking (4) and (5) into account, the variation of $\mathcal{L}$ arising from the variations of V and V' can be written

$$\begin{aligned}\delta\mathcal{L}&=\iiint\left(\frac{\partial \mathcal{L}}{\partial V}\,\delta V+\frac{\partial \mathcal{L}}{\partial V'}\,\delta V'\right)\mathrm{d}\varpi\\&=\iiint(l'\,\delta V+l\,\delta V')\,\mathrm{d}\varpi\\&=\iiint\{\delta(lV')+l'\,\delta V-V'\,\delta l\}\,\mathrm{d}\varpi\\&=\delta\mathcal{H}+\delta\mathcal{L}+\iiint(l'\,\delta V-V\,\delta l)\,\mathrm{d}\varpi\end{aligned} \tag{6}$$

The variation of $\mathcal{H}$ arising from the variations of V and l is given by

$$\delta\mathcal{H}=\iiint\left(\frac{\partial \mathcal{H}}{\partial V}\,\delta V+\frac{\partial \mathcal{H}}{\partial l}\,\delta l\right)\mathrm{d}\varpi \tag{7}$$

Furthermore, the functional derivatives of $\mathscr{H}$ may be written

$$\frac{\partial \mathscr{H}}{\partial V} = \frac{\partial H}{\partial V} - \sum_{xyz} \frac{\partial}{\partial x} \frac{\partial H}{\partial\left(\frac{\partial V}{\partial x}\right)}$$

$$\frac{\partial \mathscr{H}}{\partial l} = \frac{\partial H}{\partial l} - \sum_{xyz} \frac{\partial}{\partial x} \frac{\partial H}{\partial\left(\frac{\partial l}{\partial x}\right)}$$

Comparing (6) and (7) for arbitrary variations δV and δl yields

$$V' = \frac{\partial \mathscr{H}}{\partial l} \quad \text{and} \quad l' = -\frac{\partial \mathscr{H}}{\partial V}$$

Finally, let us write down the Poisson brackets. Consider some function $\mathscr{F}$ of V and l, which we express in terms of a density function, $F(V, l)$. We can write

$$\begin{aligned} \mathscr{F}' &= \iiint \left(\frac{\partial F}{\partial V} V' + \frac{\partial F}{\partial l} l' \right) \mathrm{d}\bar{\omega} \\ &= \iiint \left(\frac{\partial \mathscr{F}}{\partial V} \frac{\partial \mathscr{H}}{\partial l} - \frac{\partial \mathscr{F}}{\partial l} \frac{\partial \mathscr{H}}{\partial V} \right) \mathrm{d}\bar{\omega} = [\mathscr{F}, \mathscr{H}] \end{aligned}$$

[98] The General Case

We then write the Lagrangian in the form

$$L\left(V, \frac{\partial V}{\partial x}, \ldots \frac{\partial V}{\partial t}; A_x, \frac{\partial A_x}{\partial x}, \ldots \frac{\partial A_z}{\partial x}; t\right)$$

Varying each function independently, the reasoning given above leads to as many Lagrange and Hamilton equations as there are functions.

HISTORICAL AND BIBLIOGRAPHICAL NOTES

[99] Variational Methods and Field Theory

We have already mentioned in § [88] the authors who have applied these methods to Maxwell's equations for the electromagnetic field. From the very beginning of quantum mechanics, the study of the interaction between matter and radiation has made it necessary to quantize not only systems of particles (atoms, molecules) but also the electromagnetic field. The quantization rules for particles were already known, however, and to extend these rules to field calculations, the field equations had to be cast into the same form as those relating to particles.

A whole new aspect of the macroscopic theory of fields was thus developed, to provide a starting-point for quantum mechanics. These theories, which were originally produced for the electromagnetic field, are today also applied to other fields (meson fields, for example). The method described in tihs chapter is due in principle to Pauli and Heisenberg [1], and is set out in several textbooks, Schiff [7] for example. The field amplitudes are treated as coordinates, their time derivatives as velocities, and the derivatives of the Lagrangian with respect to these "velocities" as momenta.

There are other ways of obtaining the Hamiltonian expressions which I cannot include without going beyond the scope of the present work: *my object is simply to bridge the gap to the texts on quantum theory*.

As a variant upon the method given above, the Lagrangian may be expanded as a series of orthogonal functions.

A different method, due to Born and Weyl [2, 3, 4], involves treating the four derivatives of the amplitudes with respect to the four space-time coordinates as independent velocities; this leads to the introduction of four momenta. This idea has been taken up and generalized by Good [10], but I shall say no more about this here.

The reader is referred to the works of Heitler [5], Wentzel [6], L. de Broglie [8], March [9] and others for further details.

[1] W. Heisenberg and W. Pauli: Zur quanten Dynamik der Wellenfelder. *Z. Physik* **56** (1929) 1–61.

[2] M. Born: On the quantum theory of the electromagnetic field. *Proc. Roy. Soc. London* **143** (1934) 410–437.

[3] H. Weyl: Observations on Hilbert's independence theorem and Born's quantization of field equations. *Phys. Rev.* **46** (1934) 505–508.

[4] H. Weyl: Geodesic fields in the calculus of variation for multiple integrals. *Ann. Math.* **36** (1935) 607–629.

[5] W. Heitler: *The Quantum Theory of Radiation*, Oxford University Press, 3rd ed., 1954.

[6] G. Wentzel: *Quantum Theory of Fields*, Interscience, New York, London, 1949.

[7] L. I. Schiff: *Quantum Mechanics*, McGraw-Hill, 1949.

[8] L. de Broglie: *Mécanique ondulatoire du photon et théorie quantique des champs*, Gauthier-Villars, 1949.

[9] A. March: *Quantum Mechanics of Particles and Wave Fields*, Wiley, New York, 1951.

[10] R. H. Good: Hamiltonian mechanics of fields. *Phys. Rev.* **93** (1954) 239–243.

[11] H. Umezawa: *Quantum Field Theory*, North Holland, Amsterdam, 1956.

[12] N. N. Bogolyubov and D. V. Shirkov: *Introduction to the Theory of Quantized Fields*, Interscience, New York, 1959.

[13] A. Visconti: *Théorie quantique des champs*, Gauthier-Villars, Paris, 1961 (vol. I) and 1965 (vol. II).

[14] H. T. Flint: *The Quantum Equations and the Theory of Fields*, Methuen, London, 1965.

CHAPTER VIII

TRANSFER OF THE EQUATIONS OBTAINED BY VARIATIONAL METHODS INTO THE MINKOWSKI CONTINUUM

[100] The Tensor $\mathfrak{D}$, H; the Current Density Four-vector; the Ampère–Maxwell Equations in World-notation

The set of equations of § [91] may be written in the form

$$0+\frac{\partial H_z}{\partial y}-\frac{\partial H_y}{\partial z}-\frac{\partial}{\partial l}\left(\frac{ic}{c_0}\mathcal{D}_x\right) = 4\pi j_x$$

$$-\frac{\partial H_z}{\partial x}+0+\frac{\partial H_x}{\partial z}-\frac{\partial}{\partial l}\left(\frac{ic}{c_0}\mathcal{D}_y\right) = 4\pi j_y$$

$$\frac{\partial H_y}{\partial x}-\frac{\partial H_x}{\partial y}+0-\frac{\partial}{\partial l}\left(\frac{ic}{c_0}\mathcal{D}_z\right) = 4\pi j_z$$

$$\frac{\partial}{\partial x}\left(\frac{ic}{c_0}\mathcal{D}_x\right)+\frac{\partial}{\partial y}\left(\frac{ic}{c_0}\mathcal{D}_y\right)+\frac{\partial}{\partial z}\left(\frac{ic}{c_0}\mathcal{D}_z\right)+0 = 4\pi\frac{ic\varrho}{c_0}$$

The left-hand sides may be regarded as the components of the vector divergence of an *antisymmetric second-rank tensor*, $\mathcal{M}^{ij}$, with the following Cartesian components:

$$\mathcal{M} = \begin{vmatrix} 0 & H_z & -H_y & -\frac{ic}{c_0}\mathcal{D}_x \\ -H_z & 0 & H_x & -\frac{ic}{c_0}\mathcal{D}_y \\ H_y & -H_x & 0 & -\frac{ic}{c_0}\mathcal{D}_z \\ \frac{ic}{c_0}\mathcal{D}_x & \frac{ic}{c_0}\mathcal{D}_y & \frac{ic}{c_0}\mathcal{D}_z & 0 \end{vmatrix}$$

We recall that the first index characterizes the line and the second, the column; lines and columns are placed in the order x, y, z, l.

The right-hand sides of the equations are equal to (4π times) the components of a world-vector having the four components

$$P_x = j_x, \quad P_y = j_y, \quad P_z = j_z, \quad P_l = \frac{ic}{c_0}\varrho$$

We shall call these the four components of the current density. The set of equations can then be put into *tensor form:*

$$\operatorname{div} \mathfrak{M} = 4\pi \mathbf{P}$$

It is, however, better to use the following notation, which brings out more clearly the nature of the quantities $\mathfrak{M}$ and $\mathbf{P}$:

$$\partial_k \mathcal{M}^{jk} = \frac{\partial \mathcal{M}^{jk}}{\partial x^k} = 4\pi P^j$$

In arbitrary coordinates, therefore, since M is antisymmetric, we have

$$\frac{1}{\sqrt{g}} \frac{\partial \sqrt{g}\,\mathcal{M}^{jk}}{\partial x^k} = 4\pi P^j$$

[101] The tensor E, $\mathfrak{B}$; the Faraday–Maxwell equations in world-notation

The set of equations of § [70] may be written

$$0 - \frac{\partial}{\partial y}\left(\frac{ic_0}{c} E_z\right) + \frac{\partial}{\partial z}\left(\frac{ic_0}{c} E_y\right) + \frac{\partial \mathcal{B}_x}{\partial l} = 0$$

$$\frac{\partial}{\partial x}\left(\frac{ic_0}{c} E_z\right) + 0 - \frac{\partial}{\partial z}\left(\frac{ic_0}{c} E_x\right) + \frac{\partial \mathcal{B}_y}{\partial l} = 0$$

$$-\frac{\partial}{\partial x}\left(\frac{ic_0}{c} E_y\right) + \frac{\partial}{\partial y}\left(\frac{ic_0}{c} E_x\right) + 0 + \frac{\partial \mathcal{B}_z}{\partial l} = 0$$

$$\frac{\partial \mathcal{B}_x}{\partial x} + \frac{\partial \mathcal{B}_y}{\partial y} + \frac{\partial \mathcal{B}_z}{\partial z} + 0 = 0$$

These can be interpreted in two ways. Using the tensor $\mathcal{N}$ of § [49],

$$\frac{\partial \mathcal{N}_{ij}}{\partial x^k} + \frac{\partial \mathcal{N}_{jk}}{\partial x^i} + \frac{\partial \mathcal{N}_{ki}}{\partial x^j} = 0$$

We may also write

$$\frac{\partial \sqrt{g}\,\mathcal{N}^{*jk}}{\partial x^k} = 0$$

in which $\mathcal{N}^*$ is the *dual* of the tensor $\mathcal{N}$. The dual $\mathcal{N}^*$ has contravariant components:

$$\mathcal{N}^* = \begin{pmatrix} 0 & -\frac{ic_0}{c} E_z & \frac{ic_0}{c} E_y & \mathcal{B}_x \\ \frac{ic_0}{c} E_z & 0 & -\frac{ic_0}{c} E_x & \mathcal{B}_y \\ -\frac{ic_0}{c} E_y & \frac{ic_0}{c} E_x & 0 & \mathcal{B}_z \\ -\mathcal{B}_x & -\mathcal{B}_y & -\mathcal{B}_z & 0 \end{pmatrix}$$

[102] The Variational Equations of Motion of a Free Particle

(a) *The pre-relativistic formula*

We have

$$\delta \int_{t_0}^{t_1} \mathcal{L}_p \, \mathrm{d}t = \frac{1}{2} m_0 \delta \int_{t_0}^{t_1} \left(\frac{\mathrm{d}\sigma}{\mathrm{d}t}\right)^2 \mathrm{d}t = 0$$

which reduces to

$$\delta \int_{t_0}^{t_1} v \, \mathrm{d}\sigma = 0 \quad \text{or} \quad \delta \int_{t_0}^{t_1} \mathrm{d}\sigma = 0$$

if we use the fact that the velocity is constant. The motion follows the geodesics of the space through which the particle is travelling.

(b) *The relativistic formula*

We now have

$$\mathcal{L}_p = -m_0 c^2 \sqrt{1 - \frac{1}{c^2}\left(\frac{\mathrm{d}\sigma}{\mathrm{d}t}\right)^2}$$

$$= -m_0 c \frac{\sqrt{c^2 \, \mathrm{d}t^2 - \mathrm{d}\sigma^2}}{\mathrm{d}t} = -m_0 c^2 \frac{\mathrm{d}T}{\mathrm{d}t}$$

in which T denotes proper time and $\mathrm{d}\sigma$ the spatial length element.

In tensor notation

$$\mathcal{L}_p = -m_0 c \sqrt{-\mathrm{d}x_k \, \mathrm{d}x^k}$$

since

$$\mathrm{d}x_k \, \mathrm{d}x^k = -\mathrm{d}T^2$$

so that

$$-m_0 c^2 \delta \int_{M_0}^{M_1} \mathrm{d}T = 0$$

or, denoting the element of world-length by $\mathrm{d}s (= ic \, \mathrm{d}T)$,

$$\delta \int_{M_0}^{M_1} \mathrm{d}T = 0 \quad \text{or} \quad \delta \int_{M_0}^{M_1} \mathrm{d}s = 0$$

in which M_0 and M_1 are world-points. The motion follows the geodesics in space-time.

Remark. We notice that if, in Chapter VI, we had reasoned directly in four-dimensional space, the form of $\mathcal{L}_p$ would have been nearly fixed. The variational method in fact involves seeking the stationary value of a particular integral for each problem. For a free particle (no external influence) this integral cannot depend upon the system of reference. The most satisfactory quantity fulfilling this condition is the length of the world-line or the proper time. We are thus led to form the integral

$$k \int_{M_0}^{M_1} \mathrm{d}T$$

We know (*Kinematics*, § [141]) that the straight line is the longest path between two infinitesimally close points M_0 and M_1. Taking k negative will therefore give us a minimum. To obtain the value of k, we expand in series:

$$k\sqrt{1-\beta^2} = k + \frac{kv^2}{2c^2} + \dots$$

Disregarding the constant term which does not affect differentiations, we must write

$$\frac{kv^2}{2c^2} = \frac{1}{2} m_0 v^2 \quad \text{or} \quad k = m_0 c^2$$

to recover pre-relativistic dynamics.

[103] The Variational Equations of Motion in a Field with Potentials

(a) If Φ denotes the four-potential, we can show that for a point charge, the term $\mathscr{L}_i$ must be written

$$\mathscr{L}_i = \frac{\varepsilon}{c_0} \Phi_i \, \mathrm{d}x^i$$

as we have

$$\Phi_i \, \mathrm{d}x^i = A_x \, \mathrm{d}x + A_y \, \mathrm{d}y + A_z \, \mathrm{d}z + \frac{ic_0}{c} Vic \, \mathrm{d}t$$

$$= (A_x v_x + A_y v_y + A_z v_z - c_0 V) \, \mathrm{d}t$$

$$= -c_0 \left(V - \frac{\mathbf{v}}{c_0} . \mathbf{A} \right) \mathrm{d}t$$

Hence the world action integral is of the form

$$\delta \int_{M_0}^{M_1} -m_0 c^2 \, \mathrm{d}T + \frac{\varepsilon}{c_0} \Phi_i \, \mathrm{d}x^i = 0 \tag{1}$$

which is invariant under the Lorentz transformation.

We may also use the form given in § [77]. Defining a four-vector ($\mathfrak{P}$, $\mathscr{H}$) with components

$$\mathfrak{P}, \quad \frac{i}{c} \mathscr{H} = \mathcal{D}_l$$

the differential element in the Hamilton integral appears as the scalar product of this four-vector and the element of world-line. When $\mathfrak{P}$ and $\mathbf{p}$ are identical, the four-vector is the world-momentum, I; the Hamilton integral becomes

$$\int_{M_0}^{M_1} I \, \mathrm{d}s$$

Here again, we see how much four-dimensional notation simplifies the search for $\mathcal{L}_i$, for this function must characterize the field of force and the motion, by means of an invariant expression. The simplest combination is obviously the scalar product of the vectors **I** (characterizing the field) and ds (characterizing the motion).

From § [81], we obtain

$$\left(\mathcal{P}_x-\frac{\varepsilon}{c_0}\Phi_x\right)^2+\left(\mathcal{P}_y-\frac{\varepsilon}{c_0}\Phi_y\right)^2+\left(\mathcal{P}_z-\frac{\varepsilon}{c_0}\Phi_z\right)^2+\left(\mathcal{P}_l-\frac{\varepsilon}{c_0}\Phi_l\right)^2+m_0^2c^2=0$$

(b) The method is extremely useful in treating generalized electrodynamics (Born and others). To familiarize the reader with this notation, we give the variational reasoning in world-notation, starting from (1). The first term in (1) gives (with $ic\,\mathrm{d}T=\mathrm{d}s$)

$$im_0c\int_{M_0}^{M_1}\delta\sqrt{\mathrm{d}x_i^2}=im_0c\int_{M_0}^{M_1}\frac{\mathrm{d}x_i\,\delta(\mathrm{d}x_i)}{\mathrm{d}s}$$

$$=\int_{M_0}^{M_1}m_0\frac{\mathrm{d}x_i}{\mathrm{d}T}\,\delta(\mathrm{d}x_i)=\int_{M_0}^{M_1}m_0\frac{\mathrm{d}x_i}{\mathrm{d}T}\,\mathrm{d}(\delta x_i)$$

Integrating by parts, we obtain

$$-\int_{M_0}^{M_1}m_0\,\mathrm{d}V_i\,\delta x_i;$$

the integrated term vanishes since the variations δ are zero at the limits. The second term in (1) gives

$$\frac{\varepsilon}{c_0}\int_{M_0}^{M_1}\Phi_i\,\delta\,\mathrm{d}x^i+\mathrm{d}x^i\,\delta\Phi_i$$

$$=\frac{\varepsilon}{c_0}\int_{M_0}^{M_1}\Phi_i\,\mathrm{d}\,\delta x^i+\mathrm{d}x^i\,\delta\Phi_i$$

$$=\frac{\varepsilon}{c_0}\int_{M_0}^{M_1}-\mathrm{d}\Phi_i\,\delta x^i+\mathrm{d}x^i\,\delta\Phi_i$$

the last line is obtained by integration by parts (the integrated term vanishes).

Also,

$$\delta\Phi_i=\frac{\partial\Phi_i}{\partial x^k}\,\delta x^k,\quad \mathrm{d}\Phi_i=\frac{\partial\Phi_i}{\partial x^k}\,\mathrm{d}x^k$$

so that

$$0=\int_{M_0}^{M_1}-m_0\,\mathrm{d}V^i\,\delta x^i-\frac{\varepsilon}{c_0}\,\frac{\partial\Phi_i}{\partial x^k}\,\delta x^i\,\mathrm{d}x^k+\frac{\varepsilon}{c_0}\,\frac{\partial\Phi_i}{\partial x^k}\,\mathrm{d}x^i\,\delta x^k$$

The second and third terms contain summations, and the indices i and k may therefore be interchanged. For example,

$$\frac{\partial\Phi_i}{\partial x^k}\,\mathrm{d}x^i\,\delta x^k=\frac{\partial\Phi_k}{\partial x^i}\,V^k\,\delta x^i\,\mathrm{d}T$$

Condition (1) thus yields

$$0 = \int_{M_0}^{M_1} \left\{ -m_0 \frac{\mathrm{d}V^i}{\mathrm{d}T} + \frac{\varepsilon}{c_0} \left(\frac{\partial \Phi_k}{\partial x^i} - \frac{\partial \Phi^i}{\partial x^k} \right) V^k \right\} \delta x^i \, \mathrm{d}T$$

and finally

$$m_0 \frac{\mathrm{d}V^i}{\mathrm{d}T} = \frac{\varepsilon}{c_0} \left(\frac{\partial \Phi_k}{\partial x^i} - \frac{\partial \Phi^i}{\partial x^k} \right) V^k = \frac{\varepsilon}{c_0} \mathcal{H}_{ik} V^k$$

We recover the dynamics equation of § [49].

[104] Variational Methods; the Equations for Fields with Potentials

The calculations of § [90] may also be translated into four-dimensional terms. Considering, for example, the Maxwellian Lagrangian, §§ [52] and [94] suggest that we take as Lagrangian density the quantity

$$\left(\frac{\varepsilon}{c_0} \Phi_i \, \mathrm{d}x^i + \mathcal{H}_{ij}^2 \right) \mathrm{d}\varpi$$

The variation must be carried out keeping the motion of the charges unaltered, that is the currents (it is for this reason that the term in $m_0 c \, \mathrm{d}T$ does not occur here). The tensor $\mathcal{H}_{ij}$ is of course replaced by the expression for it as a function of the Φ_i.

Detailed calculation leads to the expression given in § [100].

HISTORICAL AND BIBLIOGRAPHICAL NOTES

See Chapter VI.

CHAPTER IX

ANALYSIS OF THE CONCEPTS OF FORCE AND MASS

[105] Introduction; the Necessity of Selecting a Doctrine for Defining Physical Quantities

(a) There are virtually as many definitions of force and mass as there are authors defining them; there are even some who do not take the trouble to define these concepts at all.

Confronted with such disagreement and disorder, it is only right to lay stress upon some general ideas: *to choose a doctrine.*

I assert first that *all physical quantities must be defined.* If the reader does not assent to this requirement, then for him there is no problem; to my way of thinking, however, such an attitude belongs to "primitive physics", and is no longer acceptable.

There are many aspects of definitions in physics; three of these seem to me important in this context (see Preface, p. xiv).

(i) A new physical quantity X must be defined axiomatically (or mathematically) by means of a formula containing X and other quantities already known, $A, B, C, \ldots$. Such a formula is both a definition and a postulate: it defines X through certain properties which must be confirmed by experiment.

Such a procedure cannot, of course, be applied to the first quantity to be defined; this first quantity, which is a consequence only of the second procedure, defined below, will be designated the fundamental quantity.

(ii) Every new quantity X must also be defined operationally, by describing a physical procedure which would enable it to be measured. Such a procedure may indeed be implicit in the mathematical definition. Thus every operational definition brings in some physical phenomenon, an apparatus.

(iii) Physical quantities *sometimes* interpret sensations (lapse of time, force, heat, colour, ...) and we can then conceive of definitions in terms of these sensations. This feature often plays a fundamental role when the quantity in question is first introduced into science, but it is a minor aspect which must subsequently be superceded. This is the case for the concepts of heat and temperature, for example.

These requirements, and especially the operational requirement, apply to *all physical quantities*, and hence, to space and time also. This fact is almost always overlooked, and an abstract, *a priori* character is attributed to these quantities which is incompatible with the present state of knowledge.

(b) In the first place, the fundamental quantity must be selected. The reader is reminded that, given the present state of metrology and physical theory (*Kinematics*, § [32]), the basic quantity is time-interval, a variable provided operationally by a periodic phenomenon called a clock. The basic definition is the operational one: there can be no definition in terms of a mathematical formula, for the good reason that X is now the first quantity; and there are not yet any known quantities, A, B, C. The properties of this definition are discussed elsewhere (*Gravitation*, vol. II, p. 259), and I shall not dwell here on questions concerning continuity.

After making some convention for adjusting the clocks, the distance between two clocks is defined to be the time taken by a light pulse to make the round trip from one clock to the other and back.

Geometry and kinematics are then developed, and various quantities defined, in particular velocity $\mathbf{v}$ and acceleration $\boldsymbol{\gamma}$. These developments enable us to define kinematically the concept of a *Galileian reference system* (*Kinematics*, § [32]); the latter is a reference system in which, given the selected standards of length and time, the geometry is Euclidean and the kinematics satisfies the Lorentz transformation.

(c) The definition of the quantities is connected with the question of units and the various related problems: the choice of units, systems of units, the distinction between fundamental quantities and derived quantities, the dimensions of the quantities, the idea of a coherent system and the distinction between the quantity and its measure.

[106] Definition of the Concepts of Rest-mass and Force through the Phenomenon of Gravitation (the definition of mass precedes that of force)

(a) *The concept of mass*

Consider two particles A_0 and A_1 in a Galileian reference system, A_0 stationary and A_1 travelling slowly or at rest. We *assert* that the particle A_0 communicates to the particle A_1 the acceleration

$$\boldsymbol{\gamma}_1 = \frac{K_1}{r^2}\,\frac{\mathbf{r}}{r}$$

in which K_1 is a constant, r denotes the distance A_0A_1, and $\mathbf{r}$ is the vector $\overrightarrow{A_1A_0}$. If another particle A_2 replaces A_1 in the presence of A_0, we shall have

$$\boldsymbol{\gamma}_2 = \frac{K_2}{r^2}\,\frac{\mathbf{r}}{r}$$

We *assert* that the constant K is independent of the particles A_1, A_2, ... and depends only on A_0:

$$K_1 = K_2 = m_0 .$$

This is equivalent to saying that under the action of A_0, every body experiences the same acceleration. We shall return to this very important point in § [107] (c). *By definition,*

m_0 is the *rest-mass* of A_0. The measure of m_0, that is, its operational definition, is obtained by measuring $r^2\gamma$. The experimental verification of the postulate does of course follow from a confirmation that $r^2\gamma$ is a constant. The formula

$$\gamma = \frac{m_0}{r^2}$$

defines the value of m_0 in the coherent system (coherent in the sense of § [110]). There is of course nothing to prevent us from representing the same mass in terms of a different unit, by writing

$$m_0' = km_0$$

I return to the question of systems of units in § [110].

(b) *The concept of gravitational force*

Let m_0 and m_1 be the masses of A_0 and A_1 (to obtain m_1, we measure the acceleration of an arbitrary mass in the presence of A_1, held stationary). If A_1 is present with A_0 fixed, we may write

$$\boldsymbol{\gamma}_1 = \frac{m_0}{r^2}\frac{\mathbf{r}}{r}, \quad m_1\boldsymbol{\gamma}_1 = \frac{m_1 m_0}{r^2}\frac{\mathbf{r}}{r}$$

By definition, the quantity

$$\mathbf{F}_1 = \frac{m_0 m_1}{r^2}\frac{\mathbf{r}}{r}$$

is the gravitational force exerted on A_1 by A_0, or the *weight* of A_1 in the presence of A_0. We may then write

$$\mathbf{F}_1 = m_1\boldsymbol{\gamma}_1$$

The ratio of the vectors $\mathbf{F}_1$ and $\boldsymbol{\gamma}_1$ is a constant. The measure of the gravitational force is therefore obtained with these definitions by measuring m_1 and m_0 and the distance r. From the postulate, this is the same as measuring the product $m_1\gamma_1$.

(c) *The concept of gravitational field*

Let A_0 be fixed. The values of the forces exerted on the masses m_1, m_2, ... *placed successively at the same point*, are

$$\mathbf{F}_1 = \frac{m_0 m_1}{r^2}\frac{\mathbf{r}}{r}, \quad \mathbf{F}_2 = \frac{m_0 m_2}{r^2}\frac{\mathbf{r}}{r}, \ldots$$

and we can therefore write

$$\mathbf{F}_1 = m_1\mathbf{H}, \quad \mathbf{F}_2 = m_2\mathbf{H}, \ldots$$

where

$$\mathbf{H} = \frac{m_0}{r^2}\frac{\mathbf{r}}{r}$$

By definition, $\mathbf{H}$ is the gravitational field of A_0 at the point in question.

(d) *The parallelogram postulate; the dynamics of masses experiencing gravitational interaction*

Let a mass m be placed at A, in the presence of fixed masses $m_1, m_2, \ldots$, successively, the latter placed at points $A_1, A_2, \ldots$, respectively. Let $\mathbf{r}_1, \mathbf{r}_2, \ldots$ denote the vectors $\overrightarrow{A_1A}, \overrightarrow{A_2A}, \ldots$ Under the action of m_1 alone, we have

$$\gamma_1 = \frac{m_1}{r_1^2} \frac{\mathbf{r}_1}{r_1}$$

and under the action of m_2 alone,

$$\gamma_2 = \frac{m_2}{r_2^2} \frac{\mathbf{r}_2}{r_2}$$

We *postulate* that under the action of all the masses, m acquires acceleration

$$\gamma = \gamma_1 + \gamma_2 + \ldots$$

This is the same as writing

$$\mathbf{F} = m\gamma \quad \text{with} \quad \mathbf{F} = \mathbf{F}_1 + \mathbf{F}_2 + \ldots$$

From here, we can establish the expression for the field of an arbitrary ensemble of masses, continuous or discontinuous. Statics and dynamics can be deduced, for gravitational forces and low velocities.

Remark 1. The concepts of force and field are intermediate steps in the calculation, and could be dispensed with; the basic quantity is mass.

Remark 2. I have *defined* the gravitational force to be the product m_0m_1/r^2. From the postulate, it is equal to the product $m_0\gamma$, but $m_0\gamma$ is *not* the definition of force.

(e) *Generalization of the concept of force; the dynamical law at low velocities*

Consider a point mass, m. We can set it in motion by physical means of various kinds; under the action of these, it will have an acceleration γ at each moment in time. For accelerations *other than gravitational in origin*, we define the concept of force by the relation

$$\mathbf{F} = m\gamma$$

For the moment, this definition of $\mathbf{F}$ is merely an *empty equation*. To give it physical meaning, $\mathbf{F}$ must be related to the other parameters of the phenomenon, *in every case*.

(f) *The postulate of the theory of elasticity; the idea of the dynamometer*

We state this postulate for a simple case, quite adequate for our present purposes. Consider an elastic rod of negligible mass, to which a mass m is attached (Fig. 15). Let m now experience the attraction of a mass m_0; the rod will then experience a force

$$F = \frac{m_0m_1}{r^2}$$

We *postulate* that the extension is proportional to F

$$F = k\,\Delta l$$

Hence F may be measured in terms of the extension Δl, once the instrument has been calibrated (determination of k). Such a device is called a dynamometer. With it, the force acting on m can be measured while m remains stationary.

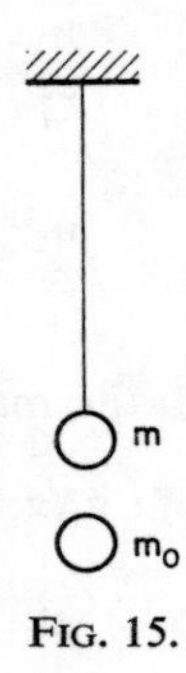

FIG. 15.

[107] Another Procedure: Definition of the Concepts of Force and Mass using the Phenomenon of Elastic Deformation (the force is defined before the mass)

(a) *Definition of force*

We mentioned in § [105] that in the early stages of a science, or again, for pedagogic reasons when the first steps of some branch of science are explained, it may be convenient to define some physical quantity in terms of a sensation. The mental sequence is as follows, if we assume for the moment that the concepts of force and mass are not yet defined. The point of departure is the sensation of effort. Observation and experiment relate this sensation to the deformation of physical bodies. If a solid experiences an effort, or let us say force, it is deformed. Hence the idea of defining force in terms of this deformation. We then define

$$\mathbf{F} = \overrightarrow{\Delta l}$$

choosing the extension of a particular rod as standard of length. This definition offers the following advantages.

For teaching, it is convenient, for it does not depend at all on the idea of motion. I believe that it should not be denied an important place. If the parallelogram postulate is added to it, the whole of statics can be constructed without recourse to dynamics. It possesses all the attributes which we require of a good physical definition, and in particular, the same attributes as the definition of mass using the phenomenon of gravitation: the choice of a fundamental phenomenon, definition of the new quantity through a formula. Its principal advantage, however, is that it brings out clearly the equivalence principle, and the possibility of defining the inertial mass and the gravitational mass separately, as we shall show.

Remark. The sensation of effort provides only a starting-point; from it, the physicist extracts only the idea that there is a physical quantity to be created and specified. We must not be surprised, therefore, if no importance is attached to the analysis of the sensation itself: this is the psychologists' affair. However interesting their work may be in other contexts, it has nothing to offer the physicist but the obvious, common or garden idea of effort.

(b) *Definition of the inertial mass*

Suppose that we apply a force to a corpuscle at rest; this may be visualized, as in (a), as an elastic rod or spring which pulls the corpuscle. As a postulate for dynamics at low velocities, we assert that the acceleration is parallel and proportional to the force; by definition, the coefficient of proportionality is the inertial mass, m:

$$\frac{F}{\gamma} = \text{const} = m$$

(c) *Definition of the gravitational mass; the equivalence principle*

For the same body, we then define a gravitational mass, M, using the attraction postulate

$$F = \frac{M^2}{r^2}$$

which gives the force between two identical masses a distance r apart. The concept of gravitational field H is then defined by

$$F = MH, \quad H = M/r^2$$

If we then consider two masses placed in a gravitational field H, they are acted on by forces

$$F_1 = M_1 H, \quad F_2 = M_2 H$$

giving them accelerations

$$\gamma_1 = \frac{F_1}{m_1} = \frac{M_1}{m_1} H$$

$$\gamma_2 = \frac{F_2}{m_2} = \frac{M_2}{m_2} H$$

These accelerations are equal only if

$$\frac{M_1}{m_1} = \frac{M_2}{m_2}$$

This proportionality between the gravitational and inertial masses is none other than the postulate of § [106] that the K are equal. If the $K_1, K_2, \ldots$ were different, the definitions of § [106] would collapse. Einstein called this property the equivalence principle (equivalence of inertia and gravitation) and upon it he erected his theory of gravitation.

The experimental checks of this postulate are studied in my *Relativité généralisée, Gravitation.*

Remark 1. We perceive that this principle is indispensable to the definition, with the procedure of § [106], and in a certain sense, this shows how fundamental is its character. The statement remains implicit, however, and for a long time it passed unnoticed. It required Einstein's genius to realize its importance and make it play a major role. With the present procedure, the principle is unnecessary: we can define the masses without it. As it has to be stated explicitly, it is all the more strikingly brought to our attention.

Remark 2. The procedure given above is none other than the reasoning employed in the definition of electric charge (§ [117]); in the electrical case, however, the attractive mass and the inertial mass are no longer proportional.

[108] What Procedure are we to Choose?

(a) Each procedure possesses advantages and drawbacks, which I have analysed Nevertheless, we may make a few remarks in this connection. With the procedure of § [107], it seems that force and mass must necessarily be defined together. With that of § [106], on the contrary, force is an intermediate quantity, with which we could, if need be, dispense. This is, however, a somewhat Platonic remark, for we should then have to do without the concept of field as well, and construct electricity without speaking of electric and magnetic fields. The opponents of the concept of force do not seem to have thought very hard about this; furthermore, they do nothing, they produce nothing based upon their preferences. The procedure of § [106] is founded on the phenomenon which, in the final analysis, is used to produce the mass standard; it is not possible to imagine using a dynamometer as standard, because of the inconstancy of the metals. This would remain true even if we rallied to the relativistic idea of a single fundamental quantity, and if we regarded the unit of mass as a derived unit.

(b) It seems clear that the best procedure, for a thorough reconstruction of physics, is that of § [106]. The procedure using elastic deformation offers solid advantages, however, and the objections made to it on grounds of principle are sometimes based on false preconceptions. In the French edition, I adopted this procedure alone. I still feel that it should be widely known and used, as a first introduction at least.

[109] Change of Galileian Reference System; the Dynamical Law at High Velocities and Transformation of the Dynamical Quantities

(a) *Ordinary notation*

We assume that the laws have been written down and the quantities defined for a Galileian system of reference, K_0, in which the velocities are *zero*. We transform to another Galileian system, K. Using the procedure described in § [38], we then obtain the law of dynamics in the new system. From this, as I have explained in detail, we obtain the definition of relativistic mass, the force transformation, ... The essential point is that the law in K is a *consequence* of the law in K_0 and of relativistic kinematics.

In the course of these calculations, we have defined the following concepts: *the four-vector force K^i; the four-tensor force $\mathcal{N}^{ij}$; the four-vector impulse* or *world-impulse, $K^i\,dT$; the four-vector momentum, m_0V^i*. We have also introduced the ideas of *transverse and longitudinal mass, and relativistic mass.*

(b) *Variants*

I have mentioned several times (§§ [5] and [38]) that the notation adopted in the tridimensional equations of relativistic dynamics is to some extent arbitrary. I have already indicated one variant (§ [25]), using a new definition of acceleration. Suppose that in K_0, the quantities m_0 and $\mathbf{F}_0$ are given by their definitions as in § [106]; we are required to write down the law of motion in K. The concepts of force and mass in K are therefore mathematical intermediaries, the choice of which is *logically* of no importance provided that the phenomenon which is measurable (the law of motion) is unaffected.

A detailed study of this question is highly illuminating for the comprehension of relativistic epistemology; conversely, it cannot be fully grasped until the fundamental points of this epistemology have been digested.

With the usual notation, we have (§§ [8] and [28])

$$X = \frac{m_0\gamma_x}{(1-\beta^2)^{3/2}} \qquad Y = \frac{m_0\gamma_y}{\sqrt{1-\beta^2}}$$

$$X = X_0 \qquad Y = Y_0\sqrt{1-\beta^2}$$

The Z component transforms like Y.

With the mass invariant, we write

$$X' = m_0\gamma_x, \quad Y' = m_0\gamma_y$$

This defines a new concept of force in K, having components

$$X' = X(1-\beta^2)^{3/2} = X_0(1-\beta^2)^{3/2}$$

$$Y' = Y\sqrt{1-\beta^2} = Y_0(1-\beta^2)$$

Energy and work in K must then be defined in such a way that an energy theorem is obtained giving the same phenomena as the ordinary notation, that is, the same increase in velocity during time dt (m_0 and F_0 being given). One of the definitions is therefore arbitrary.

In the general case, the transformation of force obviously yields formulae different from those of § [30]. For, consider two reference systems K_1 and K_2; βc is the velocity of K_2 in K_1, q_1 and q_2 are the velocities of a particle in K_1 and K_2 respectively. We have

$$X_2' = m_0\gamma_{2x} = m_0\gamma_{1x}\frac{(1-\beta^2)^{3/2}}{\left(1-\dfrac{vq_{1x}}{c^2}\right)^3}$$

or

$$X_2' = \frac{(1-\beta^2)^{3/2}}{\left(1-\frac{vq_{1x}}{c^2}\right)^3} X_1'$$

For the components normal to the relative velocity,

$$Y_2' = m_0\gamma_{2y} = \frac{m_0(1-\beta^2)}{\left(1-\frac{vq_{1x}}{c^2}\right)^2}\left(\gamma_{1y} + \frac{\frac{vq_{1y}}{c^2}}{1-\frac{vq_{1x}}{c^2}}\gamma_{1x}\right)$$

so that

$$Y_2' = \frac{1-\beta^2}{\left(1-\frac{vq_{1x}}{c^2}\right)^2}\left(Y_1' + \frac{\frac{vq_{1y}}{c^2}}{1-\frac{vq_{1x}}{c^2}}X_1'\right)$$

The formulae with the force invariant

Suppose that in K we have a new force concept, with components

$$X' = X_0 = X, \quad Y' = Y_0 = \frac{Y}{\sqrt{1-\beta^2}}$$

The equations of motion become

$$X' = \frac{m_0\gamma_x}{(1-\beta^2)^{3/2}}, \quad Y' = \frac{m_0\gamma_y}{1-\beta^2}$$

This is equivalent to taking the quantity measured by a dynamometer in K_0 as the force, irrespective of K.

This formalism is certainly seductive in theory, for only the measured quantities appear: the force $\mathbf{F}_0$ in K_0 and the velocity $\mathbf{v}$ in K; the formula gives γ which again is measurable.

Consider now two systems of references K_1 and K_2. For example, we have

$$X_2' = \frac{m_0\gamma_{2x}}{\left(1-\frac{q_2^2}{c^2}\right)^{3/2}}$$

$$= \frac{m_0}{\left(1-\frac{q_1^2}{c^2}\right)\frac{1-\beta^2}{(1-vq_{1x}/c^2)^2}} \frac{(1-\beta^2)^{3/2}}{\left(1-\frac{vq_{1x}}{c^2}\right)^3}\gamma_{1x}$$

so that

$$X_2' = \frac{\sqrt{1-\beta^2}\,X_1'}{\left(1-\frac{q_1^2}{c^2}\right)\left(1-\frac{vq_{1x}}{c^2}\right)}$$

Formalism based on measurement, in K, of the reading of a dynamometer in K_0

We measure force by means of a dynamometer fixed in K_0, that is, attached to the moving body. Let us now read the deformation on a measuring-rod fixed in K (instead of reading it on a scale fixed to the spring). Force then transforms like length

$$X' = X_0\sqrt{1-\beta^2}, \qquad Y' = Y_0$$

giving the equations

$$X' = \frac{m_0\gamma_x}{1-\beta^2}, \qquad Y' = \frac{m_0\gamma_y}{1-\beta^2}$$

Force and acceleration are parallel; the longitudinal and transverse masses are equal.

[110] The Coherent System† with a Single Fundamental Quantity, using the Procedure of § [107] (Absolutely Coherent System)

By a coherent system, I mean any system in which the definitions of the units do not introduce any arbitrary constants. I give these units, together with a comparison with the usual units.

(a) The fundamental quantity is time-interval, which we denote by the dimension T. The unit is arbitrary but once chosen, the units of all other physical quantities are fixed. Let us, for the moment, use the second.

The unit of length is the distance traversed *in vacuo* by a light signal during one second; this unit is therefore the second and its dimension is T. We thus write

$$l = ct$$

in which c is a constant which by the definition of a coherent system is taken equal to unity. In centimetres, this becomes $c = 3\times10^{10}$. Area has the dimension T^2, and the unit of area is equivalent to c^2 centimetres; volume has dimension T^3 and the unit is equivalent to c^3 centimetres. The unit of velocity is c cm/s and it is dimensionless. Acceleration has dimensions T^{-1} and its unit is c cm/s^2. Mass, defined by $r^2\gamma$, has dimension T and its unit is c^3 grammes. Force is dimensionless and the unit is c^4 dynes.

(b) *Practical remarks on the use of such a system.* The unit of time-interval would have to be chosen in such a way that for the other quantities, more convenient units were obtained. Or perhaps this is a somewhat Platonic desire. We cannot disregard the evidence: it is impossible to devise units which are convenient in all branches of physics. How can we hope to satisfy fundamental particle specialists, opticians and cosmologists simultaneously? Less exactingly, we accept that units are convenient if their multiples or sub-multiples—mega, micro, pico—are convenient.

If agreement were obtained on such a system, the names of the units could be eliminated altogether. We should simply say, for example, this rod is 25, or 25 mega, or 25 pico, long.

† See § [141].

Granted the theoretical worth of the coherent system with a single fundamental quantity (and it is *the system of the future*), it seems high time that the international commissions on metrology looked into the question.

(c) *Remark on dimensional analysis.* This adapts itself quite naturally to this new system. We notice, however, that the use of a single quantity would make its arguments far less effective. However, nothing would prevent us from calling the dimensions of force and surface F and S provisionally, in such arguments, if the phenomenon being studied involved force and surfaces. All this is very arbitrary, and is merely a question of convenience and effectiveness.

[111] Coherent Systems with Several Fundamental Quantities (Pseudo-coherent Systems)

(a) *Two fundamental quantities (time-interval, length)*

Let us retain time-interval as one fundamental quantity. Instead of defining length in terms of time, however, let us define it directly and solely by an operational method. For this, we choose as unit the length of some rod (the old definition in terms of a platinum rod), or again, some wavelength (the new definition in terms of krypton). To the concept of length we thus ascribe the same fundamental character as to that of time-interval, and we accord the former the dimension L. The relation

$$l = ct$$

now becomes a physical law, and the magnitude and dimension of the constant, c, depend upon the units chosen independently for l and t. We call a system coherent with two fundamental quantities when all other quantities are defined in terms of l and t without the introduction of arbitrary constants. I shall not dwell upon this system which, so far as I know, has never been employed. It would of course be possible to use other quantities than length and time-interval.

(b) *Three fundamental quantities (time-interval, length, mass)*

The mass is selected as the third quantity, and is defined purely operationally. This system has been and is still very widely employed, with its many variants (cgs, mks, . . .). Force has also been suggested as the third quantity, rather than mass (M, K_p, S).

(c) *Four fundamental quantities (time-interval, length, mass, the strength of an electric current)*

As we shall see later (§ [141]), to attribute a fundamental character to current or to electric charge is characteristic of the Giorgi system. We shall find that for deep-rooted theoretical reasons, *this system is an aberration.* From the present point of view, it is already clear that it is diametrically opposed to the really coherent system, with a single fundamental unit.

[112] Critical Scrutiny of Other Definitions of Mass and Force

(a) *Can the definition of the concept of energy precede those of mass and force?*

The concepts work and energy are usually defined in terms of those of force and mass. Some authors, and some teachers, feel that energy is a "primary concept", more basic than force. An examination of their work reveals that the statement is gratuitous, the consequence of a sentimental preference. For they give *no definition.* It is, for example, a vicious circle to say "I cannot define the general concept energy, but I shall give you some specific examples". In fact, the first example given is kinetic energy; if, however, mass is not used, we have merely a vague notion gleaned from everyday observations, with no discrimination between kinetic energy and momentum. I have myself tried to start with energy, but I did not get anywhere.

(b) *Is the concept of mass compatible with the inertia of the potential energy?* (See § [156])

(c) *Can we define mass as quantity of matter?*

Such a definition is of course meaningless if we have not defined "quantity of matter" beforehand. It is, however, employed by some authors, when they say "the weight of a body varies but the mass, which is a measure of the quantity of matter, is independent of position".

More serious is the suggestion that this quantity of matter should be defined, in theory at least, in terms of the number and nature of elementary particles of which the body is composed. Then, however, this concept is different from mass. Because of the equivalence of energy and mass, the mass of a body composed of identical interacting particles is not equal to the sum of the masses of the constituent particles. Furthermore, when the particles are different, the definition cannot be applied, for it assumes that some notion of mass common to all the particles has already been defined, and this is exactly the problem under consideration.

Finally, for the physicist, matter is that which possesses inertia (inertial mass), and the concept of mass is defined beforehand.

(d) *Can we state the principle of inertia (Newton's first law) before defining mass?*

Books on classical mechanics invariably begin by stating this principle in the form: an isolated particle remains in uniform rectilinear motion (or has zero acceleration).

I confess that I do not understand the physical meaning of this statement. If, by "an isolated particle", is meant a particle which is alone in the universe, then the term "isolated" is clear; but what does "motion" mean, for by hypothesis the particle is alone? If "isolated" really means "practically isolated", that is, the rest of the universe exerts a negligible influence on the particle, then I wonder how we are to verify that this influence is negligible. By the rectilinear and uniform nature of the motion? If so, then the principle is a mere definition. By measuring the forces? They have not been defined yet! Beneath its abstract and general appearance, this statement is either tautological or a vicious circle.

(e) *Can we give a quantum mechanical definition of mass?*

In the present work, we are considering only the macroscopic domain (pre-relativistically or relativistically); it is with this in mind that the definitions of force and mass that we have adopted must be understood.

In § [65] we saw that mass could be defined in terms of the equations of quantum mechanics. However, this interpretation of the parameter which is introduced requires force to be defined beforehand. More generally, we may define mass as the parameter which occurs in Schrödinger's equation

$$\nabla^2\Psi - \frac{8\pi^2 m}{h^2} V\Psi = \frac{4\pi i m}{h} \frac{\partial \Psi}{\partial t}$$

In the *Comptes rendus de l'Académie des Sciences de Paris* for 27 October 1941, on page 563, for example, I find the phrase "Considérons une particule de nom propre m_0, liée à un système de référence S ...", over the signature Roubaud-Valette.

However, this assumes that the potential function V has first been defined, which involves using the concept of force (see Appendix II).

[113] The Use of Non-Galileian Reference Systems

(a) *Pre-relativistic physics*

Let us consider the example of the rotating disc, which we take as system of reference. We place a heavy particle M of mass m at rest on a disc D; the latter rotates with constant angular velocity ω about its axis with respect to the laboratory, which is regarded as a Galileian system of reference K_0. If we assume that D is horizontal and that there are no frictional forces, M cannot remain at rest on D unless it is attached to a spring. In the system D, this spring behaves as a dynamometer, and its extension gives the static force $\mathbf{F}_s$; its magnitude here is

$$\mathbf{F}_s = m\mathbf{r}\omega^2$$

We now release the particle and refer its motion to the system K_0. The force $\mathbf{F}_0$ acting upon it is zero. We shall say that, by definition, the real or absolute force $\mathbf{F}_0$ is zero, and hence so also is the acceleration $\boldsymbol{\gamma}_0$. Let us now refer the motion to the system of the disc.

Let $\boldsymbol{\gamma}_r$ denote the acceleration with respect to D; it can no longer be calculated using the relation

$$\boldsymbol{\gamma}_r = \mathbf{F}/m$$

with $\mathbf{F}$ equal to $\mathbf{F}_0$ or $\mathbf{F}_s$. We know that now, we must write

$$m\boldsymbol{\gamma}_r = m\mathbf{r}\omega^2 + 2m\boldsymbol{\omega}\times\mathbf{v}_r$$

This relation can be interpreted in two very different fashions. We may wish to retain the simple relation

$$\gamma_r = \frac{\text{Force}}{m} = \frac{\mathbf{F}_r}{m}$$

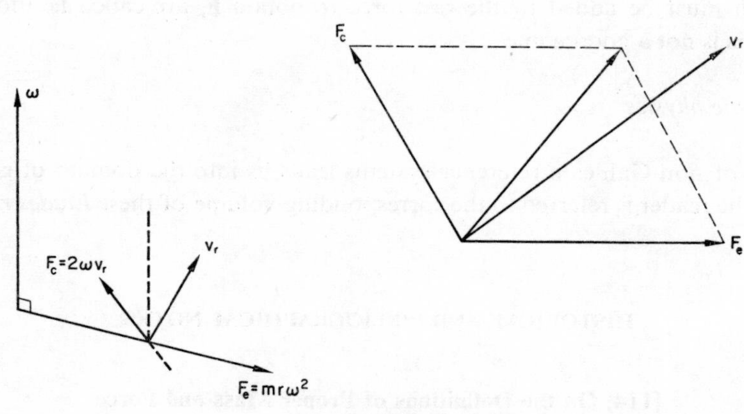

FIG. 16.

in the disc system, thus introducing fictitious forces (Fig. 16). When we use arbitrarily accelerated systems, two fictitious forces must be added to the real force $\mathbf{F}_0$ (Fig. 16):

—The inertial drag force

$$\mathbf{F}_e = m\mathbf{r}\omega^2$$

—The centrifugal force

$$\mathbf{F}_c = -2m\boldsymbol{\omega}\times\mathbf{v}_r$$

From this standpoint, we write in general

$$m\gamma_r = \mathbf{F}_0+\mathbf{F}_e+\mathbf{F}_c = \mathbf{F}_r$$

This is to say that the force $\mathbf{F}_r$ employed in the calculations is not the real force, but is deduced from $\mathbf{F}_0$ with the aid of a formula enabling us to transfer from a Galileian reference system K_0 to rotating systems.

In conclusion, therefore, we state two laws: the law of dynamics (which is the same as that for Galileian systems) and a law of transformation of force.

Alternatively, we could write

$$m(\gamma_r-\mathbf{r}\omega^2+2\boldsymbol{\omega}\times\mathbf{v}_r) = \mathbf{F}_0$$

where we have collected on to one side the cause of the motion (the force $\mathbf{F}_0$), and on the other, the characteristics of the motion. Force is then invariant, but we modify the law of dynamics.

Different though these two interpretations are, they lead to the same results. Just as in § [109], the idea of the force on a moving body in an arbitrary system of reference is open to several definitions. Only one force is well defined physically: the force $\mathbf{F}_0$ given by the extension of a spring which is extended by the moving body and produces the force (in the present case, this spring would replace the terrestrial attraction)—whence the names absolute force and real force. Nevertheless, we abandon these terms in favour of proper force or rest force, which are closer to experiment and possess none of the metaphysical overtones of the terms "absolute" and "real". In addition, the terms which must be added to the rest force to obtain $\mathbf{F}_r$ are called fictitious forces, although this is not a good name.

(b) *Relativistic physics*

The use of non-Galileian reference systems leads us into the domain of generalized relativity. The reader is referred to the corresponding volume of these *Etudes relativistes.*

HISTORICAL AND BIBLIOGRAPHICAL NOTES

[114] On the Definitions of Proper Mass and Force

(a) Since the essential part of such a bibliography is by nature pre-relativistic, I shall limit myself to a few general signposts, intended to stimulate curiosity. For a thorough study, the reader is referred to the books by Dugas [18, 21] and *in particular, by Jammer* [24, 25], *which I cannot praise too highly.*

(b) Euler [1] seems to have been the first to give exact definitions. He gives priority to the concept of force, and an analysis of his work reveals that he adopts the procedure of § [107] more or less explicitly. Also in favour of this point of view, let us mention the illustrious name of Maxwell [5]. Interesting discussion is to be found in Höfler [9, 10, 11] and Engelmeyer [8]; the latter even suggests the use of a system in which force is one of the three fundamental quantities.

(c) After d'Alembert, mechanics acquired the appearance of a purely mathematical science in the hands of many authors who claimed to use only obvious ideas and principles, or at least only those "of necessary and not contingent truth". The mathematicians have always felt a certain revulsion before the idea of force, however, which they find "metaphysical". On the contrary, they consider that the concepts of velocity and acceleration are clearer and satisfactory as primary data, *even though their origin is the same as that of the concept of force:* an intuitive quantity leads to a physical definition by a choice of standard and of measuring procedure. This fact of course escapes those thinkers *whose preoccupations are exclusively theoretical and who do not bother to specify how the quantities employed are to be measured.*

D'Alembert [2] tells us explicitly: "... sans vouloir discuter ici si ce principe est de vérité nécessaire ou contingente, nous nous contenterons de le prendre comme une définition, et d'entendre seulement par le mot de force accélératrice, la quantité à laquelle l'accroissement de vitesse est proportionnel."† The principle in question is the impulse equation. D'Alembert dismisses the forces as "êtres obscurs et métaphysiques, qui ne sont capables que de répandre les ténèbres sur une science claire par elle-même".‡ The majority of later authors adopt and aggravate this attitude.

L. Carnot also considered that force, used as a primary concept (in Euler's sense), is a "notion métaphysique et obscure". It is entertaining to observe that for these scholars, the terms "metaphysical" and "obscure" are synonymous. How say the metaphysicians?

Barré de Saint-Venant (whom Dugas quotes) cannot be too sarcastic about forces, these "sortes d'intermédiaires d'une nature occulte et métaphysique", and again "ces sortes d'êtres problématiques, ou plutôt d'adjectifs substantisés, qui ne sont ni matière ni esprit, êtres aveugles et inconscients et qu'il faut douer de la merveilleuse faculté d'apprécier les distances et d'y proportionner ponctuellement leurs in-

† "Without wishing to discuss here whether this principle is of necessary or contingent truth, we shall agree to take it as a definition, and by the term accelerating force we shall understand only the quantity to which the increase of velocity is proportional."

‡ "Obscure metaphysical entities, only capable of spreading confusion in a science in itself clear."

tensités".† Carried away by the balance of his prose, Saint-Venant does not notice that accelerations could be subjected to the same brand of irony (they too can appreciate distances ... in Newton's laws).

The work of Mach [4] and Kirchhoff [6] forms a clear milestone in this evolution of mechanics into a purely rational science; for them, too, the fundamental concepts are time, space and matter; force is a derived concept.

In his *Science et Hypothèse*, Poincaré [13] analyses the principles and concepts of mechanics at length, but I cannot reconcile myself to his conclusions. He tells us that "c'est par définition que la force est égale au produit de la masse par l'accélération". But what is mass? He examines the various ways of introducing the latter, and halts in particular at the gravitational mass, and then concludes that "les masses sont des coefficients qu'il est commode d'introduire dans les calculs". Later, he says: "On s'explique maintenant comment l'expérience a pu servir de base aux principes de la mécanique et cependant ne pourra jamais les contredire."‡ It is this point of view that has already been criticized in *Kinematics* (§ [37]).

In connection with a survey of the literature, Rocard once wrote: "Henri Poincaré, grand magicien des sciences, ... se promenait avec aisance au milieu des mirages de toutes sortes dont sont remplis ses livres, si passionnants pour l'amateur, si troublants pour le professionnel.§ This is certainly the impression that remains after reading Poincaré's physical and epistemological writings; and if we merely use the word "troublants" it is out of deference to Poincaré's name.

Let me not be misunderstood. I am full of deference to the formulae discovered by d'Alembert, Lagrange and their school. On the other hand, their ideas on epistemology, and especially on the definition of physical quantities, seem highly questionable; the reader can scarcely accuse me of *lèse-majesté* for this.

I shall not labour the "necessary" utility of the postulates of mechanics, and even of geometry. This idea has played an important role in enabling mechanics to be fitted into a remarkably axiomatic structure; there is, however, no one left to accept it.

(d) The definition of mass in terms of the number of particles was adopted at the end of the nineteenth century by various authors, in particular in the celebrated treatise of Hertz [7]. In practice, however, Hertz defined mass in terms of weight, which leads us back to the procedure of § [107].

Some Soviet theoreticians, Joffé [17] for example, ever careful to interpret physics in the light of dialectical materialism, define mass in terms of quantity of matter; this last term is not defined, however —they doubtless consider that it is common sense.

(e) Ostwald [12] claims priority for the concept of energy; according to him, mass must be defined as a function of energy by means of the relation

$$m = \frac{E}{\frac{1}{2}v^2}$$

This is all very well, but what about the definition of energy? Ostwald seems to be silent on this point; he implicitly regards this word as having an obvious meaning. This point of view is inadmissible even for the concepts of space and time, let alone that of energy.

(f) An axiomatic presentation of the postulates of dynamics has been attempted by various authors, for example McKinsey, Sugar and Suppes [20], Rubin and Suppes [22], Hermes [15], Pendse [16]. If their work is set out explicitly, I cannot see that they achieve any new physical ideas.

Remark. Finally, for the reader's amusement, this is what I find under the headings "force" and "mass" in the *Vocabulaire Philosophique* of Lalande [19].

---

† "Kinds of intermediaires, of occult metaphysical nature ... these kinds of problematical beings, or rather, adjectives turned into nouns, which are neither mind nor matter, blind oblivious beings to which we must ascribe the miraculous faculty of judging distances and fixing their intensities accordingly."

‡ "Force is equal to the product of mass and acceleration by definition ... The masses are coefficients which it is convenient to introduce into the calculations ... We can explain now how experiment was able to provide a basis for the principles of mechanics, but will, however, never be able to contradict them."

§ "Henri Poincaré, great magician of the sciences, ... strolled at his ease amid the mirages of every kind which fill his books, so compelling to the amateur, so disquieting to the professional."

"**Force:** ... En mécanique la définition usuelle de la force est celle-ci: étant admis que tout corps abandonné à lui-même persiste indéfiniment dans un mouvement rectiligne et uniforme (ou dans le repos qui peut être considéré comme un cas particulier), on appelle force tout ce qui peut modifier cet état de repos ou de mouvement rectiligne et uniforme. La force est égale au produit de la masse par l'accélération."

"**Masse:** ... Etant donné qu'une même force, appliquée à des corps différents, leur donne des accélérations inégales; et que, pour un même corps, les accélérations sont proportionnelles aux forces, on appelle masse d'un corps le rapport constant qui existe pour ce corps entre les forces qui sont appliquées et les accélérations correspondantes."†

Thus force is defined in terms of mass and mass of force! This reminds me of the pleasure I felt as a schoolboy when I found such definitions in the dictionary as *Pear:* fruit of the pear-tree; *Pear-tree:* the tree that produces pears.

I shall never accustom myself to the small importance which philosophers attach to the meanings of scientific terms. In these times of materialistic technology, it is ever more necessary to read and digest philosophical and metaphysical writings; a strong dose of patience and good temper is required to put up with and adapt them, however.

[1] L. EULER: *Mechanica sive motus scientia analytice exposita*. St. Petersburg, 1736. Published by Paul Stäckel, Leipzig and Berlin, 1912.

[2] J. D'ALEMBERT: *Traité de Dynamique*. Paris, 1743. The second edition, which appeared in 1758, has been republished by Gauthier-Villars in their Collection *Les Maîtres de la pensée scientifique*, 1921; Quotations, pp. xxvi and 26.

[3] BARRÉ DE SAINT-VENANT: *Principes de mécanique fondés sur la cinématique*. Paris, 1851.

[4] E. MACH: Ueber die Definition der Masse. *Carl's Repertorium der Experimentalphysik* **4** (1868) 355–359.

[5] J. C. MAXWELL: *Matter and Motion*. 1876. Republished by Dover (N.D.).

[6] G. KIRCHHOFF: *Vorlesungen über Mechanik*. Leipzig, 1876.

[7] A. HERTZ: *Die Prinzipien der Mechanik in neuem Zusammenhang dargestellt*. Leipzig, 1894. English translation by D. E. Jones and J. T. Walley: *The Principles of Mechanics Presented in a New Form*. London, 1899. Republished by Dover, 1956.

[8] CLÉMENTITCH DE ENGELMEYER: Sur l'origine sensorielle des grandeurs mécaniques. *Revue Philos. France* **39** (1895) 511–517.

[9] A. HÖFLER: Einige Bemerkungen über das C.S.G. System. *Z. Phys. Chem.* **11** (1898) 79.

[10] A. HÖFLER: *Kants Metaphysische Anfangsgründe der Naturwissenschaft*. Leipzig, 1900, p. 76.

[11] A. HÖFLER: Studien zur gegenwärtigen Philosophie der Mechanik. *Veröffentlichtungen der phil. Ges. Univ. Wien*. **3b** (1900).

[12] W. OSTWALD: *Vorlesungen über Naturphilosophie*, Leipzig, 1902.

[13] H. POINCARÉ: *Science et hypothèse*. 1910.

[14] CH. PLATRIER: La masse en cinématique et théorie des tenseurs du second ordre. *Act. Sci. et Ind.*, Hermann, 1936. Les axiomes de la Mécanique newtonienne. *Ibid.* In chapter III of the latter, I find that "la force ... n'est qu'une conception humaine et la cause profonde des mouvements nous est inconnue".‡ But what is *not* a mere "conception humaine" in science? To wish to eliminate physical concepts is, properly speaking, to engage in metaphysics. Is this the object of classical mechanics?

[15] H. HERMES: Eine Axiomatisierung der allgemeinen Mechanik. *Forschungen zur Logik und zur Grundlegung der exakte Wissenschaften*. Heft 3, Leipzig, 1938.

[16] C. G. PENDSE: A further note on the definition and determination of mass in Newtonian mechanics. *Phil. Mag.* **27** (1939) 55.

---

† "**Force:** ... In mechanics, the usual definition of force is as follows: granted that every body if left to itself will continue indefinitely in uniform rectilinear motion (or remain at rest, which may be regarded as a special case), we call force anything capable of modifying this state of rest or uniform rectilinear motion. Force is equal to the product of mass and acceleration.

"**Mass:** ... Given that if the same force is applied to different bodies, it will give them different accelerations, and given that for the same body, the accelerations are proportional to the forces, the mass of the body is the name given to the constant ratio, which exists for this body, of the applied forces to the corresponding accelerations."

‡ "Force ... is but a human concept, and the underlying cause of motion is unknown."

[17] A. F. JOFFÉ: *Basic Concepts in Contemporary Physics.* Translated from the Russian, Moscow, 1949.
[18] R. DUGAS: *Histoire de la Mécanique.* Dunod, 1950.
[19] LALANDE: *Vocabulaire Philosophique.* Presses universitaires, 1st ed., 1902–1923, 6th ed., 1951.
[20] J. C. C. MCKINSEY, A. C. SUGAR and P. SUPPES: Axiomatic foundations of classical particle mechanics. *J. Rat. Mechanics and Analysis* **2** (1953) 252–272.
[21] R. DUGAS: *La Mécanique au XVII siècle.* Dunod, 1954.
[22] H. RUBIN and P. SUPPES: Axioms of relativistic particle mechanics in "Transformations of systems of relativistic particle mechanics". *Pacific J. Math.* **4** (1954) 563–601.
[23] H. HERMES: Zur Axiomatisierung der Mechanik, in "The axiomatic method". *Proc. Intern. Symp.* North Holland, 1959, pp. 282–290.
[24] M. JAMMER: *Concepts of Force.* Harvard University Press, Cambridge, Mass., 1957.
[25] M. JAMMER: *Concepts of Mass.* Harvard University Press, Cambridge, Mass., 1961.

[114A] On the Idea of the Force of a Moving Body

The word force has frequently been employed to describe different concepts of force in the strict sense (static or accelerating force).

The scholars of the seventeenth century were long divided over the following problem: what is to be called the force of a moving body (not to be confused with the accelerating force acting on a moving body)? According to some, this force is proportional to the velocity, and to others, proportional to the square of the velocity. In fact, both opinions were correct. We have only to distinguish between two equally important characteristics of a body in motion: its momentum, mv and its kinetic energy, $\frac{1}{2}mv^2$. These quantities are measured by stopping the body with a constant force, for example, and measuring either the distance travelled or the time which elapses before it comes to a halt.

The term, the force of a body in motion, has been abandoned. It is to be regretted that impulse ($F\,dt$) and momentum (mv) are so often confused. The reader who desires more details about the historical arguments might read d'Alembert's *Dynamique* (ref. [2]) of § [114], see p. xxvi) and Bouasse's *Dynamique*.

Kant himself first went into print with a paper on this question (E. Kant: Pensées sur la véritable évaluation des forces vives, 1747) in which he attempts to reconcile the doctrines of Descartes and Leibnitz.

[115] On the Various Forms of the Equations of Relativistic Dynamics

Einstein [1] gives the equations with the force invariant of § [109], but his interpretation is very different. He specifies that the force employed is that of the system K_0, "measured with a spring-balance, at rest in this system". Nevertheless, he in no way assumes that the force is invariant, for in his memoir, we find the usual transformation formulae (§ [44]) of the Lorentz components. Einstein's equations are consequently those usually in fact adopted: it is strange that he should not have written them out explicitly, instead of retaining quantities measured in different systems. He does, moreover, say himself that it would be better to adopt Planck's point of view.

The idea that relativistic mechanics can be treated, regarding mass as a constant, is to be found in a *Note* by Dugas [5], together with observations by Lecornu.

Born [4] transforms the forces as lengths: this is the point of view of the last equations of § [109]. In seeking the transformation for the components of the acceleration, he forgets to transform the time, however, and thus obtains wrong formulae. Finally, in compensation, he obtains the usual formulae of dynamics. I draw attention to this fact not to criticize its author (who must long since have noticed his *lapsus calami)*, but to prevent the reader of Born's book, which is widely used, from being misled.

Esclangon [12–14] adopts the definition of the last equations of § [109] (Esclangon's equations are not exactly the same as those of § [109]). He is amazed that his equations do not coincide with "les formules classiques de la mécanique électronique",† but the reason for this disagreement, which is, as

† "The classical formulae of electron mechanics."

we have seen, purely formal, escapes him. He even speaks of "désaccord grave" and proposes that a distinction should be made between dynamometric forces (his definition) and field forces (the usual definition).

In fact, the important thing is to specify the physical problem under consideration carefully, and one is then at liberty to translate it into the terminology one prefers. Thus, for example, the problem of the motion of a particle acted upon by a constant force, as Esclangon treats it, does not correspond to the physical problem of the motion of an electron in a static electric field.

This question, incorrectly analysed on the epistemological planes naturally provides a warhorse for the non-relativistic authors; we mention, for example, Ritz [3], Le Roux [7], Prunier [8, 9, 15], Sevin [10] and Sivadjian [16]. These paper, are interesting from various points of view, but their criticisms do not affect the theory of relativity in any way.

[1] A. Einstein: Ref. [1] of Appendix I, p. 45.
[2] Ph. Frank: Die Stellung des Relativitäts-Prinzips im System der Mechanik und der Elektrodynamik. *Sitz. Akad. Wiss. Wien* **118** (1909) 373–446.
[3] W. Ritz: Recherches critiques sur l'électrodynamique générale. *Ann. Chim. Phys.* **13** (1908) 145–275.
[4] M. Born: *La Theorie de la relativité d'Einstein et ses bases physiques*. Gauthier-Villars, 1923, p. 258.
[5] R. Dugas: Sur le mouvement d'un point matériel de masse variable avec la force vive, soumis à une force centrale. *C. R. Acad. Sci. Paris* **178** (1924) 547–549.
[6] L. Lecornu: Observation sur la Communication précédente (same reference).
[7] J. Le Roux: La variation de la masse. *C. R. Acad. Sci. Paris* **180** (1925) 1470–1473.
[8] F. Prunier: *Newton, Maupertuis et Einstein. Réflexions à propos de la relativité*. Paris, 1929.
[9] F. Prunier: *Essai d'une physique de l'éther*. Blanchard, 1932.
[10] E. Sevin: *Le Temps absolu et l'espace à quatre dimensions*. Dunod, 1934, 127 pages. See, in particular, p. 39.
[11] E. Esclangon: La notion de temps: temps physique et relativité. *Bull. Astron.* **10**, fasc. I (1937) 1–72.
[12] E. Esclangon: *La Notion de temps*. Gauthier-Villars, 1938, 80 pages.
[13] E. Esclangon: Sur la définition de la force en relativité restreinte. *C. R. Acad. Sci. Paris* **208** (1939) 62–65.
[14] E. Esclangon: Sur les forces dynamométriques et les forces de champ. *C. R. Acad. Sci. Paris* **208** (1939) 685–689.
[15] F. Prunier: Quelques observations et expériences nouvelles ... *Arch. Sci.* Geneve, **1**, fasc. I (1948) 1–160.
[16] J. Sivadjian: Le coéfficient d'inertie (la masse) et le mouvement. *Arch. Sci.* Genève, **2**, fasc. I (1949) 61–86.

CHAPTER X

THE INTERACTION BETWEEN TWO ELECTRIC CHARGES. THE RELATIVISTIC RECONSTRUCTION OF ELECTROMAGNETISM

A. GENERAL REMARKS

[116] The Point of View Adopted

With the aid of the principle of relativity, it is often possible to obtain laws for systems in motion if we know the laws at rest (or for low velocities); we have used this procedure to deduce relativistic dynamics from Newtonian dynamics (§ [38]). We now apply it to electrical phenomena. We shall therefore set out from the laws of electrostatics (Coulomb's postulate) and deduce the laws of electromagnetism by purely mechanical methods. *The concepts of magnetic field and force will have the appearance of intermediate quantities, which relativistic methods allow us to eliminate if we so desire.* The result is a reduction of the number of fundamental concepts, and from the epistemological point of view, this constitutes one of the main advantages of the method.

Nevertheless, we must draw attention to one essential point from the outset: the starting-point will not be Coulomb's postulate properly speaking, but a statement which depends upon the latter but is better adapted to relativistic requirements. Of Coulomb's law, I retain only the Newtonian variation as $1/r^2$. I have of course a perfect right to assert any postulate I choose at the beginning. The present chapter may serve as an outline for a reconstruction of the whole of electromagnetism. As we shall see, two methods are possible. The first exploits fully the results of Chapters III and V; it requires a complete previous knowledge of relativistic dynamics. This point of view, wherein the vectors $\mathbf{E}$ and $\mathfrak{B}$ are regarded as purely mechanical quantities, has not yet become habitual. Meanwhile, we may use a second method, which is more direct and of which I give the essence below; the reader who desires a full text based on this method may consult my *Electricité relativiste*. Whatever the point of view adopted, we recognize that electromagnetism forms a special case of the fields with potentials, and hence a special case of Chapter V.

[117] The New Relativistic Form of Coulomb's Postulate

(a) *The Coulomb force*

Let A and B be two charged particles (Fig. 17), with which we associate two algebraic numbers ε_A and ε_B which we call the charge or electrical mass of the particles. We assume that particle A is stationary in a Galileian system of reference K_A, while *the particle B may*

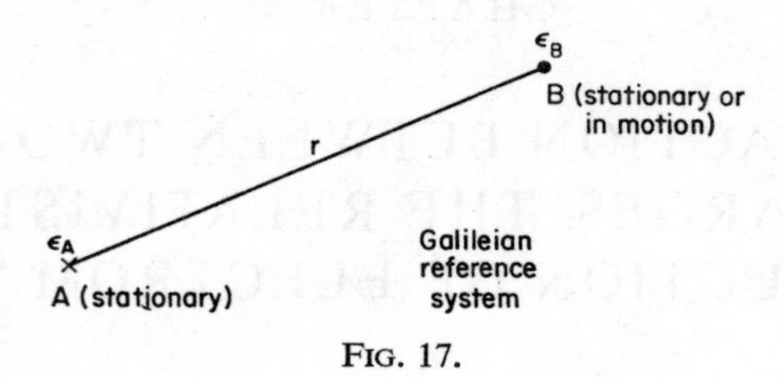

FIG. 17.

be either stationary or in motion. The action of A on B is a force, directed along $\overrightarrow{AB} = \mathbf{r}$ if the charges have the same sign (repulsive force), or along $\overrightarrow{BA} = -\mathbf{r}$ if they have opposite signs (attractive force); it is given by

$$\mathbf{F} = \frac{\varepsilon_A \varepsilon_B}{K_0} \frac{\mathbf{r}}{r^3} \tag{1}$$

K_0 is a constant of proportionality† (which is put in the denominator so that we later recover the usual expressions).

The postulate characterizes each particle individually by means of a number, ε, instead of considering the whole product $\varepsilon_A \varepsilon_B$. From this hypothesis, we extract the following consequences. Let us denote the action of A (stationary) on B (moving) by F_{AB}; first, therefore, we have

$$F_{AB} = \frac{\varepsilon_A \varepsilon_B}{K_0 r^2} = F_{BA}$$

The reader should notice that action and reaction are *not equal*: in fact, F_{AB} and F_{BA} denote the forces on B and A in different situations.

For A and some other particle C, with r unaltered,

$$F_{AC} = F_{CA} = \frac{\varepsilon_A \varepsilon_C}{K_0 r^2}$$

Hence, for B and C

$$F_{BC} = F_{CB} = \frac{\varepsilon_B \varepsilon_C}{K_0 r^2} = \frac{F_{AB} F_{AC}}{F_{AA}}$$

Important remark. Equation (1) is a *provisional* statement of the postulate, which is adequate only when the velocity and *acceleration* of the particle A vanish; we shall see later

† K_0 frequently also denotes a proper system of reference; no possible confusion can arise, however.

(§ [136A]) that the definitive statement, from which the whole of electromagnetism can be defined, brings in the potential and not the force. The statement above will suffice for the present; the general form is necessary only when retarded potentials are required.

(b) *The state of motion of the charge B is immaterial*

This point is essential for magnetostatics. Furthermore, the postulate tells us nothing about the action of B upon A; the equality of action and reaction is therefore eliminated from the law (we shall find that this is a relativistic property).

This form of the statement differs from the one adopted in the majority of electrostatics textbooks; it constitutes one of the special features of my standpoint *and I take full responsibility for it.*

All the books in fact assume that the two charges A and B are stationary (in what reference system is, moreover, not mentioned). Such a statement is inadequate. With it, we cannot even calculate the force acting on an electron B projected between the plates of a condenser.

(c) *Action at a distance; inclusion of the relativistic constant c*

In stating the postulate, we have implicitly accepted the idea of action at a distance. This idea, due to Newton, was considered inconsistent by many physicists: they could not accept that a body might act on another without contact or without action on the part of the intervening medium. This, however, is not the true meaning of the postulate. In reality, we specify as hypothesis the expression for the force, without defining what happens in the intervening medium; we are not concerned with the way in which the force is transmitted to some point, but only with the value it has at the point. So far as physical theory is concerned, this is just as legitimate. If, furthermore, the idea of contact seems clearer, this is uniquely because it arises from our everyday experience. In fact, however, only statements about action at a distance are axiomatically satisfactory; in the particle universe, there is never contact in the common meaning of the word. The Newtonian attitude thus constitutes a very considerable epistemological step forward.

We can make a similar comment upon instantaneous actions. There is no implication in the postulate that the action of A on B is instantaneous. On the contrary, the theory of relativity requires that the velocity at which this action is transmitted should not be greater than c; *we assert that it is equal to* c. If we put ε_A at A at time t, its action reaches B at time $t+r/c$. If, however, ε_A is stationary and independent of t, its magnitude and position are the same at times t and $t+r/c$; it is exactly as though the action were transmitted instantaneously. The two points of view yield different results when ε_A is moving; we shall see (§ [136A]) that if ε_A is accelerating, the hypothesis concerning c is essential. It alone is in agreement with experiment.

(d) *Limits of validity*

We shall regard particles as points. The inverse square law gives a force which increases indefinitely as r tends to zero, but Coulomb's law ceases to be valid experimentally for distances smaller than 10^{-13} cm in order of magnitude.

[118] Comparison with the Usual Methods; the Constants K_0, μ_0 and c_0

In modern textbooks on electricity, the law of Biot and Savart and the law of force on a moving charge are commonly employed. We shall show that these can be deduced from Coulomb's law with the aid of the Lorentz transformation. To simplify comparison of the usual method and the direct relativistic method, we recapitulate these formulae in the present notation.

A current element i ds produces at an arbitrary point M the magnetic field $\mathfrak{B}$ given by

$$\mathfrak{B} = \mu_0 i \, \mathrm{d}\mathbf{s} \times \frac{\mathbf{r}}{r^3}$$

in which $\mathbf{r}$ is the radius vector joining the element and the point M and μ_0 is a constant.

A *neutral* current element i ds placed in a magnetic field experiences a force

$$\mathbf{F} = i \, \mathrm{d}\mathbf{s} \times \mathfrak{B}$$

A charge travelling at velocity $\mathbf{v}$ may be regarded as a current element

$$i \, \mathrm{d}\mathbf{s} = \frac{\varepsilon \mathbf{v}}{c_0}$$

in which c_0 is a third arbitrary constant. The three constants K_0, μ_0 and c_0 which have been introduced to render all the systems of units easily accessible (§ [138]) are not independent. They are related by

$$c = \frac{c_0}{\sqrt{K_0 \mu_0}}$$

and this may be regarded as the definition of c_0.

These laws are applied to point charges, together with Coulomb's law for electrostatic interactions. We shall show that comparison of the direct relativistic method with this, the usual method, leads to the following conclusions:

—the usual method is valid to first order (in $\beta = v/c$) for calculating the interaction between point charges;

—its validity for current elements depends on the nature of the current; for currents of zero density, it is valid to second order.

I stress this, for the most varied and false ideas are still prevalent, even in standard texts.

B. THE FIRST METHOD, USING THE RESULTS OF CHAPTERS III AND V

[119] General Remarks; the Essence of the Method

(a) *Definition of the concept of the electromagnetic field created by a charge*

We use the notation of § [117], and situate ourselves in the reference system K_A of the charge A; the force on B is therefore

$$\mathbf{F} = \varepsilon_B \frac{\varepsilon_A}{K_0 r^2} \frac{\mathbf{r}}{r}$$

This force $\mathbf{F}$ does not depend upon the velocity of B. We may write

$$\mathbf{F} = \varepsilon_B \mathbf{E}_A, \qquad \mathbf{E}_A = \frac{\varepsilon_A}{K_0 r^2} \frac{\mathbf{r}}{r}$$

Comparison with the formulae of § [43] shows that, in K_A, there is no component $\mathfrak{B}$. Furthermore, the coefficient ε is, in the present case, recognized to be the electric charge. If we take another reference system K, we have the general formula

$$\mathbf{F} = \varepsilon_B \left(\mathbf{E} + \frac{\mathbf{v}}{c_0} \times \mathfrak{B} \right)$$

where $\mathbf{E}$ and $\mathfrak{B}$ are vectors, deduced from $\mathbf{E}_A$ with the aid of the transformations of § [44]. I remind the reader that the parameter ε is treated as an invariant *by convention.* (It *is* no more than a convention: it is not a physical law, as some would have us believe.) By definition, $\mathbf{E}$ and $\mathfrak{B}$ are the components of the electromagnetic field created by the charge ε_A in the reference system K; $\mathbf{E}$ is the electric field and $\mathfrak{B}$ is the magnetic field.

(b) We shall later calculate these fields $\mathbf{E}$ and $\mathfrak{B}$, when they are produced by a single particle. It will be more convenient to change the notation. We shall denote the charge, the field of which we wish to calculate, by ε; the reference system in which this charge is at rest by K_0; and every other system of reference by K. Because of the limitation on the validity of Coulomb's law, we shall denote the radial distance beyond which we are confident of the correctness of the law by a; it is often said that a is the radius of the charge, but this is an abuse of the language. The calculations will also be valid for a macroscopic spherical distribution, a charged metal sphere for example. In this case, the restrictions on validity do not interfere, and a denotes the radius of the sphere.

[120] The Electric Field of a Charge in Uniform Motion

In the system K_0 in which the charge A is stationary, the field at every point is simply the electric field (§ [119])

$$E_{0x} = \frac{\varepsilon x_0}{K_0 r_0^3}, \qquad E_{0y} = \frac{\varepsilon y_0}{K_0 r_0^3}, \qquad E_{0z} = \frac{\varepsilon z_0}{K_0 r_0^3}$$

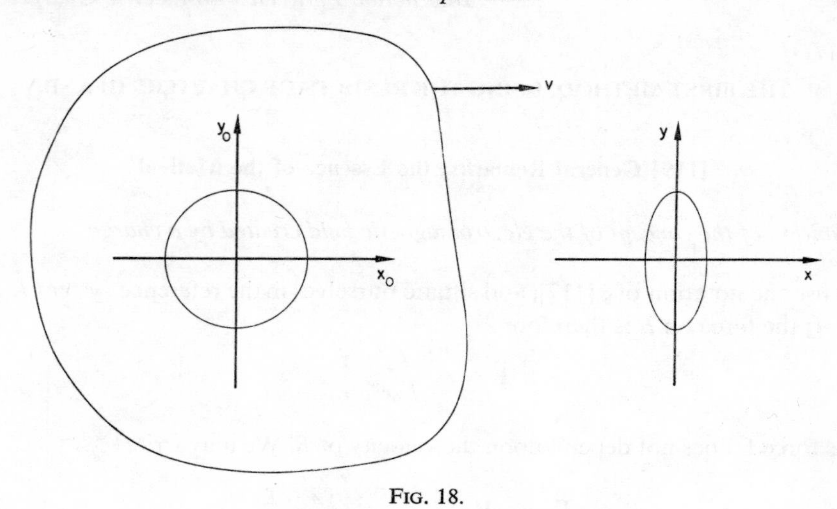

FIG. 18.

We suppose that the centre of A is placed at the origin of coordinates, O_0 (Fig. 18); r_0 denotes the distance from the origin to the point at which the field is to be calculated:

$$r_0^2 = x_0^2+y_0^2+z_0^2$$

The field lines, given by the equations

$$\frac{dx_0}{E_{0x}} = \frac{dy_0}{E_{0y}} = \frac{dz_0}{E_{0z}}$$

are here radial straight lines. In the system K_0, the field and field lines are independent of time.

Let us consider another Galileian reference system with respect to which the charge has velocity v parallel to O_x. We make an observation at time t of K, that is, *simultaneously in* K. To each point $xyzt$ of K, there corresponds a point $x_0y_0z_0t_0$ of K_0, where t_0 depends upon the point considered. *There is no simultaneity in* K_0, but this does not matter, since the field in K_0 is independent of t_0. The shape of the charge becomes an ellipsoid of revolution in K, flattened in the sense of the velocity v (by the Lorentz contraction), with semi-axes

$$a, \quad a\sqrt{1-\beta^2}$$

In the limiting case $v = c$, the charge becomes a plane disc, of zero thickness and radius a. The components of the electric field at an arbitrary point are given by

$$E_x = E_{0x} = \frac{\varepsilon x_0}{K_0(x_0^2+y_0^2+z_0^2)^{3/2}} = \frac{\varepsilon}{K_0}\,\frac{\dfrac{x-vt}{\sqrt{1-\beta^2}}}{\left\{\left(\dfrac{x-vt}{\sqrt{1-\beta^2}}\right)^2+y^2+z^2\right\}^{3/2}}$$

$$E_y = \frac{E_{0y}}{\sqrt{1-\beta^2}} = \frac{\varepsilon}{K_0\sqrt{1-\beta^2}}\,\frac{y}{\left\{\left(\dfrac{x-vt}{\sqrt{1-\beta^2}}\right)^2+y^2+z^2\right\}^{3/2}}$$

E_z transforms like E_y. The coordinates x, y, z are independent of t, and so the field only depends upon time through the explicit presence of the symbol t. We shall simplify these formulae, but we must return to the complete expressions whenever the time variable has to be introduced. *We shall perform the calculations on the assumption that the charge lies at the origin O at time zero*. The results for an arbitrary time, that is, for an arbitrary position of A, are deduced by a simple displacement of the origin. We therefore write

$$E_x = \frac{\varepsilon}{K_0} \frac{(1-\beta^2)x}{\{x^2+(1-\beta^2)(y^2+z^2)\}^{3/2}}$$

$$E_y = \frac{\varepsilon}{K_0} \frac{(1-\beta^2)y}{\{x^2+(1-\beta^2)(y^2+z^2)\}^{3/2}}$$

We should obviously obtain the same expressions for the field created at the origin by a charge placed at $(-x, -y, -z)$.

More simply

$$E_x = \frac{\varepsilon}{K_0} \frac{(1-\beta^2)x}{s^3}, \ldots$$

where

$$s = \sqrt{x^2+(1-\beta^2)(y^2+z^2)}$$

We denote the vector with components x, y, z by $\mathbf{r}$, giving

$$\mathbf{E} = \frac{\varepsilon}{K_0} \frac{1-\beta^2}{s^3} \mathbf{r}$$

If α denotes the angle between the radius vector and the velocity, we may write the resulting field in the form

$$E = \frac{\varepsilon}{K_0} \frac{1-\beta^2}{r^2(1-\beta^2 \sin^2 \alpha)^{3/2}}$$

Just as in K_0, the electric field in K is directed along the straight line joining the charge to the point at which the field is calculated. It is odd that the sense of the velocity v is immaterial. The general equations for the field lines,

$$\frac{dx}{E_x} = \frac{dy}{E_y} = \frac{dz}{E_z},$$

in which the expressions for F_x, ... are substituted, confirm that these lines are radial straight lines, as in the system K_0.

A diagram makes all this far more easily comprehensible. We display the results for the two reference systems by making the two systems of axes coincide (only the plane xOy is shown in Fig. 19). A given point in space will be represented by two points, M_0 and M, corresponding to the systems K_0 and K. Thus the point at which the axis Ox intersects the surface of the charge will be denoted by A_0 and A; the Lorentz transformation takes us from one to the other.

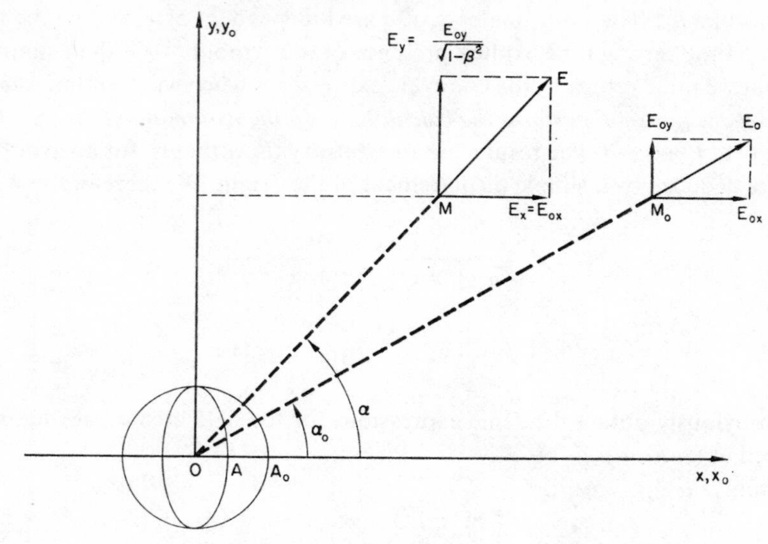

FIG. 19.

Let us plot the field at a point M_0 with respect to the system K_0. With respect to K, this point is transferred to M. The components of the new electric field are

$$E_{0x}, \qquad \frac{E_{0y}}{\sqrt{1-\beta^2}}$$

We can verify that the field directions are indeed OM_0 and OM. In fact

$$\tan \alpha_0 = \frac{y_0}{x_0} = \frac{E_{0y}}{E_{0x}}$$

$$\tan \alpha = \frac{y}{x} = \frac{y}{x_0\sqrt{1-\beta^2}} = \frac{\tan \alpha_0}{\sqrt{1-\beta^2}}$$

$$= \frac{E_{0y}}{\sqrt{1-\beta^2}E_{0x}} = \frac{E_y}{E_x}$$

The pattern of field lines in the system K is thus deduced graphically from the pattern in K_0 by means of the Lorentz contraction.

[121] The Magnetic Field of a Charge in Uniform Motion

In the system K_0, this field is zero. In K, it has components

$$\mathcal{B}_x = 0, \quad \mathcal{B}_y = -\frac{c_0\beta}{c}\frac{E_{0z}}{\sqrt{1-\beta^2}}, \quad \mathcal{B}_z = \frac{c_0\beta}{c}\frac{E_{0y}}{\sqrt{1-\beta^2}}$$

Comparison with the expression for **E** as a function of $\mathbf{E}_0$ shows that $\mathfrak{B}$ *is proportional to* *the vector product of the velocity and electric field vectors:*

$$\mathfrak{B} = \frac{c_0}{c^2}\,\mathbf{v}\times\mathbf{E}$$

In terms of the variables of the system K, the components may be written

$$\mathcal{B}_y = -\frac{c_0\varepsilon\beta}{K_0 c}\,\frac{(1-\beta^2)z}{s^3}, \qquad \mathcal{B}_z = \frac{c_0\varepsilon\beta}{K_0 c}\,\frac{(1-\beta^2)y}{s^3}$$

and the magnitude of the resultant is

$$\mathcal{B} = \frac{c_0\varepsilon\beta}{K_0 c}\,\frac{(1-\beta^2)\sin\alpha}{r^2(1-\beta^2\sin^2\alpha)^{3/2}}$$

The field lines of $\mathfrak{B}$ *are circles lying in a plane normal to the velocity* and centred on the direction Ox of this velocity. The field $\mathfrak{B}$ at a point M is therefore *normal to the plane containing the velocity* $\mathbf{v}$ *and the line* OM. Its direction is given by Maxwell's left-hand rule. Irrespective of $\mathbf{v}$, therefore, the orientation of $\mathfrak{B}$ is given by the familiar elementary rules.

Only when v *is negligibly small in comparison with* c *does the quantity* $\mathfrak{B}$ *reduce to the quantity given by the Biot and Savart law, however.*

[122] The Ultra-relativistic Case: Velocities Close to c

Let us now investigate the limiting case, $v = c$. We have already seen that the charge is flattened into a disc of zero thickness, lying in the plane yOz.

The component E_x is zero everywhere. The components E_y and E_z are only non-zero in the plane yOz, normal to the velocity and passing through the charge. The field is

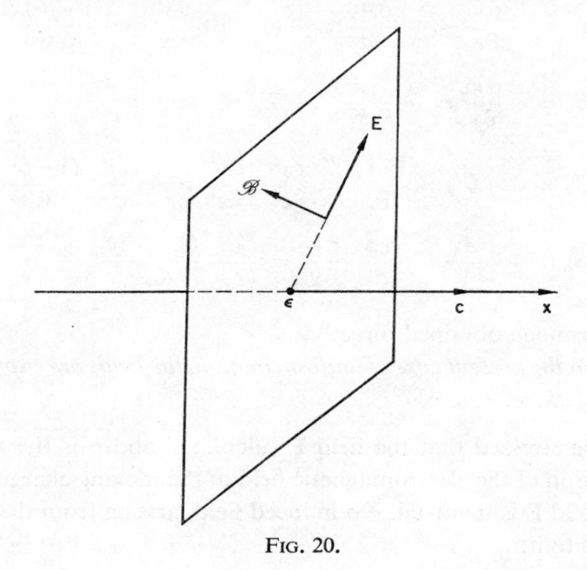

FIG. 20.

therefore "concentrated" in this plane (Fig. 20). It can be shown that all the lines of force lie in this plane, and that the transformed equipotential surfaces are so flattened as to lie in this plane.

The field $\mathfrak{B}$ has analogous properties, with the relation

$$\mathfrak{B} = \frac{c_0}{c^2}\,\mathbf{c}\times\mathbf{E}$$

[123] The Potentials of a Charge in Uniform Motion

In the rest system K_0, the Coulomb field is derived from the scalar potential

$$V(x_0, y_0, z_0) = \frac{\varepsilon}{K_0 r_0} = \frac{\varepsilon}{K_0\sqrt{x_0^2+y_0^2+z_0^2}}$$

The components of the electric field are then given by

$$\mathbf{E}_0 = -\text{grad } V_0$$

We can then apply the theory of fields, given in Chapter V. In the system K, we shall have scalar and vector potentials (§ [69])

$$V = \frac{V_0}{\sqrt{1-\beta^2}} = \frac{\varepsilon}{K_0\sqrt{(x-vt)^2+(1-\beta^2)(y^2+z^2)}}$$

$$A_x = \frac{c_0\beta}{c}\,\frac{V_0}{\sqrt{1-\beta^2}} = \frac{c_0\beta}{c}V, \qquad A_y = A_z = 0$$

For the field, therefore, we find

$$E_x = -\frac{\partial V}{\partial x} - \frac{1}{c_0}\frac{\partial A_x}{\partial t} = -(1-\beta^2)\frac{\partial V}{\partial x} = \frac{\varepsilon(1-\beta^2)x}{K_0 s^3}$$

$$E_y = -\frac{\partial V}{\partial y} = \frac{\varepsilon(1-\beta^2)y}{K_0 s^3}, \qquad E_z = \ldots$$

$$\mathcal{B}_x = 0 \quad \mathcal{B}_y = \frac{\partial A_x}{\partial z} = \frac{c_0\beta}{c}\frac{\partial V}{\partial z} = -\frac{c_0\beta}{c}\frac{(1-\beta^2)z}{K_0 s^3}$$

$$\mathcal{B}_z = -\frac{\partial A_x}{\partial y} = \frac{c_0\beta}{c}\frac{\partial V}{\partial y} = \frac{c_0\beta}{c}\frac{(1-\beta^2)y}{K_0 s^3}$$

We recover the formulae obtained directly.

We notice that in the present case of uniform motion, the fields are expressed as functions of a scalar potential alone.

Remark. It is to be stressed that the field **E** calculated above is the *total electric field.* If we derive the action of the electromagnetic field of the moving charge on the stationary charge, only this field **E** is involved. No induced field, arising from the time variation of $\mathfrak{B}$, has to be added to it.

[124] Surfaces of Equal Phase and Equipotential Surfaces†

In the proper system of reference of the charge, K_0, the surfaces of equal phase and the equipotential surfaces are spheres centred on the charge (Fig. 21). In K, the formulae of § [123] show that the equipotential surfaces are ellipsoids, centred at each instant of time t on the charge (Fig. 22(b)). The surfaces of equal phase at the same instant t are illustrated in Fig. 22(a). This means that when the charge is at O (time t), the sphere O_1

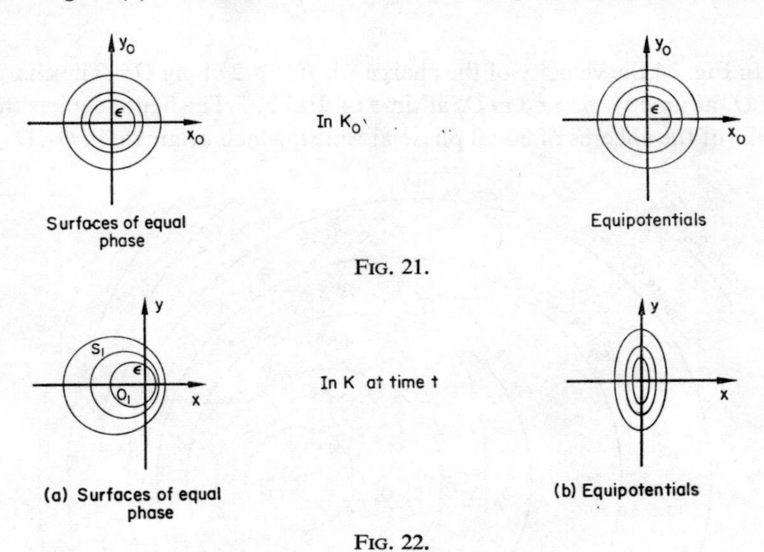

FIG. 21.

FIG. 22.

is reached by the action which was transmitted from an earlier position O_1 (with $r_1 = (c/v)\overline{OO_1}$). If the action were transmitted like a wave in a medium, the spheres would be equipotential surfaces and the two figures would be incompatible. Here, however, the action which begins at O_1 is not transmitted isotropically, as it would be for a medium stimulated at O_1. The spheres are not equipotentials because of the coefficient $1+v_r/c$ which appears in the Liénard formulae. If the equipotential points of Fig. 22(a) are joined (Fig. 23), the ellipsoids of Fig. 22(b) are obtained. This can be understood qualitatively.

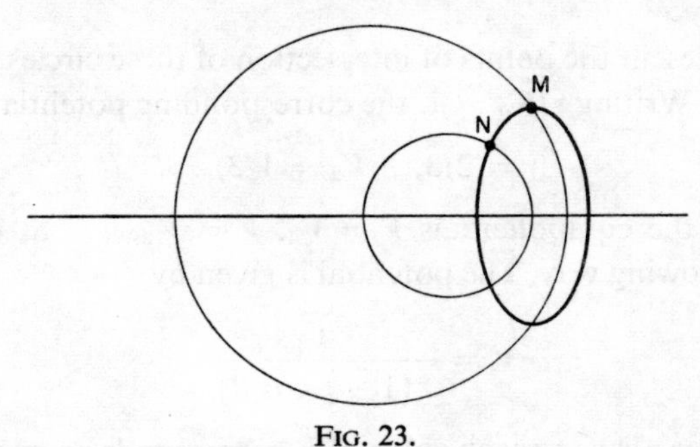

FIG. 23.

† § [136A] should be read before this section.

In passing from M to N, r is reduced but θ is increased, whence the possibility of compensation.

Remark 1. A statement in this connection in the French edition of 1957 requires correction. On p. 206, l. 3, the words "ou même onde lumineuse ou hertzienne" must be deleted. They would be acceptable if only uniform motion were considered; this would give rise to an insufficiently general point of view (to which I restricted myself in 1957), which would be incompatible with the form I have adopted for the fundamental postulate (§ [136A]).

Remark 2. In Fig. 24 the velocity of the charge ε is $v = c/2$ along Ox. The charge is at M at time t, at O_1 at time $t - t/5$ and at O_2 at time $t - 2t/5$ .... The figure depicts the circular cross-sections of the spheres of equal phase at time t, which originate in $O_1, O_2, \ldots$. The

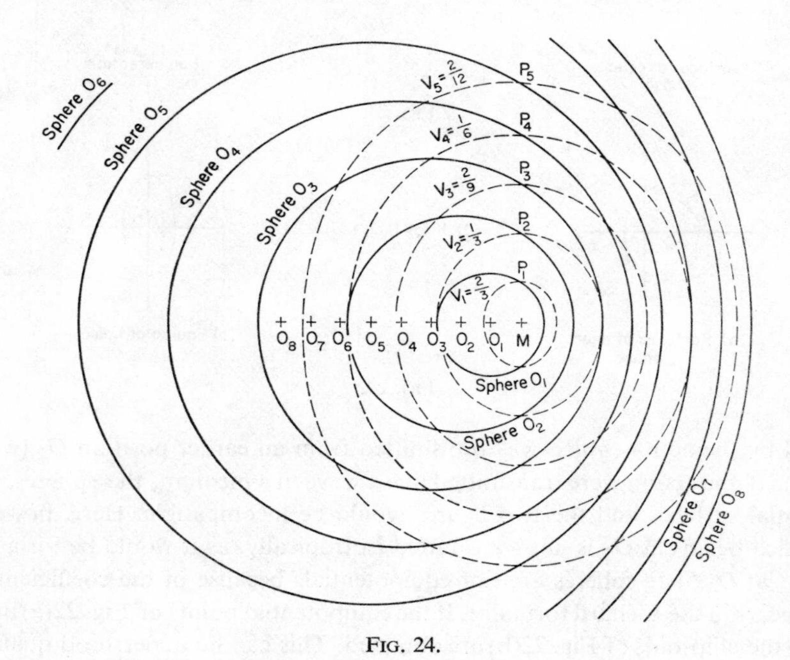

FIG. 24.

points $P_1, P_2, \ldots$ represent the points of intersection of these circles and the line through M perpendicular to $\mathbf{v}$. Writing $\varepsilon/K_0 = 1$, the corresponding potentials are

$$V_1 = 2/3, \quad V_2 = 1/3, \ldots$$

The figure illustrates the equipotentials $V = V_1$, $V = V_2$, ... at time t. They can be constructed in the following way. The potential is given by

$$V = \frac{1}{r(1 - \frac{1}{2}\cos\theta)}$$

we select a circle, radius r say: to each value of V corresponds an angle θ, and hence, two points on the circle.

[125] The Force Exerted on a Charge ε_2 (at B) by a Charge ε_1 (at A)

(a) *The velocities of the two charges are arbitrary*

Let u be the velocity of B and v be that of A; we take the Ox axis parallel to v. Then

$$X = \varepsilon_2\left(E_x + \frac{u_y}{c_0}\mathcal{B}_z\right) = \varepsilon_2\left(E_x + \frac{vu_y}{c^2}E_y\right)$$

$$= \frac{\varepsilon_1\varepsilon_2}{K_0 s^3}(1-\beta^2)\left(x + \frac{vu_y}{c^2}y\right)$$

$$Y = \varepsilon_2\left(E_y - \frac{u_x\mathcal{B}_z}{c_0}\right) = \varepsilon_2\left(1 - \frac{vu_x}{c^2}\right)E_y$$

$$= \frac{\varepsilon_1\varepsilon_2}{K_0 s^3}(1-\beta^2)\left(1 - \frac{vu_x}{c^2}\right)y$$

(b) *The charge B is stationary*

The force is simply given by

$$\mathbf{F} = \varepsilon_2\mathbf{E}$$

(c) *The velocities of A and B are the same*

It is necessary to examine this special case here, in order to clarify the concept of dynamic potential, the use of which can lead to confusion.

In K the components of the force exerted on B are

$$F_x = F_{0x} = \varepsilon_2 E_{0x} = -\varepsilon_2\frac{\partial V_0}{\partial x_0}$$

$$F_y = \sqrt{1-\beta^2}F_{0y} = \varepsilon_2\sqrt{1-\beta^2}E_{0y} = -\varepsilon_2\sqrt{1-\beta^2}\frac{\partial V_0}{\partial y_0}$$

We now express the derivatives in terms of the K variables:

$$V_0 = \frac{\varepsilon_1}{K_0}\frac{1}{\sqrt{x_0^2+y_0^2+z_0^2}} = \frac{\varepsilon_1}{K_0}\frac{\sqrt{1-\beta^2}}{s}$$

$$\frac{\partial V_0}{\partial x_0} = \frac{\partial V_0}{\partial x}\frac{\partial x}{\partial x_0} = \sqrt{1-\beta^2}\frac{\partial V_0}{\partial x}$$

The components of the force are therefore

$$F_x = -\varepsilon_2\sqrt{1-\beta^2}\frac{\partial V_0}{\partial x}$$

$$F_y = -\varepsilon_2\sqrt{1-\beta^2}\frac{\partial V_0}{\partial y}$$

This force has potential

$$V_F(x, y, z) = \sqrt{1-\beta^2}V_0(x, y, z)$$

and by various authors, von Laue for example, this is called the dynamic potential. The corresponding equipotential ellipsoids are the transformed surfaces of the equipotential spheres of K_0. The force is normal to these ellipsoids. If we consider a charged sphere in motion, the boundary surface in K is a flattened ellipsoid. We see that the charge distribution is indeed in equilibrium, since the force is normal to the limiting surface.

It is, however, essential to realize that the dynamic potential only yields the force in the present special case. We draw attention too to a possible source of confusion. It might be thought that, since this function is the transform of $V_0(x_0, y_0, z_0)$, the field **E** can be deduced from it. This is obviously not true, for V_F is not the same as the electric potential in K.

Remark. We can also calculate the force **F** using the Lorentz formulae; making use of the fact that **u** and **v** are equal, we have

$$F_y = \varepsilon_2\left(E_y - \frac{v}{c_0}\mathcal{B}_z\right) = \varepsilon_2(1-\beta^2)E_y$$

$$F_x = \varepsilon_2 E_x$$

[126] Application of the Foregoing Remarks to Current Elements

(a) *The position of the problem*

We consider a conducting medium in the form of a wire, stationary in K, and we suppose that it carries a current. We picture the situation as moving charges sliding without friction along the interior of a stationary tube. Crude though this model may be, it leads to a good interpretation of the ordinary experimental results. Furthermore, it is no cruder than the common model which leads to Ohm's law. The problem is now to determine the interaction between such tubes (this interaction can be measured with a dynamometer). We can say that the force F which the charges exert on one another is transmitted to the tubes (no friction) and that the dynamometer will then indicate F. In fact the charges strike the walls and tend to transfer momentum to the tube. We shall see in Part Three that the conservation of momentum is retained in relativistic dynamics, if the theory is used in its ordinary form. Hence the product $F\,dt$ is conserved, and the dynamometer does measure F.

(b) *Metallic conductors (Ohm's law satisfied)*

The nature of the conduction current. Present-day ideas on metallic conductors lead us to accept that current corresponds to the displacement of negative charges (electrons). In addition, there are fixed positive charges distributed in such a way as to preserve the overall neutrality of the medium. The conductor behaves merely as a support or guide for the current; its substance does not intervene, which enables us to consider this question here.

The formula for the interaction. Figure 25 represents two of the schematic tubes (dotted lines), and a pair of charges in each tube. Let us calculate the force on the pair B_1,B_2, for example. The charge A_1 produces a purely electric field at B, leading to equal and opposite forces on B_1 and B_2; these forces therefore cancel out.

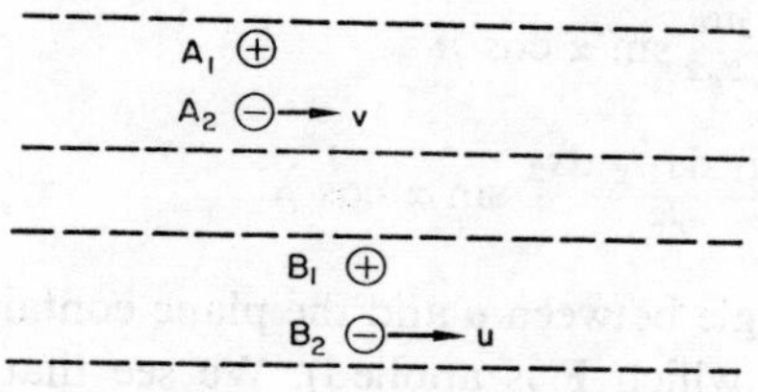

FIG. 25.

The action of the moving charge A_2 on the moving charge B_2 is given by the formula of § [125]:

$$X = \frac{\varepsilon_1 \varepsilon_2}{K_0 s^3}(1-\beta^2)\left(x + \frac{vu_y}{c^2}y\right)$$

$$Y = \frac{\varepsilon_1 \varepsilon_2}{K_0 s^3}(1-\beta^2)\left(1 - \frac{vu_x}{c^2}\right)y$$

The action of A_2 on stationary B_1 is obtained from these formulae by reversing the sign of ε_2 and setting u equal to zero.

The total force exerted by the element A on the element B therefore has components

$$X = \frac{\varepsilon_1 \varepsilon_2}{K_0} \frac{1-\beta^2}{s^3} \frac{vu_y}{c^2} y$$

$$Y = -\frac{\varepsilon_1 \varepsilon_2}{K_0} \frac{1-\beta^2}{s^3} \frac{vu_x}{c^2} y$$

In the general case, X and Y cannot be expressed solely as functions of the current elements; the velocity v appears explicitly. If v is small enough for β to be negligible in comparison with unity,

$$X = \frac{\varepsilon_1 \varepsilon_2}{K_0 r^3} \frac{vu_y}{c^2} y, \qquad Y = -\frac{\varepsilon_1 \varepsilon_2}{K_0 r^3} \frac{vu_x}{c^2} y$$

These formulae can obviously be obtained by the ordinary method (§ [118]).

In conclusion, therefore, these are the correct formulae for conductors when the velocities are known to be effectively low.

Within these media, the elementary charges possess random motion (thermal agitation), at velocities of the order of a *kilometre per second.* When the average of these velocities possesses a non-zero component in some direction, a macroscopic current flows. This component is far smaller than the real instantaneous velocity, and is in general smaller than a *millimetre per second.*

The Ampère–Grassmann–Reynard formula for low velocities. The preceding approximate formulae give a force **F** of magnitude

$$F = \frac{\varepsilon_1\varepsilon_2}{K_0 r^3}\,\frac{vy}{c^2}\sqrt{u_x^2+u_y^2} = \frac{\varepsilon_1}{K_0}\,\frac{u}{c^2}\,\frac{\varepsilon_2 v}{r^2}\,\frac{y}{r}\sqrt{1-\frac{u_z^2}{u^2}}$$

$$= \frac{\varepsilon_1\varepsilon_2 uv}{K_0 c^2 r^2}\sin\alpha\cos\mu$$

$$= \frac{\mu_0 i_1\,\mathrm{d}s_1 i_2\,\mathrm{d}s_2}{r^2}\sin\alpha\cos\mu$$

in which μ denotes the angle between **u** and the plane containing **r** and **v** (**u** denotes the velocity of the charge to which **F** is applied). We see that only the component of **u** lying in the plane containing **r** and **v** is involved; the force **F** is zero if **u** is normal to this plane.

A special case

Suppose that we have the arrangement illustrated in Fig. 26. The element $\mathrm{d}s_1$ is placed at O_1 in the plane yOz and is inclined at angle θ_1 to Oy; the element $\mathrm{d}s_2$ is at O_2 in the plane xOy and makes an angle θ_2 with Oy. Figure 26 shows the fields $\mathcal{B}_1$ and $\mathcal{B}_2$ at O_1 and O_2 and the forces exerted on $\mathrm{d}s_1$ and $\mathrm{d}s_2$. The following values are obtained:

$$\mathcal{B}_1 = \frac{\mu_0 i_2\,\mathrm{d}s_2}{r^2}\sin\theta_2, \quad F_1 = \mu_0\frac{i_1 i_2\,\mathrm{d}s_1\,\mathrm{d}s_2}{r^2}\sin\theta_2\cos\theta_1$$

$$\mathcal{B}_2 = \frac{\mu_0 i_1\,\mathrm{d}s_1}{r^2}\sin\theta_1, \quad F_2 = \mu_0\frac{i_1 i_2\,\mathrm{d}s_1\,\mathrm{d}s_2}{r^2}\sin\theta_1\cos\theta_2$$

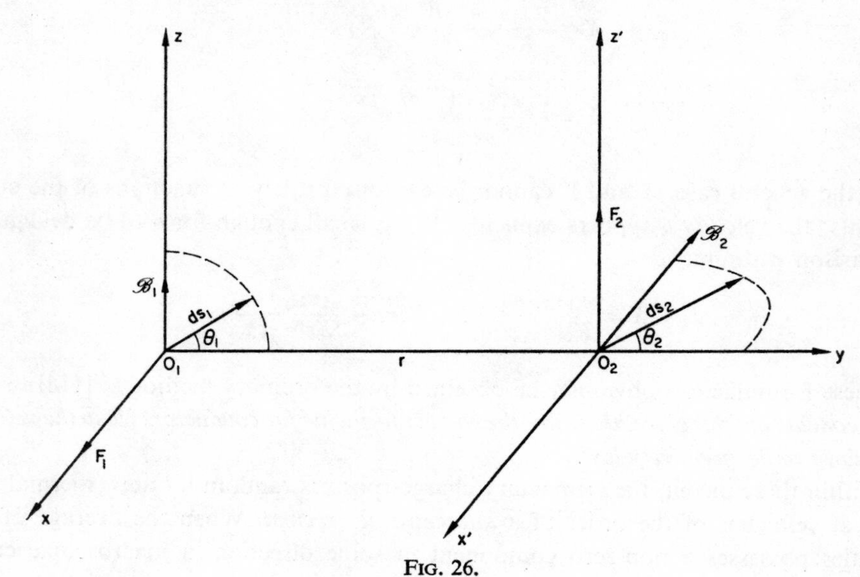

FIG. 26.

F_1 and F_2 are perpendicular and unequal. The element ds_1 can act on ds_2 without ds_2 acting on ds_1: we have only to choose θ_1 equal to 90°.

The field of a current element

The resulting electric field for a fixed charge $+\varepsilon$ and a moving charge $-\varepsilon$ travelling with velocity v parallel to Ox is

$$E_x = -\frac{\varepsilon}{K_0}\frac{(1-\beta^2)x}{s^3}+\frac{\varepsilon}{K_0}\frac{x}{r^3}$$

At low velocities, the electric field is zero; this is not the case for arbitrary velocities, however. The same is true of the electric potential V.

The components of the field $\mathfrak{B}$ are those given in § [121]. Vectorially

$$\mathfrak{B} = \frac{1-\beta^2}{s^3}\mu_0 i\,\mathbf{dl}\times\mathbf{r}$$

and at low velocities, we have the Biot–Savart law

$$\mathfrak{B} = \mu_0 i\,\mathbf{dl}\times\frac{\mathbf{r}}{r^3}$$

The vector potential is given by

$$A_x = \frac{c_0\beta}{K_0 c}\frac{\varepsilon}{r_0\sqrt{1-\beta^2}}, \qquad A_y = A_z = 0$$

To first order

$$A_x = \frac{\mu_0 \varepsilon v}{c_0 r} = \frac{\mu_0 i\,dl}{r}$$

and vectorially

$$\mathbf{A} = \mu_0 i\frac{\mathbf{dl}}{r}$$

(c) *Electrolytic conductors*

The current consists of positive and negative charges in motion. As an experimental model, we must imagine insulating tubes containing electrolyte and measure the interaction between the tubes. This case is studied in § [135].

[127] Arbitrary Distribution of Electric Charges in Uniform Overall Motion

(a) *Calculation of the field*

We consider a body of arbitrary shape, electrostatically charged, at rest in a system K_0. In this system, it produces a purely electric field, with components E_{0x}, E_{0y}, E_{0z}. We now consider a new system K, the axes of which are parallel to those of K_0, which moves with uniform velocity $-v$ along Ox. This is a generalization of §§ [120] and [121]. In the system K, the charged body produces an electromagnetic field with components

$$E_x = E_{0x} \qquad E_y = \frac{E_{0y}}{\sqrt{1-\beta^2}} \qquad E_z = \frac{E_{0z}}{\sqrt{1-\beta^2}}$$

$$\mathcal{B}_x = 0 \qquad \mathcal{B}_y = -\frac{c_0\beta}{c}\frac{E_{0z}}{\sqrt{1-\beta^2}} \qquad \mathcal{B}_z = \frac{c_0\beta}{c}\frac{E_{0y}}{\sqrt{1-\beta^2}}$$

The three relations for $\mathfrak{B}$ can be written vectorially

$$\mathfrak{B} = \frac{c_0}{c^2}\,\mathbf{v}\times\mathbf{E}$$

Let δ denote the angle between the electric field and the velocity in the system K, and δ_0 the same angle in K_0; then

$$\tan\delta = \frac{\tan\delta_0}{\sqrt{1-\beta^2}}$$

This expression and those given earlier for $\mathbf{E}$ lead us to the following conclusions. Let us consider the network of electrostatic field lines in K_0. To deduce the corresponding network in K (the axes $Ox_0y_0z_0$ and $Oxyz$ coincide at the instant in question), we have only to deform the network in K_0 by shortening all lengths parallel to v in the ratio $\sqrt{1-\beta^2}$. The collection of field lines behaves like a body undergoing the Lorentz contraction; in applications, therefore, we need only consider the assembly of charged body and network of lines of force and apply this contraction to it.

Important remark. The observations are assumed to be taken simultaneously in K and are hence not simultaneous in K_0 (but this does not matter in K_0, as mentioned in § [120]).

(b) *The scalar potential and the vector potential*

In K_0, there is only the scalar potential V_0 by hypothesis; hence, in K,

$$V = \frac{V_0}{\sqrt{1-\beta^2}}, \qquad A_x = \frac{c_0\beta}{c}\frac{V_0}{\sqrt{1-\beta^2}}, \qquad A_y = A_z = 0$$

and so, as in § [123],

$$E_x = -\frac{\partial V}{\partial x} - \frac{1}{c_0}\frac{\partial A_x}{\partial t} = -(1-\beta^2)\frac{\partial V}{\partial x}$$

$$E_y = -\frac{\partial V}{\partial y}, \qquad E_z = -\frac{\partial V}{\partial z}$$

$$\mathcal{B}_x = 0, \quad \mathcal{B}_y = \frac{\partial A_x}{\partial z} = \frac{c_0\beta}{c}\frac{\partial V}{\partial z}, \qquad \mathcal{B}_z = -\frac{\partial A_x}{\partial y} = \frac{c_0\beta}{c}\frac{\partial V}{\partial y}$$

[128] The Straight-line Charge in Uniform Motion along its Length; the Unbounded Rectilinear Current

We shall discuss this topic first by means of the field method; we then compare this with the calculations of the usual method in certain cases.

We shall consider either a line charge of continuous charge density ϱ_0 or a sequence of sufficiently closely spaced particles indiscriminately. If there are n_0 particles, each of charge ε, per unit length, we have

$$\varrho_0 = n_0 \varepsilon$$

(a) *Electromagnetic field*

In its rest system K_0, the line charge which we assume to lie along the axis Ox_0 creates an electric field

$$E_{0x} = 0, \quad E_{0y} = \frac{2\varrho_0}{K_0} \frac{y_0}{y_0^2+z_0^2}, \quad E_{0z} = \frac{2\varrho_0}{K_0} \frac{z_0}{y_0^2+z_0^2}$$

Hence, in the system K, where the coordinates y_0 and z_0 transform into y and z respectively,

$$E_x = 0, \quad E_y = \frac{2\varrho_0}{K_0\sqrt{1-\beta^2}} \frac{y}{y^2+z^2}$$

$$E_z = \frac{2\varrho_0}{K_0\sqrt{1-\beta^2}} \frac{z}{y^2+z^2}$$

$$\mathcal{B}_x = 0, \quad \mathcal{B}_y = -\frac{2c_0}{K_0 c^2} \frac{\varrho_0 v}{\sqrt{1-\beta^2}} \frac{z}{y^2+z^2}$$

$$\mathcal{B}_z = -\frac{2c_0}{K_0 c^2} \frac{\varrho_0 v}{\sqrt{1-\beta^2}} \frac{y}{y^2+z^2}$$

We thus have a radial electric field of magnitude

$$E = \frac{2\varrho_0}{K_0 r\sqrt{1-\beta^2}},$$

and a magnetic field, directed according to the classical rules, of magnitude

$$\mathcal{B} = \frac{2c_0}{K_0 c^2} \frac{\varrho_0 v}{r\sqrt{1-\beta^2}}$$

In K the charge density is given by†

$$\varrho = \frac{\varrho_0}{\sqrt{1-\beta^2}}$$

† This formula may here be regarded simply as the *definition* of ϱ in K; see also § [93].

and the current density in K is therefore given by

$$j = \frac{\varrho v}{c_0} = \frac{\varrho_0 v}{c_0\sqrt{1-\beta^2}}$$

Finally, *for all velocities*, we have

$$E = \frac{2\varrho}{K_0 r}, \quad \mathcal{B} = \frac{c_0^2}{K_0 c^2}\frac{2j}{r} = \mu_0\frac{2j}{r}$$

(b) *Action on a particle having the same velocity*

This is the problem of the interaction between two beams of cathode rays, in particular. We obviously have to calculate the action of all the particles of one beam on *one particle* of the other. This must not be confused with the interaction between two currents (see below).

We consider charges of the same sign. The electric field will produce a repulsive force

$$\frac{2\varrho\varepsilon}{K_0 r} = \frac{2n\varepsilon^2}{K_0 r}$$

in which n is the number of charges per unit length in K. The field $\mathfrak{B}$ produces the attraction

$$\frac{2n\varepsilon^2\beta^2}{K_0 r}$$

giving a total repulsive force

$$\frac{2n\varepsilon^2}{K_0 r}(1-\beta^2)$$

We can obtain this result without bringing in the magnetic field. In K_0, the repulsive force (normal to Ox) is

$$\frac{2n_0\varepsilon^2}{K_0 r_0}$$

which leads to the expression found above when transformed into K.

For charges ε_1 and ε_2 travelling parallel to one another at different velocities, the formula may be generalized to

$$\frac{2n\varepsilon_1\varepsilon_2}{K_0 r}\left(1-\frac{v_1 v_2}{c^2}\right)$$

(c) *Interaction between parallel currents of non-zero current density*

We desire to know the force exerted by an unbounded rectilinear current on a length l of parallel current. The preceding formulae immediately give the repulsive force:

$$\frac{2n\varepsilon(1-\beta^2)}{K_0 r}n\varepsilon l = \frac{2\varrho^2 l}{K_0 r}(1-\beta^2)$$

(d) *Interaction between currents of zero current density*

This is the common case of currents in metallic conductors. We assume that the current consists of two or more distributions such that the charge density in K is zero. The electric field is then zero.

(e) *On a paradox*

With pre-relativistic thinking, it is incomprehensible that a line charge travelling along its length and a stationary line charge should have different effects on another charge. There is no way in which an observer can distinguish between two positions of the moving line charge and we cannot conceive how this motion can have any effect. In relativity, on the other hand, the density ϱ is modified by the Lorentz contraction. It is exactly as though the density were fixed, but altered in magnitude. The interaction may be calculated from this point of view; we have only to be careful to use the correct symbols ϱ and n in each reference system.

(f) *Other calculations*

The task of calculating the transformation for a line charge travelling perpendicular to its length and, more generally, in any direction, is left to the reader.

[129] The Plane in Uniform Motion Parallel to Itself

(a) *Electromagnetic field*

Let the charged plane be the plane xOy. In its proper system K_0, we have the electric field

$$E_{0x} = 0, \quad E_{0y} = 0, \quad E_{0z} = \frac{2\pi\sigma_0}{K_0}$$

In K,

$$E_x = 0, \quad E_y = 0$$

$$E_z = \frac{E_{0z}}{\sqrt{1-\beta^2}} = \frac{2\pi\sigma_0}{K_0\sqrt{1-\beta^2}} = \frac{2\pi\sigma}{K_0}$$

$$\mathcal{B}_x = 0, \quad \mathcal{B}_z = 0$$

$$\mathcal{B}_y = -\frac{K_0\mu_0}{c_0}\frac{vE_{0z}}{\sqrt{1-\beta^2}} = -\frac{2\pi\mu_0}{c_0}\sigma v = -2\pi\mu_0 j$$

in which j is the current density *per unit length*.

(b) *Action on a positive particle travelling at the same velocity*

The force is independent of the distance of the particle from the plane. The electric field produces the repulsive force

$$\frac{2\pi\sigma\varepsilon}{K_0}$$

and the field $\mathfrak{B}$, the attractive force

$$\frac{2\pi\mu_0\sigma v}{c_0}\,\frac{\varepsilon v}{c_0}$$

leading to an overall repulsive force

$$\frac{2\pi\varepsilon\sigma}{K_0}(1-\beta^2)$$

This problem may also be solved as follows. The repulsive force is

—in K_0:

$$X_0 = Y_0 = 0, \quad Z_0 = \frac{2\pi\sigma_0\varepsilon}{K_0}$$

—in K:

$$X = Y = 0, \quad Z = Z_0\sqrt{1-\beta^2} = \frac{2\pi\varepsilon\sigma}{K_0}(1-\beta^2)$$

(c) *Other calculations*

As an exercise, the reader may care to examine the case of the circular cylinder with either a volume or a surface charge distribution.

[130] Maxwell's Equations

The equations of § [70] for curl **E** and div $\mathfrak{B}$ are obviously valid when **E** and $\mathfrak{B}$ are the components of an electromagnetic field, since they are applicable to all fields. The equations of § [91] for curl **H** and div $\mathfrak{D}$ likewise apply here, if we use the Lagrangian

$$L = -(\varrho V - \mathbf{j}\,.\,\mathbf{A}) + \frac{1}{8\pi}\left(K_0E^2 - \frac{\mathcal{B}^2}{\mu_0}\right)$$

as I mentioned in § [94].

I shall dwell no more on this: the reader is referred to my electricity textbook for a detailed account.

C. THE SECOND METHOD, IN WHICH THE RESULTS OF CHAPTERS III AND V ARE NOT EMPLOYED

[131] General remarks

Method B is a very general one, but requires a detailed knowledge of relativistic dynamics. To regard **E** and $\mathfrak{B}$ as purely mechanical quantities, to accept that Maxwell's equations are equations of mechanics, these depend upon an attitude of mind that has not yet been widely adopted. For use until it is adopted, and for normal teaching requirements, I also give a direct method. The method is of course still a relativistic one, but only the transformation of force is required. This method is, moreover, of interest in its own right, far in excess of its convenience for teaching purposes. It draws attention to the fact that it is possible to formulate electromagnetism without using the concept of a field, and in particular, without using the concept of magnetic field. We can thus obtain electromagnetism without magnetism.

We shall first show how the interactions between charges can be calculated without the introduction of the concept of magnetic field. We shall then see how the concept of field can be introduced directly, without invoking the general results of Chapters III and V.

For all complementary details, the reader may consult my textbook of electricity.

[132] The Interaction between Two Charges Travelling at the Same Velocity; Calculation without the Introduction of the Concept of Field

We first treat the very simple special case in which two charges are at rest with respect to one another, for this case enables us to explain the fundamentals of the method with only elementary calculations. This is not mere padding, however, for which certain readers might kindly reproach me: since the various authors disagree over the results, a detailed analysis is worth while.

(a) *The line joining the two charges is perpendicular to the velocity*

In the system K_0 in which the charges are at rest, they repel one another with a force

$$X_0 = 0, \quad Y_0 = \frac{\varepsilon_1\varepsilon_2}{K_0 r_0^2}$$

when the axes are suitably chosen (Fig. 27). Thus, in the system K in which the charges

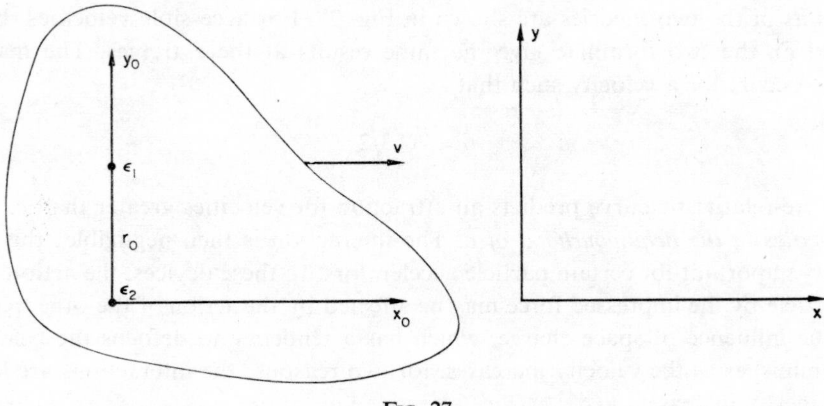

FIG. 27.

have the common velocity $v = \beta c$, we have

$$r = r_0, \quad X = X_0 = 0, \quad Y = \sqrt{1-\beta^2}Y_0 = \frac{\varepsilon_1\varepsilon_2}{K_0 r_0^2}\sqrt{1-\beta^2}$$

This elementary example shows that the result is obtained immediately by purely mechanical calculations with a change of system, without introducing the ideas of current element and magnetic field.

The case of low velocities; the second-order calculation. Expanding in series and considering only the first two terms, we have the repulsive force

$$Y \simeq \frac{\varepsilon_1\varepsilon_2}{K_0 r^2}\left(1-\frac{1}{2}\beta^2\right)$$

With the older method, the charge ε_1 creates a magnetic field at the point where ε_1 is situated, given by

$$\mathcal{B}_z = \frac{\mu_0}{c_0}\frac{\varepsilon_2 v}{r^2}$$

Thus, using the law of force on a moving charge, the attractive force on the charge ε_1 is given by

$$Y = \frac{\mu_0 \varepsilon_2 v}{c_0 r^2}\frac{\varepsilon_1 v}{c_0} = \frac{\mu_0}{c_0^2}\frac{\varepsilon_1\varepsilon_2}{r^2}v^2 = \frac{\varepsilon_1\varepsilon_2}{K_0 r^2}\beta^2$$

Adding the Coulomb term, we obtain the total force

$$Y = \frac{\varepsilon_1\varepsilon_2}{K_0 r^2}(1-\beta^2)$$

Even at low velocities, therefore, the two theories disagree. To bring them into agreement, we should have to neglect the terms in β^2, which is equivalent to calculating with stationary charges; in this case, however, the Biot–Savart law would not intervene. The results of the two theories are shown in Fig. 28. For accessible velocities (between zero and c), the two formulae give the same results at the extremes. The maximum disparity occurs for a velocity such that

$$\beta = \sqrt{3}/2$$

The pre-relativistic curve predicts an attraction for velocities greater than c.

Velocities in the neighbourhood of c. The interaction is then negligible; this case is extremely important for certain particle accelerators. In these devices, the action exerted on a particle by the impressed force may be affected by the action of the other particles; this is the influence of space charge, which has a tendency to defocus the beam. This effect diminishes as the velocity increases, for two reasons: the interactions are lessened and the inertia increases.

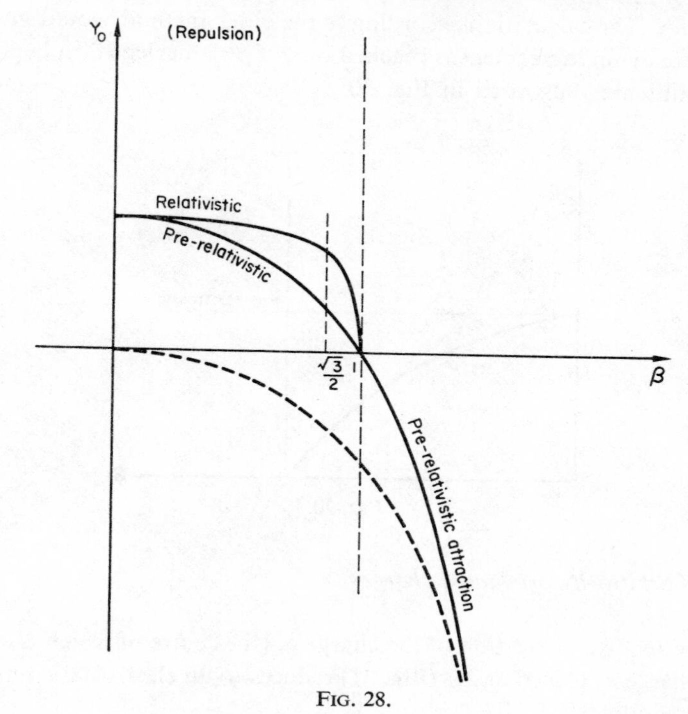

FIG. 28.

(b) *The line joining the two charges is parallel to the velocity*

In the system K_0 (Fig. 29),

$$X_0 = \frac{\varepsilon_1 \varepsilon_2}{K_0 r_0^2}, \qquad Y_0 = 0$$

so that in K,

$$r = r_0 \sqrt{1-\beta^2}$$

$$X = X_0 = \frac{\varepsilon_1 \varepsilon_2}{K_0 r^2} (1-\beta^2)$$

$$Y = Y_0 \sqrt{1-\beta^2} = 0$$

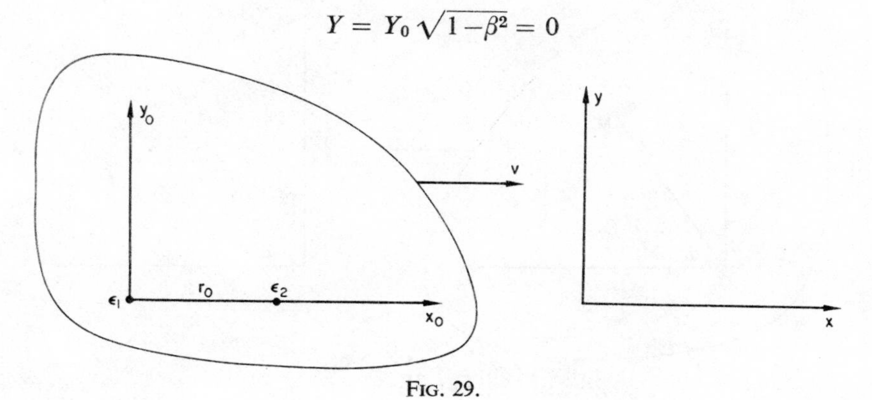

FIG. 29.

Low velocities. The calculation according to the older method would give zero magnetic force. Here again, agreement is reached only if β^2 is neglected in comparison with unity. The results are illustrated in Fig. 30.

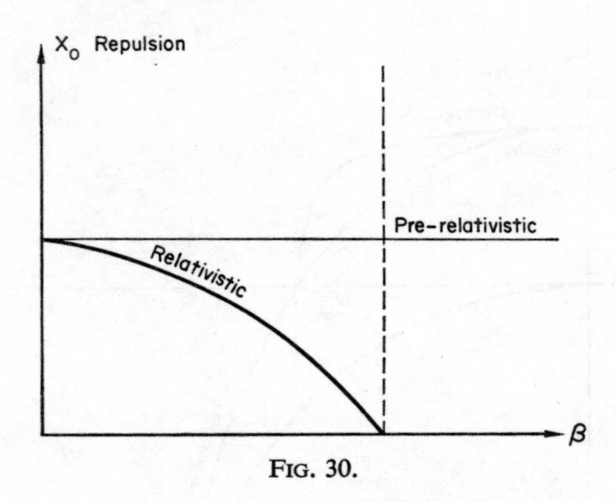

FIG. 30.

(c) *The case of arbitrarily orientated charges*

(i) In the system K_0, the action of the charge ε_1 (the centre of which is assumed to be at O) on the charge ε_2 placed at M_0 (Fig. 31) reduces to an electrostatic repulsion, along OM_0, with components

$$X_0 = \frac{\varepsilon_1\varepsilon_2}{K_0}\,\frac{x_0}{(x_0^2+y_0^2)^{3/2}}, \qquad Y_0 = \frac{\varepsilon_1\varepsilon_2}{K_0}\,\frac{y_0}{(x_0^2+y_0^2)^{3/2}}$$

The angle α_0 between the force and Ox_0 is given by

$$\tan\alpha_0 = \frac{Y_0}{X_0} = \frac{y_0}{x_0}$$

and the force is therefore directed along the line joining the charges. To transform into

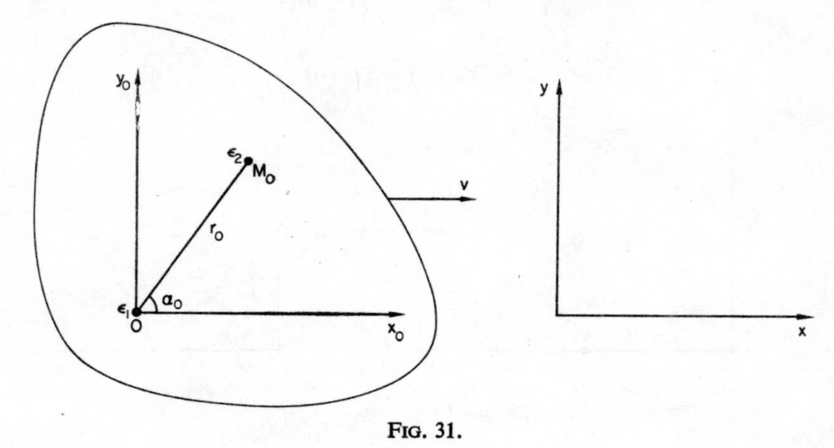

FIG. 31.

the system K, we must make the transformations

$$X_0 = X, \quad Y_0 = \frac{Y}{\sqrt{1-\beta^2}}$$

$$x_0 = \frac{x}{\sqrt{1-\beta^2}}, \quad y_0 = y$$

so that

$$X = \frac{\varepsilon_1\varepsilon_2}{K_0} \frac{(1-\beta^2)x}{\{x^2+(1-\beta^2)y^2\}^{3/2}}$$

$$Y = \frac{\varepsilon_1\varepsilon_2}{K_0} \frac{(1-\beta^2)y}{\{x^2+(1-\beta^2)y^2\}^{3/2}}(1-\beta^2)$$

$$\tan\alpha = \frac{Y}{X} = (1-\beta^2)\frac{y}{x}$$

(ii) Except in the special cases studied above in (a) and (b), the force is not along the line joining ε_1 and ε_2. We should obtain the same result by writing directly

$$\tan\alpha = \frac{Y}{X} = \frac{\sqrt{1-\beta^2}Y_0}{X_0} = \sqrt{1-\beta^2}\frac{y_0}{x_0} = (1-\beta^2)\frac{y}{x}$$

We must beware of going too hastily, and applying to $\tan\alpha_0$ the kinematics formula for the transformation of angles (*Kinematics*, § [68]); this formula is valid for lines but *not* for vectors representing forces (§ [30]).

Let us suppose that the two charges are connected rigidly (in the relativistic sense whereby the distance between them remains constant in a given system). It seems that since the forces are not collinear, there is a couple. The question of non-simultaneity must be considered, however, and we then find that the couple can vanish. We shall reconsider this at great length in § [179].

(iii) We can put the results into the following form which is more convenient:

—component along **r** (repulsion)

$$\frac{\varepsilon_1\varepsilon_2}{K_0} \frac{1-\beta^2}{r^2(1-\beta^2\sin^2\alpha)^{3/2}}$$

—component normal to **r**

$$\frac{\varepsilon_1\varepsilon_2}{K_0} \frac{(1-\beta^2)^2\sin\alpha}{r^2(1-\beta^2\sin^2\alpha)^{3/2}}$$

Low velocities; second-order approximation. We expand X and Y in series, and retain only terms in β^2

$$X = \frac{\varepsilon_1\varepsilon_2}{K_0} \frac{x}{(x^2+y^2)^{3/2}}\left(1+\frac{\beta^2}{2}\frac{y^2-2x^2}{x^2+y^2}\right)$$

$$Y = \frac{\varepsilon_1\varepsilon_2}{K_0} \frac{y}{(x^2+y^2)^{3/2}}\left(1-\frac{\beta^2}{2}\frac{y^2+4x^2}{x^2+y^2}\right)$$

These expressions are generalizations of those given in (a) and (b). To this approximation, the expression for tan α is unaffected. Let us investigate what we obtain with the formula of Biot and Savart and the force on a moving charge. The charge ε_1 produces a Coulomb repulsion at ε_2, with components

$$X = \frac{\varepsilon_1\varepsilon_2}{K_0}\frac{x}{(x^2+y^2)^{3/2}}, \qquad Y = \frac{\varepsilon_1\varepsilon_2}{K_0}\frac{y}{(x^2+y^2)^{3/2}}$$

and magnetic field

$$\mathcal{B}_x = 0, \quad \mathcal{B}_y = 0, \quad \mathcal{B}_z = \frac{\mu_0\varepsilon_1 v}{c_0}\frac{y}{(x^2+y^2)^{3/2}}$$

which produces a magnetic force with components

$$X = 0, \quad Y = \frac{\mu_0\varepsilon_1 v}{c_0}\frac{y}{(x^2+y^2)^{3/2}}\varepsilon_2 v$$

so that the total force is

$$X = \frac{\varepsilon_1\varepsilon_2}{K_0}\frac{x}{(x^2+y^2)^{3/2}}, \qquad Y = \frac{\varepsilon_1\varepsilon_2}{K_0}\frac{y}{(x^2+y^2)^{3/2}}(1-\beta^2)$$

with

$$\tan\alpha = \frac{Y}{X} = (1-\beta^2)\frac{y}{x}$$

Apart from the expression for tan α, the results of the two theories are different.

[133] The Interaction between Two Charges Travelling at Arbitrary Velocities; Calculation not Introducing the Concept of Field

(a) *General remarks*

Let two charges ε_A and ε_B, situated at different points A and B, be travelling with arbitrary uniform velocity with respect to a Galileian reference system K (Fig. 32). We denote by K_A the Galileian system of reference in which the charge ε_A is at rest. We seek the force exerted on the charge ε_B.

The calculation is performed in two stages: first the force exerted by ε_A on ε_B is calculated in K_A; this force is then calculated in the system K, using the relativistic formulae for the transformation of force.

(b) *The force exerted on ε_B in K_A*

In K_A, we take ε_A as the origin of coordinates, A_A; the charge B is situated at the point B_A. We take the x-axis parallel to the velocity $\mathbf{v}$ of ε_A with respect to K and the plane $(xAy)_A$ contains $\mathbf{v}$ and the line A_AB_A. These conventions do not limit the generality of the results in any way. Let (x, y) be the coordinates of ε_B in K and (x_A, y_A) be the

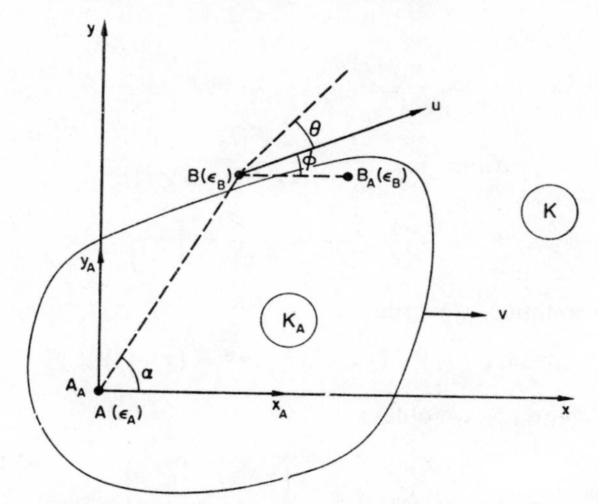

FIG. 32. At time $t = 0$, the points representing ε_A coincide (these are A_A in K_A and A in K); those representing ε_B are separated.

corresponding coordinates in K_A. Let r_A be the distance A_AB_A, at the time t of K, in K_A. The Cartesian components of the force acting on ε_B are

$$X_A = \frac{\varepsilon_A\varepsilon_B}{K_0 r_A^2}\frac{x_A}{r_A}, \qquad Y_A = \frac{\varepsilon_A\varepsilon_B}{K_0 r_A^2}\frac{y_A}{r_A}, \qquad Z_A = 0 \tag{1}$$

where

$$r_A^2 = x_A^2 + y_A^2$$

We recollect that the force exerted by a stationary charge ε_A on a moving charge ε_B is *independent of the motion* of ε_B. *By convention,* we treat the charge as an invariant. I stress the fact that this is a mere convention of the notation, and not a physical law, as is often stated. Here the charge may be thought of as playing the role of rest mass m_0 in dynamics. Nothing prevents us from selecting another *convention*, but the definitions and transformations of the vectors **E** and $\mathfrak{B}$ (see later) would be modified.

(c) *The force* **F** *exerted on* ε_B *in K*

The Cartesian components of **F**, *X and Y*. Let the axes in K, Ax, Ay and Az be parallel to Ax_A, Ay_A, Az_A respectively at the moment in question. At time $t = 0$ in K, the charges lie at A (which coincides with A_A) and B (which is not the same as B_A). Let **r** denote the vector $\overrightarrow{AB}$ and $\mathbf{u}(u_x, u_y, u_z)$ the velocity of ε_B with respect to K. The force acting on ε_B in K therefore has the components (§ [30])

$$X = X_A + \frac{vu_y/c^2}{\sqrt{1-\beta^2}} Y_A, \qquad Y = \frac{1 - vu_x/c^2}{\sqrt{1-\beta^2}} Y_A, \qquad Z = 0 \tag{2}$$

in which $\beta = v/c$.

This force lies in the plane xOy, which is the plane defined by **r** and **v** (irrespective of **u**). We must replace X_A and Y_A by their expressions in terms of the K variables;

at arbitrary time t, we have

$$x_A = \frac{x-vt}{\sqrt{1-\beta^2}}, \qquad y_A = y$$

$$X_A = \frac{\varepsilon_A\varepsilon_B}{K_0}\,\frac{(1-\beta^2)(x-vt)}{\{(x-vt)^2+(1-\beta^2)y^2\}^{3/2}} \tag{3}$$

$$Y_A = \frac{\varepsilon_A\varepsilon_B}{K_0}\,\frac{(1-\beta^2)^{3/2}\,y}{\{(x-vt)^2+(1-\beta^2)y^2\}^{3/2}}$$

To simplify the notation, we write

$$s^2 = (x-vt)^2+(1-\beta^2)y^2, \quad r^2 = (x-vt)^2+y^2$$

and substituting (3) into (2), we obtain

$$X = \frac{\varepsilon_A\varepsilon_B}{K_0}\,\frac{1-\beta^2}{s^3}\left(x-vt+\frac{vu_y}{c^2}\,y\right)$$

$$Y = \frac{\varepsilon_A\varepsilon_B}{K_0}\,\frac{1-\beta^2}{s^3}\left(1-\frac{vu_x}{c^2}\right)y \tag{4}$$

The expressions for **F**. We transform X and Y as follows:

$$X = \frac{\varepsilon_A\varepsilon_B}{K_0}\,\frac{1-\beta^2}{s^3}\left[x-vt+\frac{v}{c^2}\{(x-vt)u_x+yu_y\}-\frac{v}{c^2}(x-vt)u_x\right]$$

$$= \frac{\varepsilon_A\varepsilon_B}{K_0}\,\frac{1-\beta^2}{s^3}\left[(x-vt)\left(1-\frac{vu_x}{c^2}\right)+\frac{v}{c^2}\{(x-vt)u_x+yu_y\}\right]$$

$$= \frac{\varepsilon_A\varepsilon_B}{K_0}\,\frac{1-\beta^2}{s^3}\left[(x-vt)\left(1-\frac{\mathbf{v}\cdot\mathbf{u}}{c^2}\right)+\frac{v}{c^2}\,\mathbf{r}\cdot\mathbf{u}\right]$$

$$Y = \frac{\varepsilon_A\varepsilon_B}{K_0}\,\frac{1-\beta^2}{s^3}\left(1-\frac{\mathbf{v}\cdot\mathbf{u}}{c^2}\right)y$$

This gives an expression for **F** containing only physical quantities:

$$\mathbf{F} = \frac{\varepsilon_A\varepsilon_B}{K_0}\,\frac{1-\beta^2}{s^3}\left\{\left(1-\frac{\mathbf{v}\cdot\mathbf{u}}{c^2}\right)\mathbf{r}+\frac{\mathbf{v}}{c^2}(\mathbf{r}\cdot\mathbf{u})\right\} \tag{5}$$

In this formula, $\mathbf{r}$ is the radius vector $\overrightarrow{AB}$ and s is given by

$$s^2 = (x-vt)^2+(1-\beta^2)y^2$$

$$= \left(\frac{\mathbf{v}\cdot\mathbf{r}}{v}\right)^2+(1-\beta^2)\left(\frac{\mathbf{v}}{v}\times\mathbf{r}\right)^2$$

We introduce the following notation:

α: the angle between $\mathbf{v}$ and $\mathbf{r}$ (or Ax and AB);

θ: the angle between $\mathbf{u}$ and $\mathbf{r}$;

ϕ: the angle between $\mathbf{u}$ and $\mathbf{v}$.

We may write

$$\mathbf{F} = \frac{\varepsilon_A \varepsilon_B}{K_0} \frac{(1-\beta^2)r}{s^3} \left\{ \left(1 - \frac{vu \cos\phi}{c^2}\right) \frac{\mathbf{r}}{r} + \frac{u \cos\theta}{c^2} \mathbf{v} \right\} \tag{6}$$

The angle δ between the force and Ax is given by

$$\tan\delta = \frac{Y}{X} = \frac{(1-vu\cos\phi/c^2)\sin\alpha}{(1-vu\cos\phi/c^2)\cos\alpha + uv\cos\theta/c^2} \tag{7}$$

Remark 1. To obtain the force exerted on ε_A by ε_B, we have only to exchange the roles of **u** and **v** and reverse the sense of **r**.

Remark 2. If the charge ε_B does not lie in the plane $x_A A y_A$, the quantity s will be given by

$$s^2 = x^2 + (1-\beta^2)(y^2+z^2)$$

Remark 3. In formula (5), all the quantities are taken at the same instant t and the two charges ε_A and ε_B are both considered at this time t. The result of this is that in K_A, the two charges are considered at times between which the difference is not in general equal to r/c. This does not create any difficulties, however, since ε_A is stationary and independent of time; it is exactly as though ε_A and ε_B were considered at the same time. Formulae (1) do not involve simultaneity.

(d) *Action and reaction*

In the present case, the two forces in K are neither in the same direction nor of equal magnitude. Action and reaction are not equal. Nevertheless, we must not immediately conclude that there is a couple, as certain authors do (myself included, in the 1957 edition). The question of simultaneity obliges us to look more closely into this (§ [179]). Here we shall demonstrate that *the inequality arises because of the definition adopted above for force in K, and that it disappears if the forces on each charge are each measured in the system of the corresponding charge* (that is, with a dynamometer).

For the force on charge ε_B in K_B, we have

$$X_B = X_A = \frac{\varepsilon_A \varepsilon_B}{K_0} \frac{x_0}{r_0^3}, \qquad Y_B = \frac{Y_A}{\sqrt{1-\beta^2}} = \frac{\varepsilon_A \varepsilon_B}{K_0 \sqrt{1-\beta^2}} \frac{y_0}{r_0^3}$$

$$\tan\delta_B = \frac{Y_B}{X_B} = \frac{y_0}{x_0\sqrt{1-\beta'^2}}$$

where β' is calculated using the *velocity of charge ε_A in K_B*. The same expression would be obtained, apart from the sign, if the force exerted on the charge ε_A in K_A were calculated. The proper forces are therefore equal, and make the same proper angle with the relative velocity of the two charges.

[134] Special Cases

(a) If we set u and v equal to zero, the formula does reduce to Coulomb's formula. If we make u and v equal but not to zero, we recover the formulae of § [132] (c). Thus

$$\phi = 0, \qquad \theta = \alpha$$

If v is so small that β^2 is negligible in comparison with unity, r and s become the same, and we obtain the formulae

$$\mathbf{F} = \frac{\varepsilon_A \varepsilon_B}{K_0 r^2}\left\{\left(1 - \frac{vu \cos\phi}{c^2}\right)\frac{\mathbf{r}}{r} + \frac{u\cos\theta}{c^2}\mathbf{v}\right\}$$

$$X = \frac{\varepsilon_A \varepsilon_B}{K_0 r^3}\left(x + \frac{vu_y}{c^2}y\right), \qquad Y = \frac{\varepsilon_1 \varepsilon_2}{K_0 r^3}\left(1 - \frac{vu_x}{c^2}\right)y$$

In these formulae, it is assumed that uv is large enough not to be negligible in comparison with c^2; otherwise the other terms in β^2 would have to be retained. The product of $\varepsilon_A v$ and $\varepsilon_B u$ appears, which is not true of the general case. The expression for δ is unaltered (action is not equal to reaction).

If *both velocities* u and v are small and of the same order of magnitude, these formulae clearly yield the Coulomb force.

(b) *Comparison with the usual argument.* The charge ε_B first experiences the Coulomb force, with components

$$\frac{\varepsilon_A \varepsilon_B}{K_0}\frac{x}{(x^2+y^2)^{3/2}}, \qquad \frac{\varepsilon_A \varepsilon_B}{K_0}\frac{y}{(x^2+y^2)^{3/2}}$$

The charge ε_A creates magnetic field at charge B with components

$$\mathcal{B}_x = 0, \qquad \mathcal{B}_y = 0, \qquad \mathcal{B}_z = \frac{\mu_0}{c_0}\frac{\varepsilon_A v}{r^2}\frac{y}{r}$$

giving a force with components

$$\frac{\mu_0}{c_0}\frac{\varepsilon_A v y}{r^3}\frac{\varepsilon_B u_y}{c_0}, \qquad -\frac{\mu_0}{c_0}\frac{\varepsilon_A v y}{r^3}\frac{\varepsilon_B u_x}{c_0}, \qquad 0$$

The total force is thus

$$X = \frac{\varepsilon_A \varepsilon_B}{K_0 r^3}\left(x + \frac{vu_y}{c^2}y\right)$$

$$Y = \frac{\varepsilon_A \varepsilon_B}{K_0 r^3}\left(1 - \frac{vu_x}{c^2}\right)y$$

We obtain the same result as in (a), but with no indication as to its range of validity. We might at most think that the velocities u and v must be small, that is, that the expressions are valid in second order, but this we have seen to be wrong; in fact, the formulae are valid only in first order—they require u to be large.

It is for this reason that I have treated certain special cases in detail; the agreement between the formulae of (b) and (a) could be misleading.

The origin of these differences lies in the fact that the analysis above brings in u and v symmetrically, whereas in the relativistic calculation, these two velocities play very different physical roles.

[135] Current Elements: the Case of Electrolytic Conductors

We now resume the calculations of § [126]. By way of variety, we treat here the case of electrolytic conductors. As an example, I shall consider only parallel current elements, perpendicular to the line connecting them. This is the case shown in Fig. 33. We have first

—the action of A_1 on B_1:

$$\frac{\varepsilon_1\varepsilon_2}{K_0 r^2}\sqrt{1-\frac{v_1^2}{c^2}}$$

—the action of A_2 on B_2:

$$\frac{\varepsilon_1\varepsilon_2}{K_0 r^2}\sqrt{1-\frac{v_2^2}{c^2}}$$

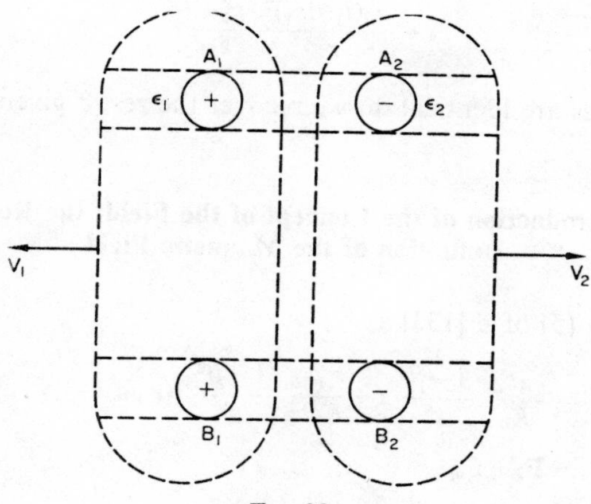

FIG. 33.

To obtain the action of A_1 on B_2, we select the system of A_1, in which we find the Coulomb force

$$Y_1 = -\frac{\varepsilon_1\varepsilon_2}{K_0 r^2}$$

or, transforming into the system of our measurements,

$$Y = \frac{Y_1}{\sqrt{1-\dfrac{v_1^2}{c^2}}} = -\frac{\varepsilon_1\varepsilon_2}{K_0 r^2}\,\frac{1}{\sqrt{1-\dfrac{v_1^2}{c^2}}}$$

Likewise, the action of A_2 on B_1 is

$$-\frac{\varepsilon_1\varepsilon_2}{K_0 r^2}\frac{1}{\sqrt{1-\dfrac{v^2}{c^2}}}$$

giving the total interaction

$$\frac{\varepsilon_1\varepsilon_2}{K_0 r^2}\left(\sqrt{1-\frac{v_1^2}{c^2}}+\sqrt{1-\frac{v_2^2}{c^2}}-\frac{1}{\sqrt{1-\dfrac{v_1^2}{c^2}}}-\frac{1}{\sqrt{1-\dfrac{v_2^2}{c^2}}}\right)$$

In the special case of low velocities, we have

$$-\frac{\varepsilon_1\varepsilon_2}{K_0 r^2}\frac{v_1^2+v_2^2}{c^2}=\frac{1}{K_0 r^2 c^2}\left(\frac{\varepsilon_2}{\varepsilon_1}\varepsilon_1^2 v_1^2+\frac{\varepsilon_1}{\varepsilon_2}\varepsilon_2^2 v_2^2\right)$$

$$=-\frac{\mu_0}{r^2}\left\{\frac{\varepsilon_2}{\varepsilon_1}(i_1\,\mathrm{d}s_1)^2+\frac{\varepsilon_1}{\varepsilon_2}(i_2\,\mathrm{d}s_2)^2\right\}.$$

If the current is such that the medium remains neutral,

$$\varepsilon_1 v_1+\varepsilon_2 v_2=0, \quad i_1\,\mathrm{d}s_1+i_2\,\mathrm{d}s_2=0$$

and hence the interaction becomes

$$-\frac{\mu_0(i_1\,\mathrm{d}s_1)^2}{r^2}\frac{\varepsilon_1^2+\varepsilon_2^2}{\varepsilon_1\varepsilon_2}$$

Only if the charges are identical do we recover the result given by the usual theory.

[136] Introduction of the Concept of the Field; the Relativistic Definition of the Magnetic Field

We write formula (5) of § [133] as

$$\mathbf{F}=\frac{\varepsilon_A\varepsilon_B}{K_0}\frac{1-\beta^2}{s^3}\mathbf{r}+\frac{\varepsilon_A\varepsilon_B}{K_0 c^2}\frac{1-\beta^2}{s^3}\{(\mathbf{r}\cdot\mathbf{u})\mathbf{v}-(\mathbf{v}\cdot\mathbf{u})\mathbf{r}\}$$

$$=\mathbf{F}_e+\mathbf{F}_m$$

We have

$$\mathbf{F}_e=\frac{\varepsilon_A\varepsilon_B}{K_0}\frac{1-\beta^2}{s^3}\mathbf{r}=\varepsilon_B\mathbf{E}$$

The vector $\mathbf{E}$, which gives the force on unit charge ε_B at rest, is by definition the electric field produced at B by ε_A. If ε_B is not at rest, the total force contains an extra component

$$\mathbf{F}_m=\frac{\varepsilon_A\varepsilon_B}{K_0 c^2}\frac{1-\beta^2}{s^3}\{(\mathbf{r}\cdot\mathbf{u})\mathbf{v}-(\mathbf{v}\cdot\mathbf{u})\mathbf{r}\}$$

which can be written

$$\mathbf{F}_m=\varepsilon_B\mathbf{u}\times\left(\frac{1-\beta^2}{K_0 c^2}\frac{\varepsilon_A\mathbf{v}\times\mathbf{r}}{s^3}\right)$$

with the aid of the identity

$$(\mathbf{r}\,.\,\mathbf{u})\mathbf{v}-(\mathbf{v}\,.\,\mathbf{u})\mathbf{r} = \mathbf{u}\times(\mathbf{v}\times\mathbf{r})$$

We thus define a vector $\mathfrak{B}$ by the relations

$$\mathfrak{B} = \frac{c_0}{K_0 c^2}\,\frac{1-\beta^2}{s^3}\,\varepsilon_A \mathbf{v}\times\mathbf{r}, \quad \mathbf{F}_m = \frac{\varepsilon_B \mathbf{u}}{c_0}\times\mathfrak{B}$$

The vector $\mathfrak{B}$ is by definition the magnetic field created at B by ε_A. We thus recover the formulae of § [121]. From here, we can define and calculate potentials, and deduce Maxwell's equations. I refer the reader to my treatise on electricity.

[136A] The Definitive Form of the Fundamental Postulate and the Relativistic Calculation of the Retarded Potentials of Liénard–Wiechert; the Interaction between Accelerating Charges

(a) *The fundamental postulate*

As I mentioned in § [117], the force postulate does not allow us to construct the whole of electromagnetism. For this reason, I suggest that the fundamental postulate should be as follows: if a charge ε is stationary at a point A at time t and has arbitrary acceleration, it creates an electrical potential

$$V = \varepsilon/K_0 r$$

at a point B at time $t+r/c$; there is no arbitrary constant, for reasons associated with the inertial energy (§ [155]).

(b) *The Liénard–Wiechert potentials*

In the calculations of § [133], we dealt with uniform motion; for the reasons stated (Remark 3), questions of simultaneity do not arise. These calculations are no longer valid for a charge in accelerated motion, however, for at successive positions the charge has different characteristics (velocity, acceleration). The charge no longer remains in the same proper system K_0. Furthermore, the field of a stationary but accelerating charge is not given by Coulomb's law. Nevertheless, the Coulomb potential remains valid. These two points lead us to start from the retarded potential in K_0.

Consider therefore a charge ε which at time $t_0 = 0$ is situated at the origin O of K_0, with zero velocity but arbitrary acceleration. At time $t_0 = r_0/c$, it creates potentials at a point M ($OM = r_0$) given by

$$V_0 = \frac{\varepsilon}{K_0 r_0}, \qquad \mathbf{A}_0 = 0 \tag{1}$$

Consider now another system K (with axes parallel to those of K_0) with respect to which K_0 has velocity v parallel to Ox; the common time origin is such that O and O_0

coincide. The potentials at M at the time corresponding to t_0 are given by

$$V = \frac{V_0}{\sqrt{1-\beta^2}} = \frac{\varepsilon}{K_0 r_0 \sqrt{1-\beta^2}}$$

$$A_x = \frac{K_0 \mu_0}{c_0} \frac{v V_0}{\sqrt{1-\beta^2}}, \qquad A_y = A_z = 0 \tag{2}$$

We have only to express r_0 as a function of the variables of K. The variables r_0 and t_0 are connected by

$$r_0 = ct_0, \qquad x_0^2+y_0^2+z_0^2-c^2t_0^2 = 0 \tag{3}$$

To r_0 and t_0 correspond variables r and t, given by

$$r = ct, \qquad (x+vt)^2+y^2+z^2-c^2t^2 = 0 \tag{4}$$

Using the time transformation

$$t_0 = \frac{t-vx/c^2}{\sqrt{1-\beta^2}}$$

formulae (3) and (4) give

$$r_0 = ct_0 = -c\frac{t-vx/c^2}{\sqrt{1-\beta^2}} = \frac{r-vx/c}{\sqrt{1-\beta^2}}$$

or

$$r_0\sqrt{1-\beta^2} = r-vx/c = r(1-vx/cr) = r(1+v_r/c)$$

in which v_r denotes radial velocity.

Finally,

$$V = \frac{\varepsilon}{K_0 r(1+v_r/c)}, \qquad A_x = \frac{\mu_0 \varepsilon v}{c_0 r(1+v_r/c)}$$

In general, there are of course three components of vector potential, so that we should write

$$A_x = \frac{\mu_0 \varepsilon v_x}{c_0 r(1+v_r/c)}, \ldots$$

In these formulae, V, A_x, A_y, A_z are the potentials at M at time t; r, v_r, v_x are quantities calculated for $t' = t-r/c$.

The radial velocity v_r is positive when the velocity increases the separation.

In vector form, the preceding formulae may be written

$$V = \frac{\varepsilon}{K_0(r-\mathbf{r}.\mathbf{v}/c)} \qquad \mathbf{A} = \frac{\mu_0 \varepsilon \mathbf{v}}{c_0(r-\mathbf{r}.\mathbf{v}/c)}$$

in which $\mathbf{r}$ *is the vector from the charge to the point at which the field is calculated.*

The formulae for the retarded potentials may be interpreted as follows (Fig. 34). The potential V at an arbitrary point at time t can be calculated by applying Coulomb's law to a charge

$$\frac{\varepsilon}{1+v_r/c}$$

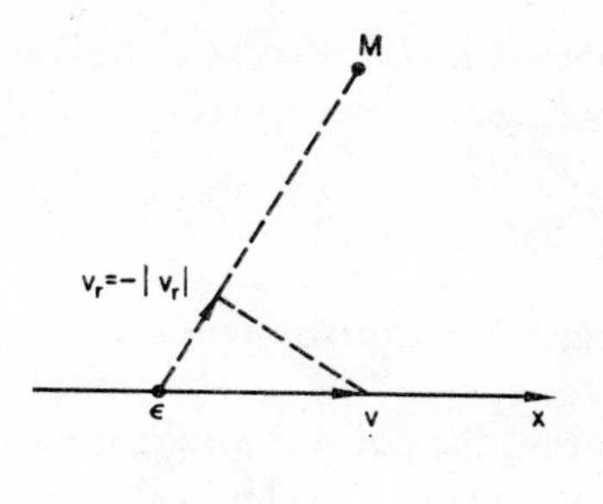

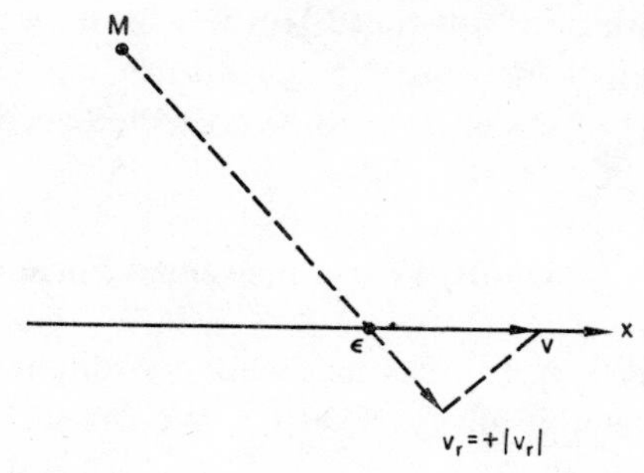

FIG. 34.

(Left). The charge appears to increase:

$$\frac{\varepsilon}{1-|v_r|/c} = \frac{\varepsilon}{1+v_r/c}$$

(Right). The charge appears to be reduced:

$$\frac{\varepsilon}{1+|v_r|/c} = \frac{\varepsilon}{1+v_r/c}$$

which is assumed to be at rest at the point where ε had been at an earlier time t'. According as the charge is approaching or moving away, it is as though its magnitude were increased or diminished in the ratio

$$1+v_r/c$$

This is a similar phenomenon to the Doppler effect.

Likewise, the potential **A** at time t may be calculated by means of the Biot and Savart law, for a current element given by position and velocity of the charge at an earlier time t', if a coefficient of increase or decrease is introduced.

From the potentials V and **A** above, the fields **E** and $\mathfrak{B}$ of an accelerating charge can be calculated, and hence, the interaction between two accelerating charges.

(c) *Relationship with the expressions given in* § [123]

In § [123], an expression giving the potential V at M at time t is set out. At this time, the charge lies on the positive x-axis, distance vt from the origin. If we wish to express V as a function of the retarded quantities, t must be replaced by r/c, with $r = OM$. Thus, the denominator becomes

$$\sqrt{(x-vr/c)^2+(1-\beta^2)(y^2+z^2)} = \sqrt{r^2+v^2x^2/c^2-2vrx/c}$$

leading to the form already obtained in $r-vx/c$.

(d) *A remark on the concept of advanced potential*

Let us provisionally accept this idea, as some authors have (and not the least distinguished: Dirac, . . .). Instead of the statement made in (a), we must assert that the potential V is created at the point B at times $t-r/c$ and $t+r/c$. The reasoning of (b) remains valid for $t-r/c$, and only the relevant signs need to be changed. We then obtain the Liénard potentials in "advanced" form, with $r(1-v_r/c)$ in the denominator. I reject this advanced solution (this rejection constitutes one of the main alterations to the 1957 edition; § [151] of the latter is to be corrected accordingly).

[136B] Four-dimensional Form of the Liénard–Wiechert Potentials

Let ξ, η, ζ, t' be the earlier coordinates of the charge; the point at which the field is calculated is always x, y, z, t. We denote the world-coordinates ξ^i and x^i, with

$$\xi^4 = ict' \quad \text{and} \quad x^4 = ict$$

The four-potential is then

$$\Phi^i = \frac{c_0}{K_0 c} \frac{\varepsilon \xi'^i}{\xi'_k(x^k - \xi^k)} = \frac{c_0}{K_0 c} \frac{\varepsilon \xi'^i}{\xi'_k r^k}$$

in which the ξ' are the derivatives of the ξ with respect to t'. In this formula, r_j is the four-vector representing the distance in space-time between the events: position of the charge at time t', field at x, y, z at time t. The components of r^j are

$$x-\xi, \quad y-\eta, \quad z-\zeta, \quad ic(t-t')$$

and its length is zero:

$$(r^j)^2 = 0$$

It can indeed be shown that the denominator of Φ^i is of the form

$$\frac{1}{\sqrt{1-\beta^2}} \left\{(x-\xi)\frac{d\xi}{dt'} + (y-\eta)\frac{d\eta}{dt'} + (z-\zeta)\frac{d\zeta}{dt'} + ic.ic(t-t')\right\}$$

$$= \frac{c}{\sqrt{1-\beta^2}} \left\{\frac{\mathbf{r}.\mathbf{v}}{c} - c(t-t')\right\}$$

To recover V, we recall that (§ [66])

$$\Phi^l = \frac{ic_0}{c} V.$$

C. CONSEQUENCES FOR THE SYSTEMS OF UNITS: THE ARTIFICIAL AND OUTDATED NATURE OF THE GIORGI SYSTEM AND THE GAUSSIAN SYSTEM

[137] Preliminary Remark on Vocabulary

The names of the vectors of electromagnetism have given rise to numerous arguments which are not yet finished. To adopt a reasonable attitude, let us separate the chaff of words from the grain of facts. What, in short, is it all about?

In the general case of polarizable media, four vectors are necessary:

— two vectors characterizing the force exerted on a charge moving through the medium (the vectors that I have called **E** and $\mathfrak{B}$);

— two vectors characterizing the state of electrical polarization of the medium (which we call $\mathfrak{J}_e$ and $\mathfrak{J}_m$).

The theory of electromagnetism leads us to use not $\mathfrak{J}_e$ and $\mathfrak{J}_m$ but the vectors

$$\mathfrak{D} = K_0\mathbf{E}+4\pi\mathfrak{J}_e, \quad \mathbf{H} = \frac{\mathfrak{B}-4\pi\mathfrak{J}_m}{\mu_0}$$

Virtually everyone agrees that these are the physical meanings of the vectors. Furthermore, **E** and $\mathfrak{B}$ on one hand and **H** and $\mathfrak{D}$ on the other combine to form the tensors $\mathcal{N}_{ij}$ and $\mathcal{M}_{ij}$ in Minkowski space. It is therefore perfectly natural to call **E** and $\mathfrak{B}$ the electric and magnetic fields, respectively ($\mathcal{N}_{ij}$ being the four-field) and $\mathfrak{D}$ and **H** the electric and magnetic induction ($\mathcal{M}_{ij}$ is then the four-induction). It may perhaps be objected that with this nomenclature, the expressions for the inductions are not symmetrical (there is a minus sign in **H**). However, this asymmetry reflects the fact that electric polarization and magnetic polarization are very different physical states. The symmetry to which we have become accustomed by the use of magnetic poles has falsified our idea of the nature of the vectors: it is **H** that characterizes the magnetic polarization (that is, the distribution of Ampère currents) and not $\mathfrak{B}$.

Remark 1. In many textbooks, the opposite names are used for **H** and $\mathfrak{B}$. This arises because the concept of magnetic pole has been used.

Remark 2. In the French edition of 1957, I gave the preceding arguments but retained the old names so as not to depart too abruptly from the familiar. I announced, however, that if I were to publish a textbook of electricity, I should change the names; I have!

[138] General Remarks on Systems of Units

(a) The systems of units in use are legion. Many authors have striven to find a system possessing enough advantages over the others to eliminate them. *A priori*, it seems difficult to create an ideal system, for the same electrical quantity may have very different orders of magnitude in different applications.† There is a widespread feeling that the Giorgi (mks) system should be adopted, and many textbooks of electricity are written using this system exclusively. I have not followed this example, to avoid giving the reader tiresome transformations to make. Whether we like it or not, we have behind us a century and a half of texts and mental attitudes. It seems to me premature to adopt a single system, in a work such as this at least. I have thought it preferable to set out the formulae

† In the French edition of 1957, I added that the arguments about this question are often interesting, but that I could not understand the passion that some scientists feel for it. The importance of the scientific and above all philosophical attitudes escaped me.

We shall see later, however, that the problem of units can be put upon a sound theoretical footing.

in a form valid for the majority of systems, and I have therefore used three arbitrary constants K_0, μ_0 and c_0 (only two of which are independent); these give access to most extant systems. I give below a table with the aid of which the various notations may easily be extracted.

(b) We first recall the general formula

$$c = \frac{c_0}{\sqrt{K_0 \mu_0}}$$

(The relativistic constant has the dimensions of a velocity.)

The constant c (*not to be confused with* c_0) is equal to the velocity of light; it is nearly equal to 300,000 km/s. In more exact calculations, we must use 299,792 km/s. The coefficients below will then be modified accordingly.

[139] C.G.S. Systems

These systems use three fundamental units: the second, the centimetre and the gramme mass. For all these systems, we have

$$c = 3 \times 10^{10} \text{ cm/s} \quad (\mathrm{LT}^{-1})$$

The numerical values and dimensions of the constants in the four cgs systems in common use are as follows:

Electrostatic system (e.s.u.):

$$c_0 = 1\ (0), \quad K_0 = 1\ (0), \quad \mu_0 = 1/c^2\ (\mathrm{L}^{-2}\mathrm{T}^2)$$

Electromagnetic system (e.m.u.):

$$c_0 = 1\ (0), \quad K_0 = 1/c^2\ (\mathrm{L}^{-2}\mathrm{T}^2), \quad \mu_0 = 1\ (0)$$

Mixed system (de Broglie, Bergmann, Lichnerowicz, ...):

$$c_0 = c\ (\mathrm{LT}^{-1}), \quad K_0 = 1\ (0), \quad \mu_0 = 1\ (0)$$

Rationalized mixed system (von Laue, Tolmann, Møller):

$$c_0 = 4\pi c\ (\mathrm{LT}^{-1}), \quad K_0 = 4\pi\ (0), \quad \mu_0 = 4\pi\ (0)$$

Gaussian system:

$$c_0 = 1\ (0), \quad K_0 = 1/c\ (\mathrm{L}^{-1}\mathrm{T}), \quad \mu_0 = 1/c\ (\mathrm{L}^{-1}\mathrm{T})$$

[140] MKSA Systems (Giorgi)

These systems are characterized by the use of the mks geometrical and mechanical units: the metre, the kilogramme mass and the second, to which a fourth unit, the coulomb (or the ampère), is added.

The velocity c is therefore

$$c = 3\times10^8 \text{ m/s}$$

In the unrationalized system, we write

$$c_0 = 1\ (0),\quad \mu_0 = 10^{-7}\ (0),\quad K_0 = \frac{1}{9\times10^9}\ (\mathrm{L^{-2}T^2})$$

More exactly,

$$K_0 = 4\pi\times8{\cdot}85434\times10^{-12}\ (\mathrm{L^{-3}T^2M^{-1}Q^2})$$

The rationalized Giorgi system is obtained from this system by replacing K_0 and μ_0 everywhere by the new constants

$$K_0' = K_0/4\pi \quad\text{and}\quad \mu_0' = 4\pi\mu_0$$

respectively. As a result, the induction vectors are not the same in rationalized and non-rationalized units. The vectors $\mathfrak{D}$ and $\mathbf{H}$ are connected with the vectors $\mathbf{E}$ and $\mathfrak{B}$ by

$$\mathfrak{D} = K_0\mathbf{E},\quad \mathbf{H} = \mathfrak{B}/\mu_0$$

With the new constants K_0' and μ_0', we define two new vectors, $\mathfrak{D}'$ and $\mathbf{H}'$, by means of the relations

$$\mathfrak{D}' = K_0'\mathbf{E},\quad \mathbf{H}' = \mathfrak{B}/\mu_0'$$

We therefore have

$$\mathfrak{D}' = \mathfrak{D}/4\pi,\quad \mathbf{H}' = \mathbf{H}/4\pi$$

which of course affects the appearance of many of the formulae. Thus in electrostatics, for example, we have

$$\operatorname{div} \mathfrak{D}' = \varrho$$

instead of

$$\operatorname{div} \mathfrak{D} = 4\pi\varrho$$

Remark. By using my constants, K_0, μ_0 and c_0, all the common systems of units can be employed. More general formulae are to be found in an article by H. Gelman, *Amer. J. Phys.* **34** (1966) 291–295.

[141] Coherent System with Three Fundamental Quantities; Criticism of the Systems of Giorgi and Gauss

In § [110], we designated systems coherent if no arbitrary constants appear in the definitions of the units. Let us make this idea more exact. The definition of a physical quantity has two essential features. A new physical quantity X may be defined mathematically by means of a formula containing X and other quantities already known. The quantity X must also be defined operationally by describing a physical procedure with which X can be measured against a standard. The first quantity can, of course, be defined by the second process only. Two quantities are said to be independent or fundamental if they are defined only operationally using independent standards. A system of units is said to be independent with n fundamental quantities if every other quantity (called a derived quantity) is defined mathematically by the theory in terms of the fundamental quantities and without introducing arbitrary coefficients. Thus the system of geometrical quantities with one fundamental quantity—length—is said to be coherent if the area, S, of a rectangle of sides a and b is $S = ab$; it is incoherent if we introduce an arbitrary coefficient k and write $S = kab$.

In the present state of physics, the systems used in mechanics possess three fundamental quantities: time, length and mass. *This will be our starting-point.* I stress, however, that the figure 3 is wholly arbitrary and provisional. The advent of relativity and the evolution of metrology are tending to reduce it to 2 in the near future; I myself believe that the system of the future will contain only one fundamental quantity, time.

For electromagnetism, therefore, we seek a coherent system using the three fundamental quantities of mechanics and regulated naturally by the theory deduced from the Coulomb postulate. The first quantity we encounter in electromagnetism is electric charge, ε. It is defined by Coulomb's postulate

$$F = \varepsilon^2/r^2 \tag{1}$$

but if we write this with an arbitrary constant K_0,

$$F = \varepsilon^2/K_0 r^2 \tag{2}$$

we obtain an incoherent system. In fact, formula (2) can only define ε^2/K_0; one of these quantities is necessarily arbitrary. With (1), the quantity ε is a derived quantity, and we shall obtain a system containing only the three fundamental quantities of mechanics. The Giorgi system, on the contrary, requires the charge to be a fourth fundamental quantity. We can certainly obtain a coherent system with four quantities in this way, provided we define the charge (or the current which may replace it) operationally, and independently of the mechanical concepts. I have never encountered such an operational definition. One merely uses (2) with ε and K_0: the Giorgi system is in fact an incoherent system with three quantities. As the theory is developed, all the various authors introduce the concept of magnetic field with the aid of a supplementary postulate, the laws of Biot and Savart and the force on a moving charge. This is a logical break in theory, at which a second arbitrary constant μ_0 is introduced. *In my reconstruction of electromagnetism, this break is removed.* I extract the expression for $\mathfrak{B}$ from the (generalized) Coulomb postulate and the Lorentz transformation. If a coherent system (free of arbitrary constants) is desired, we write

$$\mathfrak{B} = \frac{1-\beta^2}{c^2 s^3}\, \varepsilon \mathbf{v} \times \mathbf{r}$$

The whole of electricity follows from this. Maxwell's equations *in vacuo* in the coherent system become

$$\text{curl } \mathbf{E} = -\frac{\partial \mathfrak{B}}{\partial t} \quad \text{and} \quad \text{curl } \mathfrak{B} = \frac{1}{c^2}\frac{\partial E}{\partial t}$$

and the relations between field and induction,

$$\mathfrak{D} = \mathbf{E} \quad \text{and} \quad \mathbf{H} = c^2 \mathfrak{B}$$

Only the relativistic constant c occurs, and *it arises from the Lorentz transformation (and not from an arbitrary convention)*. These considerations are very damning† to the Giorgi and Gaussian systems especially. In the former, the constant c does not figure explicitly, but two constants replace it. In the latter, c does indeed appear, but it is introduced quite arbitrarily for aesthetic reasons, and is not in the right place. We must never forget that c is a fundamental constant with a physical meaning: we cannot deal with it arbitrarily.

[142] Practical Consequences

In theoretical work, two attitudes are possible. If we wish to bring out the physical meaning of the formulae, and to decide whether or not they are relativistic, they must be written in the coherent system with the single constant, c; this is the truly scientific attitude. For purely practical reasons of textual convenience, however, it may be desirable to have formulae which can be adapted to any text. In this case, I advise the reader to retain the three constants K_0, μ_0 and c_0. *In all circumstances, I advise very strongly against the use of the Giorgi and Gaussian systems* for they are sources of misconceptions. For numerical calculations, it is a matter of indifference which system the formulae are expressed in, for no system employs units which suit every possible situation. It is rare for all the data of a problem to be expressed in terms of the units of a single system; transformation is almost always necessary. One can, of course, state a problem and solve it wholly in Giorgi units. This is an artificial exercise, however, divorced from real problems; this is the attitude of the ostrich. For a detailed commentary, see Appendix II of my *Electricité*.

HISTORICAL AND BIBLIOGRAPHICAL NOTES

[142A] Preliminary Remarks

(a) *Differences between the French edition of 1957 and the present edition*

Since 1957 I have pursued my studies in electromagnetism and published three books dealing with it. The essential feature of my thinking—the construction of electromagnetism from relativistic mechanics and formula (5) of § [133]—has not altered. My subsequent research has merely confirmed me in this point of view, expanded it and found applications for it. Nevertheless, I have had to modify or abandon certain arguments which I had originally taken over from earlier relativistic authors. The evolution has therefore taken place along the following lines: rejection of classical ideas in 1957 in favour of relativistic ones, subsequent improvement of this relativistic approach (leading up to a complete reformation). It is a continuous progress of the relativistic interpretation, and never a return to the pre-relativistic point of view.

The important changes are as follows—I refer to the section numbers of the 1957 edition.

Firstly, because of its relevance in electromagnetism, the problem of the jointed lever and the stretched rod: the theory given here is very different from that of 1957 (this is discussed in § [186]).

† The Giorgi system is not particularly precious from the practical point of view either (see my *Electricité*).

In § [124] of the 1957 edition, I adopted the Coulomb postulate involving force and with no propagation. It must be replaced (this is a fundamental amelioration) by a postulate involving potential, and propagation must be introduced. Without propagation, the retarded potentials cannot be derived correctly.

Of §§ [129] and [132], I retain the inequality of action and reaction. The deductions concerning the existence of a couple and an energy current must be wholly reconsidered, however. § [145] has been entirely rewritten, for the distinction between surfaces of equal phase and equipotential surfaces had escaped me.

The main features of § [150] remain, but the proof needed to be improved. In particular, the statement that the values of V_0 and $\mathbf{A}_0$ are independent of time and valid for any value of t_0 is to be deleted; it is wrong (and serves no purpose in the calculations in any case). See above, § [136A].

§§ [160] and [169], in which couples and energy currents are also involved, also require modification as I show in §§ [180–182].

(b) *The object of the following sections*

I give below a historical survey of work dealing with the formulae for the interactions between electric charges. I make no claim to completeness, for almost two centuries of literature are involved, and the publications are very numerous. I shall try above all to trace the evolution of ideas. Quite apart from giving credit to our precursors, this survey will give me an opportunity to discuss various contemporary opinions in the light of relativistic attitudes which are often misunderstood; despite its lacunae and imperfections, it will also reveal that relativistic physicists are not unaware of the earlier interaction formulae, which are offered from time to time by certain authors as new discoveries.

[143] On the Electrostatic Interactions and the Concept of the Electric Field

The first writings in which the inverse square law of distance for electric charges is stated are, so far as I know, those of Priestley [1] and Robison [2]. The basic work is that of Coulomb [4]; the measurements made by this latter author were incomparably more sensitive and accurate than those of his predecessors, thanks to his use of the torsion balance. Coulomb does not mention [1] and [2]; it must be said that the idea of this theoretical law would have come to mind naturally, at this period when the ideas of Newton were all the rage. The real problem was to show experimentally that this law also applied to electricity; it is therefore with justice that we call it Coulomb's law.

I have not been able to consult the works of Priestley and Robison myself: I have taken the references from Whittaker (ref. [29] of Appendix I). Indeed there are many others, sometimes of great originality (Cavendish's memoir [3] merits special mention), punctuating the long period between Thalès de Milet and Coulomb, although they in no way lessen the credit due to the latter. Newton likewise was preceded by many authors, in the formulation of the law of gravity, Boulliau, Borelli and Hooke, for example; no one today would think of contesting his right to be regarded as the father of this law.

Poisson [5, 6] developed the mathematical theory of electrostatic phenomena very considerably. Basing his work on the hypotheses of Coulomb (two kinds of electricity, the inverse square law) and on the work of Lagrange and Laplace on gravitation, he introduced the notion of potential into electricity. Laplace had shown, about 1782, that the potential satisfies the relation

$$\nabla^2 V = 0$$

at every external point. Poisson generalized the formula to include internal points thus:

$$\nabla^2 V = 4\pi\varrho/K_0$$

The Memoirs published by Poisson contain the theory of conductors and of equilibrium distributions in the form still used today. Prior to him, however, despite Coulomb's work, writings on electrostatics retained their eighteenth-century appearance and were thickly studded with false statements. It is therefore fair to say that after Coulomb, Poisson is the founder of modern electrostatics.

Let us finally mention the name of Gauss, whose well-known theorem was published in a paper on the attraction between ellipsoids and is constantly in use in electrostatics.

[1] J. PRIESTLEY: *The History and Present State of Electricity, with Original Experiments*, London, 1767, p. 732.
[2] JOHN ROBISON: *Mechanical philosophy*, Edinburgh, Vol. IV, 1769, p. 73.
[3] *The Electrical Researches of the Hon. Henry Cavendish, written between 1771 and 1781*, ed. J. Clark Maxwell, Cambridge U.P., 1879. The French reader will find a discussion of this in M. Brillouin: *Propagation de l'électricité, Histoire et théorie*, Hermann, 1904.
[4] C. A. COULOMB: Où l'on détermine suivant quelles lois le fluide électrique ainsi que le fluide magnétique agissent soit par répulsion, soit par attraction. *Mémoire* communicated to the Académie Royale des Sciences in 1785, and reproduced in *Collection des Mémoires publiés par la Société française de Physique*, Gauthier-Villars, Vol. I, 1884, pp. 116–146.
[5] S. D. POISSON: Sur la distribution de l'électricité à la surface des corps conducteurs. *Mém. de l'Institut* **1** (1811) 1–274.
[6] S. D. POISSON: Remarques sur une équation qui se présente dans la théorie des attractions des sphéroïdes. *Bull. Soc. Phil.* **3** (1812) 388–392.

[144] The Interaction Between Current Elements (in the form $i\,ds$)

(a) In order to explain his experimental results on the interactions between current-carrying wires (which he performed after learning of Oersted's observations, ref. [1] of § [146], Ampère tried to find a law of interaction between current elements without worrying about the structure of the currents: he characterized them by their strength alone.

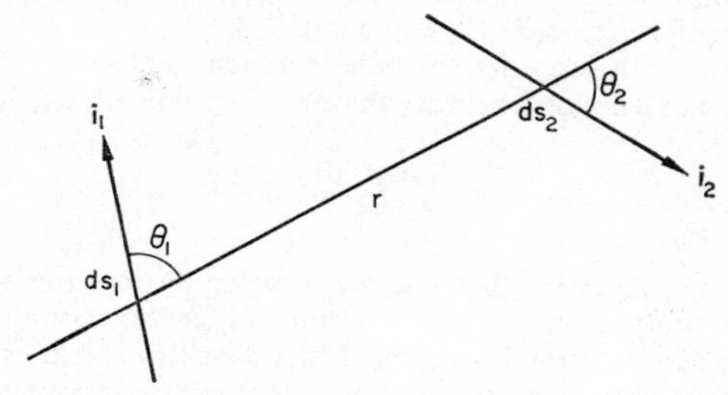

FIG. 35.

Setting out with a certain number of experimental results, which he regarded as rigorously valid theoretically, he showed that *if the law to be established is required to satisfy Newton's third law* (action and reaction must be equal) the following expression for the force is obtained (Fig. 35):

$$\begin{aligned}\frac{d\mathbf{F}}{\mu_0} &= \frac{2i_1i_2\,ds_1\,ds_2}{r^3}\left(\sin\theta_1\sin\theta_2\cos\omega-\frac{\cos\theta_1\cos\theta_2}{2}\right)\mathbf{r}\\ &= \frac{2i_1i_2\,ds_1\,ds_2}{r^3}\left(\cos\varepsilon-\frac{3}{2}\cos\theta_1\cos\theta_2\right)\mathbf{r}\\ &= \frac{i_1i_2\,ds_1\,ds_2}{r^3}\left(\frac{\partial r}{\partial s_1}\frac{\partial r}{\partial s_2}-2r\frac{\partial^2 r}{ds_1\,\partial s_2}\right)\mathbf{r}\\ &= \frac{4i_1i_2\,ds_1\,ds_2}{\sqrt{r}}\frac{\partial^2\sqrt{r}}{\partial s_1\,\partial s_2}\frac{\mathbf{r}}{r}\end{aligned}\qquad(1)$$

To establish this sequence, we use

$$\cos\varepsilon = \cos\theta_1\cos\theta_2+\sin\theta_1\sin\theta_2\cos\omega$$

$$\frac{\partial r}{\partial s_1} = -\cos\theta_1,\qquad \frac{\partial r}{\partial s_2} = \cos\theta_2$$

$$r\frac{\partial^2 r}{\partial s_1\,\partial s_2} = \cos\theta_1\cos\theta_2-\cos\varepsilon$$

dF denotes the force exerted by ds_1 on ds_2; $i_1\, ds_1$ and $i_2\, ds_2$ are the two current elements, distance r apart and **r** is the corresponding vector from ds_1 to ds_2; θ_1 and θ_2 are the angles between the currents $\mathbf{i}_1$ and $\mathbf{i}_2$ and **r**, and ω is the angle between the planes containing r and ds_1 on one hand and r and ds_2 on the other; ε is the angle between the two elements ds_1 and ds_2.

The force is directed along the line joining the elements and in addition, the force exerted on ds_2 by ds_1 is equal and opposite to the force exerted on ds_1 and ds_2. The principle of "action and reaction" is satisfied, hence the reason for the prominence formerly accorded this formula: today, we know that this is one of its vulnerable points.

The expression above gives correct results for closed circuits, but we shall see that there are simpler formulae for this task.

(b) Many textbooks of electricity point out that an infinite number of elementary formulae exist for current elements, all leading to the same result when applied to closed circuits. In fact, any total differential may be added which gives zero when integrated round a closed loop. Ampère was led to adopt the formula above by adding a supplementary mechanical condition (action equal to reaction).

If, however, influenced by Maxwell and his displacement currents, we accept that all currents do flow in closed circuits, we see that although we can observe the force exerted on a current element, we cannot observe that exerted on one current element by another. In these conditions, we cannot but attach little importance and even deny all meaning to the problem of whether the elementary formulae are correct. Which is chosen is a matter of convenience only.

I consider that this attitude (still adopted by many authors) *is no longer compatible with the present state of knowledge*. The basic problem now is not to find a formula for the interaction between current elements but between moving charges. The "elementary" formula therefore has an exact meaning and must be unique: we have given it in § [133] for weakly accelerated charges.

The same is true in consequence of the formula for current elements, once their nature has been specified (metal conductors, for Ampère's experiments).

We now give an example of a proposed formula in which the forces $d\mathbf{F}_1$ (the action of ds_1 on ds_2) and $d\mathbf{F}_2$ (the action of ds_2 on ds_1) no longer satisfy the law of action and reaction. Ampère himself pointed out that the formula

$$d\mathbf{F}_1 = \mu_0 \frac{i_1\, ds_1 i_2\, ds_2}{r^2} \sin\theta_1 \cos\mu \tag{2}$$

as equivalent to (1) for closed circuits (μ is the angle between ds_2 and the plane containing r and ds_1). He, however, preferred formula (1) because of the action and reaction condition. In the older textbooks, formula (2) is often called the formula of Grassmann, Hankel or Reynard; these authors gave it in 1845 [2], 1865 [3] and 1870 [4], much later than Ampère but with new and important comments. In the first place, they suggested that it should be adopted and not, as did Ampère, rejected. This led them to seek an explanation of the inequality of action and reaction, and not purely and simply to reject it as an absurdity. I shall not go into their explanations, but it seemed interesting to draw attention to this new attitude.

We know today that (2) is the only correct formula for metal conductors; it is deduced from Biot and Savart's law and the law of force on a moving charge. For the reduction of formula (2) to formula (1) by ꞁidding total differentials, the reader is referred to Mascart's treatise ([6], p. 508).

I shall not discuss other types of formulae; the reader who is interested in this subject will find abundant details and references in Duhem [5] and Whittaker [8] concerning the formulae suggested by Helmholtz, Neumann and Clausius; we shall, moreover, obtain formulae of the same type by a different method (§ [145]).

The formulae for elements have been virtually abandoned. Nevertheless, we mention that in a recent treatise [9] the formula for the interaction between two closed circuits is used as the basic postulate of magnetostatics, and this is equivalent to stating the law between current elements, apart from the definite integral.

[1] M. A. Ampère: *Collection de Mémoires publiés par la Société française de Physique*, Gauthier-Villars, Vols. II and III, 1887: Articles on electrodynamics. Reynard's formula is given in Vol. III, p. 123. We recall—for this is rather a sensational fact—that Ampère's first communication before the Académie was on 18th September 1820, just a week after Arago's communication on Oersted's experiment (ref. [1], § [146]).

[2] H. Grassmann: Neue Theorie der Elektrodynamik. *Ann. Phys. Chem.* **64** (1845) 1–18.

[3] Hankel: Neue Theorie der elektrischen Erscheinungen. *Ann. Phys. Chem.* (1865) 440–466; (1867) 607–621.

[4] M. REYNARD: Nouvelle théorie des actions électrodynamiques. *Ann. Chim. Phys.* **19** (1870) 272–328.
[5] P. DUHEM: *Leçons sur l'électricité et le magnétisme.* Gauthier-Villars, Vol. III, 1892. This book is undoubtedly the best guide for anyone wishing to make a thorough historical study of the question. The various formulae are studied and compared in a most painstaking way.
[6] E. MASCART: *Leçons sur l'électricité et le magnétisme,* 2nd ed., Masson and Gauthier-Villars, Vol. I, 1896. The interactions between elements are studied on pp. 503–523 and 655–665.
[7] H. POINCARÉ: *Electricité et optique,* 2nd ed., Gauthier-Villars, 1901. Republished, 1954. The theories of Ampère, Weber and Helmholtz are described on pp. 229–310. This book and that of Massart are the most suitable for a rapid survey.
[8] E. T. WHITTAKER: ref. [29] of Appendix I.
[9] E. DURAND: *Magnétostatique,* Masson, 1968.
[10] R. A. R. TRICKER: *Early Electrodynamics,* Pergamon, Oxford, 1965.

[145] The Interaction Between Moving Charges; Arbitrary Velocities, Small Accelerations

(a) *Work before 1905*

The search for postulates about interaction at a distance. The authors who followed the route indicated by Ampère attempted to construct the formulae for current elements from two categories of hypotheses:

— hypotheses involving the nature of the current, charges with opposite signs and velocities, for example;
— hypotheses about the interaction between moving charges.

The phenomenon to be explained was thus more finely analysed, and this line of attack on the question was to prove very fruitful. Examination of the simplest possible case immediately leads to the result that the action between two moving charges is not the same as that between two stationary charges at the same points. From this emerged the idea of the delay caused by the time of propagation. From it might have emerged the idea of the relativity of force and hence the necessity for a new dynamics!

The first known expression is probably due to Gauss [1]; it is equivalent to the formula

$$\mathbf{F} = \frac{\varepsilon_1 \varepsilon_2}{K_0 r^3}\left\{1+\frac{v^2}{c^2}\left(1-\frac{3}{2}\cos^2\theta\right)\right\}\mathbf{r}$$

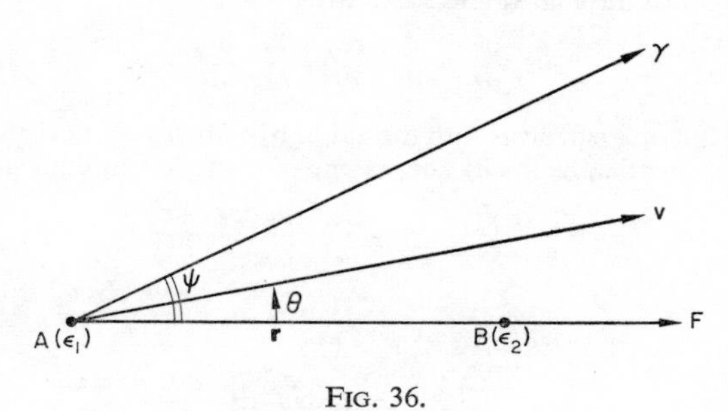

FIG. 36.

The symbols are defined in Fig. 36. At a given time with respect to the system of reference employed, the charges ε_1 and ε_2 are at positions A and B respectively; the angle between their relative velocity v and AB is θ. Hence

$$\frac{\mathrm{d}r}{\mathrm{d}t} = v\cos\theta$$

It is I who add "with respect to the system of reference employed"; Gauss and especially his followers insisted, on the contrary, on the fact that only the velocity of one charge relative to the other would matter. It is clear, however, that a reference system reduced to a point does not allow us to define v and θ, but only $\mathrm{d}r/\mathrm{d}t$. The formula satisfies the law of action and reaction.

Weber [2] gives the following formula, which takes possible accelerations into account but which differs from Gauss's formula even for small accelerations:

$$\mathbf{F} = \frac{\varepsilon_1\varepsilon_2}{K_0 r^3}\left\{1+A\frac{\mathrm{d}^2 r}{\mathrm{d}t^2}+B\left(\frac{\mathrm{d}r}{\mathrm{d}t}\right)^2\right\}\mathbf{r}$$

in which A and B are functions of r only and must be selected in such a way as to yield Ampère's formula,

$$\frac{i_1 i_2\,\mathrm{d}s_1\,\mathrm{d}s_2}{r^3}\left(\frac{\partial r}{\partial s_1}\frac{\partial r}{\partial s_2}-2r\frac{\partial^2 r}{\partial s_1\,\partial s_2}\right)\mathbf{r}$$

Weber's formula was of an exploratory kind, based upon the following general conditions; the required expression must:

—contain the parameters of the motion (velocity and acceleration) as simply as possible; the velocity must appear squared, so that the same force is obtained when the direction of motion of the two charges is reversed;
—reduce to Coulomb's law in equilibrium;
—yield a force along **r**.

The reader can acquire a rapid idea of the calculation of the constants from the works of Poincaré or Mascart, mentioned in the preceding section. The full formula, using the notation of Fig. 36, is

$$\mathbf{F} = \frac{\varepsilon_1\varepsilon_2}{K_0 r^3}\left\{1+r\frac{\mathrm{d}^2 r}{\mathrm{d}t^2}-\frac{1}{2}\left(\frac{\mathrm{d}r}{\mathrm{d}t}\right)^2\right\}\mathbf{r}$$

$$= \frac{\varepsilon_1\varepsilon_2}{K_0 r^3}\left(1+r\gamma\cos\psi-\frac{v^2\cos^2\theta}{2}\right)\mathbf{r}$$

Action and reaction are equal. The total force has potential

$$\frac{\varepsilon_1\varepsilon_2}{K_0 r}\left\{1-\frac{1}{2}\left(\frac{\mathrm{d}r}{\mathrm{d}t}\right)^2\right\};$$

the magnetic force (the total force less the electrostatic term, that is, the force between current elements) is obtained from the dynamical potential

$$-\frac{\varepsilon_1\varepsilon_2}{2K_0 r}\left(\frac{\mathrm{d}r}{\mathrm{d}t}\right)^2$$

Since r is a function of t through s_1 and s_2 we write

$$\frac{\mathrm{d}r}{\mathrm{d}t} = \frac{\partial r}{\mathrm{d}s_1}\frac{\mathrm{d}s_1}{\mathrm{d}t}+\frac{\partial r}{\partial s_2}\frac{\mathrm{d}r_2}{\mathrm{d}t}$$

Weber then eliminates the squared terms with the aid of hypotheses about the structure of the currents; for the electrodynamic potential, he finally adopts one of the following equivalent expressions:

$$\frac{\varepsilon_1\varepsilon_2}{K_0}\frac{v_1 v_2}{r}\frac{\partial r}{\partial s_1}\frac{\partial r}{\partial s_2} = \frac{i_1\,\mathrm{d}s_1 i_2\,\mathrm{d}s_2}{r}\frac{\partial r}{\partial s_1}\frac{\partial r}{\partial s_2}$$

$$= i_1\,\mathrm{d}s_1 i_2\,\mathrm{d}s_2\frac{\cos\theta_1\cos\theta_2}{r}$$

$$= i_1\,\mathrm{d}s_1 i_2\,\mathrm{d}s_2\frac{\partial\sqrt{r}}{\partial s_1}\frac{\partial\sqrt{r}}{\partial s_2}$$

Some later authors retain this idea of electrodynamic potential.

The formulae of Carl Neumann and von Helmholtz, the existence of which was mentioned in the preceding section, essentially depend upon the existence of dynamical potentials:

Neumann

$$i_1\,\mathrm{d}s_1 i_2\,\mathrm{d}s_2\frac{\cos\varepsilon}{r}$$

Helmholtz

$$i_1\,\mathrm{d}s_1 i_2\,\mathrm{d}s_2\left(A\frac{\cos\varepsilon}{r}+B\frac{\cos\theta_1\cos\theta_2}{r}\right)$$

the last expression is a generalization of the two preceding formulae.

These potentials give different laws of interaction from that of Ampère. For the latter, there is a potential only for the force between two closed circuits, and not for the interaction between two elements nor for the interaction between a closed circuit and an element. I shall not discuss how Weber and the others obtain agreement (all these authors regard Ampère's formula as the only really true one); for a preliminary historical account, the reader should consult Poincaré's book. I merely mention that all these theories agree for closed metallic circuits.

In this case, the reader will have recognized in the foregoing expressions the classical form of the mutual energy of two closed circuits.

Riemann [6] suggested

$$\frac{\varepsilon_1\varepsilon_2}{K_0 r^2}\left\{\left(1+\frac{v^2}{c^2}\right)\frac{\mathbf{r}}{r}-\frac{2v}{c^2}\,\mathbf{v}\cos\theta+\frac{2r}{c^2}\boldsymbol{\gamma}\right\}$$

in which $\boldsymbol{\gamma}$ is the acceleration vector. He also introduced the idea that the velocity of propagation of the action ought to appear in the formulae; thus for a variable charge, for example, he pointed out that

$$\varepsilon(t-r/c)$$

should be used instead of

$$\varepsilon(t)$$

The interaction calculated by means of a field

Research into these questions took on a new direction with the use of the idea of electromagnetic field, developed principally under the guidance of Faraday and Maxwell. Instead of looking for postulates which would give the interaction between two charges directly, these authors sought to elucidate the way in which the action is transmitted between the charges. The interaction formulae are no longer basic postulates, but consequences of the field theory. This point of view enabled them to establish in the pre-relativistic era the correct formulae for arbitrary velocities and small accelerations. The historical study of this is postponed until §§ [146] ff.

(b) *Work after 1905*

The theory of relativity enables one to adopt either standpoint at will. If we wish to discuss the interaction, a simple change of system allows us to pass from the Coulomb force to the general case. Alternatively the idea of field may be used, with the same result of course. We have shown this above, in the present chapter, for weakly accelerated charges. This double point of view also applies to the general case of arbitrary accelerations. As many authors have shown, modern electromagnetism can be developed in a coherent and equivalent fashion, either using the field concept or using only the concept of action at a distance. For a detailed bibliography, see my *Rayonnement et Dynamique*. As a strikingly typical article, I mention here that of Wheeler and Feynman (ref. [2] of the Complementary Remarks below), which follows upon the work of Schwarzschild, Tetrode and Fokker. We might therefore expect this historical review to end here. On the contrary, however, there is a whole group of papers trying to cast doubt on the whole question.

In a broad critical survey of the foundations of electrodynamics, Ritz [7] suggests that the concept of field should be discarded, for he regards it as a useless mathematical intermediary. He finds it indispensable to base this science uniquely upon the interaction postulate, as did the earliest authors; this, however, is an underestimate of the worth of the field method, taking only one side of the question into account. Furthermore, Ritz makes indefensible conditions at the outset: equality of action and reaction for charges in uniform relative motion; the formulae are to contain only the relative velocities between the charges; relativistic kinematics is to be rejected (for, he says, it would be regrettable for our mental economy to have to accept such complications). He first sets out in a most clear and interesting way the Lorentz theory and the formulae that we regard as correct, but he then rejects them since they do not satisfy his prerequisites. He then offers a new theory, which yields the formula

$$\frac{\varepsilon_1\varepsilon_2}{K_0 r^2}\left[\left\{1+\frac{3-k}{4}\,\frac{v^2}{c^2}-\frac{3(1-k)}{4}\,\frac{v^2}{c^2}\cos^2\theta-\frac{r}{2c^2}\gamma\cos\psi\right\}\frac{\mathbf{r}}{r}-\frac{k+1}{r}\,\frac{v\cos\theta}{c^2}\,\mathbf{v}-\frac{r}{c^2}\boldsymbol{\gamma}\right]$$

in which k is an arbitrary constant to be determined by experiment. If it is determined from the condition that the formula must reduce to that of Gauss for negligible $\boldsymbol{\gamma}$, we obtain $k=-1$; the formula becomes

$$\frac{\varepsilon_1\varepsilon_2}{K_0 r^2}\left[\left\{1+\frac{v^2}{c^2}\left(1-\frac{3}{2}\cos^2\theta\right)\right\}\frac{\mathbf{r}}{r}-\frac{r}{c^2}\boldsymbol{\gamma}\right]$$

Action and reaction are different, but the difference contains only the acceleration; for uniform velocities, therefore, it vanishes. As Ritz himself pointed out, this formula is a general linear combination of the laws of Weber and Riemann. Moreover, Ritz stressed the fact that his Memoir was only a critical study of the theories of Lorentz and Maxwell, and that his own theory was as yet still in a tentative form. His early death (at 31 years of age) prevented him from pursuing it further.

His ideas—essentially, the search for a theory not using relativistic mechanics—were taken up afresh 25 years later by a few authors. In 1934 Hovgaard disinterred Ritz's theory which, suitably developed and transformed, should allow one to avoid the complications of relativity, according to him.

Bush [9] rediscovered Gauss's formula independently, and applied it to various problems.

Warburton [11, 12], having stated the potential energy, obtained a formula which was a generalization of those of Gauss, Weber, Riemann and Ritz. Moon and Spencer [13–20] again proposed to rebuild electrodynamics upon formulae for the interaction between charges. They first studied the case of uniform velocities and used the unacceptable postulates already mentioned (on action and reaction, ...). For two charges in uniform motion, they find Gauss' formula and for two current elements, the classical formula of Ampère. I shall not go into the more general formulae they suggest for the case of accelerated charges or time-varying charges. In connection with these two authors, I draw attention to a historical article [14], a new way of introducing Maxwell's equations [16] and finally, a polemic with Warburton (both points of view seem to me equally wrong).

Conclusion

Let us glance back over all the preceding formulae, neglecting the acceleration terms and ignoring those calculated by means of the field, the historical study of which is postponed until § [147]. They are all different from the formulae given in the present text, in § [133]. I therefore regard them all as wrong, despite the warning inherent in the recent works that I have just quoted, and despite the fact that in certain cases, they lead to the correct results (for closed metallic circuits, for example).

It seems obvious that they cannot be absolutely satisfactory because according to their authors (who insist upon this time and again), only the relative velocity v of the two charges (the velocity of one with respect to the other) occurs. If the velocity v is zero (as well as the acceleration), they reduce to Coulomb's law, and this is clearly wrong (see my calculation § [132]). The velocity of the charges *with respect to the system of reference* must necessarily be introduced (even for low velocities). I say nothing of high velocities, with which the authors I have mentioned do not deal. Also, from the epistemological point of view, the general attitude that they adopt is hardly satisfactory. They wish to state a law for moving charges which is to be an initial postulate, in isolation. We have seen, however, that the interaction formulae are given by a simple application of particle dynamics, provided that the correct mechanics is employed. Here some authors have gone astray over the fact that relativistic dynamics must be used *even at low velocities;* to others this dynamics is wholly repugnant. Why should one wish to dispense with the principle of relativity at all costs, when one ought to be exploiting its astounding possibilities to the full? These remarks are directed at the papers by Moon and Spencer in particular. These authors ought to have reflected upon the recent article by Sard (ref. [11] of § [147]) which gives a good digest of the subject.

I am not at all unfamiliar with the state of mind of the non-relativistic authors; for several years, I worked in the intellectual atmosphere of Bouasse, who was perhaps the most violent of the anti-relativistic physicists. These prejudices must be abandoned; it should no longer be lightly asserted that the relativistic physicists (Einstein foremost among them) hold naïve beliefs: the theory must be *studied* and *used* before it is judged.

In conclusion, the only formulae which agree with our present knowledge are those given by field theory or by the direct relativistic theory that I have developed. In particular, *the Reynard–Ampère formula is obtained* for low velocities and current elements. What detours and what labour, only to return finally to one of the starting-points! Two steps forward have, nevertheless, been taken: the choice of formula is no longer arbitrary and the formula is attached to dynamics.

Remark. For the older papers, earlier than 1900 say, a better bibliographic starting-point is to be found in the textbooks of the period (Duhem in particular). Here I list only a few striking works, to set off the principal names. A complete bibliography would fill 20 or 30 pages.

[1] C. F. Gauss: *Zur Mathematischen Theorie der Elektrodynamischen Wirkung. Werke*, Göttingen,

Vol. V, 1867, pp. 601–629. The formula given on p. 617 dates from 1835, but was published only in 1867; it was found among Gauss's papers after his death.

[2] W. WEBER: *Elektrodynamische Massbestimmungen über ein Allgemeines Grundgesetz der Elektrischen Wirkung*, Julius Springer, Berlin, 1893. The first publication dates from 1846 (in the *Abhandlungen Leibnitzens Gesellschaft* in Leipzig). One of his papers is to be found in Vol. III of *Mémoires sur l'Electrodynamique*, ref. [1] of § [144]; Poincaré, Mascart and Duhem studied his ideas in great detail.

[3] CARL NEUMANN: *Die elektrischen Kräfte. Darlegung und Erweiterung der von A. Ampère, F. Neumann, W. Weber, G. Kirchhoff, entwickelten mathematischen Theorien*, Leipzig, 1873. In this text, the work of F. E. Neumann on closed circuits, written in 1845 and 1847, is studied in particular, in which the law for elements is used implicitly.

[4] H. HELMHOLTZ: Ueber die Theorie der Elektrodynamik: Die elektrodynamischen Kräfte in bewegten Leitern. *Borchardt's J. Reine Angew. Math.* **78** (1874) 273–324.

[5] R. CLAUSIUS: *Die Mechanische Wärmetheorie.*

[6] B. RIEMANN: Ein Beitrag zur Elektrodynamik. *Ann. Physik* **131** (1867) 237–243. A contribution to electrodynamics, *Phil. Mag.* **34** (1867) 368–372; see also *Gesammelte Mathematischen Werke*, Teubner, Leipzig, 1892, p. 288.

[7] W. RITZ: Recherches critiques sur l'électrodynamique générale. *Ann. Chim. Phys.* **13** (1908) 145–275.

[8] W. RITZ: Recherches critiques sur les théories électrodynamiques de Maxwell et Lorentz. *Arch. Genève* **26** (1908). This paper, which is shorter than [7], is a good introduction to the ideas of Ritz. An English translation is given in the article by Hovgaard [10].

[9] V. BUSH: The force between moving charges. *J. Math. Phys.* **5** (1926) 129–157.

[10] W. HOVGAARD: Ritz's electrodynamic theory. *J. Math. Phys.* **11** (1932) 218–254.

[11] F. W. WARBURTON: Reciprocal electric force. *Phys. Rev.* **69** (1946) 40.

[12] F. W. WARBURTON: Relative electrodynamics. *Bull. Amer. Phys. Soc.* **23** (1948) 4.

[13] P. MOON and D. E. SPENCER: Interpretation of the Ampère experiments. *J. Frank. Inst.* **257** (1954) 203–220.

[14] P. MOON and D. E. SPENCER: The Coulomb force and the Ampère force. *J. Frank. Inst.* **257** (1954) 305–315.

[15] P. MOON and D. E. SPENCER: Electromagnetism without magnetism: an historical sketch. *Amer. J. Phys.* **22** (1954) 120–124.

[16] D. E. SPENCER: Maxwell's equations and the new electrodynamics. *Bull. Amer. Math. Soc.* **60** (1954) 167.

[17] P. MOON and D. E. SPENCER: A new electrodynamics. *J. Frank. Inst.* **258** (1954) 395–398.

[18] This reference in the French edition omitted here.

[19] P. MOON and D. E. SPENCER: Electromagnetism, old and new; a reply. *J. Frank. Inst.* **258** (1954) 398–400.

[20] P. MOON and D. E. SPENCER: A postulational approach to electromagnetism. *J. Frank. Inst.* **259** (1955) 293–305.

[21] P. MOON and D. E. SPENCER: On the Ampère force. *J. Frank. Inst.* **260** (1955) 295–311.

Complementary Remarks, prepared for the present edition

The relativistic calculations of the interactions between charges in uniform motion has been given by Tolman [1], a fact of which I was unaware when composing the French edition of 1957. This remarkable paper has passed unnoticed, and had hardly any influence; indeed, Tolman himself seems subsequently to have forgotten it when he was writing his treatise! Tolman's and my formulae are identical, and it is curious, in view of their wholly independent derivations, that the same notation, s, is found in both.

The problem has also been investigated by Alsina in his thesis [3, 4] and in recent papers [8, 10]. This author has just written to tell me that since 1962, the Electricity lectures in Caracas University employ these basic relativistic principles. Finally, the work of Rosser [6, 7] may be consulted.

To my certain knowledge, therefore, the same train of thought has been followed independently by Tolman, Alsina, Dacos (§ [147]), Rosser and myself. Yet another striking example of the remark in § [74] of my *Kinematics*. Nevertheless, I should like to emphasize the following, rather deep-rooted differences between my account and those of my predecessors:

— The basic postulate that I adopt is concerned not with the field but with the potential; furthermore it contains the delay conditions with the velocity c from the outset.

— The postulate is unique; the concept of magnetic field and Maxwell's equations (which are no longer treated independently) are deduced from it.
— The method has been exploited to good effect in a treatise on general electricity.

Calculations on the interactions between parallel currents, generalizing Searle's results (ref. [17] of § [146]), have been pursued by Malatesta (1a).

In Blanc's thesis [9], the general question of the interaction between moving charges is reconsidered by asserting postulates directly; I agree neither with his method nor, generally speaking, with his results, but his work nevertheless makes instructive and suggestive reading.

[1] R. C. Tolman: Non-Newtonian mechanics; some transformation equations. *Phil. Mag.* **25** (1913) 150–157.
[1a] S. Malatesta: Interpretazione relativistica delle azioni elettrodinamiche. *Acta Pontificia Acad Sci.* **9** (1945) No. 3, 21–26.
[2] J. A. Wheeler and R. P. Feynman: Classical electrodynamics in terms of direct interparticle action. *Rev. Mod. Phys.* **21** (1949) 425–433.
[3] F. Alsina: Relaciones entre electromagnetismo y relatividad. Thesis, La Plata, 1951.
[3a] F. Alsina: Acciones entre conductores paralelos. *An. Soc. Cient. Arg.*, Enero, 1951.
[4] F. Alsina: Fuerzas entre cargas y conductores. *An. Soc. Cient. Arg.*, Febr. 1952.
[5] P. Moon and D. E. Spencer: The new electrodynamics and its bearing on relativity. *Kritik und Fortbildung der Relativitätstheorie*, Ed. Karl Sapper, Akad. Verlag, Graz, Austria, n.d. (approx. 1960) pp. 144–159.
[6] W. G. V. Rosser: Electromagnetism as a second order effect. *Contemp. Phys.* **1** (1960) 453–466.
[7] W. G. V. Rosser: Electromagnetism as a second order effect. *Contemp. Phys.* **3** (1961) 28–44.
[8] F. Alsina: La electrotecnica como consecuencia relativista. *Rev. Un. Mat. Arg.* **22** (1962) 85–90.
[9] P. Blanc: Analyse des fondements logiques et expérimentaux des lois de l'électromagnétisme classique. Thesis, Paris, 1963, Imprimerie Nationale.
[10] F. Alsina: La electrotecnica como consecuencia relativista, II. Corrientes variables. *Ciencia y Tecnica*, **133** (March–April 1964) No. 671, 77–80.

To these authors, I must add R. Mendez and chiefly J. Henry, who worked with me when I was preparing the 1957 edition.

[146] The Electromagnetic Field of a Current or of a Charge in Uniform Motion

We have sketched the history of the concept of electric field in § [143]. The magnetic field first appeared as a field acting on a magnetized needle. The majority of textbooks still define it in this way.

The topic of which we now study the history is thus concerned with the action of an electric current or of charges in motion (the two cases must be distinguished, for the two problems were not originally treated together) on a magnetic needle. We then study the interactions between currents or moving particles using the field as intermediary.

Ever since the opening decades of the eighteenth century, we find papers on the relations between electricity and magnetism. They dealt with unrelated theories and observations, however (thus Desormes suspended an insulated battery to see whether it would align itself in the earth's magnetic field). Not until 1820 was an exact experiment described, which displayed the phenomenon clearly. This was Oersted's paper [1], dealing with the field of a rectilinear current, and by virtue of the immediate reactions it provoked among the physicists, it deserves to be regarded as the effective starting-point (*there are few examples of such rapid progress, after more than a century of stagnation*).

The quantitative law was given by Biot and Savart, first for the action of an infinite rectilinear current [2] and later for a current element [3] in the form still used today.

The first positive experiment, showing the magnetic action of moving charges, is due to Rowland [5]; he observed the action of an electrostatically charged rotating disc on a magnetized pointer. This is described in most electricity textbooks.

This experiment was repeated by Crémieu [7] using a different method (measurement by induction), and as a result of parasitic effects arising from his experimental arrangement, he obtained negative results. Numerous arguments ensued, the echoes of which are still to be heard in Poincaré's work [8].

Pender [9] repeated Crémieu's experiments in America, with some improvements. He, however, obtained positive results, and a very good approximation (he measured the ratio of electrostatic and magnetic units).

Thanks to a fortunate and most rare conjunction of circumstances,† the two physicists were able to work together in Paris and come to an agreement, thus furnishing a remarkable example of scientific collaboration. The final outcome was complete quantitative agreement between experiment and field calculation using Biot and Savart's law.

It is instructive to glance over the arguments of the period, to understand properly that the identification of a charge in motion with a current element was *far from obvious*. The full study of this question involves the electrodynamics of moving media, which is treated in another book [24].

Crookes, Goldstein and others studied electrical discharge in rarefied gases and thus observed charged particles at high velocities. They studied the interactions among these particles, the action of a magnetic field on the beam, and so on. This was the experimental starting-point from which relativistic mechanics was to emerge.

Using Maxwell's theory [4], J. J. Thomson [11] calculated the electromagnetic field created by a charged sphere in uniform motion. He agreed that the electric field at each moment is the same as for a stationary charge (correct for low velocities). From the variation of this field, he deduced the magnitude of Maxwell's displacement currents. He then treated these currents as real currents, of which he calculated the vector potential, **A**. The magnetic field is given by curl **A**.

This method is not absolutely correct, since it does not take the convection current formed by the moving charge into account. Fitzgerald [12] noticed this shortcoming, and showed that for a point charge, ε, the magnetic field is the same as that of a current element $\varepsilon\mathbf{v}$. He calculated along the same lines as Thomson, but included the convection current; the magnetic field arising from the displacement currents is zero.

In modern terminology, the reasoning is as follows: we shall not use the vector potential, which is here a source of confusion (we shall return to this in connection with Heaviside). We set out from Maxwell's equation, which we write

$$\frac{K_0}{c_0}\frac{\partial \mathbf{E}}{\partial t} = \text{curl } \mathbf{H}$$

or, for a closed circuit,

$$\frac{K_0}{c_0}\frac{\partial}{\partial t}\iint \mathbf{E}\,.\,\mathrm{d}\mathbf{S} = \frac{K_0}{c_0}\frac{\partial \mathcal{F}}{\partial t} = \iint \text{curl } \mathbf{H}\,.\,\mathrm{d}\mathbf{S}$$

$$= \oint \mathbf{H}\,.\,\mathrm{d}\mathbf{l}$$

Consider a charge ε travelling along Ox with velocity v (Fig. 37). At a point M, the flux of **E** through a circle radius R and centred on Ox is

$$\mathcal{F} = \frac{2\pi\varepsilon}{K_0}(1-\cos\alpha)$$

It was this flux which, at that period, was called the flux of displacement current; hence

$$\frac{\partial \mathcal{F}}{\partial t} = \frac{\partial \mathcal{F}}{\partial x}\frac{\mathrm{d}x}{\mathrm{d}t} = \frac{2\pi\varepsilon v}{K_0}\frac{\partial}{\partial x}\left(1-\frac{x}{r}\right)$$

where

$$r^2 = x^2+R^2, \qquad r\,\mathrm{d}r = x\,\mathrm{d}x$$

Finally,

$$\frac{\partial \mathcal{F}}{\partial t} = -\frac{2\pi\varepsilon v}{K_0}\frac{R^2}{r^3}, \qquad H = -\frac{\varepsilon v}{c_0}\frac{\sin\alpha}{r}$$

O. Heaviside [13, 14] gives the solution for a point moving at an arbitrary velocity. On this occasion, he used his celebrated operator calculus, the techniques of which were still obscure and full of pitfalls. However, his argument can be transposed with some changes of terminology and retained today, in an account based upon Maxwell's equations. The following reasoning is indeed to be found almost word for word in Searle (see below), who kept in touch with Heaviside in scientific matters.

---

† Nowadays, this kind of collaboration is fortunately becoming common.

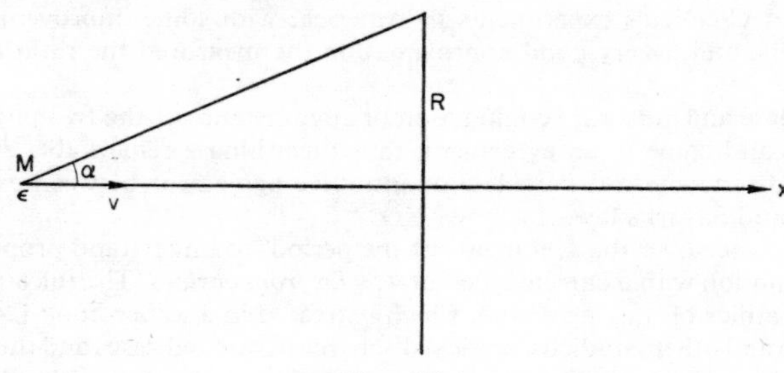

FIG. 37.

They set out from the relations which we accept today for the potentials:

$$\nabla^2\mathbf{A} - \frac{K_0\mu_0}{c_0^2}\frac{\partial^2\mathbf{A}}{\partial t^2} = -\frac{4\pi\mu_0}{c_0}\varrho\mathbf{v}$$

$$\nabla^2 V - \frac{K_0\mu_0}{c_0^2}\frac{\partial^2 V}{\partial t^2} = -\frac{4\pi}{K_0}\varrho$$

We notice in passing that Thomson's error, mentioned above, was to write the equation for **A** without the right-hand side. In the case of uniform motion, the field of the electron moves with it. The quantities V and **A** are therefore unaltered when t is increased by dt and x by $dx = v\,dt$ simultaneously; thus

$$\frac{\partial V}{\partial t}dt + \frac{\partial V}{\partial x}v\,dt = 0$$

so that

$$\frac{\partial V}{\partial t} = -v\frac{\partial V}{\partial x}, \quad \frac{\partial^2 V}{\partial t^2} = -v^2\frac{\partial^2 V}{\partial x^2}$$

and hence, for the potentials,

$$(1-\beta^2)\frac{\partial^2 A_x}{\partial x^2} + \frac{\partial^2 A_x}{\partial y^2} + \frac{\partial^2 A_x}{\partial z^2} = -\frac{4\pi\mu_0}{c_0}\varrho v$$

$$(1-\beta^2)\frac{\partial^2 V}{\partial x^2} + \frac{\partial^2 V}{\partial y^2} + \frac{\partial^2 V}{\partial z^2} = -\frac{4\pi}{K_0}\varrho$$

Only the component A_x is non-zero, since the velocity $v = v_x$ is along Ox. Comparison of these two relations shows that

$$A_x = \frac{K_0\mu_0}{c_0}vV = \sqrt{K_0\mu_0}\beta V$$

Thus only V need be determined. The fact that the fields are given by a single function is stated explicitly by Thomson (1889). We then make the change of variables

$$x = x_0\sqrt{1-\beta^2}, \quad y = y_0, \quad z = z_0$$

With the variables x_0, y_0, z_0, the density ϱ_0 is given by

$$\varrho\,dx\,dy\,dz = \frac{\varrho_0}{\sqrt{1-\beta^2}}dx_0\,dy_0\,dz_0$$

giving the relation

$$\frac{\partial^2 V}{\partial x_0^2} + \frac{\partial^2 V}{\partial y_0^2} + \frac{\partial^2 V}{\partial z_0^2} = -\frac{4\pi}{K_0}\frac{\varrho_0}{\sqrt{1-\beta^2}}$$

We thus obtain the same equation as for a charge at rest in the system (x_0, y_0, z_0) by writing

$$\frac{\partial^2 V_0}{\partial x_0^2} + \frac{\partial^2 V_0}{\partial y_0^2} + \frac{\partial^2 V_0}{\partial z_0^2} = -\frac{4\pi}{K_0}\varrho_0$$

with

$$V = \frac{V_0}{\sqrt{1-\beta^2}}$$

Using the results of electrostatics, we have

$$V = \frac{\varepsilon}{K_0} \frac{1}{\sqrt{\frac{x^2}{1-\beta^2}+y^2+z^2}}$$

To calculate $\mathfrak{B}$ and $\mathbf{E}$ is then a trivial matter.

From this we conclude in particular that the electric field of a point charge is radial. Heaviside believed that he could prove that the same solution applied to the exterior of a uniformly charged sphere.

Thomson [15, 16] then reconsidered the problem by means of a new method. He assumed that the electromagnetic field arose because of the displacement of the lines of electric force in the aether; I shall not go into this.

Heaviside's conclusions about the sphere were contested by Searle [17, 19]; after some discussion, the latter and Heaviside himself established that the field of a point charge is the same as that of a flattened ellipsoid of revolution (with the ratio $\sqrt{1-\beta^2}$), and they stated that the total force is normal to this surface. It was thus that the ellipsoid appeared. We should not, however, regard this as the Lorentz contraction; for, according to these authors, the ellipsoid in question would remain ellipsoidal even at rest. In other words, they associated the field of a point charge with that of an ellipsoid in uniform motion.

Nevertheless, the use of the variable x does seem to suggest that we obtain a sphere at rest; this is because we are arguing with hindsight, however. At that time, the real shape was the ellipsoid and the sphere was a fictitious shape, obtained by introducing auxiliary coordinates. One other point also still needed modification. The possibility of velocities greater than c was generally accepted. The corresponding solution is pursued in several papers, for example those of Heaviside and Th. des Coudres [22].

After the appearance of Einstein's fundamental memoir, the whole question took on a very different aspect. Henceforward, no Galileian system was privileged *in vacuo*, so that the field of a moving charge was calculated by simply transforming the electric field of the system at rest. This is the method explained in the foregoing chapter, and which is to be found in the earliest relativistic texts, in particular that of von Laue [23]. Further references are to be found in *Rayonnement et Dynamique*. [24].

[1] J. Chr. Oersted: *Experimenta circa conflictus electrici in acum magneticum*, Copenhagen, Hafniae, 1820. German trans.: *Schweigger's J. Phys. Chem.* **29** (1820) 275. English trans.: *Thomson's Annals of Philosophy* **16** (1820) 273. French trans: Expériences relatives à l'effet du conflit électrique sur l'aiguille aimantée. *Ann. Chim. Phys.* **14** (1820) 417–425; *J. Phys. Rad.* **91,** 72. This text is also reproduced in ref. [1] of § [144]. Oersted's experiment was described at a meeting of the *Académie des Sciences* on 11 September 1820, by Arago.

[2] J. B. Biot and F. Savart: Note sur le magnétisme de la pile de Volta. *Ann. Chim. Phys.* **15** (1820) 222–223. This *Note* was read on 30 October 1820. The principal passage is as follows: "Par le point où réside cette molécule (de magnétisme austral ou boréal) menez une perpendiculaire à l'axe du fil: la force qui sollicite la molecule est perpendiculaire à cette ligne et à l'axe du fil. Son intensité est réciproque à la simple distance."† The sense of the force is also given.

[3] J. B. Biot and F. Savart: Sur l'aimantation imposée aux métaux par l'électricité en mouvement. *Précis élémentaire de physique*, Vol. II, 2nd ed. 1821, p. 117; 3rd ed. 1823, p. 704. Text reproduced in ref. [1] of § [144], pp. 80–127.

[4] J. C. Maxwell: *A Treatise of Electricity and Magnetism*, 1873; 3rd ed., 1892.

[5] H. A. Rowland: Sur l'action électromagnétique de la convection électrique. *Ann. Chim. Phys.* **12** (1877) 119–124.

[6] H. Helmholtz: Bericht betreffend Versuche über die elektromagnetische Wirkung elektrischer Convection ausgeführt von Hrn. Henry A. Rowland der J. Hopkins Universität in Baltimore. *Ann. Phys. Chem.* (1876) 487–493; *Monats. Kön. Preus. Akad. Berlin* (1876) 211–216.

[7] V. Crémieu: Recherches sur l'existence du champ magnétique produit par le mouvement d'un corps électrisé. *C. R. Acad. Sci. Paris* **130** (1900) 1544; *Thèse de doctorat*, Paris, 1901. For further references, see [8].

[8] H. Poincaré: A propos des expériences de M. Crémieu. *Rev. Gén. Sci. Pures Appl.* **12** (1901) 994–1107; *Œuvres complètes*, Gauthier-Villars, Vol. X, 1954, pp. 391–420; Sur les expériences de

† "Through the point at which this molecule is situated (with either northern or southern polarity), drop a perpendicular on to the axis of the wire: the force exerted on the molecule is perpendicular to this line and to the axis of the wire. Its intensity is inversely proportional to the distance."

M. Crémieu et une objection de M. Wilson. *L'Eclairage électrique* **31** (1902) 83–93; *Œuvres complètes*, Vol. X, pp. 421–437. The latter work contains detailed references in connection with Crémieu; see also the notes, p. 626.
[9] H. PENDER: Doctoral Thesis, Johns Hopkins University. See also *Phys. Rev.* **13** (1901) 203 and **15** (1902) 291.
[10] V. CRÉMIEU and H. PENDER: Recherches contradictoires sur l'effet magnétique de la convection électrique. *Bull. Soc. Franç. Phys.* (1903) 136–162.
[11] J. J. THOMSON: On the electromagnetic effects produced by the motion of electrified bodies. *Phil. Mag.* **11** (1881) 229–249.
[12] C. F. FITZGERALD: Note on Mr. J. J. Thomson's investigation of the electromagnetic action of a moving electrified sphere. *Proc. Roy. Dublin Soc.* **3** (1883) (but the Communication is dated Nov. 1881), 250–254; *Scientific Writings*, p. 102.
[13] O. HEAVISIDE: *Electrician*, 23 Nov. 1888.
[14] O. HEAVISIDE: On the electromagnetic effects due to the motion of electrification through a dielectric. *Phil. Mag.* **27** (1889) 324–339.
[15] J. J. THOMSON: On the magnetic effects produced by motion in the electric field. *Phil. Mag.* **27** (1889) 1–14.
[16] J. J. THOMSON: On the illustration of the properties of the electric field by means of tubes of electrostatic induction. *Phil. Mag.* **31** (1891) 149–171.
[17] F. C. SEARLE: Problems in electric convection. *Phil. Trans. London,* **187A** (1896) 675–713. This very remarkable paper deserves to be rescued from oblivion.
[18] W. V. MORTON: Notes on the electromagnetic theory of moving charges. *Proc. Phys. Soc.* **14** (1896) 180–187.
[19] F. C. SEARLE: On the steady motion of an electrified ellipsoid. *Phil. Mag.* **44** (1897) 329–343.
[20] A. SCHUSTER: On the magnetic force acting on moving electrified spheres. *Phil. Mag.* **43** (1897) 1–11.
[21] W. B. MORTON: Notes on the electro-magnetic theory of moving charges. *Phil. Mag.* **41** (1896) 488–494.
[22] TH. DES COUDRES: Zur Theorie der Kraftfeldes elektrischer Ladungen, die sich mit Überlichtgeschwindigkeit bewegen. *Arch. Néerl. Sci.*, series II, **5** (1900) 652–664.
[23] VON LAUE: ref. [2] of Appendix I.
[24] H. ARZELIÈS: ref. [34] of Appendix I.

[147] On the Interactions Between Moving Charges, Calculated via the Field

I find the problem studied for the first time in Searle [1]. The two charges are assumed to have the same velocity (the problem of § [132]); Searle gives the correct result and analyses it fully, bringing out the inequality of action and reaction.

We see that for the case of uniform motion, *the problem was solved in 1896*. We have already mentioned in § [145] that it has often been reconsidered and solved wrongly by authors who start with different assumptions as their direct postulates about the interaction. In field theory, too, we encounter some questionable recent publications in this connection.

We first mention the ideas of Hering [2] and their refutation by Liénard [3]. Litman [4] has considered the case of § [132]; as he did not take relativity into account, he found it strange that an attractive force should exist in one system and not in the other, and he wondered whether this force really exists. Tripp [5 and 10] replied in the negative, concluding that certain concepts at the foundations of electromagnetism require radical alteration. In a very sensible article, Sard [11] explained this by bringing out the role of the Lorentz transformation. Nevertheless, he seems to have gone unnoticed, witness the article by Robertson [12], whose historical study [6] I also mention. Han Kong-Chi [13] considers the same problem, wholly unaware that it has long since been solved. Hughes [14] puts it to use in teaching the force transformation formulae. Page and Adams [7] study the interactions to second order, using the pre-relativistic formulae; to explain the inequality of action and reaction, they introduce the energy properties of the field. I cannot agree with their way of solving the problem in their book on electrodynamics [8]. Despite this, and several other minor criticisms, this book is on the whole *a very remarkable piece of work*. O'Leary [9] begins with these results and suggests adding a third term to the Heaviside–Thomson force (which he names after Lorentz).

In the same year as the French edition of the present work (1957), a book by Dacos appeared [16], in which the author also proposed to obtain electromagnetism from the equations of electrostatics, with the aid of relativity. Our methods are very different, however. Dacos first treats electrostatics in the usual way, using Coulomb's postulate with respect to stationary charges. He then states a new postulate (p. 86), the invariance of electric flux, or the invariance of the equation div $\mathbf{E} = 4\pi\varepsilon$, ε being invariant. From this he extracts the transformation of the electric field of a particle from K_0 to K, where K_0 is the proper system of reference (which he subsequently appears to forget). He then uses this transformation to calculate the interaction between charges in uniform motion. Comparison of our results shows hat they disagree. I do not understand his formula V. 2 on p. 87:

$$E_y'' = \frac{E_y}{\sqrt{1-\beta_2^2}}$$

It seems to me that in the system of reference S, the charge q_1 also creates a field $\mathcal{B}_z$ which ought to figure in the expression for E_y''. His transformation formulae seem incomplete.

[1] SEARLE: ref. [19] of § [146].

[2] C. HERING: Electromagnetic forces; a search for more rational fundamentals; a proposed revision of the laws. *J. Amer. Inst. El. Eng.* **42** (1923) 140–154.

[3] A. LIÉNARD: Examen des idées de Carl Hering sur la nécessité d'une révision des lois de l'électromagnétisme. *Rev. Gen. Electr.* **16** (1924) 811–859.

[4] B. LITMAN: Letter to the Editor. *Electr. Eng.* **64** (1945) 381.

[5] W. A. TRIPP: An analysis of electromagnetic forces. *Electr. Eng.* **64** (1945) 351–356.

[6] I. A. ROBERTSON: An historical note on a paradox in electrodynamics. *Phil. Mag.* **36** (1945) 32–43.

[7] L. PAGE and N. I. ADAMS: Action and reaction between moving charges. *Amer. J. Phys.* **13,** No. 3 (1945) 141–147.

[8] L. PAGE and N. I. ADAMS: *Electrodynamics,* Van Nostrand, New York, 1941.

[9] AUSTIN J. O'LEARY: Addition of a third term in the Lorentz force equation. *Amer. J. Phys.* **14** (1946) 63.

[10] W. A. TRIPP: A paradox in electrodynamics. *Phil. Mag.* **38** (1947) 61–65.

[11] R. D. SARD: The forces between moving charges. *Electr. Eng.* (1947) 61–65.

[12] I. A. ROBERTSON: A paradox in electrodynamics. *Phil. Mag.* **39** (1948) 162–163.

[13] HAN KONG-CHI: A problem about moving charges. *Amer. J. Phys.* **16** (1948) 398–399.

[14] J. V. HUGHES: A problem about moving charges. *Amer. J. Phys.* **17** (1949) 319–320.

[15] J. BACKUS: An alternative interpretation of magnetic forces between moving charges. *Bull. Amer. Phys. Soc.* **30,** No. 8 (1955) 24.

[16] F. DACOS: *Conception actuelle de l'électromagnétisme théorique.* Dunod, 1957.

[148] The Retarded Potentials of Liénard–Wiechert; the Interaction between Accelerating Charges

For the calculation of the field of an accelerated charge, setting out from the Liénard–Wiechert potentials, see my *Electricité*, for example. A historical study of these retarded potentials is to be found in my *Rayonnement et Dynamique*.

PART TWO

THE DYNAMICS OF PARTICLES OF VARIABLE PROPER MASS

CHAPTER XI

RECONSIDERATION† OF THE EQUIVALENCE OF MASS AND ENERGY; VARIOUS ASPECTS

[149] The Equivalence of Mass and Energy in the Dynamics of Particles of Constant Proper Mass

In this section, we consider an invariant particle, which does not alter its nature (an electron, for example); this is the case considered in Part One.

In § [13] we saw that the energy variation is

$$dW = d\left(\frac{m_0c^2}{\sqrt{1-v^2/c^2}}\right) \tag{1}$$

whence

$$W = \frac{m_0c^2}{\sqrt{1-v^2/c^2}} + \text{const} \tag{2}$$

and the constant is independent of velocity. At rest,

$$W_0 = m_0c^2 + \text{const}$$

The kinetic energy is given by

$$E = W - W_0 = m_0c^2\left(\frac{1}{\sqrt{1-v^2/c^2}} - 1\right)$$

and the constant has dropped out; the expression for E thus follows from the impulse postulate, without supplementary postulates. Let us now make the supplementary postulate that the constant is zero:

$$W = \frac{m_0c^2}{\sqrt{1-v^2/c^2}} \tag{3}$$

so that the rest energy is

$$W_0 = m_0c^2 \tag{4}$$

In calculating the motion of an invariant particle, with constant proper mass, equations (2) and (3) are equivalent, for only the variation dW is involved.

We might nevertheless point out that when the laws of motion are written in four-dimensional form, we are led to introduce the quantity (3); the vector four-impulse cannot be written in terms of E.

† See § [13].

[150] The Equivalence of Mass and Energy in the Transmutation of Elementary Particles

We shall encounter examples of the transmutations of such particles in Part Three; thus, for example, an electron positron pair can interact to give two photons.

In these phenomena, expression (3) of § [149] plays the vital role; we write down formulae stating that the total energy (and the impulse) are the same before and after the interaction. The essential fact is that *the rest energy W_0 can yield kinetic energy*. This is the case, for example, when a neutral particle, of rest mass m_0 in its proper reference frame, decays into two photons in this same reference system (§ [236]):

$$m_0 = \frac{2h\nu_0}{c^2}$$

[151] Mass and Kinetic Energy for a Complex Particle

We consider an ensemble of free, non-interacting particles, and refer them to two systems K_1 and K_2. We shall show that their kinetic energy can be regarded as mass. We assume that the geometrical centre of mass G (in the ordinary sense of the term) is stationary in K_2, and that the particles are in random motion about this point. In K_2 the velocity of one particle is inclined at an angle α to the axis Ox_2 at time t_2. With respect to K_1, the energy is given by

$$W_1 = m_0c^2 \frac{1+\dfrac{uv}{c^2}\cos\alpha}{\sqrt{1-u^2/c^2}\,\sqrt{1-v^2/c^2}}$$

u is the velocity of the particles in K_2.

This can be seen by writing $W_1 = mc^2$ in K_1 and transforming the velocity of the particle into K_1. For an arbitrary number of particles, the total energy at time t_1 of K_1 is given by

$$W_1 = \frac{1}{\sqrt{1-v^2/c^2}} \sum \frac{m_0c^2}{\sqrt{1-u^2/c^2}} + \frac{v}{\sqrt{1-v^2/c^2}} \sum \frac{m_0 u \cos\alpha}{\sqrt{1-u^2/c^2}}$$

In K_2, the particles are considered at different times; we assume that the statistical distribution of velocities is stationary in K_2. Since G is fixed, we can then write

$$\sum \frac{m_0u_x}{\sqrt{1-u^2/c^2}} = 0, \quad \sum \frac{m_0u_y}{\sqrt{1-u^2/c^2}} = 0, \quad \sum \frac{m_0u_z}{\sqrt{1-u^2/c^2}} = 0$$

The expression for W_1 now reduces to

$$W_1 = \frac{1}{\sqrt{1-v^2/c^2}} \sum \frac{m_0c^2}{\sqrt{1-u^2/c^2}} = \frac{W_2}{\sqrt{1-v^2/c^2}}$$

Seen from K_1, therefore, it is exactly as though the system were a single corpuscle with proper mass

$$M_0 = W_2/c^2$$

The proper mass of the system is *greater* than the sum of the masses of its constituents, because of the "internal" motions.

Remark. Whenever we are dealing with an extended system, we must be on our guard against the relativity of simultaneity, when we change reference system.

[152] Mass and Energy in Thermodynamics

(a) *Mass and thermal energy (heat)*

Warming a body is equivalent to transferring kinetic energy to it on the molecular scale. According to §[151], warming a body should also be accompanied by an increase in its mass. This observation forms one of the basic tenets of relativistic thermodynamics (§§ [183] ff.).

(b) *Mass and rest energy (or internal energy)*

We shall study an example in which changes of mass arising from changes of velocity are excluded; the mass must then change like the rest energy. In the calculation, we shall employ the expression for the kinetic energy, which is the only one fixed by the impulse postulate (§ [149]). We obtained in *Kinematics* (§ [80]) the relation

$$\frac{1}{\sqrt{1-q_1^2/c^2}} = \frac{1+vq_{2x}/c^2}{\sqrt{1-q_2^2/c^2}\sqrt{1-v^2/c^2}}$$

whence

$$\frac{q_{1x}}{\sqrt{1-q_1^2/c^2}} = \frac{q_{1x}+v}{\sqrt{1-q_2^2/c^2}\sqrt{1-v^2/c^2}},$$

$$\frac{q_{1y}}{\sqrt{1-q_1^2/c^2}} = \frac{q_{2y}}{\sqrt{1-q_2^2/c^2}}, \quad \frac{q_{1z}}{\sqrt{1-q_1^2/c^2}} = \frac{q_{2z}}{\sqrt{1-q_2^2/c^2}}$$

Consider two identical particles travelling with velocities $\mathbf{q}_{2+}$ and $\mathbf{q}_{1-}$, equal and opposite with respect to a system of reference K_2. The signs $+$ and $-$ are merely labels to distinguish the two particles; for this pair ($q_{2+} = q_{2-}$), we have

$$\frac{1}{\sqrt{1+q_1^2/c^2}} + \frac{1}{\sqrt{1-q_1^2/c^2}} = \frac{2}{\sqrt{1-q_2^2/c^2}\sqrt{1-v^2/c^2}}$$

Suppose that they collide. If this occurs perfectly symmetrically, which we are at liberty to assume, the centre of mass G will remain stationary in K_2 and the change in W will be the same for both corpuscles. We write down the condition that energy is conserved in K_1 and in K_2. This energy consists of the kinetic energy and an energy W, the nature of which we leave unspecified. This will be all the internal energy of the corpuscle (partly kinetic, if the corpuscle is complex), that is, the energy necessary to create it in its present state. Indicating quantities after collision by a bar, we have in K_2

$$2W_2+2m_0c^2\left(\frac{1}{\sqrt{1-q_2^2/c^2}}-1\right) = 2\overline{W}_2+2\overline{m}_0c^2\left(\frac{1}{\sqrt{1-\overline{q}_2^2/c^2}}-1\right)$$

and in K_1

$$2W_2+m_0c^2\left(\frac{1}{\sqrt{1-q_{1+}^2/c^2}}-1\right)+m_0c^2\left(\frac{1}{\sqrt{1-q_{1-}^2/c^2}}-1\right)$$

$$= 2\overline{W}_2+\overline{m}_0c^2\left(\frac{1}{\sqrt{1-\overline{q}_{1+}^2/c^2}}-1\right)+\overline{m}_0c^2\left(\frac{1}{\sqrt{1-\overline{q}_{1-}^2/c^2}}-1\right)$$

Using a relation indicated earlier, the last equation can be written

$$W_2-m_0c^2+\frac{m_0c^2}{\sqrt{1-q_2^2/c^2}\sqrt{1-v^2/c^2}}$$

$$= \overline{W}_2-\overline{m}_0c^2+\frac{\overline{m}_0c^2}{\sqrt{1-\overline{q}_2^2/c^2}\sqrt{1-v^2/c^2}}$$

The first equation is written

$$W_2-m_0c^2+\frac{m_0c^2}{\sqrt{1-q_1^2/c^2}} = W_2-\overline{m}_0c^2+\frac{\overline{m}_0c^2}{\sqrt{1-\overline{q}_2^2/c^2}}$$

and on multiplying it by $1/\sqrt{1-v^2/c^2}$ and subtracting it from the preceding equation, we obtain

$$(W_2-m_0c^2)\left(1-\frac{1}{\sqrt{1-v^2/c^2}}\right) = (\overline{W}_2-\overline{m}_0c^2)\left(1-\frac{1}{\sqrt{1-v^2/c^2}}\right)$$

As $v \neq 0$, this gives

$$W_2-\overline{W}_2 = (m_0-\overline{m}_0)c^2$$

The change in the energy W is thus proportional to the change in the mass m_0.

[153] Mass and Potential Energy in the Dynamics of Compressible and Extensible Media

(a) Consider a medium at rest, which is compressed by a uniform, constant pressure p, say. If the volume decreases by $d\mathcal{V}$, the work done is $p\, d\mathcal{V}$; for an elastic medium, this work is stored as elastic potential energy and can be recovered. We realize that in virtue of this, the mass of the body increases by an amount

$$dm = \frac{p\, d\mathcal{V}}{c^2}$$

It is upon this that the relativistic theory of continuous media is based.

(b) We reconsider the calculations of § [179]. If we did not take the increase of mass of the compressed body into account, we should obtain an impulse in K that was different from zero, and hence a constant momentum.

[154] Mass and Potential Energy in Atoms

(a) *Phenomena associated with the planetary electrons*

Consider first the simplest possible case, the Bohr model of the hydrogen atom, with circular electron trajectory of radius r. The potential energy is the energy required to construct the atom when its two components are brought in from infinity (where they do not interact). If ε is the absolute value of the common charge of electron and nucleus, this energy will be negative:

$$E_p = -\varepsilon^2/K_0 r$$

The kinetic energy is

$$E_k = \varepsilon^2/2K_0 r$$

and the total energy is

$$W = E_p + E_k = -\varepsilon^2/2K_0 r$$

When the electron moves to a different energy level, a photon is emitted or absorbed, of frequency ν:

$$W_1 - W_2 = h\nu$$

The difference between the energies W is wholly transformed into the kinetic energy of the photon. We shall therefore treat potential energy in exactly the same way as kinetic energy, and attribute mass to it.

With elliptic trajectories, potential energy is continuously being transformed into kinetic energy and conversely.

If, finally, we abandon the Bohr model and adopt Schrödinger's theory, the two types of energy are no longer discriminated.

(b) *Nuclear energy; fission and fusion*

Generally speaking, the mass of a nucleus is the sum of the proper masses of the constituent particles, their kinetic energies and their interaction energies. For stationary atoms, the last two terms give a negative sum, W_i, and we write

$$M_0 = \Sigma m_0 - \frac{W_i}{c^2}$$

The quantity W_i/c^2 is called the mass defect.

When fission occurs, the nucleus turns into other nuclei together with elementary particles. A part of the mass energy $M_0 c^2$ is converted into kinetic energy which can do work. The vast part of the energy $M_0 c^2$ goes into the mass energy of the new nuclei. Fusion is a special case of fission; only the elementary particles remain, and we can extract much more energy than by fission.

[155] Mass and Potential Energy in Electromagnetism

(a) *An ensemble of stationary point charges*

In this case, the potential energy is, by definition, the energy required to bring up all the charges from infinity (where they do not interact) to their present positions. In electricity, it is shown that this energy is given by

$$W = \tfrac{1}{2}\Sigma Q_i V_i \tag{1}$$

in which V_i is the potential created at the point where Q_i is located by all the charges *except* Q_i. This energy W can be positive or negative. If we associate mass with this energy, the proper mass of the ensemble will be given by

$$M_0 = \Sigma m_{0i} + \frac{W}{c^2} \tag{2}$$

The ensemble may be regarded as a complex particle. The mass M_0 can be measured (by deflection) and hence has a well-determined value. The same is true of W and so the definition of the latter is fixed; we can no longer say, as we did in Newtonian mechanics, that W is defined apart from an arbitrary constant. As a result of this, the potential V too has a well-determined value, *without any arbitrary constant*; for one particle the potential is $\varepsilon/K_0 r$. It is for this reason that we have adopted the statement of § [136A].

(b) *Charged conductors; arbitrary macroscopic distributions of stationary charges*

For a single charged conductor of capacity C, we have

$$W = \tfrac{1}{2}VQ = \tfrac{1}{2}CV^2 \tag{3}$$

in which V is its potential. In consequence, $W > 0$ whatever the sign of V. This result is as we should expect. For, to charge a neutral body positively, at least one positive charge must be brought up to it to begin with; further charges of the same sign will then be repelled. The conclusion is the same if a negative charge is removed; work will have to be done to remove subsequent charges. If the body is to be charged negatively, the same conclusion is again reached.

Let us denote by $\pm\Sigma m_0$ the change of rest mass which occurs when charges are added or removed. When a conductor passes from a neutral state to a charged state, its mass increases by

$$\tfrac{1}{2}CV^2 \pm \Sigma m_0 \tag{4}$$

For a single conductor or for an arbitrary number of conductors, it can be shown that the energy W can be put into the equivalent form

$$W = \frac{K_0}{8\pi}\iiint E^2\, d\varpi \tag{5}$$

(see, for example, my *Electricité*, § [142]); E is the electric field within the element of volume $d\varpi$ and the integration is taken over all space. In formulae (3) and (5), V and E are the real values of potential and field respectively. The equivalence of these formulae depends upon the assumption that the distribution contains a very large number of elementary charges, so that the field created at a point by all the charges except one is virtually the same as the effective field. *This point is fundamental.* We have only to bear it in mind to see just *how deceptive is the interpretation in which an energy density is associated with the field*—to every field, the density $K_0E^2/8\pi$ is said to correspond. Equation (5) is only a mathematical equivalent of (1) in the case of a large number of charges; the differential elements being integrated have no physical meaning. If the field E is

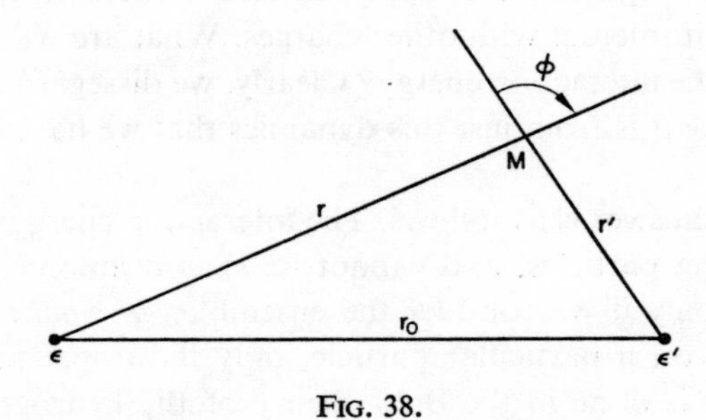

FIG. 38.

created by a small number of charges, expression (5) for W is wrong. Consider, for example, two charges ε and ε'. If (5) were generally valid, we should write (Fig. 38)

$$W = \frac{1}{8\pi K_0} \iiint \left(\frac{\varepsilon^2}{r^4} + \frac{\varepsilon'^2}{r'^4} + \frac{2\varepsilon\varepsilon'}{r^2 r'^2} \cos\phi \right) d\varpi$$

in which $d\varpi$ is the element of volume surrounding M. This expression for W is *incorrect*; we must write

$$W = \frac{\varepsilon\varepsilon'}{K_0 r_0}$$

This value is, moreover, quite simply the energy necessary to bring one charge into the field of the other; it is obtained from the integral

$$W = \frac{1}{8\pi K_0} \iiint \frac{2\varepsilon\varepsilon'}{r^2 r'^2} \cos\phi \, d\varpi$$

(c) *The general concept of electromagnetic energy; the Poynting vector*

We have just said that the concept of the energy density of a field E must be discarded, for it arises from a faulty interpretation of formula (5). Nevertheless, this concept has been generalized to include the case of an electromagnetic field E and $\mathcal{B}$; the density is taken as

$$\frac{K_0E^2 + \mu_0 H^2}{8\pi} \quad \text{with} \quad H = \mathcal{B}/\mu_0$$

This expression, which is still usually to be found in the majority of treatises on electricity, is *unacceptable*; the same is true of the corresponding interpretation of the Poynting vector. I refer the reader to my *Electricité*, or to *Rayonnement et Dynamique*.

[156] On a Conceptual Difficulty

(a) We associate mass with interaction energy. We have, however, built up dynamics by characterizing a material particle by a well-determined proper mass, which is independent of the surroundings of the particle. An electron, for example, always has proper mass $0{\cdot}90 \times 10^{-27}$ g. We now apply an electric field to this electron, which is equivalent to making it interact with other charges. What are we then to say about the mass associated with the interaction energy? Clearly, we disregard it. Is this not an internal inconsistency, since it is from just this dynamics that we have extracted the concept of mass energy.

(b) As I see it, the answer is as follows. The interaction energy is connected with the ensemble of interacting particles, and cannot be shared among the various particles. Its mass is involved only if we consider the ensemble, *en bloc*, as a complex particle. If we focus attention on a particular particle, only its proper mass will be involved. This is certainly what is done in the Bohr theory of the hydrogen atom. The motion of the electron is calculated using its proper mass. The interaction energy is not taken into account unless we are considering the complete atom, to calculate its radiation, for example.

(c) Some authors believe in a different interpretation, and correct the proper mass of a corpuscle for its interaction energy. They see this correction as related to the mass renormalization that one finds in field theory. At first sight, this idea seems attractive. I do not, however, see how this interaction energy is to be shared, and a well-determined fraction of it attributed to each particle. One can, of course, make supplementary hypotheses, but the validity of the usual value of the mass of the electron, for example, then requires explanation. Let us compare the proper energy of the electron,

$$m_0c^2 = 8{\cdot}20 \times 10^{-7} \text{ erg} = 0{\cdot}51028 \text{ MeV}$$

with the interaction energy between electron and nucleus in the hydrogen atom. Taking r of the same order of magnitude as the radius of the atom, say 10^{-13} cm, this energy is given by

$$\frac{\varepsilon^2}{r} = \frac{(4{\cdot}80)^2 \times 10^{-20}}{10^{-13}} = 22{\cdot}84 \times 10^{-7} \text{ erg}$$

The two energies are of the same order of magnitude. It is difficult to see, therefore, how the success of Bohr's calculation can be reconciled with the present point of view. This observation seems to me to favour the interpretation of (b).

HISTORICAL AND BIBLIOGRAPHICAL NOTES

The history of the topics dealt with in this chapter is very fascinating; it marks a turning-point, not only in the progress of physics, but also in the evolution of humanity. It is to be found in my book *Rayonnement et dynamique*..., but because of its length, there can be no question of reproducing it here.

CHAPTER XII

FUNDAMENTAL POSTULATES AND GENERAL THEOREMS†

A. THE IMPULSE POSTULATE AND ITS FIRST CONSEQUENCES‡

[157] The Postulate

(a) As in § [5], we shall call the momentum **p** and the relativistic mass m the quantities defined by

$$\mathbf{p} = \frac{m_0\mathbf{v}}{\sqrt{1-v^2/c^2}} = m\mathbf{v} \tag{1}$$

where **v** is the velocity and m_0 is the rest mass, which in the preceding chapters has been assumed constant. As a generalization, *we now assume that m_0 is no longer constant*; by hypothesis, it is a quantity that depends upon time. The mass m_0 may be expressed as a function of the position variables but since these depend upon time through the law of motion, we may always consider that m is a function of a single variable, t (as we do for velocity, for example).

We adopt the impulse postulate

$$\mathrm{d}\mathbf{p} = \mathrm{d}\mathbf{I} = \mathbf{F}\,\mathrm{d}t \tag{2}$$

F is the resultant of all the forces acting on the corpuscle, d**I** is the impulse of **F** during time dt and d**p** is the variation of the momentum. The force **F** arises in two ways: external actions (electric field and gravity, for example) and actions connected with the variation of proper mass (the reaction of a rocket, for example). We shall be more specific later; it is convenient to study the formalism of equation (2) first.

(b) *Justification for the postulate*. I make a general statement, so as to obtain a general formalism; asserted thus, *ex abrupto*, the statement takes the form of a postulate. We shall see, however, in studying the various special cases, that the statement is in fact a consequence of the equations of dynamics for constant proper mass. The proof will always proceed as follows. Consider two instants of time t and $t+\mathrm{d}t$ at which the proper mass has values m_0 and $m_0+\mathrm{d}m_0$; we can always assume that the proper mass is constant

† The proper mass is assumed to vary continuously; abrupt changes are studied in Part Three (collisions).

‡ It is possible to use other formalisms; we shall return to this point later, and demonstrate (in § [177]) the advantage of the postulate adopted here.

before t and after $t+\mathrm{d}t$. At the two ends of the interval $\mathrm{d}t$, the formulae of ordinary dynamics can be applied and in particular, the conservation equations for momentum and energy.

[158] The Relation between Mass and Acceleration

Let us write out the derivative of the momentum, using the notation of § [8]:

$$\mathbf{F} = \frac{\mathrm{d}}{\mathrm{d}t}\left(\frac{m_0 v\mathbf{t}}{\sqrt{1-\beta^2}}\right)$$

$$= \frac{m_0\mathbf{t}}{(1-\beta^2)^{3/2}}\frac{\mathrm{d}v}{\mathrm{d}t}+\frac{m_0\mathbf{n}}{\sqrt{1-\beta^2}}\frac{v^2}{\varrho}+\frac{v\mathbf{t}}{\sqrt{1-\beta^2}}\frac{\mathrm{d}m_0}{\mathrm{d}t}$$

$$= m_t\gamma_t+m_n\gamma_n+\frac{1}{\sqrt{1-\beta^2}}\frac{\mathrm{d}m_0}{\mathrm{d}t}\mathbf{v}$$

The third term is characteristic of variable proper mass.

[159] The Energy Theorem

From the relationship given in § [158], we deduce

$$\mathbf{F}\,.\,\mathbf{v} = \frac{m_0 v}{(1-\beta^2)^{3/2}}\frac{\mathrm{d}v}{\mathrm{d}t}+\frac{v^2}{\sqrt{1-\beta^2}}\frac{\mathrm{d}m_0}{\mathrm{d}t}$$

$$= \frac{m_0 v}{(1-\beta^2)^{3/2}}\frac{\mathrm{d}v}{\mathrm{d}t}+\frac{\beta^2}{\sqrt{1-\beta^2}}\frac{\mathrm{d}(m_0c^2)}{\mathrm{d}t}$$

$$= \frac{\mathrm{d}}{\mathrm{d}t}(mc^2)-\frac{c^2}{\sqrt{1-\beta^2}}\frac{\mathrm{d}m_0}{\mathrm{d}t}+\frac{\beta^2}{\sqrt{1-\beta^2}}\frac{\mathrm{d}(m_0c^2)}{\mathrm{d}t}$$

$$= \frac{\mathrm{d}}{\mathrm{d}t}(mc^2)-\sqrt{1-\beta^2}\frac{\mathrm{d}(m_0c^2)}{\mathrm{d}t}$$

$$\mathbf{F}\,.\,\mathbf{v}\,\mathrm{d}t+c^2\sqrt{1-\beta^2}\,\mathrm{d}m_0 = \mathrm{d}(mc^2) \qquad (1)$$

This relation reduces to an identity in the rest system.

Retaining the usual definitions of energy W and work $\mathcal{T}$, we have

$$\mathrm{d}W = \mathrm{d}(mc^2) \qquad \mathrm{d}\mathcal{T} = \mathbf{F}\,.\,\mathbf{v}\,\mathrm{d}t$$

$$\mathrm{d}W = \mathrm{d}\mathcal{T}+c^2\sqrt{1-\beta^2}\,\mathrm{d}m_0$$

The total energy variation can originate in only two ways: work done by external forces, $\mathrm{d}\mathcal{T}_e$, and the energy change arising from the change of proper mass. We cannot be more specific for the time being, however; in fact, the change of mass can alter the energy by acting both on m_0 and on v. If this variation does not affect v, and only in this case, we can write

$$\mathrm{d}W = \mathrm{d}\mathcal{T}_e+c^2\frac{\mathrm{d}m_0}{\sqrt{1-\beta^2}}$$

We discuss this further, in § [174] for instance.

[160] Velocities Small in Comparison with c

The equation of motion becomes

$$\mathbf{F} = m_0(\boldsymbol{\gamma}_t + \boldsymbol{\gamma}_n) + \frac{dm_0}{dt_0}\mathbf{v} = m_0\boldsymbol{\gamma} + \frac{dm_0}{dt_0}\mathbf{v}$$

The energy equation is now written

$$\mathbf{F} \cdot \mathbf{v}\, dt - \frac{v^2}{2}\, dm_0 = d\left(\frac{1}{2} m_0 v^2\right)$$

[161] Another Formulation of the Fundamental Equations

In K_0, the variation of proper mass entails variations $d\mathbf{I}_0$ and $dW_0 = c\, dm_0$; whence, in K, variations $d\mathbf{I}_i$ and dW_i. Taking the external force $\mathbf{F}_e$ into account, so that

$$d\left(\frac{m_0 \mathbf{v}}{\sqrt{1-\beta^2}}\right) = \mathbf{F}_e\, dt + d\mathbf{I}_i$$

and

$$d\left(\frac{m_0 c^2}{\sqrt{1-\beta^2}}\right) = \mathbf{F}_e \cdot \mathbf{v}\, dt + dW_i$$

we can obviously write

$$d\left(\frac{m_0 \mathbf{v}}{\sqrt{1-\beta^2}}\right) = \mathbf{F}\, dt, \quad \mathbf{F} = \mathbf{F}_e + \frac{d\mathbf{I}_i}{dt}$$

Then, however, we have

$$d\left(\frac{m_0 c^2}{\sqrt{1-\beta^2}}\right) = \left(\mathbf{F} - \frac{d\mathbf{I}_i}{dt}\right) \cdot \mathbf{v}\, dt + dW_i$$

B. TRANSFORMATION OF THE PRINCIPAL QUANTITIES

[162] Passage from the rest system K_0 to another system, K

We examine the force transformation, taking up the reasoning of § [28] again. Here we have

$$X = \frac{m_0 \gamma_x}{(1-\beta^2)^{3/2}} + \frac{v}{\sqrt{1-\beta^2}} \frac{dm_0}{dt}$$
$$Y = \frac{m_0}{\sqrt{1-\beta^2}} \gamma_y \tag{1}$$

In the rest system, (1) reduces to

$$X_0 = m_0 \gamma_{0x}, \qquad Y_0 = m_0 \gamma_{0y} \tag{2}$$

Furthermore,

$$\gamma_{0x} = \frac{\gamma_x}{(1-\beta^2)^{3/2}}, \quad \gamma_{0y} = \frac{\gamma_y}{1-\beta^2} \tag{3}$$

Comparing (1) and (2) and taking (3) into account, we obtain

$$X = X_0 + \frac{v}{\sqrt{1-\beta^2}}\frac{dm_0}{dt} = X_0 + v\frac{dm_0}{dt_0}$$

$$Y = Y_0\sqrt{1-\beta^2}$$

Two important results emerge:

—*even at low velocities, the force is transformed (term in* dm_0/dt_0*);*
—*a non-zero force in K corresponds to a zero force in* K_0.

[163] General Transformations of Masses and Momenta

The formulae are the same as those for constant m_0:

$$m_1 = m_2\frac{1+vu_{2x}/c^2}{\sqrt{1-\beta^2}}$$

$$p_{1x} = \frac{p_{2x}+m_2v}{\sqrt{1-\beta^2}}$$

$$p_{1y} = p_{2y}, \qquad p_{1z} = p_{2z}$$

We have in fact changed nothing in the definitions of m and p.

Remark. Although the mass m_0 depends upon the time, it is treated *by convention* at each moment as an invariant with respect to all changes of reference system.

[164] General Transformation of Force†

The component X_1

We have

$$X_1 = \frac{dp_{1x}}{dt_1} = \frac{d}{dt_2}\left(\frac{p_{2x}+m_2v}{\sqrt{1-\frac{v^2}{c^2}}}\right)\frac{dt_2}{dt_1}$$

$$= \frac{1}{1+\frac{vu_{2x}}{c^2}}\left(\frac{dp_{2x}}{dt_2}+v\frac{dm_2}{dt_2}\right) \tag{1}$$

$$= \frac{1}{1+\frac{vu_{2x}}{c^2}}\left(X_2+v\frac{dm_2}{dt_2}\right)$$

† The remainder of the present chapter has been prepared in collaboration with A. Guessous (ref. [52] of § [172]).

In K_2, the energy theorem takes the form

$$X_2u_{2x}+Y_2u_{2y}+Z_2u_{2z} = c^2\frac{\mathrm{d}m_2}{\mathrm{d}t_2}-\sqrt{1-\frac{u_2^2}{c^2}}\,c^2\frac{\mathrm{d}m_0}{\mathrm{d}t_2}$$

so that

$$\frac{\mathrm{d}m_2}{\mathrm{d}t_2} = \frac{1}{c^2}(X_2u_{2x}+Y_2u_{2y}+Z_2u_{2z})+\sqrt{1-\frac{u_2^2}{c^2}}\,\frac{\mathrm{d}m_0}{\mathrm{d}t_2}$$

$$= \frac{1}{c^2}(X_2u_{2x}+Y_2u_{2y}+Z_2u_{2z})+\left(1-\frac{u_2^2}{c^2}\right)\frac{\mathrm{d}m_0}{\mathrm{d}t_0}$$

Substituting in (1),

$$X_1 = \frac{1}{1+\dfrac{vu_{2x}}{c^2}}\left\{X_2+\frac{v}{c^2}(X_2u_{2x}+\ \ldots)+v\left(1-\frac{u_2^2}{c^2}\right)\frac{\mathrm{d}m_0}{\mathrm{d}t_0}\right\}$$

Finally

$$X_1 = X_2+\frac{v/c^2}{1+\dfrac{vu_{2x}}{c^2}}(Y_2u_{2y}+Z_2u_{2z})+\frac{v\left(1-\dfrac{u_2^2}{c^2}\right)}{1+\dfrac{vu_{2x}}{c^2}}\frac{\mathrm{d}m_0}{\mathrm{d}t_0}$$

We may also write

$$X_1 = \frac{1}{1+\dfrac{vu_{2x}}{c^2}}\left\{X_2+\frac{v}{c^2}\mathbf{u}_2\cdot\mathbf{F}_2+v\left(1-\frac{u_2^2}{c^2}\right)\frac{\mathrm{d}m_0}{\mathrm{d}t_0}\right\}$$

The components Y_1 and Z_1

The result here is the same as that for m_0 constant. Thus

$$Y_1 = \frac{\mathrm{d}p_{1y}}{\mathrm{d}t_1} = \frac{\mathrm{d}p_{2y}}{\mathrm{d}t_1} = \frac{\mathrm{d}p_{2y}}{\mathrm{d}t_2}\frac{\mathrm{d}t_2}{\mathrm{d}t_1} = Y_2\frac{\mathrm{d}t_2}{\mathrm{d}t_1}$$

or

$$Y_1 = Y_2\frac{\sqrt{1-\dfrac{v^2}{c^2}}}{1+\dfrac{vu_{2x}}{c^2}}$$

and likewise for Z_1.

Other expressions

In the expression for X_1, let us replace u_{2y} and u_{2z} by the expressions for them (*Kinematics*, § [80]):

$$u_{2y} = u_{1y}\frac{1+\dfrac{vu_{2x}}{c^2}}{\sqrt{1-\beta^2}}$$

We obtain

$$X_1 = X_2 + \frac{vu_{1y}/c^2}{\sqrt{1-\frac{v^2}{c^2}}} Y_2 + \frac{vu_{1z}/c^2}{\sqrt{1-\frac{v^2}{c^2}}} Z_2 + \frac{v\left(1-\frac{u_2^2}{c^2}\right)}{1+\frac{vu_{2x}}{c^2}} \frac{dm_0}{dt_0}$$

Likewise

$$Y_1 = Y_2 \frac{1-\frac{vu_{1x}}{c^2}}{\sqrt{1-\beta^2}}$$

C. USE OF THE VECTORS E AND 𝔅

[165] Definition of E and 𝔅; the transformation Laws

(a) The form of the force transformation suggests that, as a generalization of the relation for constant proper masses, we should write

$$\mathbf{F} = \varepsilon\left(\mathbf{E} + \frac{\mathbf{u}}{c_0} \times \mathfrak{B}\right) + \mathbf{G}(u) \frac{dm_0}{dt_0} \tag{1}$$

where **E** and 𝔅 are independent of **u** and **G** is independent of **E** and 𝔅. Substituting into the relation (§ [164]),

$$Y_1 = Y_2 \frac{\sqrt{1-v^2/c^2}}{1+vu_{2x}/c^2} \tag{2}$$

we obtain

$$\varepsilon E_{1y} + \frac{\varepsilon}{c_0}(u_{1z}\mathcal{B}_{1x} - u_{1x}\mathcal{B}_{1z}) + G_{1y}\frac{dm_0}{dt_0}$$

$$= \frac{\sqrt{1-v^2/c^2}}{1+vu_{2x}/c^2}\left\{\varepsilon E_{2y} + \frac{\varepsilon}{c_0}(u_{2z}\mathcal{B}_{2x} - u_{2x}\mathcal{B}_{2z}) + G_{2y}\frac{dm_0}{dt_0}\right\} \tag{3}$$

Since $\mathbf{u}_1$, $\mathbf{u}_2$, and sometimes **E** and 𝔅 also, are independent of dm_0/dt_0, we must have

$$G_{1y} = G_{2y} \frac{\sqrt{1-v^2/c^2}}{1+vu_{2x}/c^2} \tag{4}$$

Since this is the transformation formula for u_y, let us take $\mathbf{G} = \mathbf{u}$ and check whether this is acceptable. Taking (4) into account, equation (3) reduces to the case of constant m_0, and yields (see the calculation in § [44])

$$E_{1y} = \frac{E_{2y} + \frac{v}{c_0}\mathcal{B}_{2z}}{\sqrt{1-v^2/c^2}}, \quad \mathcal{B}_{1z} = \frac{\mathcal{B}_{2z} + \frac{c_0 v}{c^2} E_{2y}}{\sqrt{1-v^2/c^2}}$$

Analogous calculations can be performed on the transformations for Z_1 and X_1. We can finally write

$$\boxed{\mathbf{F} = \varepsilon\left(\mathbf{E}+\frac{\mathbf{u}}{c_0}\times\mathfrak{B}\right)+\mathbf{u}\,\frac{\mathrm{d}m_0}{\mathrm{d}t_0}}$$

The quantities $\mathbf{E}$ *and* $\mathfrak{B}$ *transform just as for constant proper mass* m_0.

[166] Another Form of the Impulse Equation

Substituting into the equation of motion, we obtain

$$\boxed{m_0\,\frac{\mathrm{d}}{\mathrm{d}t}\left(\frac{v}{\sqrt{1-v^2/c^2}}\right) = \varepsilon\left(\mathbf{E}+\frac{\mathbf{v}}{c_0}\times\mathfrak{B}\right)} \tag{1}$$

On the right-hand side, we have the usual expression for the force. We must be on our guard, however, for contrary to what (1) might lead us to believe, *this force is not always the external force*. We can only say that it reduces to the external force in certain special cases. It can include actions created by changes of mass. This equation could have been taken as the basic postulate, but its origin would have seemed obscure. It is better to deduce it from the impulse postulate.

[167] Another Form of the Energy Equation

For the work, we find

$$\mathbf{F}\,.\,\mathbf{v}\;\mathrm{d}t = \varepsilon\mathbf{E}\,.\,\mathbf{v}\;\mathrm{d}t+\frac{v^2\,\mathrm{d}m_0}{\sqrt{1-v^2/c^2}}$$

Substituting into equation (1) of § [159], we obtain

$$\mathbf{F}\,.\,\mathbf{v}\;\mathrm{d}t+c^2\sqrt{1-\beta^2}\;\mathrm{d}m_0 = \varepsilon\mathbf{E}\,.\,\mathbf{v}\;\mathrm{d}t+\frac{c^2\,\mathrm{d}m_0}{\sqrt{1-\beta^2}}$$

$$= \mathrm{d}\left(\frac{m_0c^2}{\sqrt{1-\beta^2}}\right) = \frac{c^2\,\mathrm{d}m_0}{\sqrt{1-\beta^2}}+m_0c^2\,\mathrm{d}\left(\frac{1}{\sqrt{1-\beta^2}}\right)$$

Hence

$$\varepsilon\mathbf{E}\,.\,\mathbf{v}\;\mathrm{d}t = m_0c^2\,\mathrm{d}\left(\frac{1}{\sqrt{1-\beta^2}}\right)$$

On the left-hand side is the work done by $\mathbf{E}$ (which is not the total work).

[167A] Special Cases

(a) *The case* $\mathbf{E} = \mathfrak{B} = 0$

We then have a uniform or zero velocity, whether m_0 is constant or variable. The two forms of dynamics cannot differ unless $\mathbf{E}$ or $\mathfrak{B}$ is not equal to zero.

(b) *The case of rectilinear motion with* $\mathbf{v}\times\mathfrak{B} = 0$ *($\mathfrak{B}$ zero or normal to* $\mathbf{v}$*)*

Relation (2) reduces to

$$\varepsilon \mathbf{E} = m_0 \frac{\mathrm{d}}{\mathrm{d}t}\left(\frac{v}{\sqrt{1-\beta^2}}\right) = \frac{m_0}{(1-\beta^2)^{3/2}} \frac{\mathrm{d}v}{\mathrm{d}t}$$

$$= \frac{m_0}{1-\beta^2} \frac{\mathrm{d}v}{\mathrm{d}t_0}$$

$$\int \frac{\mathrm{d}v}{1-v^2/c^2} = \int \frac{\varepsilon E}{m_0} \mathrm{d}t_0$$

The problem would be solved if $m(t_0)$ and $E(t_0)$ were known.

D. THE FOUR-DIMENSIONAL FORMALISM

[168] Definition of the Four-vectors Π_i and K_i

(a) We define the four-vectors momentum Π_i and force K_i by means of the relations

$$\Pi_\alpha = \frac{m_0 v_\alpha}{\sqrt{1-\beta^2}} = p_\alpha, \quad \Pi_l = \frac{i m_0 c}{\sqrt{1-\beta^2}} = \frac{iW}{c}$$
$$K_\alpha = \frac{F_\alpha}{\sqrt{1-\beta^2}}, \quad K_l = \frac{i}{c} \frac{\mathbf{v}\cdot\mathbf{F}}{\sqrt{1-\beta^2}} + ic \frac{\mathrm{d}m_0}{\mathrm{d}t} \tag{1}$$

v_α and F_α denote velocity and force in three dimensions.

The definition of the four-vector Π_i is the same as in dynamics with constant proper mass; in the four-vector K_i, the expression for K_l is new.

Instead of postulating relations (1), we can define K_i and Π_i to be the four-vectors which, in the proper system K_0, have the components

$$\mathbf{F}_0, \quad ic\frac{\mathrm{d}m_0}{\mathrm{d}t_0} \quad \text{and} \quad \mathbf{p}_0, \quad \frac{i}{c} m_0 c^2 \tag{2}$$

Using the simple Lorentz transformation, the components of the four-vector transform as follows:

$$K_x = \frac{F_x}{\sqrt{1-\beta^2}} = \frac{K_{0x}-i\beta K_{0l}}{\sqrt{1-\beta^2}}$$

$$K_l = \frac{F_l}{\sqrt{1-\beta^2}} = \frac{K_{0l}+i\beta K_{0x}}{\sqrt{1-\beta^2}}$$

Hence

$$F_x = F_{0x}+v\,\frac{\mathrm{d}m_0}{\mathrm{d}t_0}$$

$$F_l = \frac{i}{c}\,v_x F_{0x}+ic\,\frac{\mathrm{d}m_0}{\mathrm{d}t_0}$$

$$= \frac{i}{c}\,v_x\left(F_x-v\,\frac{\mathrm{d}m_0}{\mathrm{d}t_0}\right)+ic\,\frac{\mathrm{d}m_0}{\mathrm{d}t_0}$$

$$= \frac{i}{c}\left\{\mathbf{v}.\mathbf{F}+(c^2-v^2)\,\frac{\mathrm{d}m_0}{\mathrm{d}t_0}\right\}$$

and the expression for K_l is that given in (1).

[169] The Equations of Dynamics

(a) We shall show that we recover the equations of §§ [157] and [159] by writing the law (as in § [38]) in the form

$$K_i = \frac{\mathrm{d}\Pi_i}{\mathrm{d}T} \tag{1}$$

$T = t_0$ is the proper time, related to the variable t by

$$\frac{\mathrm{d}t}{\mathrm{d}T} = \frac{1}{\sqrt{1-\beta^2}}$$

Thus for the spatial components, we have

$$\frac{F_x}{\sqrt{1-\beta^2}} = \frac{\mathrm{d}}{\mathrm{d}t}\left(\frac{m_0 v_x}{\sqrt{1-\beta^2}}\right)\frac{\mathrm{d}t}{\mathrm{d}T}$$

so that

$$F_x = \frac{\mathrm{d}}{\mathrm{d}t}\left(\frac{m_0 v_x}{\sqrt{1-\beta^2}}\right)$$

or

$$\mathbf{F} = \frac{\mathrm{d}}{\mathrm{d}t}\left(\frac{m_0\mathbf{v}}{\sqrt{1-\beta^2}}\right)$$

Furthermore, the time component gives

$$\frac{F_l}{\sqrt{1-\beta^2}} = \frac{\mathrm{d}}{\mathrm{d}t}\left(\frac{im_0 c}{\sqrt{1-\beta^2}}\right)\frac{\mathrm{d}t}{\mathrm{d}T}$$

or

$$F_l = \frac{\mathrm{d}}{\mathrm{d}t}\left(\frac{m_0 c^2}{\sqrt{1-\beta^2}}\right),$$

$$\mathbf{v} \,.\, \mathbf{F} + (c^2 - v^2)\frac{\mathrm{d}m_0}{\mathrm{d}t_0} = \frac{\mathrm{d}}{\mathrm{d}t}(mc^2)$$

We do indeed recover the relation of § [159]:

$$\mathbf{v} \,.\, \mathbf{F}\,\mathrm{d}t + c^2\sqrt{1-\beta^2}\,\mathrm{d}m_0 = \mathrm{d}(mc^2)$$

(b) Expanding (1),

$$K_i = \frac{\mathrm{d}}{\mathrm{d}T}(m_0 V_i) = m_0\frac{\mathrm{d}V_i}{\mathrm{d}T} + V_i\frac{\mathrm{d}m_0}{\mathrm{d}T}$$

$$= m_0 \Gamma_i + V_i\frac{\mathrm{d}m_0}{\mathrm{d}T}$$

[170] The Tensor N^{ij}

(a) We attempt to determine a tensor N^{ij} with which we can calculate the force K using the same relation† as in the case of constant proper masses, namely

$$K^i = \frac{\varepsilon}{c_0} N^{ij} V_j \tag{1}$$

The desired matrix N^{ij} must reduce to

$$N_0^{ij} \to \begin{pmatrix} 0 & \mathcal{B}_{0z} & -\mathcal{B}_{0y} & -\frac{ic_0}{c}E_{0x} \\ -\mathcal{B}_{0z} & 0 & \mathcal{B}_{0x} & -\frac{ic_0}{c}E_{0y} \\ \mathcal{B}_{0y} & -\mathcal{B}_{0x} & 0 & -\frac{ic_0}{c}E_{0z} \\ \frac{ic_0}{c}E_{0x} & \frac{ic_0}{c}E_{0y} & \frac{ic_0}{c}E_{0z} & \frac{c_0}{\varepsilon}\frac{\mathrm{d}m_0}{\mathrm{d}t_0} \end{pmatrix}$$

in the proper system K_0. In fact, the force in K_0 must reduce to

$$K^\alpha = F^\alpha = \varepsilon E^\alpha, \qquad K^4 = ic\frac{\mathrm{d}m_0}{\mathrm{d}t_0} = \frac{\varepsilon}{c_0}N^{44}V_4$$

The components in another system K are given by

$$\{N^{ij}\} = \{L\}\{N_0^{ij}\}\{L\}$$

This matrix is as follows:

† We could also retain the tensor $\mathcal{N}_{ij}$ for constant proper masses, and alter the form of the law.

$$\begin{pmatrix} \dfrac{-\dfrac{c_0v^2}{c^2\varepsilon}\dfrac{dm_0}{dt_0}}{1-\beta^2} & \dfrac{\mathcal{B}_{0z}+\dfrac{c_0v}{c^2}E_{0y}}{\sqrt{1-\beta^2}} & -\dfrac{\mathcal{B}_{0y}-\dfrac{c_0v}{c^2}E_{0z}}{\sqrt{1-\beta^2}} & -\dfrac{ic_0}{c}\left(E_{0x}+\dfrac{\dfrac{v}{\varepsilon}\dfrac{dm_0}{dt_0}}{1-\beta^2}\right) \\ -\dfrac{\mathcal{B}_{0z}+\dfrac{c_0v}{c^2}E_{0y}}{\sqrt{1-\beta^2}} & 0 & \mathcal{B}_{0x} & -\dfrac{ic_0}{c}\dfrac{E_{0y}+\dfrac{v}{c_0}\mathcal{B}_{0z}}{\sqrt{1-\beta^2}} \\ \dfrac{\mathcal{B}_{0y}-\dfrac{c_0v}{c^2}E_{0z}}{\sqrt{1-\beta^2}} & -\mathcal{B}_{0x} & 0 & -\dfrac{ic_0}{c}\dfrac{E_{0z}-\dfrac{v}{c_0}\mathcal{B}_{0y}}{\sqrt{1-\beta^2}} \\ \dfrac{ic_0}{c}\left(E_{0x}-\dfrac{\dfrac{v}{\varepsilon}\dfrac{dm_0}{dt_0}}{1-\beta^2}\right) & \dfrac{ic_0}{c}\dfrac{E_{0y}+\dfrac{v}{c_0}\mathcal{B}_{0z}}{\sqrt{1-\beta^2}} & \dfrac{ic_0}{c}\dfrac{E_{0z}-\dfrac{v}{c_0}\mathcal{B}_{0y}}{\sqrt{1-\beta^2}} & \dfrac{\dfrac{c_0}{\varepsilon}\dfrac{dm_0}{dt_0}}{1-\beta^2} \end{pmatrix}$$

It is perfectly natural to write

$$N^{ij} \to \begin{pmatrix} \dfrac{-\dfrac{c_0v^2}{c^2\varepsilon}\dfrac{dm_0}{dt_0}}{1-\beta^2} & \mathcal{B}_z & -\mathcal{B}_y & -\dfrac{ic_0}{c}\left(E_x+\dfrac{\dfrac{v}{\varepsilon}\dfrac{dm_0}{dt_0}}{1-\beta^2}\right) \\ -\mathcal{B}_z & 0 & \mathcal{B}_x & -\dfrac{ic_0}{c}E_y \\ \mathcal{B}_y & -\mathcal{B}_x & 0 & -\dfrac{ic_0}{c}E_z \\ \dfrac{ic_0}{c}\left(E_x-\dfrac{\dfrac{v}{\varepsilon}\dfrac{dm_0}{dt_0}}{1-\beta^2}\right) & \dfrac{ic_0}{c}E_y & \dfrac{ic_0}{c}E_z & \dfrac{\dfrac{c}{\varepsilon}\dfrac{dm_0}{dt_0}}{1-\beta^2} \end{pmatrix}$$

The fields **E** and $\mathfrak{B}$ thus transform just as for m_0 constant.

(b) The tensor N^{ij} can be decomposed into two tensors,

$$N^{ij} = \mathcal{N}^{ij}+\mathcal{S}^{ij}$$

$\mathcal{N}^{ij}$ is the usual antisymmetric tensor for constant proper mass, m_0:

$$\mathcal{N}^{ij} \to \begin{pmatrix} 0 & \mathcal{B}_z & -\mathcal{B}_y & -\dfrac{ic_0}{c}E_x \\ -\mathcal{B}_z & 0 & \mathcal{B}_x & -\dfrac{ic_0}{c}E_y \\ \mathcal{B}_y & -\mathcal{B}_x & 0 & -\dfrac{ic_0}{c}E_z \\ \dfrac{ic_0}{c}E_x & \dfrac{ic_0}{c}E_y & \dfrac{ic_0}{c}E_z & 0 \end{pmatrix}$$

$\mathcal{S}^{ij}$ is a symmetric tensor, given by

$$\mathcal{S}^{ij} \rightarrow \begin{pmatrix} \dfrac{-\dfrac{c_0 v^2}{c^2 \varepsilon} \dfrac{dm_0}{dt_0}}{1-\beta^2} & 0 & 0 & \dfrac{-\dfrac{ic_0 v}{c\varepsilon} \dfrac{dm_0}{dt_0}}{1-\beta^2} \\ 0 & 0 & 0 & 0 \\ 0 & 0 & 0 & 0 \\ \dfrac{-\dfrac{ic_0 v}{c\varepsilon} \dfrac{dm_0}{dt_0}}{1-\beta^2} & 0 & 0 & \dfrac{\dfrac{c_0}{\varepsilon} \dfrac{dm_0}{dt_0}}{1-\beta^2} \end{pmatrix}$$

(c) It can be shown that, on expanding (1), the spatial components are indeed

$$X = \varepsilon E_x + v \frac{dm_0}{dt_0}$$

$$Y = \varepsilon E_y - \frac{\varepsilon v}{c_0} \mathcal{B}_z$$

$$Z = \varepsilon E_z + \frac{\varepsilon v}{c_0} \mathcal{B}_y$$

or in vector form

$$\mathbf{F} = \varepsilon\left(\mathbf{E} + \frac{\mathbf{v}}{c_0} \times \mathfrak{B}\right) + \mathbf{v} \frac{dm_0}{dt}$$

The time component is an expression of the energy theorem:

$$\varepsilon \mathbf{E} . \mathbf{v}\, dt = m_0 c^2\, d\left(\frac{1}{\sqrt{1-\beta^2}}\right)$$

[171] Another Tensor N^{ij}

The physical quantities are the four-velocity and the four-force; the relation

$$K_i = \mathcal{N}_{ij} V^j$$

does not define $\mathcal{N}_{ij}$ uniquely. Setting out from the expression for the three-dimensional force (§ [165]), we see that the following tensor is just as apt as that of § [170]:

$$N^{ij} = \begin{pmatrix} \dfrac{c_0}{\varepsilon} \dfrac{dm_0}{dt_0} & \mathcal{B}_z & -\mathcal{B}_y & -\dfrac{ic_0}{c} E_x \\ -\mathcal{B}_z & \dfrac{c_0}{\varepsilon} \dfrac{dm_0}{dt_0} & \mathcal{B}_x & -\dfrac{ic_0}{c} E_y \\ \mathcal{B}_y & -\mathcal{B}_x & \dfrac{c_0}{\varepsilon} \dfrac{dm_0}{dt_0} & -\dfrac{ic_0}{c} E_z \\ \dfrac{ic_0}{c} E_x & \dfrac{ic_0}{c} E_y & \dfrac{ic_0}{c} E_z & \dfrac{c_0}{\varepsilon} \dfrac{dm_0}{dt_0} \end{pmatrix}$$

This tensor seems to be more satisfactory, for it does not contain v but only the quantities characteristic of the field and of the variation of m_0.

HISTORICAL AND BIBLIOGRAPHICAL NOTES

[172] On Point Dynamics with Variable Proper Mass; Relativistic Rockets and Astronautics

(a) *The general equations*

The dynamics of the particle having variable proper mass is still a marginal topic in textbooks on mechanics; in the French edition of the present work, I devoted only a short chapter to the question of rockets (in the second volume). This branch of mechanics is becoming more and more important, however, because of its applications in astronautics, in the theory of continuous media and in thermodynamics, for example. I have taken note of this development by promoting the fundamental equations to the first volume.

The general equations of motion, both Newtonian and relativistic, are to be found in the works by Tolman [2], Relton [4], Sommerfeld [11], Mandelker [19], Szamosi [20], Marx [24], Kalitzin [27, 29, 32], Bradeanu [48], Oliveri [51], Gadsen [55c], Pomeranz [56] and van den Akker [57]. Not all these are correct. Some authors—Tolman, for example, and myself in the first edition (§ [5])—believe that it is enough to employ the usual equation, merely allowing for the variation of m_0. The formalism adopted in §§ [157] and [159] is my own. At my request, Monsieur Guessous who was then a pupil of mine worked out some of the consequences of it. His diploma work extends my point of view on the general significance of the force expressed as **E** and $\mathfrak{B}$ to the present case (§ [165]); it contains studies of various problems of motion.

The equations in a gravitational field have been studied by Grodzovsky, Ivanov and Tokarev [47, 49] and Krzywoblocki [59].

(b) *Rockets*

The study of relativistic rockets was inaugurated by the fundamental paper by Ackerett [5]; the latter gives the differential equation of the motion, and the expression for the final velocity. Publications subsequently multiplied and it is a characteristic fact that this was not because relativistic physicists were at a loose end, but was one of the activities of the astronautical centres; the papers were published in the technical journals (quite often, in *Astronautica Acta*) or given at Congresses. In connection with "*material*" *propulsion* rockets, we mention the work of Seifert, Mills and Supperfield [6], Bade [12], Sänger [13], Krause [18], Kooy [28], Kalitzin [32, 55], Huth [41] and Rhee [60].

Photon rockets are studied in some of the preceding papers, and especially in the articles by Sänger [14, 15, 17, 23, 35, 36, 37], Kooy [28], Stuhlinger [34], Peschka [44] and Burcev [58].

Multiple-stage rockets are considered by Peschka [16], Subotowicz [33], Evette and Wangeness [40], Spencer and Jaffe [55b] and Mišoň [62].

(c) *Astronautics and relativity*

In some respects, this question is an application of the clock paradox; the reader is therefore invited to refer first to the Notes to Chapter VIII of *Kinematics*. In his work of 1911, Langevin already foresaw the practical conditions, but he limited himself to a few brief remarks about the energy necessary for firing a bullet; it is a curious fact that he pointed out the principle of the rocket, only to deny its possible realization (doubtless he had been too avid a reader of Jules Verne in his childhood).

The real precursor in this domain, and indeed, in everything related to astronautics, seems to me to be Esnault-Pelterie [1, 3]. We must not forget that two books are involved, published in 1930 and 1935 and containing many technical calculations. Only after the Second World War was the problem fully attacked. There are many papers in which the possibility of travelling to the stars, and even to other galaxies, is very seriously studied. Apart from the purely kinematic questions, the various means of propulsion are studied (jets of matter, of ions and of photons), and the general conclusion is that the long distance vehicles will be photon rockets. The nature of the fuel is also studied thoroughly; nuclear reactions are obligatorily chosen, and some authors even speak of anti-matter motors (although no one really knows what this involves). The difficulty (or perhaps, impossibility) of carrrying enough fuel has led to the suggestion that interstellar gas might be employed. For all these problems, we mention the works of Seifert [6], Hoyle [7], Shepherd [8, 25], Maughin [9, 10], Peschka [16], Sänger [21, 30, 42], Kooy [22, 26], Karlovitz and Lewis [31], Bussard [38], Marx [39, 55a], Tsu [43], Krause [45], Boneff [46], Mocckel [50], von Hoerner [53], Spencer and Jaffe [55b] and Stan and Tóth [61].

The conclusions drawn by these authors vary widely. Some (Marx, Tsu) are very pessimistic; they believe that access to even the nearest stars would be practically impossible. They say that anti-matter would have to be used, and that even so, voyages of 100 years and more would be necessary. Others (Spencer and Jaffe, for example) are moderately optimistic, shall we say. Using multiple-stage rockets, with fission or fusion energy, it should be possible to travel to α in Centaur and back in 29 years. Finally, there are some firmly optimistic authors (Sänger). During a human lifetime, not only the nearest stars but all the stars in our galaxy and all the galaxies beyond, up to the limits of the universe, should be accessible, they believe.

I cannot really express any opinion, for a thorough technical study would be necessary. Nevertheless, *I have a deep-rooted conviction that the optimists are right.* I am convinced that the only limits to practical achievement lie in the human imagination. Sooner or later, man will perform what he has been capable of conceiving, be it but in his dreams. Hitherto, humanity has dwelled in a little corner of the universe, leading a static life with only very weak sources of energy. We are at the beginning of a new era, dominated, so far as our material existence† is concerned, by three magic words: energy, space, time.

[1] R. Esnault-Pelterie: *L'Astronautique*. Gauthier-Villars, 1930, pp. 228–241.

[2] R. C. Tolman: Ref. [4] of Appendix I, 1934, p. 46 (footnote).

[3] R. Esnault-Pelterie: *L'Astronautique; compléments*. Gauthier-Villars, 1935, pp. 87–93.

[4] F. E. Relton: *Applied Bessel Functions*. Blackie, London, 1946 (new ed. 1949), pp. 82–85.

[5] J. Ackerett: Zur Theorie der Raketen. *Helv. Phys. Acta* **19** (1946) 103–112. English trans.: L. Gilbert, *J. Brit. Interplan. Soc.* **6** (1947) 116–123.

[6] H. S. Seifert, M. W. Mills and M. Supperfield: Physics of rockets: dynamics of long range rockets. *Amer. J. Phys.* **15** (1947) 255–272.

[7] F. Hoyle: Some scientific aspects of interplanetary travel. *The Times*, Oct. 1950, p. 7.

[8] L. R. Shepherd: Interstellar flight. *J. Brit. Interplan. Soc.* **11** (1952) 149–167.

[9] Ch. Maughin: Astronautique et relativité. A l'assaut de l'espace-temps. *C. R. Acad. Sci. Paris* **234** (1952) 1004–1007.

[10] Ch. Maughin: A propos de ma Note: "Astronautique et relativité". *C. R. Acad. Sci. Paris*, **234** (1952) 1329.

[11] A. Sommerfeld: *Mechanics, Lectures on Theoretical Physics.* Academic Press, New York, 1952, Vol. 1, pp. 28–29.

[12] W. L. Bade: Relativistic rocket theory. *Amer. J. Phys.* **21** (1953) 310–312.

[13] E. Sänger: Die physikalischen Grundlagen der Strahlantriebtechnik. *V. D. I. Forschungsheft* **19** (1953) Heft 437, 5–25.

[14] E. Sänger: Zur Theorie der Photonenraketen. *Ing. Arch.* **21** (1953) 213–226.

[15] E. Sänger: Zur Theorie der Photonenraketen. Probleme der Weltraumforschung. *Proc. IVth Internat. Astron. Congr. Zürich 1953*, Biel, Laubscher, 1954.

[16] W. Peschka: Ueber die Ueberbrückung interstellarer Entfernungen. *Astr. Acta* **2** (1956) 191–200.

[17] E. Sänger: Zur Mechanik der Photonenstrahlantriebe. *Mitt. Forschungsinst. Phys. Strahlantriebe, Stuttgart*, No. 5, München, R. Oldenbourg, 1956.

[18] H. Krause: Relativistische Raketenmechanik. *Astr. Acta* **11** (1956) 30–47.

[19] J. Mandelker: Le mouvement classique et relativiste d'une particule dont la masse croit avec la vitesse. *Book of Abstr.*, Sect. II, 1956, p. 9.

[20] G. Szamosi: Die relativistische Bewegung des Massenpunktes bei einer allgemeinen Kraftannahme. *Acta Phys. Hung.* **5** (1956) 463–469.

[21] E. Sänger: Die Erreichbarkeit der Fixsterne. Abbild. 2, *Kongressvortrag*, Rome, 1956.

[22] J. M. J. Kooy: Space travel and future research into the structure of the universe. *J. Brit. Interplan. Soc.* **15** (1956) 248–259.

[23] E. Sänger: Zur Flugmechanik der Photonenraketen. *Astr. Acta* **3** (1957) 89–99.

[24] G. Marx: Innere Arbeit in der relativistischen Dynamik. *Acta Phys. Hung.* **6** (1957) 353–379.

† Let us not forget what the Gospels tell us, though: "Man cannot live by bread alone . . ." All our problems will not be solved by a trip to the galaxies.

[25] L. R. SHEPHERD: Interstellar flight, chap. 24 of *Realities of Space Travel,* Edited by Carter, New York, McGraw-Hill, 1957.
[26] J. M. J. KOOY: *Proc. VIIIth Intern. Astr. Congr., Barcelona 1957,* Vienna, Springer, 1958, p. 569.
[27] N. S. KALITZIN: Grundgleichungen der relativistischen Mechanik eines materiellen Punktes mit veränderlicher Masse. *Nuovo Cimento* **8** (1958) 843–849.
[28] J. M. J. KOOY: On relativistic rocket mechanics. *Astr. Acta* **4** (1958) 31–58.
[29] N. S. KALITZIN: Grundgleichungen der relativistischen Mechanik eines materiellen Punktes mit veränderlicher Masse. *C. R. Acad. Sci. Bulgarie* (1958) pp. 185–188.
[30] E. SÄNGER: *Proc. IXth Intern. Astr. Congr. Amsterdam 1958,* Vienna, Springer, 1959, p. 817.
[31] B. KARLOVITZ and B. LEWIS: Space propulsion by interstellar gas. *Proc. IXth Intern. Astr. Congr. Amsterdam, 1958,* p. 307.
[32] N. S. KALITZIN: Ueber eine Verallgemeinerung der Grundformel der Raketendynamik. *Nuovo Cimento* **11** (1959) 298–299.
[33] M. SUBOTOWICZ: Theorie der relativistischen n-Stufenrakete. *Proc. Xth Intern. Astr. Congr. London 1959,* Vienna, Springer, 1960, Vol. 2, pp. 852–864.
[34] E. STUHLINGER: Photon rocket propulsion. *Astronautics* **4** (1959), p. 36–; *ibid.*, p. 69–; *ibid.*, p. 72–; *ibid.*, p. 74–; *ibid.*, p. 76–; *ibid.*, p. 78–.
[35] E. SÄNGER: Strahlungsquellen für Photonenstrahlantriebe. *Astr. Acta* **5** (1959) 15.
[36] E. SÄNGER: Ueber das Richten intensiver Photonenstrahlen mittels Elektronengasspiegel. *Astr. Acta* **5** (1959) 266.
[37] E. SÄNGER: On the directing of intense photonic beams by means of electron gas mirrors. *Proc. Xth Intern. Astr. Congr. London 1959,* Vienna, Springer, 1960, p. 828.
[38] R. W. BUSSARD: Galactic matter and interstellar flight. *Astr. Acta* **6** (1960) 179–194.
[39] G. MARX: Ueber Energieprobleme der interstellarer Raumfahrt. *Astr. Acta* **6** (1960) 366–372.
[40] A. A. EVETTE and B. R. WANGENESS: Note on the separation of relativistically moving rockets. *Amer. J. Phys.* **80** (1960) 566.
[41] J. HUTH: Relativistic theory of rocket flight with advanced propulsion systems. *A.R.S. Journ.* **30** (1960) 250–253.
[42] E. SÄNGER: Atomraketen für Raumfahrt. *Astr. Acta* **6** (1960) 3–15.
[43] T. C. TSU: Requirements of interstellar flight. *Astr. Acta* **6** (1960) 247–255.
[44] W. PESCHKA: Beitrag zur inneren Ballistik der Photonenraketen. *Proc. XIth Intern. Astr. Congr. Stockholm 1960,* Vienna, Springer, 1961, p. 30.
[45] H. G. L. KRAUSE: Astrorelativity. *Proc. XIIth Intern. Astr. Congr. Washington 1961,* Vienna, Springer, 1963, pp. 131–160.
[46] N. BONEFF: Le problème des jumeaux dans la théorie de la relativité et en astronautique. *Ibid.*, pp. 191–195.
[47] G. L. GRODZOVSKY, Y. N. IVANOV and V. V. TOKAREV: On the motion of a body of variable mass with constant power consumption in a gravitational field. *Ibid.*, pp. 196–202.
[48] P. BRADEANU: Sur un problème variationnel relatif au mouvement d'un point de masse variable; application aux fusées. *Studia Univ. Babes-Bolyai,* fasc. 1, 1961, pp. 193–205.
[49] G. L. GRODZOVSKY, Y. N. IVANOV and V. V. TOKAREV: Mouvement d'un corps de masse variable avec dépense constante de puissance dans un champ de gravitation. *C. R. Acad. Sci. U.S.S.R.* **137** (1961) 1082–1085.
[50] W. E. MOCCKEL: Interplanetary trajectories for electrically propelled space vehicles. *Astr. Acta* **7** (1961) 431–444.
[51] E. OLIVERI: Equazione relativistica del moto della massa variabile. *Boll. Accad. Catania* **74** (1962) 47–50.
[52] A. GUESSOUS: Contribution à la dynamique du point à masse propre variable. *Dipl. Etud. Sup., Rabat,* 30 Oct. 1962 (directed by H. Arzeliès).
[53] S. V. HOERNER: The general limits of space travel. *Science* **137** (1962) 18.
[54] R. L. HALFMAN: *Dynamics.* Addison Wesley, 1962, pp. 542–543 and 553–554.
[55] H. KALITZIN: *Dynamik der relativistischen Raketen und einiger astronomischen Objekte.* Sofia, 1963, Vol. I.
[55a] G. MARX: The mechanical efficiency of interstellar vehicles. *Astr. Acta* **9** (1963) 131–138.
[55b] D. F. SPENCER and L. D. JAFFE: Feasibility of interstellar travel. *Astr. Acta* **9** (1963) 49–58.
[55c] C. P. GADSEN: Newton's law with variable mass. *Amer. J. Phys.* **32** (1964) 61.
[56] K. B. POMERANZ: The equation of motion for relativistic particles and systems with variable rest mass. *Amer. J. Phys.* (1964) 955–958.
[57] J. A. VAN DEN AKKER: Newton's law with variable mass. *Amer. J. Phys.* **32** (1964) 387.

[58] P. Burcev: On the mechanics of photon rockets in the general theory of relativity. *Czechoslovak J. Phys.* **5** (1964) 294–301.
[59] M. Z. v. Krzywoblocki: On the equations of motion of a body with variable mass in the general theory of relativity. *Acta Phys. Austr.* **17** (1964) 159–185.
[60] J. W. Rhee: Relativistic rocket motion. *Amer. J. Phys.* **33** (1965) 587.
[61] I. Stan and A. Tóth: Sint posibile calatoriile interstelare. *Rev. Fiz. Chim.* **2** (1965) 99–110.
[62] K. Mišoň: Rockets with impulsively separated structural mass. *Astr. Acta* **2** (1965) 306–311 (in Russian).

See Appendix III for complementary references.

CHAPTER XIII

THE ACTION THAT MODIFIES THE PROPER MASS DOES NOT AFFECT THE VELOCITY†

A. GENERAL EQUATIONS

[173] The Position of the Problem; the Various Cases

Let us consider a particle of proper mass m_0 at time t_0, at rest in a Galileian system of reference, K_0. Between times t_0 and t_0+dt_0, we add to or remove from the particle proper mass dm_0 (we count an increase of mass as positive) *without altering its state of rest in* K_0. The momentum in K_0 remains zero. By hypothesis, therefore, if we transfer to some other frame of reference K, the velocity $\mathbf{v}$ of the body is unaffected by the mass change but the momentum does vary by virtue of the variation of m_0. As well as the modification to m_0, we can add an external force, $\mathbf{F}_e$; in this case, of course, the action of the force is added to that of the change dm_0.

Numerous very important cases are included within the preceding hypothesis. The variation of proper mass is obtained in different ways, some examples of which are as follows.

Deformed elastic body

Consider an elastic body at rest in K_0, which we compress between times t_0 and t_0+dt_0 with the aid of a uniform pressure. The body remains at rest in K_0, but the work done, $d\mathcal{T}_0$, is identified with an increase of proper mass by hypothesis (the mass thus increases without receiving additional matter). We have the same situation for a gas or liquid, within a container, compressed by a piston. In other cases, the body may be deformed under the action of localized forces (the jointed lever, the extended rod), but we always assume that it remains at rest in K_0.

Continuous media in random motion

We consider a fluid, and examine the matter within an element of volume $d\varpi_0$, of which we trace the motion. We then consider this matter as a corpuscle, and we apply the

† This case is very important. The results obtained are fundamental for relativistic thermodynamics and the relativistic dynamics of continuous media; they also yield some conclusions about solids It is because the properties of the present special case have not been clearly disentangled that these branches of science have, until recently, remained studded with errors.

equations of particle dynamics, taking all the forces into account. Apart from the mass of the matter, the rest mass of this particle contains the mass-energy of the elastic deformation; we are thus dealing with a case covered by the hypothesis.

Heated or cooled bodies

The proper mass can also be increased without adding matter by adding heat; we shall remain within the case considered in the present chapter if the exchange of heat takes place by isotropic convection.

Isotropic gain or loss of matter

This case seems to be of no practical importance, but it gives a picture of the problem. We can think of a sphere full of holes, through which it loses material.

Radiating or absorbing bodies

The conclusions are the same, since radiation is a transfer of energy, and hence of mass.

Particle deformed by a field

Consider a particle placed in a force field, and in consequence, deformed. Thus we can imagine that an electric charge moving in a field $\mathbf{E}$, $\mathfrak{B}$ undergoes internal deformations if the field is high enough or varies rapidly. A change of proper mass would result. Is this perhaps an idea worth delving into in the dynamics of large accelerations?

[174] The Impulse Equation

(a) During time $\mathrm{d}t_0$ of K_0 (or $\mathrm{d}t$ of K), the body receives mass $\mathrm{d}m_0$ (which may, for example, be provided in the form of work, $\mathrm{d}\mathcal{T}_0$, or heat $\mathrm{d}Q_0$). *By hypothesis*, its state of rest in K_0 is unaffected; if it has velocity v in system K, it continues to have this velocity. At constant velocity, the change of momentum in K is

$$\frac{v\,\mathrm{d}m_0}{\sqrt{1-\beta^2}},$$

and this change occurs without any external force. If there is a non-zero resultant external force, $\mathbf{F}_e$, we must take as the impulse equation:

$$\begin{aligned}\frac{\mathrm{d}\mathbf{p}}{\mathrm{d}t} &= \frac{\mathrm{d}}{\mathrm{d}t}\left(\frac{m_0\mathbf{v}}{\sqrt{1-\beta^2}}\right) = \mathbf{F}_e + \frac{\mathbf{v}\,\mathrm{d}m_0}{\mathrm{d}t\sqrt{1-\beta^2}} \\ &= \mathbf{F}_e + \frac{\mathrm{d}m_0}{\mathrm{d}t_0}\mathbf{v} = \mathbf{F}_e + \mathbf{F}_i = \mathbf{F}\end{aligned} \qquad (1)$$

We denote the component of **F** associated with the change of proper mass by $\mathbf{F}_i$. Equation (1) is more conveniently written

$$\mathbf{F}_e = \frac{\mathbf{v}}{\sqrt{1-\beta^2}}\frac{dm_0}{dt}+m_0\frac{d}{dt}\left(\frac{\mathbf{v}}{\sqrt{1-\beta^2}}\right)-\mathbf{v}\frac{dm_0}{dt_0}$$

or

$$\mathbf{F}_e = m_0\frac{d}{dt}\left(\frac{\mathbf{v}}{\sqrt{1-\beta^2}}\right)$$

If there is no force $\mathbf{F}_e$, the motion of the body in K is a uniform translation. In this case, an extended body behaves like a point corpuscle, and its particular features can be studied here; this is the justification for §§ [178] ff.

(b) The energy equation is written

$$\mathbf{F}_e\,.\,\mathbf{v}\,dt+\mathbf{v}\,.\,\frac{dm_0}{dt_0}\,\mathbf{v}\,dt+c^2\sqrt{1-\beta^2}\,dm_0$$

$$= d\left(\frac{m_0c^2}{\sqrt{1-\beta^2}}\right) = c^2\frac{dm_0}{\sqrt{1-\beta^2}}+c^2m_0\,d\left(\frac{1}{\sqrt{1-\beta^2}}\right)$$

or

$$\mathbf{F}_e\,.\,\mathbf{v}\,dt = m_0c^2\,d\left(\frac{1}{\sqrt{1-\beta^2}}\right)$$

Remark. We may also write (§ [159])

$$dW = d\left(\frac{m_0c^2}{\sqrt{1-\beta^2}}\right) = c^2\frac{dm_0}{\sqrt{1-\beta^2}}+m_0c^2\,d\left(\frac{1}{\sqrt{1-\beta^2}}\right)$$

$$= d\mathcal{T}_e+c^2\frac{dm_0}{\sqrt{1-\beta^2}}$$

and so

$$d\mathcal{T}_e = m_0c^2\,d\left(\frac{1}{\sqrt{1-\beta^2}}\right)$$

Thus in the special case of the present chapter, the work done by the external force is equal to the fraction of the energy variation arising from the velocity change. To this fraction must be added the part relating to the change of proper mass. The two terms separate.

(c) *Other arguments*. The foregoing reasoning uses the general postulate of § [157] (a). Equation (1) can be obtained by arguing directly in the dynamics of invariable proper masses, as we mentioned in § [157] (b).

First argument. In the system K_0, the mass acquired, dm_0, has momentum dp_0 equal to zero. In K, however,

$$d\mathbf{p} = \frac{dm_0}{\sqrt{1-\beta^2}}\mathbf{v}$$

At time t, the total momentum is

$$\frac{m_0\mathbf{v}}{\sqrt{1-\beta^2}}+\frac{dm_0}{\sqrt{1-\beta^2}}\mathbf{v}$$

and must be the same at time $t+dt$; arguing in one dimension, therefore, we must have

$$\frac{m_0 v}{\sqrt{1-\beta^2}}+\frac{dm_0}{\sqrt{1-\beta^2}}v = \frac{m_0 v}{\sqrt{1-\beta^2}}+d\left(\frac{m_0 v}{\sqrt{1-\beta^2}}\right)$$

This does indeed imply that $dv = 0$.

Argument related to rocket theory (§ [191]). At time t, the momentum of the body is $m\mathbf{v}$; the masses dm' which it receives at time $t+dt$ are still outside it and each has velocity $\mathbf{w}$. Thus at time t, the total momentum is

$$m\mathbf{v}+\Sigma\mathbf{w}\, dm'$$

As, by hypothesis, the velocity $\mathbf{v}$ does not change, the total momentum at time $t+dt$ must be $(m+dm)\mathbf{v}$. Conservation of momentum gives

$$\mathbf{v}\, dm = \Sigma\mathbf{w}\, dm'$$

To ensure the constancy of v, we assume in addition that in K_0, the masses gained can be grouped into pairs with equal and opposite momenta. Consider first the simplest case of two equal masses dm_0', with velocities $\pm w_0$ parallel to v; thus

$$\begin{aligned}\Sigma\mathbf{w}\, dm' &= w_1\, dm_1' + w_2\, dm_2' \\ &= \frac{w_0+v}{1+\dfrac{vw_0}{c^2}}\, dm_0'\, \frac{1+\dfrac{vw_0}{c^2}}{\sqrt{1-\beta^2}} \\ &\quad + \frac{-w_0+v}{1-\dfrac{vw_0}{c^2}}\, dm_0'\, \frac{1-\dfrac{vw_0}{c^2}}{\sqrt{1-\beta^2}} = \frac{2v\, dm_0'}{\sqrt{1-\beta^2}}\end{aligned}$$

In addition, conservation of energy gives $dm_0 = 2dm_0'$, since dm_0 and dm_0' are the proper masses of the same substance in K_0. Finally

$$d\mathbf{p} = d(m\mathbf{v}) = \mathbf{v}\, dm = \frac{\mathbf{v}\, dm_0}{\sqrt{1-\beta^2}}$$

If, in K_0, the velocities $\mathbf{w}_0$ are normal to $\mathbf{v}$, in K they have the two components

$$w_{1x} = w_{2x} = v; \qquad w_{1y} = w_0\sqrt{1-\beta^2}; \qquad w_{2y} = -w_0\sqrt{1-\beta^2}.$$

$$\Sigma w_x\, dm' = \frac{2v\, dm_0'}{\sqrt{1-\beta^2}} = \frac{v\, dm_0}{\sqrt{1-\beta^2}}; \qquad \Sigma w_y\, dm' = 0.$$

We recover the desired equation, which is valid for an arbitrary number of pairs dm_0' with velocities $\pm w_0$ in arbitrary orientations.

[175] First Variant of the Formalism; Another Definition of Force in K

(a) In the formalism adopted above and in § [157], the impulse of the force is equal to the change of momentum, whatever the system of reference and for every problem.

In the present chapter, and in the absence of a component $\mathbf{F}_e$, the force $\mathbf{F}$ is zero in K_0 and different from zero in K. Since the definition of $\mathbf{F}$ in K is arbitrary, there is nothing to prevent us from deciding that only the component $\mathbf{F}_e$ deserves to be called force. We should then write

$$\mathbf{F}_e = \frac{\mathrm{d}}{\mathrm{d}t}\left(\frac{m_0\mathbf{v}}{\sqrt{1-\beta^2}}\right) - \mathbf{v}\frac{\mathrm{d}m_0}{\mathrm{d}t_0} \tag{1}$$

or

$$\mathbf{F} = \frac{\mathrm{d}\mathbf{p}}{\mathrm{d}t} - \mathbf{v}\frac{\mathrm{d}m_0}{\mathrm{d}t_0}$$

To avoid any confusion, however, we shall reserve the symbol $\mathbf{F}$ for the force defined in § [157].

Physical laws that can be checked are, of course, not modified: only a change of language is involved. Since certain authors adopt it, however, it is as well to display the consequences and to bring out the conventional nature of the laws.

b) *Longitudinal and transverse mass*

We denote the unit vectors along the tangent and principal normal by $\mathbf{t}$ and $\mathbf{n}$, respectively, and the radius of curvature by R. Equation (1) is then written

$$\begin{aligned} \mathbf{F}_e &= \frac{\mathrm{d}}{\mathrm{d}t}\left(\frac{m_0 v\mathbf{t}}{\sqrt{1-\beta^2}}\right) - \frac{v\mathbf{t}}{\sqrt{1-\beta^2}}\frac{\mathrm{d}m_0}{\mathrm{d}t} \\ &= \frac{m_0\mathbf{t}}{(1-\beta^2)^{3/2}}\frac{\mathrm{d}v}{\mathrm{d}t} + \frac{m_0\mathbf{n}}{\sqrt{1-\beta^2}}\frac{v^2}{R} \\ &= m_t\gamma_t + m_n\gamma_n \end{aligned} \tag{2}$$

The result is therefore the same as for fixed proper masses.

(c) *The energy theorem*

We scalar multiply the two sides of equation (2) by $\mathbf{v}\,\mathrm{d}t$:

$$\mathbf{F}_e\,.\,\mathbf{v}\,\mathrm{d}t = \frac{m_0 v\,\mathrm{d}v}{(1-\beta^2)^{3/2}} = \mathrm{d}(mc^2) - c^2\frac{\mathrm{d}m_0}{\sqrt{1-\beta^2}}$$

Introducing the work done by the external force, $\mathrm{d}\mathcal{T}_e$,

$$\mathrm{d}\mathcal{T}_e = \mathbf{F}_e\,.\,\mathbf{v}\,\mathrm{d}t$$

the energy equation becomes

$$\mathrm{d}(mc^2) = \mathrm{d}\mathcal{T}_e + c^2\frac{\mathrm{d}m_0}{\sqrt{1-\beta^2}} \tag{3}$$

The total energy changes because the external force does work and proper mass is gained; we can also write

$$\mathrm{d}\mathcal{T}_e = c^2 m_0\,\mathrm{d}\left(\frac{1}{\sqrt{1-\beta^2}}\right)$$

(d) *Transformation of the force* $\mathbf{F}_e$

From K_0 *to* K. In K we take the Ox axis parallel to the velocity at the time in question, and we denote the components of $\mathbf{F}_e$ by $X_e, \ldots$; then

$$X_e = \frac{m_0\gamma_x}{(1-\beta^2)^{3/2}}, \quad Y_e = \frac{m_0\gamma_y}{\sqrt{1-\beta^2}}, \quad Z_e = \frac{m_0\gamma_z}{\sqrt{1-\beta^2}}$$

In K_0

$$X_{0e} = m_0\gamma_{0x}, \quad Y_{0e} = m_0\gamma_{0y}, \quad Z_{0e} = m_0\gamma_{0z}$$

From kinematics,

$$\gamma_{0x} = \frac{\gamma_x}{(1-\beta^2)^{3/2}}, \quad \gamma_{0y} = \gamma_y, \quad \gamma_{0z} = \gamma_z$$

so that

$$X_e = X_{0e}, \quad Y_e = Y_{0e}\sqrt{1-\beta^2}, \quad Z_e = Z_{0e}\sqrt{1-\beta^2}$$

This is the same as the transformation for fixed proper masses.

From K_1 *to* K_2. Let v be the velocity of K_2 with respect to K_1 and u the velocity of the corpuscle. The impulse equation gives

$$X_{e1} = \frac{dp_{1x}}{dt_1} - \frac{u_{1x}\, dm_0}{dt_0}$$

$$\frac{d}{dt_2}\, \frac{p_{2x}+m_2 v}{\sqrt{1-\dfrac{v^2}{c^2}}}\, \frac{dt_2}{dt_1} - \frac{u_{1x}\, dm_0}{dt_0}$$

$$= \left(\frac{dp_{2x}}{dt_2} + v\frac{dm_2}{dt_2}\right) \frac{1}{1+\dfrac{vu_{2x}}{c^2}} - u_{1x}\frac{dm_0}{dt_0}$$

The energy theorem gives

$$c^2\, dm_2 = X_{e2}\, dx_2 + Y_{e2}\, dy_2 + Z_{e2}\, dz_2 + c^2 \frac{dm_0}{\sqrt{1-\dfrac{u_2^2}{c^2}}}$$

and hence

$$X_{e1} = \frac{1}{1+\dfrac{vu_{2x}}{c^2}} \left\{\frac{dp_{2x}}{dt_2} + \frac{v}{c^2}\left(X_{e2}\frac{dx_2}{dt_2} + Y_{e2}\frac{dy_2}{dt_2} + Z_{e2}\frac{dz_2}{dt_2} + \frac{dm_0}{dt_0}\right)\right\} - u_{1x}\frac{dm_0}{dt_0}$$

$$= \frac{1}{1+\dfrac{vu_{2x}}{c^2}} \left\{X_{e2} + \frac{u_{2x}\, dm_0}{dt_0} + \frac{v}{c^2}\left(u_{2x}X_{e2} + u_{2y}Y_{e2} + u_{2z}Z_{e2} + \frac{dm_0}{dt_0}\right)\right\} - u_{1x}\frac{dm_0}{dt_0}$$

which reduces to

$$X_{e1} = X_{e2} + \frac{\dfrac{v}{c^2}(Y_{e2}u_{2y} + Z_{e2}u_{2z})}{1+\dfrac{vu_{2x}}{c^2}}$$

because

$$\frac{u_{2x}+v}{1+\dfrac{vu_{2x}}{c^2}}-u_{1x}=0$$

Likewise

$$Y_{e1}=\frac{dp_{1y}}{dt_1}-\frac{u_{1y}\,dm_0}{dt_0}=\frac{dp_{2y}}{dt_2}\,\frac{dt_2}{dt_1}-u_{1y}\,\frac{dm_0}{dt_0}$$

$$=\frac{dp_{2y}}{dt_2}\,\frac{\sqrt{1-v^2/c^2}}{1+\dfrac{vu_{2x}}{c^2}}-u_{1y}\,\frac{dm_0}{dt_0}$$

so that

$$Y_{e1}=\frac{\sqrt{1-v^2/c^2}}{1+\dfrac{vu_{2x}}{c^2}}\,Y_{e2}$$

with a similar transformation for Z_{e1}.

For the components of $\mathbf{F}_e$, we obtain the same formulae as in the dynamics of invariant proper masses.

(e) *Use of the vectors* **E** *and* $\mathfrak{B}$

Using the results of §§ [165] and [174], we have

$$\mathbf{F}_e=\mathbf{F}-\mathbf{v}\,\frac{dm_0}{dt_0}=\varepsilon\left(\mathbf{E}+\frac{\mathbf{v}}{c}\times\mathfrak{B}\right)$$

In this case, the external force has its usual expression in **E** and $\mathfrak{B}$.

(f) *Four-vector notation*

As in ordinary dynamics, let us define the four-vectors momentum Π_i and force K_{ei} by means of the relations ($\alpha = x, y, z$)

$$\Pi_\alpha=\frac{m_0v_\alpha}{\sqrt{1-\beta^2}}=p_\alpha \qquad \Pi_l=\frac{im_0c}{\sqrt{1-\beta^2}}=\frac{iW}{c}$$

$$K_{e\alpha}=\frac{F_{e\alpha}}{\sqrt{1-\beta^2}} \qquad K_{el}=\frac{i}{c}\,\frac{\mathbf{v}.\mathbf{F}_e}{\sqrt{1-\beta^2}}$$

We show that equations (1) and (3) are equivalent to

$$\frac{d\Pi_i}{dt_0}=K_{ei}+V_i\,\frac{dm_0}{dt_0}$$

In fact, for the spatial components,

$$\frac{F_{ex}}{\sqrt{1-\beta^2}}+\frac{v_x}{\sqrt{1-\beta^2}}\,\frac{dm_0}{dt_0}=\frac{dp_\alpha}{dt_0}$$

or

$$\frac{\mathrm{d}p_x}{\mathrm{d}t} = F_{ex} + v_x \frac{\mathrm{d}m_0}{\mathrm{d}t_0}$$

For the time component

$$\frac{\mathrm{d}}{\mathrm{d}t_0}\left(\frac{im_0c}{\sqrt{1-\beta^2}}\right) = \frac{i}{c}\,\frac{\mathbf{v}.\mathbf{F}_e}{\sqrt{1-\beta^2}} + \frac{ic}{\sqrt{1-\beta^2}}\,\frac{\mathrm{d}m_0}{\mathrm{d}t_0}$$

and

$$\mathrm{d}\left(\frac{m_0c^2}{\sqrt{1-\beta^2}}\right) = \mathbf{F}_e.\mathbf{v}\,\mathrm{d}t + c^2\frac{\mathrm{d}m_0}{\sqrt{1-\beta^2}}$$

$$= \mathrm{d}\mathcal{T}_e + \frac{c^2\,\mathrm{d}m_0}{\sqrt{1-\beta^2}}$$

[176] Second Variant of the Formalism; Another Definition of Momentum

Equation (1) of § [174] contains three terms:

$$\mathbf{F}_e \qquad \frac{\mathrm{d}\mathbf{p}}{\mathrm{d}t} \qquad \mathbf{v}\,\frac{\mathrm{d}m_0}{\mathrm{d}t_0}$$

With the formalism of § [175], these three terms are kept separate:

$$\mathbf{F}_e = \frac{\mathrm{d}\mathbf{p}}{\mathrm{d}t} - \mathbf{v}\,\frac{\mathrm{d}m_0}{\mathrm{d}t_0}$$

In the general formalism of Chapter XII, $\mathbf{v}\,\mathrm{d}m_0/\mathrm{d}t_0$ is regarded as a force and $\mathbf{F}_e + \mathbf{v}\,\mathrm{d}m_0/\mathrm{d}t_0$ is treated as a single unit and called total force, $\mathbf{F}$; thus

$$\mathbf{F} = \frac{\mathrm{d}\mathbf{p}}{\mathrm{d}t}$$

We could also write

$$\mathbf{F}_e = \frac{\mathrm{d}\boldsymbol{\pi}}{\mathrm{d}t}$$

thus defining a new momentum vector, $\boldsymbol{\pi}$, such that

$$\frac{\mathrm{d}\boldsymbol{\pi}}{\mathrm{d}t} = \frac{\mathrm{d}\mathbf{p}}{\mathrm{d}t} - \mathbf{v}\,\frac{\mathrm{d}m_0}{\mathrm{d}t_0} = \frac{\mathrm{d}}{\mathrm{d}t}\left(\frac{m_0\mathbf{v}}{\sqrt{1-\beta^2}}\right) - \mathbf{v}\,\frac{\mathrm{d}m_0}{\mathrm{d}t_0} = m_0\,\frac{\mathrm{d}}{\mathrm{d}t}\left(\frac{\mathbf{v}}{\sqrt{1-\beta^2}}\right)$$

The vector $\boldsymbol{\pi}$ is no longer necessarily collinear with the velocity. We may then encounter momenta with transverse components. We must, however, beware of slipshod terminology, and never forget that two different vectors, $\boldsymbol{\pi}$ and $\mathbf{p}$, are called by the same name.

[177] On the Choice of Formalism

The third formalism seems to be of no interest for it requires the introduction of the new quantity $\boldsymbol{\pi}$ which is inconvenient to handle.

The first two formalisms, on the contrary, appear to be practically equivalent. From the conceptual point of view, because of the arbitrariness in the definition of **F**, both are acceptable. The second formalism possesses some advantages: it retains the usual force transformation formulae, as well as the expression for the force in terms of **E** and $\mathfrak{B}$ and it gives a physically very clear statement of the energy theorem. However, these advantages are a feature of the special case considered here; in the following chapters (see § [194], for example), they vanish. With this second formalism, we have to write down different equations according to the case studied. These observations have led me to adopt the formalism of Chapter XII as the general formalism, for it presents, among others, the following advantages:

—the impulse equation takes the simple form

$$d\mathbf{p} = \mathbf{F}\, dt$$

and this equation is valid *in every case*; this enables us to develop a general formalism *once and for all*;

—the vectors **E** and $\mathfrak{B}$ transform as in the dynamics of invariant proper masses, which avoids any rewriting of electromagnetic theory.

B. SOLIDS IN UNIFORM TRANSLATION, WITH OR WITHOUT ANY TENDENCY TO ROTATE; COUPLES

[178] The Right-angled Lever

(a) *The position of the problem*

Consider a right-angled lever (Fig. 39), consisting of two perpendicular arms, OA_0 and OB_0, each of length l_0. This lever is free to rotate in its own plane about the axis Ω which is stationary in the Galileian reference system K_0. In K_0, at time $t_0 = 0$, the lever is at rest. At this time, forces F_{OA} and F_{OB} are applied to the ends A_0 and B_0 respectively; the forces are both equal in magnitude to F_0 and act at right angles to the arms of the lever. We assume that at the point Ω, the reaction of the axis simultaneously applies a force having components $-F_{OA}$ and $-F_{OB}$ to the lever. In K_0, therefore, we have, at each instant of time, zero resultant external force $\mathbf{F}_{0e}$ and also, zero moment. The lever remains at rest in K_0, on the average; it is, however, deformed elastically and its various points acquire kinetic energy. To simplify the problem and reduce it to its fundamentals, we assume that the kinetic energy is negligible in comparison with the potential energy

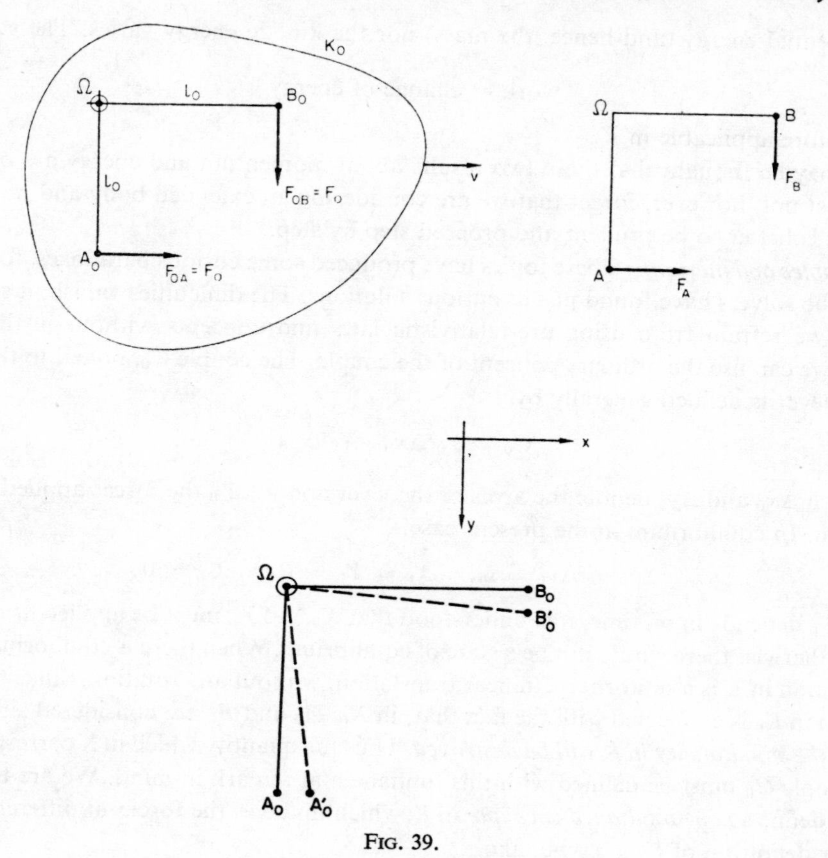

FIG. 39.

(which is stored as elastic energy). After time $t_0 = T_0$, the ends, A_0 and B_0 will have reached A_0' and B_0', and thenceforward all the parts of the lever remain stationary in K_0, under the action of the forces F_0.

(b) *Permanent state of tension*

The impulse theorem. The lever is at rest in K_0, and hence has an overall velocity v in K. The impulses $\mathbf{F}_A\,dt$ and $\mathbf{F}_B\,dt$ are cancelled by the equal and opposite impulses of the forces exerted by the axis. This is in agreement with the fact that the momentum of the lever does not vary. The equation

$$\text{total impulse} = \text{change of momentum}$$

is therefore applicable in K.

The energy theorem. The force F_B does no work. The force F_A does work $F_A v$ per unit time on the lever. The force exerted by the axis does work $-F_A v$ per unit time. The sum of these forces does no work on the lever; this is in agreement with the fact that neither the potential energy (and hence, the mass) nor the kinetic energy varies. The equation

$$\text{work} = \text{change of energy}$$

is therefore applicable in K.

It may be thought that these two results about momentum and energy are obvious. We must not, however, forget that we are considering an extended body and not a particle; it is better to be prudent and proceed step by step.

Couples and moments. These topics have produced some curious paradoxes, for which would-be solvers have found just as curious solutions. The difficulties vanish, it seems to me, if we refrain from using pre-relativistic laws and concepts without justification. In K_0, we can use the ordinary concept of the couple. The couple C_0, nomal to the plane of the lever is defined generally by

$$C_0 = Y_0 \,\Delta x_0 - X_0 \,\Delta y_0$$

in which Δx_0 and Δy_0 denote the arms of the lever and X_0, Y_0 the forces applied normal to them. In equilibrium, in the present case,

$$\Delta x_0 = \Delta y_0 = l_0, \quad X_0 = Y_0 = F_0, \quad C_0 = 0$$

If F_0 depends upon time, it is understood that X_0 and Y_0 must be applied *at the same time*, otherwise there would not be a state of equilibrium. When there is equilibrium in K_0, the motion in K is a uniform rectilinear translation, without any rotation. Since the equilibrium in K_0 is connected with the fact that, in K_0, A_0 and B_0 are considered simultaneously, *the simultaneity in K will be destroyed.* Thus the quantity which in K corresponds to the couple C_0 must be defined with this fundamental remark in mind. We are therefore led to define a *non-simultaneous coupie* in K, which involves the forces at different times. As our definition of C in K, we take

$$C = Y\,\Delta x - X\,\Delta y$$

The Lorentz transformation,

$$x = \frac{x_0 + vt_0}{\sqrt{1-\beta^2}}, \qquad y = y_0$$

gives for the time t_0 (simultaneity in K_0)

$$\Delta x = \frac{\Delta x_0}{\sqrt{1-\beta^2}}, \qquad \Delta y = \Delta y_0$$

Furthermore (see the relevant *Remark* in (c)), using the usual force transformation (in the present situation, the proper mass does not change), we have

$$X = X_0, \qquad Y = Y_0\sqrt{1-\beta^2}$$

Thus

$$C = \frac{\Delta x_0}{\sqrt{1-\beta^2}}\, Y_0\sqrt{1-\beta^2} - y_0 X_0 = 0$$

The couple defined in this way is therefore zero both in K and in K_0. With such a couple, the absence of rotation in K is associated with a zero couple.

We can also define a couple C' in K, by considering the forces X and Y at a time t of K. In this case, we must use the Lorentz contraction and we write

$$\Delta x = \Delta x_0 \sqrt{1-\beta^2}, \qquad \Delta y = \Delta y_0$$

so that

$$C' = Y\,\Delta x - X\,\Delta y = Y_0\,\Delta x_0(1-\beta^2) - X_0\,\Delta y_0$$

which is here equal to

$$C' = -\beta^2 l_0 F_0$$

In pre-relativistic thinking, such a couple would tend to rotate the lever about Ω in such a direction that Oy approaches Ox (Fig. 40). In fact, however, this rotation does not occur. For a long time this was found paradoxical (§ [186]) because it was implicitly

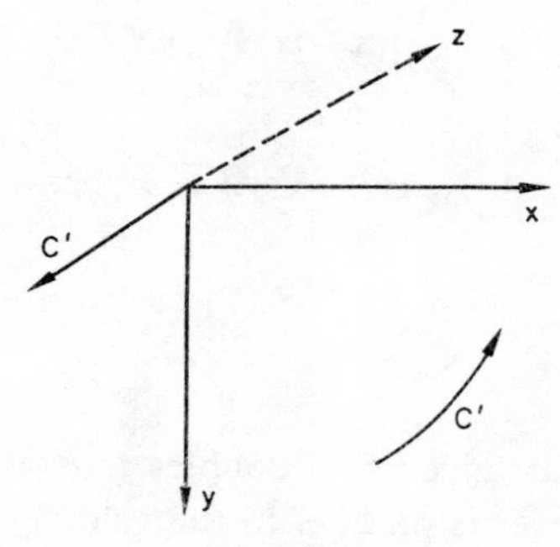

FIG. 40.

accepted that the ordinary pre-relativistic law would apply to the couple C'. We can see that such an extrapolation is unjustifiable except in the limit $\beta = 0$. If the couple is calculated with simultaneity in K, a couple $C' \neq 0$ is necessary if we are to have no rotation. The absence of this couple would lead to rotation. The quantitative aspect of this statement is straightforward. For, let us apply two equal forces at A and B for a very short time, simultaneously in K. In K_0, then, only one of the forces will act to begin with and we can therefore anticipate rotation in K_0. The problem of determining the relativistic law of rotating bodies remains to be solved, as the foregoing remarks are concerned only with equilibrium; they show that even in this very simple case, we must be wary of extrapolations of the pre-relativistic laws.

Remark 1. Another form of the calculation of C'. We may write

$$\begin{aligned} C' &= (x-vt)Y - yX \\ &= \left(\frac{x_0+vt_0}{\sqrt{1-\beta^2}} - v\,\frac{t_0+\dfrac{v}{c^2}x_0}{\sqrt{1-\beta^2}}\right) Y_0\sqrt{1-\beta^2} - y_0 X_0 \\ &= x_0(1-\beta^2)Y_0 - y_0X_0 \\ &= x_0Y_0 - y_0X_0 - \beta^2 x_0 Y_0 \end{aligned}$$

and finally

$$C' = C_0 - \beta^2 x_0 Y_0$$

With the present assumptions,

$$C_0 = 0, \quad C' = -\beta^2 l_0 F_0$$

We can also calculate in the opposite sense:

$$\begin{aligned} C_0 &= x_0 Y_0 - y_0 X_0 \\ &= \frac{x - vt}{\sqrt{1-\beta^2}} \frac{Y}{\sqrt{1-\beta^2}} - yX \\ &= \frac{1}{1-\beta^2} \{(x-vt)Y - yX + \beta^2 yX\} \\ &= \frac{1}{1-\beta^2} (C' + \beta^2 yX) \end{aligned}$$

which does indeed give

$$C' = C_0 - \beta^2 x_0 Y_0$$

Remark 2. It can be seen that, given two couples that are equal in K_0, but are such that one of the arms of the lever is parallel to the velocity and the other perpendicular to it, the couples obtained in K are equal using the first definition and different using the second. The second result arises from the Lorentz contraction, and also affects the measurement of a ruler. Here, however, it leads to a picture of a couple that is difficult to reconcile with the intuitive physical concept. In fact, the orientation of the couple in K_0 does not influence any kinematic or dynamic properties; the same should be true in K and this is a fresh argument for using the first definition (for the present problem, at any rate).

(c) *Study of the transitional situation, while the tension is building up*

The energy theorem. If the displacements are very small, we can assume that they are parallel to the axes; we write

$$A_0 A_0' = B_0 B_0' = a_0$$

The work done in K_0 between $t_0 = 0$ and $t_0 = T_0$ is

$$d\mathcal{T}_0 = 2F_0 a_0$$

corresponding to an increase of proper mass

$$dm_0 = \frac{2F_0 a_0}{c^2}$$

We now transform to the reference system K; since the body is in translational motion and v does not alter, we must have

$$dm = \frac{dm_0}{\sqrt{1-\beta^2}}$$

We now calculate the work $d\mathcal{T}$ with respect to K, and show that the preceding relation is compatible with the energy equation

$$d\mathcal{T} = c^2\, dm$$

if we transform the forces in the usual way. I write "if": it must not be forgotten that so far, the force transformation has only been established for particle dynamics; it is a consequence of the equation of motion. We know nothing about forces of the type F_A, applied to a point with which no mass concept is associated. We must therefore proceed with caution, and say that we shall use the known transformation *tentatively* (or if we prefer, we use it as a *definition*). We shall later see how the dynamics of particles with variable proper mass can be reconciled with the results that we are going to obtain.

Having made this clear, if we wish to transfer to the system K, we must now transform the point events which have the following coordinates in K_0:

application of the forces at A_0: $x_0 = 0, \quad y_0 = l_0, \quad t_0 = 0$

at B_0: $x_0 = l_0, \quad y_0 = 0, \quad t_0 = 0$

position of tension reached at $A_0(A_0')$: $x_0 = a_0, \quad y_0 = l_0, \quad t_0 = T_0$

at $B_0(B_0')$: $x_0 = l_0, \quad y_0 = a_0, \quad t_0 = T_0$

We transform into K, using

$$x = \frac{x_0 + vt_0}{\sqrt{1-\beta^2}}, \quad y = y_0, \quad t = \frac{t_0 + \frac{v}{c^2}x_0}{\sqrt{1-\beta^2}}$$

We thus have

application of the force at A: $x = 0, \quad y = l_0, \quad t = 0$

$$\text{at } B\text{:} \quad x = \frac{l_0}{\sqrt{1-\beta^2}}, \quad y = 0, \quad t = \frac{\frac{v}{c^2}l_0}{\sqrt{1-\beta^2}}$$

position of tension reached at $A(A')$:

$$x = \frac{a_0 + vT_0}{\sqrt{1-\beta^2}}, \quad y = l_0, \quad t = \frac{T_0 + \frac{v}{c^2}a_0}{\sqrt{1-\beta^2}}$$

position of tension reached at $B(B')$:

$$x = \frac{l_0 + vT_0}{\sqrt{1-\beta^2}}, \quad y = a_0, \quad t = \frac{T_0 + \frac{v}{c^2}l_0}{\sqrt{1-\beta^2}}$$

We then transform the forces F_{OA} and F_{OB}. For this, we must use the relations (§ [30])

$$X = X_0 + \frac{\frac{vu_{0y}}{c^2} Y_0}{1+vu_{0x}/c^2}, \qquad Y = \frac{\sqrt{1-\beta^2}}{1+vu_{0x}/c^2} Y_0$$

We begin with F_{OA}, which in K_0 has the components

$$(F_{OA})_{x_0} = F_0, \quad (F_{OA})_{y_0} = 0$$

For this force,

$$u_{0x} = a_0/T_0, \qquad u_{0y} = 0$$

Hence F_A has components

$$(F_A)_x = F_0, \quad (F_A)_y = 0$$

For the force F_{OB},

$$(F_{OB})_{x_0} = 0, \qquad (F_{OB})_{y_0} = F_0$$

$$u_{0x} = 0, \qquad u_{0y} = a_0/T_0$$

$$(F_B)_x = \frac{va_0}{c^2T_0} F_0, \quad (F_B)_y = \sqrt{1-\beta^2}\, F_0$$

The important fact is the existence of a component $(F_B)_x$ which is non-zero in K. At A, the work done is

$$\mathrm{d}\mathcal{T}_A = F_0 \frac{a_0 + vT_0}{\sqrt{1-\beta^2}}$$

and at B,

$$\mathrm{d}\mathcal{T}_B = \frac{va_0F_0}{c^2T_0} \frac{vT_0}{\sqrt{1-\beta^2}} + \sqrt{1-\beta^2}\, F_0a_0$$

$$= \frac{a_0F_0}{\sqrt{1-\beta^2}}$$

We then transform the force at Ω. Since this point is stationary in K_0, we use the relations

$$X = X_0, \quad Y = Y_0\sqrt{1-\beta^2}$$

and hence at Ω, in K, we have

$$(F_\Omega)_x = -F_0, \quad (F_\Omega)_y = -F_0\sqrt{1-\beta^2}$$

The work is therefore

$$\mathrm{d}\mathcal{T}_\Omega = -\frac{F_0vT_0}{\sqrt{1-\beta^2}}$$

The total work done on the lever is therefore

$$\mathrm{d}\mathcal{T} = \mathrm{d}\mathcal{T}_A + \mathrm{d}\mathcal{T}_B + \mathrm{d}\mathcal{T}_\Omega = \frac{2a_0F_0}{\sqrt{1-\beta^2}} = \frac{\mathrm{d}\mathcal{T}_0}{\sqrt{1-\beta^2}}$$

$$= c^2\,\mathrm{d}m$$

This is the anticipated result.

The impulse theorem. During time T_0, the momentum is increased by

$$\mathrm{d}p = v\,\mathrm{d}m = v\frac{\mathrm{d}m_0}{\sqrt{1-\beta^2}} = \frac{2va_0F_0}{c^2\sqrt{1-\beta^2}}$$

We shall confirm that we have

$$\mathrm{d}p = \Sigma F\,\mathrm{d}t;$$

this equation is a generalization of that for a particle. At A, we have impulse

$$(I_A)_x = F_0\frac{T_0+\dfrac{va_0}{c^2}}{\sqrt{1-\beta^2}}, \qquad (I_A)_y = 0$$

At B,

$$(I_B)_x = \frac{va_0F_0}{c^2\sqrt{1-\beta^2}}, \qquad (I_B)_y = F_0T_0$$

At Ω,

$$(I_\Omega)_x = -\frac{F_0T_0}{\sqrt{1-\beta^2}}, \qquad (I_\Omega)_y = -F_0T_0$$

The total impulse is therefore

$$I_x = \frac{2va_0T_0}{c^2\sqrt{1-\beta^2}}, \qquad I_y = 0$$

If we compare this with the observed change of momentum (parallel to Ox), we see that the equation

change of momentum = impulse

is applicable here. We notice in particular that *neither for the impulse nor for the momentum is there a transverse component*. Nevertheless, two important comments are indispensable. The total impulse I, calculated above, is *not* of the form

force×time

as in the case of a moving particle. Here, the impulses I_A and I_B are applied during *different* times, namely

$$\frac{T_0+va_0/c^2}{\sqrt{1-\beta^2}} \quad \text{and} \quad \frac{T_0}{\sqrt{1-\beta^2}}$$

Furthermore, notice that we do not introduce the impulse of the resultant force. This concept of resultant force is of no value here, for, I repeat, the forces F_A and F_B are applied for different lengths of time.

The connection with the dynamics of particles of variable proper mass; the equivalent point corpuscle. We have just shown that the sum of the impulses, applied for different times, is given by

$$\sum F\,\mathrm{d}t = \frac{2va_0F_0}{c^2\sqrt{1-\beta^2}}$$

Furthermore,

$$dm_0 = \frac{2F_0 a_0}{c^2}$$

whence

$$\sum F\, dt = v \frac{dm_0}{\sqrt{1-\beta^2}}$$

$$d\mathbf{p} = \mathbf{v} \frac{dm_0}{\sqrt{1-\beta^2}}$$

We now replace the extended body by a point corpuscle having the same impulse, and seek the force **F** which, applied to this equivalent corpuscle during time $dt = dt_0/\sqrt{1-\beta^2}$, would give the same $d\mathbf{p}$. We therefore write

$$d\mathbf{p} = \mathbf{v} \frac{dm_0}{\sqrt{1-\beta^2}} = \mathbf{F} \frac{dt_0}{\sqrt{1-\beta^2}}$$

so that

$$d\mathbf{p} = \mathbf{F}\, dt \quad \text{where} \quad \mathbf{F} = \mathbf{v} \frac{dm_0}{dt_0}$$

This force vanishes in K_0, and in K has the above value; furthermore, it acts in the direction Ox. It therefore does transform as required by the formalism of § [162].

This force, which is applied to the equivalent corpuscle, must not be confused with the real forces applied to the lever; indeed, the transformation formulae are different.

Remark. For this equivalent point corpuscle, the external force is zero. This is natural, since the external force $\mathbf{F}_e$ is the resultant of the forces at A, B and Ω.

With the general formalism, we say that the change of momentum is equal to the impulse of the force **F**. With the formalism of § [175], and in the absence of an impulse, the change of momentum is $\mathbf{v}\, dm_0/dt_0$.

[179] The Extended or Compressed Rod

(a) *The position of the problem*

Consider a rod A_0B_0 of length l_0 (Fig. 41), at rest in the system K_0 (and making an angle θ_0 with Ox_0). At time $t_0 = 0$, forces are applied to the ends of the rod, equal in magnitude but in opposite directions and both along the length of the rod. The rod therefore remains at rest in K_0, on the average, although it is elastically deformed. After a time $t_0 = T_0$, the extremities reach A_0' and B_0', and thenceforward the whole rod remains stationary in K_0. Let dl_0 be the shifts of the ends, so that $\lambda_0 = l_0 - 2\, dl_0$ is the length in the deformed state.

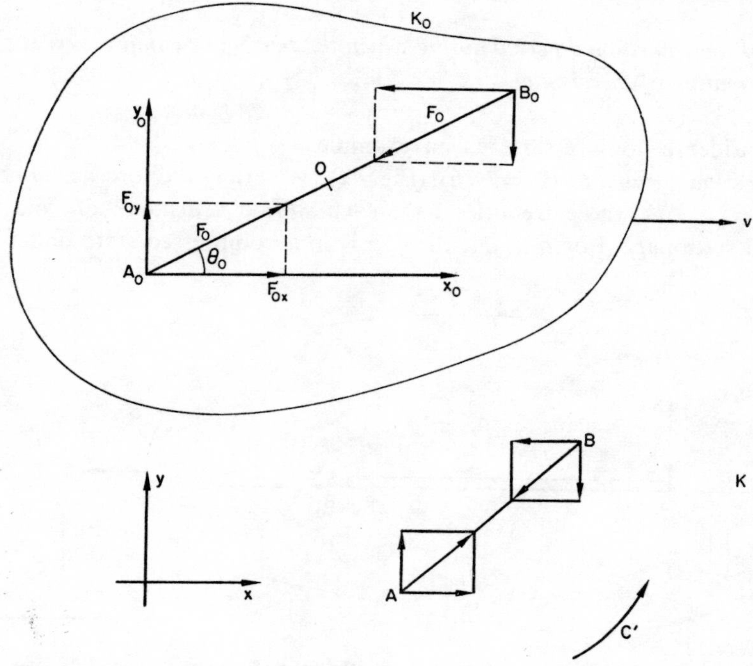

FIG. 41.

(b) *Permanently deformed state*

Let us transpose the results of § [178]. To calculate the couple, with respect to B_0 for example, we have

$$X_0 = F_{0x} = F_0 \cos \theta_0, \quad Y_0 = F_0 \sin \theta_0$$
$$\Delta x_0 = \lambda_0 \cos \theta_0, \quad \Delta y_0 = \lambda_0 \sin \theta_0$$

In K_0, $C_0 = 0$. To transfer to K, the two possibilities of § [178] are open to us. With the couple C,

$$\Delta x = \frac{\Delta x_0}{\sqrt{1-\beta^2}}, \quad \Delta y = \Delta y_0$$

so that

$$\lambda \sin \theta = \lambda_0 \sin \theta_0, \quad \lambda \cos \theta = \frac{\lambda_0 \cos \theta_0}{\sqrt{1-\beta^2}}$$

This gives C equal to nought. With the couple C',

$$\Delta x = \Delta x_0 \sqrt{1-\beta^2}, \quad \Delta y = \Delta y_0$$

so that

$$\lambda \sin \theta = \lambda_0 \sin \theta_0, \quad \lambda \cos \theta = \lambda_0 \cos \theta_0 \sqrt{1-\beta^2}$$
$$C' = F_y \, \Delta x - F_x \, \Delta y = -\beta^2 F_{0x} \, \lambda_0 \sin \theta_0$$

The comments of § [178] apply here also.

(c) *Study of the transitional period during which the tension or compression is established; transformation of the elastic work*

We consider a compression, as an example.

Compression parallel to the velocity (Fig. 42). We apply the forces at time $t_0 = 0$ in K_0; at time $t_0 = T_0$, the extremities have each shifted a distance dl_0 and from then on, remain stationary. For $t_0 > T_0$, the rod is in a compressed state under the action

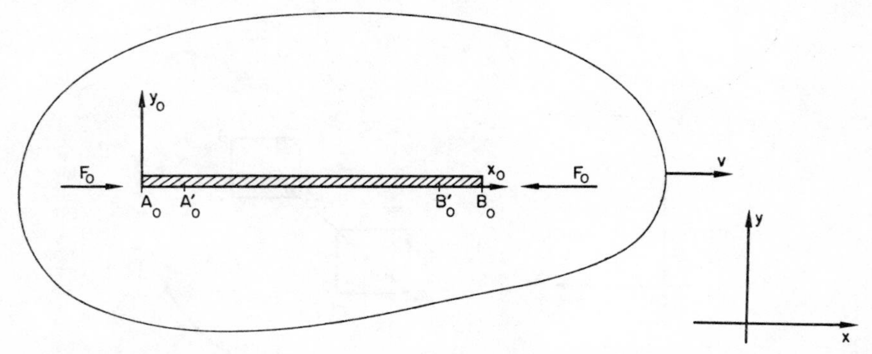

FIG. 42. The axes Ox and Ox_0, shown separately for the sake of clarity, in fact slide over one another.

of forces. The work done between $t_0 = 0$ and $t_0 = T_0$ is $d\mathcal{T}_0 = 2F_0\, dl_0$. The impulse is zero. We now refer this situation to a reference frame K, with respect to which the rod has velocity v parallel to its length. In K_0, the coordinates of the events to be transformed are as follows:

When the compression begins at A_0: $x_0 = 0, \quad t_0 = 0$

at B_0: $x_0 = l_0, \quad t_0 = 0$

When the compression is established at $A_0(A_0')$: $x_0 = dl_0, \quad t_0 = T_0$

at $B_0(B_0')$: $x_0 = l_0 - dl_0, \quad t_0 = T_0$

We transform these coordinates, using the formulae

$$x = \frac{x_0 + vt_0}{\sqrt{1-\beta^2}}, \qquad t = \frac{t_0 + \dfrac{v}{c^2}x_0}{\sqrt{1-\beta^2}}$$

and in K, they become:

when the compression begins at A: $x = 0, \quad t = 0$

$$\text{at } B: \quad x = \frac{l_0}{\sqrt{1-\beta^2}}, \qquad t = \frac{\dfrac{v}{c^2}l_0}{\sqrt{1-\beta^2}}$$

when it is established at $A(A')$:

$$x = \frac{\mathrm{d}l_0 + vt_0}{\sqrt{1-\beta^2}}, \qquad t = \frac{T_0 + \frac{v}{c^2}\,\mathrm{d}l_0}{\sqrt{1-\beta^2}}$$

at $B(B')$:

$$x = \frac{l_0 - \mathrm{d}l_0 + vT_0}{\sqrt{1-\beta^2}}, \qquad t = \frac{T_0 + \frac{v}{c^2}(l_0 - \mathrm{d}l_0)}{\sqrt{1-\beta^2}}$$

Let us calculate the work. To transform the forces, we employ the formulae

$$X = X_0 + \frac{\frac{vu_{0y}}{c^2}Y_0}{1 + \frac{vu_{0x}}{c^2}}, \qquad Y = \frac{\sqrt{1-\beta^2}}{1 + \frac{vu_{0x}}{c^2}}Y_0$$

where $\mathbf{u}_0$ denotes the velocity of the point of application of the forces. In the present case

$$u_{0y} = 0, \qquad Y_0 = 0$$
$$X = X_0, \qquad Y = 0$$

The force exerted at A and B in K therefore has the value F_0. At A, the work is

$$\mathrm{d}\mathcal{T}_A = F_0 \frac{\mathrm{d}l_0 + vT_0}{\sqrt{1-\beta^2}}$$

and at B

$$\mathrm{d}\mathcal{T}_B = -F_0 \frac{-\mathrm{d}l_0 + vT_0}{\sqrt{1-\beta^2}}$$

The total work in K is therefore

$$\mathrm{d}\mathcal{T} = \frac{2F_0\,\mathrm{d}l_0}{\sqrt{1-\beta^2}}$$

The transformation relation is hence

$$\mathrm{d}\mathcal{T} = \frac{\mathrm{d}\mathcal{T}_0}{\sqrt{1-\beta^2}}$$

This result agrees with the relations

$$\mathrm{d}\mathcal{T}_0 = c^2\,\mathrm{d}m_0, \quad \mathrm{d}\mathcal{T} = c^2\,\mathrm{d}m, \quad \mathrm{d}m = \frac{\mathrm{d}m_0}{\sqrt{1-\beta^2}}$$

which express the energy balance.

For the impulse theorem, we notice that the change of momentum is

$$\mathrm{d}p = v\,\mathrm{d}m = v\frac{\mathrm{d}m_0}{\sqrt{1-\beta^2}} = \frac{2v}{c^2}\frac{F_0\,\mathrm{d}l_0}{\sqrt{1-\beta^2}}$$

Let us now confirm that the relation

$$\mathrm{d}p = \sum F\,\mathrm{d}t$$

is satisfied. At A, we have impulse

$$I_A = F_0 \frac{T_0 + \frac{v}{c^2}\,\mathrm{d}l_0}{\sqrt{1-\beta^2}}$$

and at B,

$$I_B = -F_0 \frac{T_0 - \frac{v}{c^2}\,\mathrm{d}l_0}{\sqrt{1-\beta^2}}$$

so that

$$\sum F\,\mathrm{d}t = I = I_A + I_B = \frac{2v}{c^2}\frac{F_0\,\mathrm{d}l_0}{\sqrt{1-\beta^2}}$$

or $\mathrm{d}p$.

To relate this to the dynamics of particles of variable proper mass, we set out from the relations

$$\sum F\,\mathrm{d}t = \frac{2v}{c^2}\frac{F_0\,\mathrm{d}l_0}{\sqrt{1-\beta^2}}, \qquad \mathrm{d}m_0 = \frac{2F_0\,\mathrm{d}l_0}{c^2}$$

whence

$$\sum F\,\mathrm{d}t = v\frac{\mathrm{d}m_0}{\sqrt{1-\beta^2}}$$

We can therefore define an equivalent point corpuscle as in § [178], using the two formalisms. A force is or is not required to give the change $\mathrm{d}p$, according to the formalism adopted. The external force F_e is zero.

Compression normal to the velocity. We consider the case shown in Fig. 43, where the velocity is normal to the rod. In K_0, we still have

$$\mathrm{d}\mathcal{T}_0 = 2F_0\,\mathrm{d}l_0$$

For the various events, we have in K_0:

compression begins at A_0: $x_0 = 0, \quad y_0 = 0, \quad t_0 = 0$

at B_0: $x_0 = 0, \quad y_0 = l_0, \quad t_0 = 0$

it is established at $A_0(A_0')$: $x_0 = 0, \quad y_0 = \mathrm{d}l_0, \quad t_0 = T_0$

at $B_0(B_0')$: $x_0 = 0, \quad y_0 = l_0 - \mathrm{d}l_0, \quad t_0 = T_0$

In the reference system K:

compression begins at A: $x = 0, \quad y = 0, \quad t = 0$

at B: $x = 0, \quad y = l_0, \quad t = 0$

it is established at $A(A')$:

$$x = \frac{vT_0}{\sqrt{1-\beta^2}}, \quad y = \mathrm{d}l_0, \quad t = \frac{T_0}{\sqrt{1-\beta^2}}$$

at $B(B')$:

$$x = \frac{vT_0}{\sqrt{1-\beta^2}}, \quad y = l_0 - \mathrm{d}l_0, \quad t = \frac{T_0}{\sqrt{1-\beta^2}}$$

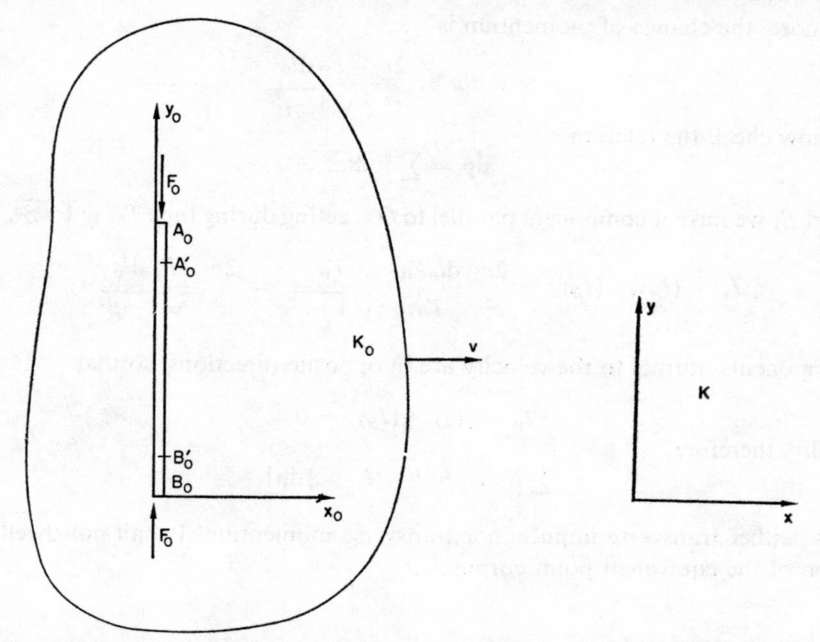

FIG. 43. The axes Ox and Ox_0, shown separately for the sake of clarity, in fact slide over one another.

Let us calculate the work. In the formulae for the transformation of the force components, we have here

$$u_{0x} = 0, \quad u_{0y} = \frac{\mathrm{d}l_0}{T_0}, \quad X_0 = 0$$

so that

$$X = \frac{v}{c^2}\frac{\mathrm{d}l_0}{T_0}Y_0, \quad Y = Y_0\sqrt{1-\beta^2}$$

The forces normal to the velocity do work

$$2Y_0\sqrt{1-\beta^2}\,\mathrm{d}l_0$$

The forces parallel to the velocity do work

$$\frac{2v}{c^2}\frac{\mathrm{d}l_0}{T_0}Y_0\frac{vT_0}{\sqrt{1-\beta^2}}$$

The total work is thus

$$\mathrm{d}\mathcal{T} = 2F_0\,\mathrm{d}l_0\left(\sqrt{1-\beta^2}+\frac{\beta^2}{\sqrt{1-\beta^2}}\right)$$

and finally,

$$\mathrm{d}\mathcal{T} = \frac{\mathrm{d}\mathcal{T}_0}{\sqrt{1-\beta^2}}$$

This result is in agreement with the energy balance,

$$\mathrm{d}\mathcal{T}_0 = c^2\,\mathrm{d}m_0, \quad \mathrm{d}\mathcal{T} = c^2\,\mathrm{d}m, \quad \mathrm{d}m = \frac{\mathrm{d}m_0}{\sqrt{1-\beta^2}}$$

Furthermore, the change of momentum is

$$dp = v\,dm = \frac{2v}{c^2}\,\frac{F_0\,dl_0}{\sqrt{1-\beta^2}}$$

We now check the relation

$$d\mathbf{p} = \sum \mathbf{F}\,dt$$

At A and B, we have a component parallel to Ox, acting during time $T_0/\sqrt{1-\beta^2}$, so that

$$I_x = (I_A)_x + (I_B)_x = \frac{2v}{c^2}\,\frac{dl_0 F_0}{T_0}\,\frac{T_0}{\sqrt{1-\beta^2}} = \frac{2v}{c^2}\,\frac{F_0\,dl_0}{\sqrt{1-\beta^2}}$$

The components normal to the velocity are in opposite directions, so that

$$I_y = (I_A)_y + (I_B)_y = 0$$

Finally, therefore,

$$\left|\sum \mathbf{F}\,dt\right| = I = I_x = |d\mathbf{p}|$$

There is neither transverse impulse nor transverse momentum. I shall not dwell on the definition of the equivalent point corpuscle.

[180] An Indefinite Straight Ohmic Current in Translational Motion through a Magnetic Field

(a) *The position of the question*

Suppose that we have an electric field in K. The electric field has no effect upon a current of zero charge density travelling in a conductor fixed in K. We shall now examine what happens when the current of zero *proper* charge is travelling with an overall uniform velocity v with respect to K. The pre-relativistic theories would reply that the action is still zero but the relativistic theory, on the contrary, predicts the existence of an effect.

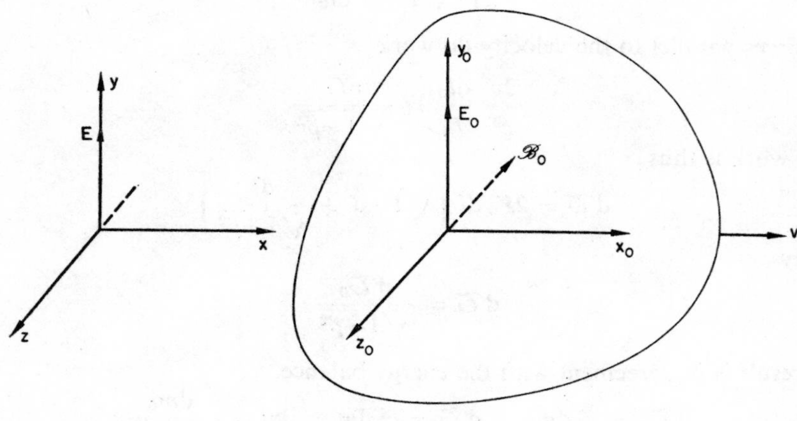

FIG. 44.

(b) We assume that the conductor is moving parallel to Ox with velocity v in its own direction. In K_0, the electric field **E** gives a component $\mathfrak{B}_0$ (Fig. 44). The conductor will therefore have a tendency to move in the direction Oy_0 in K_0. If it is free to move, there will be a measurable displacement along Oy_0 and hence along Oy also.

At first sight, this fact is paradoxical. One might think that since the current consists of pairs of equal charges, it should experience no force in K arising from **E**. This overlooks the fact that in K, however, the current is no longer neutral (§ [181 A]): if **v** and i

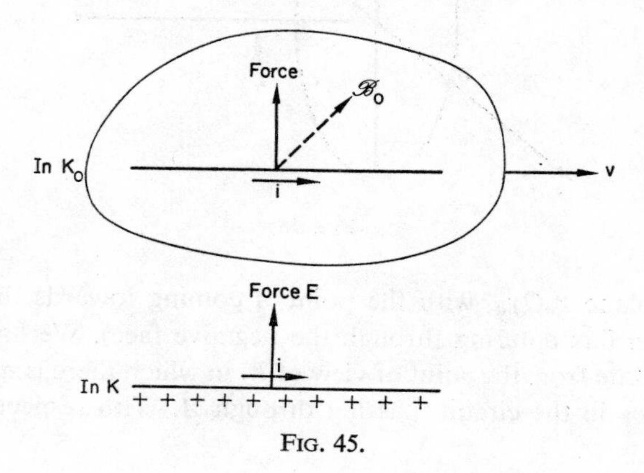

FIG. 45.

are in the same direction, it is positively charged (Fig. 45), whence the desired force and displacement. The force is ascribed to the magnetic field or to the electric field, according to the system of reference adopted.

[181] A Circular Ohmic Current, in Translational Motion in an Electric Field E

We assume that the field **E** is normal to the velocity of translation. Initially, the current is assumed to be at rest in its proper frame of reference (no rotation).

(a) *The plane of the current is initially normal to the velocity*

Taking the velocity parallel to Ox, we set the field **E** along Oy and the current in the plane yOz (Fig. 46). In K, therefore

$$E_x = E_z = 0, \quad E_y = E$$

and hence in K_0

$$E_{0x} = E_{0z} = 0, \qquad E_{0y} = E_0 = \frac{E}{\sqrt{1-\beta^2}}$$

$$\mathcal{B}_{0x} = \mathcal{B}_{0y} = 0, \qquad \mathcal{B}_{0z} = \mathcal{B}_0 = -\frac{c_0\beta}{c}\frac{E}{\sqrt{1-\beta^2}} = -\frac{c_0\beta}{c}E_{0y}$$

On applying the usual laws for the action of a field $\mathfrak{B}$ on a current, in K_0, we predict that the circuit will have a tendency to rotate about Oy_0; its position of stable equi-

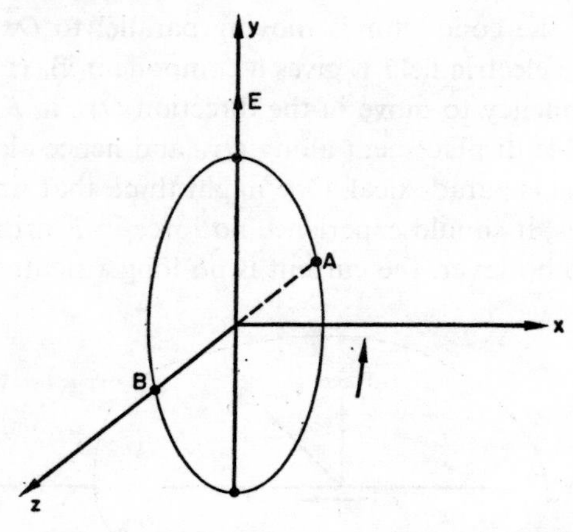

FIG. 46.

librium is in the plane x_0Oy_0, with the point A coming towards the negative part of Ox_0 (the maximum flux entering through the negative face). We have now to explain this tendency to rotate from the point of view of K, in which there is no magnetic field $\mathfrak{B}$. Consider a charge ε in the circuit, passing through A. With respect to K_0, the charge has velocity

$$u_{0x} = u_{0z} = 0, \qquad u_{0y} = u_0$$

and hence experiences a force

$$X_0 = \frac{\varepsilon u_0}{c_0}\mathcal{B}_0, \qquad Y_0 = \varepsilon E_0$$

In K, the force must be

$$X = 0, \qquad Y = \varepsilon E$$

and we recover these values by applying the transformation formulae (Fig. 47). For, in K

$$u_x = v, \qquad u_y = u_0\sqrt{1-\beta^2}$$

$$Y = Y_0\sqrt{1-\beta^2} = \varepsilon E_0\sqrt{1-\beta^2} = \varepsilon E$$

$$X = X_0 + \frac{vu_{0y}}{c^2}Y_0 = -\frac{c_0\beta}{c}\frac{E}{\sqrt{1-\beta^2}}\frac{\varepsilon u_0}{c_0} + \frac{vu_0}{c^2}\frac{\varepsilon E}{\sqrt{1-\beta^2}} = 0$$

If the charge were sliding without friction in a vertical tube (Fig. 48) stationary in K_0, the tube would experience an impulse towards negative x_0. With a circular tube (Fig. 49), the tube will tend to rotate about Oy_0 so that Oz_0 moves towards Ox_0. The impulsive couple is

$$l_0X_0\,dt_0 = \frac{\varepsilon u_0}{c_0}\mathcal{B}_0\,dt_0 = \varepsilon E\,dt_0\frac{vu_0}{c^2\sqrt{1-\beta^2}}$$

In K, the vertical tube has velocity v parallel to Ox. If no force acted on the charge, it would travel with uniform velocity having components u_x, u_y. The charge would

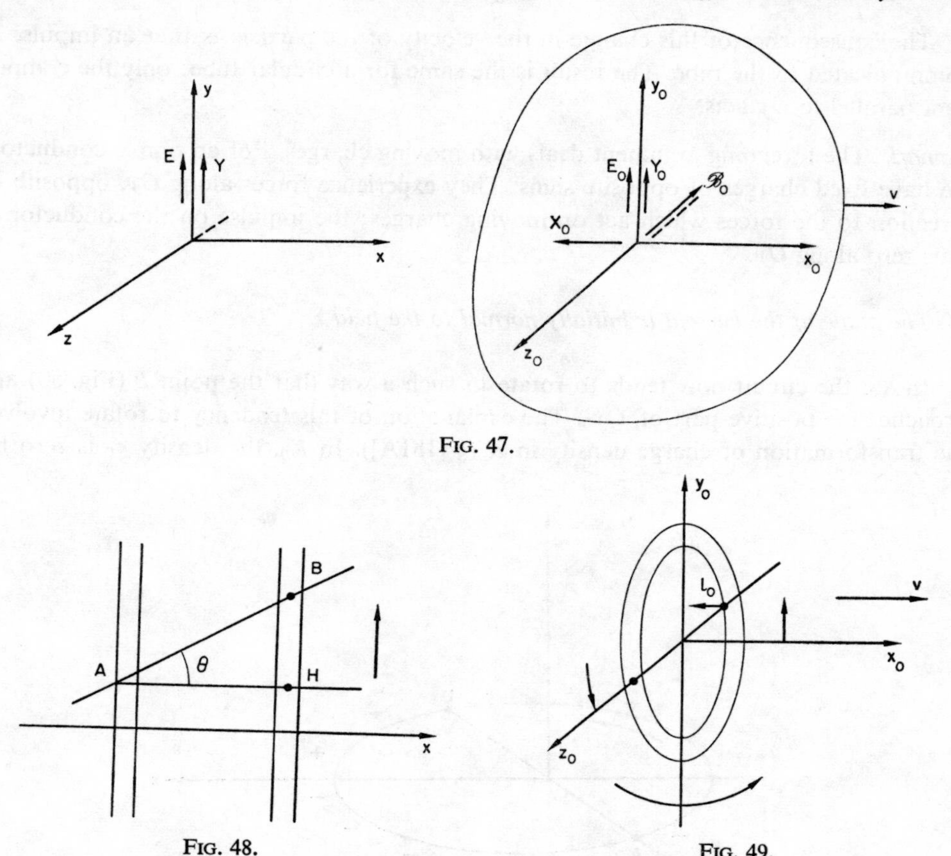

FIG. 47.

FIG. 48.

FIG. 49.

move along the tube without touching it. For its positions at times $\Delta t = 1$ apart, we should have

$$AH = v, \qquad BH = u_0\sqrt{1-\beta^2}, \qquad \tan\theta = \frac{u_0\sqrt{1-\beta^2}}{v}$$

In fact, however, there is a force, εE, and the charge accelerates. The component u_y tends to increase while the tube continues to move with velocity v. If the particle remained at velocity v along the Ox direction, the tube would experience no impulse. The velocity along Ox increases, however, for a particle travelling upwards through A because its mass increases (for a particle travelling downwards, it would decrease); *this is a specifically relativistic effect*. In consequence

$$\mathrm{d}p_x = \mathrm{d}\frac{m_0 u_x}{\sqrt{1-u^2/c^2}} = 0$$

$$m_0\frac{\mathrm{d}u_x}{\sqrt{1-u^2/c^2}} = -m_0 u_x\,\mathrm{d}\left(\frac{1}{\sqrt{1-u^2/c^2}}\right)$$

$$= -\frac{v}{c^2}\,\mathrm{d}\left(\frac{m_0c^2}{\sqrt{1-u^2/c^2}}\right) = -\frac{v}{c^2}\,\mathrm{d}W = -\frac{v}{c^2}\,\varepsilon E\,\mathrm{d}t$$

The consequence of this change in the velocity of the particle is that an impulse is communicated to the tube. The result is the same for a circular tube: only the component parallel to Oy acts.

Remark. The foregoing argument deals with moving charges. For an ohmic conductor, we have fixed charges of opposite signs. They experience forces along Oy, opposite in direction to the forces which act on moving charges; the impulse on the conductor is thus zero along Oy.

(b) *The plane of the current is initially normal to the field* **E**

In K_0, the circuit now tends to rotate in such a way that the point B (Fig. 50) approaches the positive part of Oy_0. The explanation of this tendency to rotate involves the transformation of charge density in K (§ [181A]). In K_0, the density ϱ_0 is zero by

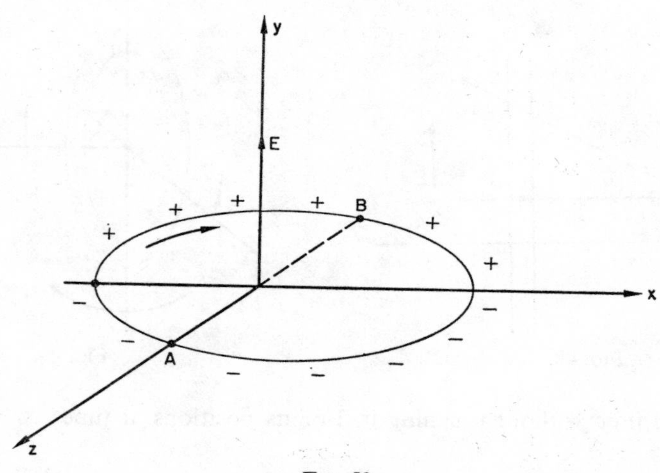

FIG. 50.

hypothesis (ohmic current). In K, the current densities are those shown in Fig. 50 (it remains neutral in the case studied in (a), Fig. 46). In K, therefore, there is a couple arising from the action of the field **E** on these charges.

(c) *The plane of the current initially contains the field* **E** *and the velocity* (Fig. 51)

There is no tendency to rotate in K_0, and in K, the action of **E** on the charges produces no couple; all the forces lie in the plane of the circuit.

(d) *Arbitrary initial position*

In studying various special cases above, we have revealed two different effects; for an arbitrary initial position, these two effects will combine.

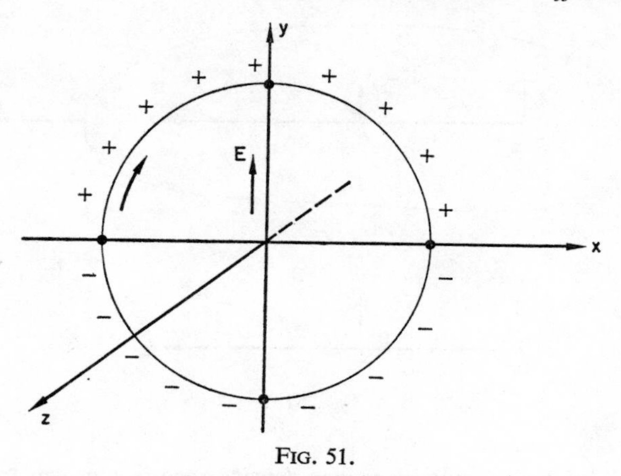

FIG. 51.

[181A] Transformation of the Charge in a Finite Volume

(a) *The volume is not insulated*

We consider, for example, a current in which the moving charges are neutralized by charges at rest in K_2. An arbitrary volume can remain neutral only if the number of charges entering at a given moment is equal to the number leaving. The number of moving charges present in the volume at a given time must remain equal to the number of neutralizing charges stationary in the medium, since the charge of a particle is independent of the system of reference. Thus, according to pre-relativistic thinking, it seems that the number of charges in a given volume at any instant must likewise be independent of the reference system; we should have a neutral volume in all systems, therefore. This reasoning overlooks the relativity of simultaneity, however. *The number of stationary charges within a given volume is independent of the reference system, but the same is not true of the moving charges*; the arrival and departure of charges are in fact two events which occur at different points in space. The example we have chosen is virtually realized in practice when currents flow in ordinary metal conductors if we assume that the conducting material, defining the proper systems, has no other electrical properties.

We consider a straight segment of conductor A_2B_2, stationary in the system K_2 (Fig. 51A). At time $t_0 = 0$, a current is set up in this conductor. The moving charges are negative charges with positive velocity u_2; they are assumed to be neutralized by positive charges at rest. The number of charges entering at A_2 is equal to the number leaving at B_2, at time $t_2 = 0$.

We now transfer to the system K_1, with respect to which K_2 has a velocity v along Ox. The readings of the clocks in K_1, corresponding to t_2, are shown in Fig. 51A. If we wish to speak of simultaneous events at A_1 and B_1, we must wait at the end-point B_1 until the clock shows the time $t_1 = 0$. We have already shown (*Kinematics*, § [72](d)) that herein lay the explanation of the Lorentz contraction. We see that here, extra negative charges will enter the segment A_1B_1 during the delay. In the conditions described, therefore, it will be negatively charged from the point of view of K_1.

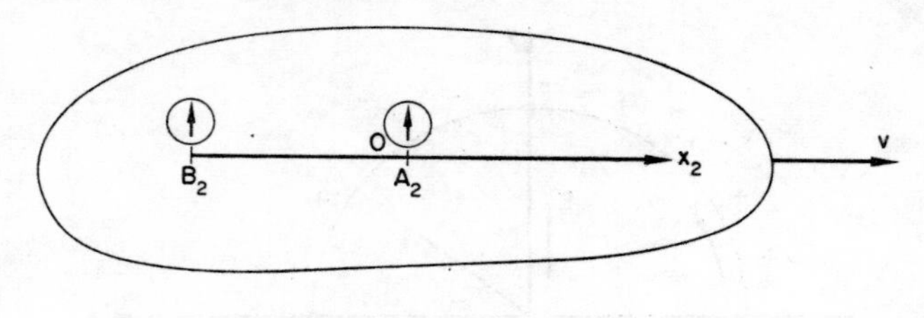

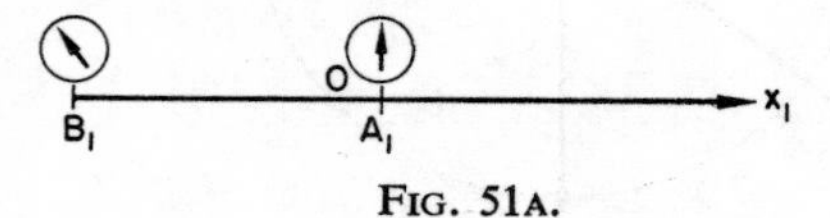

FIG. 51A.

(b) *The volume is insulated; a ring of proper density zero in uniform translational motion*

The number of charges within such a volume cannot vary; for all systems of reference, therefore, the total charge is the same. If the volume is neutral in one system, it will be neutral in every system. This statement is concerned only with the total volume, however. As in the preceding case, there is nothing to prevent regions of the volume from appearing to be charged, provided that they are compensated elsewhere.

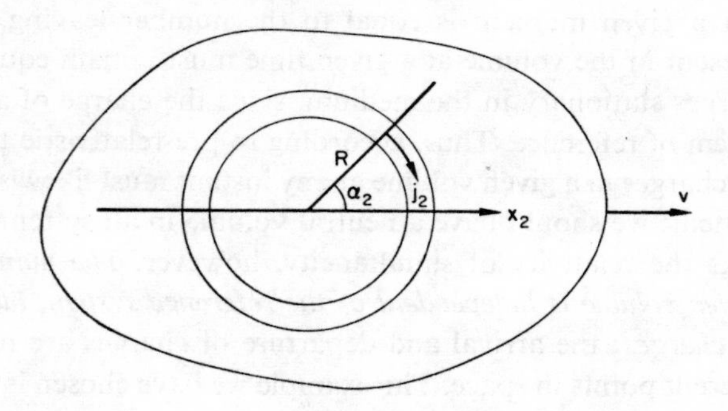

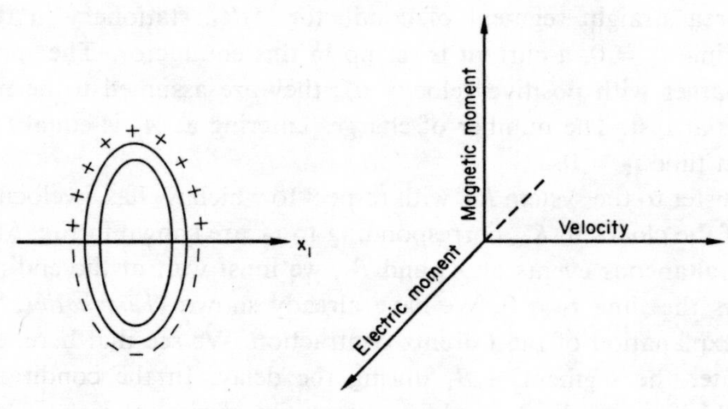

FIG. 51B.

Consider, for example, a ring of radius R through which a current j_2 flows, in the rest system K_2 of the ring, consisting of negative charges in motion (Fig. 51B); in this system K_2, stationary positive charges neutralize the negative charges. We now consider a system K_1, with respect to which the ring has a velocity v parallel to the plane containing it. The diameter parallel to v divides the ring into two parts in which the current is either in the same sense as v or in the opposite sense. Following the same argument as in (a) at each point, we see that in the system K_1, the lower part is negatively charged and the upper part is positively charged. The ring therefore possesses a resultant electrical moment. The density is clearly greatest along the diameter normal to Ox, and vanishes on Ox.

If the velocity of the ring is perpendicular to the plane of the latter, no charge appears.

Remark. I shall not dwell on the quantitative aspect of this topic here (involving calculation of the densities and of the moments).

[182] The Charged Plane Condenser in Uniform Translational Motion

(a) We take the Oz axis in the plane of the plates, assumed to be rectangular (Fig. 52), and the Ox axis is parallel to the velocity; let α be the angle between Oy and the plates. In the proper frame of reference, K_0, of the condenser, symmetry arguments show that each plate exerts a force $\mathbf{F}_0$ through the centre of the other plate and normal to it; the resultant and the couple are clearly zero. We take the centre of one of the plates as origin of coordinates. The components X_0 and Y_0 of $\mathbf{F}_0$ satisfy the relation.

$$Y_0 OM_0 \cos\alpha_0 - X_0 OM_0 \sin\alpha_0 = 0, \qquad \frac{Y_0}{X_0} = \tan\alpha_0$$

Figure 52 shows the condenser in the reference system K, with respect to which it has velocity v; to make the mechanical construction perfectly clear, I have given it with a different setting of the axes in Fig. 53. Since there is no rotation in K, the motion in K is a uniform translation. Let us examine what happens to the concept of a couple. We shall obtain the two types of couple, C and C', already introduced in § [178].

(b) The couple C is defined with respect to simultaneous observation in K_0. Under these conditions, the distance between two points is

$$\Delta x = \frac{\Delta x_0}{\sqrt{1-\beta^2}}, \qquad \Delta y = \Delta y_0$$

which here becomes

$$OM\cos\alpha = \frac{OM_0\cos\alpha_0}{\sqrt{1-\beta^2}}, \qquad OM\sin\alpha = OM_0\sin\alpha_0$$

With the transformation formulae

$$X = X_0, \qquad Y = Y_0\sqrt{1-\beta^2}$$

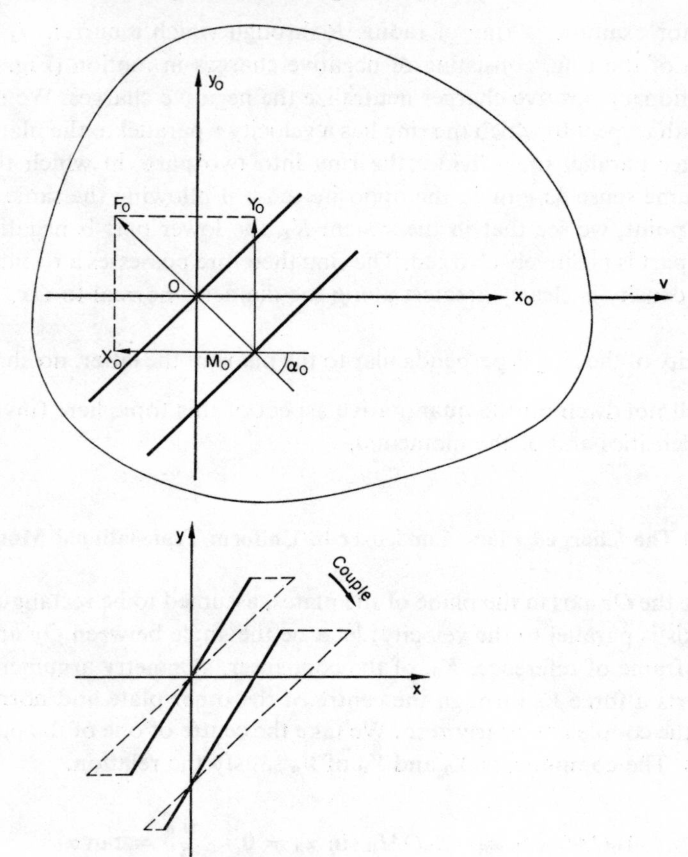

FIG. 52.

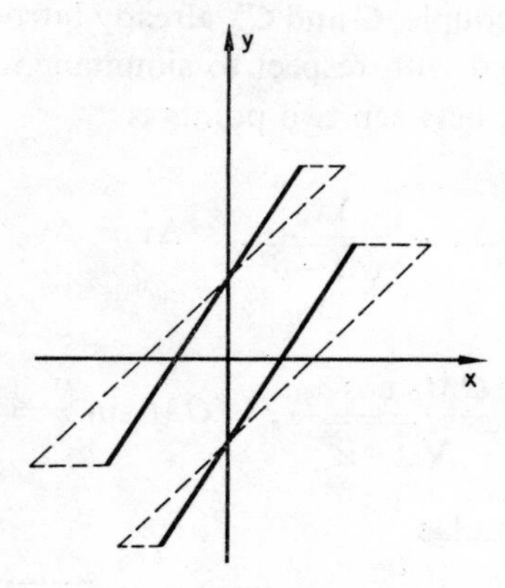

FIG. 53

we obtain the couple

$$C = Y.OM \cos\alpha - X.OM \sin\alpha$$
$$= Y_0.OM_0 \cos\alpha_0 - X_0.OM_0 \sin\alpha_0$$
$$= 0$$

The absence of rotation in K is associated with a vanishing couple C.

(c) The couple C' is defined with respect to simultaneous observation in K. We now have

$$\Delta x = \Delta x_0 \sqrt{1-\beta^2}, \qquad \Delta y = \Delta y_0$$
$$OM \cos\alpha = OM_0 \cos\alpha_0 \sqrt{1-\beta^2}, \qquad OM \sin\alpha = OM_0 \sin\alpha_0$$
$$C' = Y.OM \cos\alpha - X.OM \sin\alpha$$
$$= -\beta^2 F_0 OM_0 \sin\alpha_0 \cos\alpha_0$$

To a first approximation, we may use the field for infinite plates:

$$C' = \tfrac{1}{2} W_0 \beta^2 \sin 2\alpha_0 \quad \text{in which} \quad W_0 = \tfrac{1}{2}\gamma(V_A - V_B)^2;$$

here γ, V_A and V_B denote the capacity and potentials of the plates measured in K_0.

For discussion of the choice between C and C', see § [178].

Remark 1. The couple C' can also be obtained by calculating the field created by one of the plates and hence the action of this field on the other plate.

Remark 2. The calculation can be performed, taking into account the edge effects. In view of the present interpretation, it is of no interest, but in the past the exact calculation of the couple C' has been of considerable importance (§ [190]).

C. APPLICATION TO THERMODYNAMICS

[183] The Transformation of Heat

A body is supplied isotropically with heat dQ_0, and this increases its proper mass by an amount $dm_0 = dQ_0/c^2$. The body remains at rest in K_0 and hence continues travelling at velocity v in K. In K_0, the changes of energy and impulse are

$$dW_0 = dQ_0, \qquad dp_0 = 0$$

and in K

$$dW = \frac{dW_0 + v\,dp_0}{\sqrt{1-\beta^2}} = \frac{dQ_0}{\sqrt{1-\beta^2}}$$

$$dp = \frac{dp_0 + \frac{v}{c^2}\,dW_0}{\sqrt{1-\beta^2}} = \frac{dQ_0}{\sqrt{1-\beta^2}}\,\frac{v}{c^2}$$

If we wish the expression for the equivalence of heat and work to remain invariant, we must define the heat, dQ, in K by

$$dQ = dW$$

Hence

$$dQ = \frac{dQ_0}{\sqrt{1-\beta^2}}, \qquad dp = \frac{dQ}{c^2}\,v$$

In conformity with the dynamical equation (§ [175]), we have

$$\frac{dp}{dt} = \frac{dQ_0}{dt_0}\,\frac{v}{c^2} = v\,\frac{dm_0}{dt_0}$$

[184] Compressed Gases

Consider a gas in a cylinder, compressed by a piston. By assuming that we have two pistons working in opposite directions, we shall be able to take over the results obtained for the rod directly. The case of only one piston is obviously identical provided we take into account the work done in K by the base of the cylinder. We shall first establish a result concerning the transformation of pressure.

(a) *The invariance of a uniform pressure under certain conditions*

Consider a body, *stationary* in K_0, upon which a uniform pressure p_0 acts; we transfer to another reference frame K, recalling that the forces are transformed according to the formulae of § [28]. For an element of surface dS_{0x} perpendicular to Ox, we have a force parallel to Ox in K_0 given by

$$dF_{0x} = p_0\,dS_{0x}$$

In K

$$dS_x = dS_{0x}, \qquad dF_x = dF_{0x}$$

so that for the pressure,

$$dF_x = p\,dS_x, \qquad p = p_0$$

For an element of surface dS_{0y} normal to Oy, we have

$$dS_y = dS_{0y}\sqrt{1-\beta^2}, \qquad dF_y = dF_{0y}\sqrt{1-\beta^2}$$

$$dF_y = p\,dS_y = p\,dS_{0y}\sqrt{1-\beta^2} = dF_{0y}\sqrt{1-\beta^2}$$

so that once again, $p = p_0$.

If a pressure is uniform in the proper system of the surface to which it is applied, then it is invariant. *This requires simultaneity in K.*

(b) *Cylinder parallel to the velocity*

In K_0, we have

$$X_0 = F_0 = p_{0x}\,dS_0 = p_0\,dS_0, \qquad p_{0x} = p_0$$

$$Y_0 = p_{0y}\,dS_0 = 0, \qquad p_{0y} = 0$$

and in K,

$$X = p_x \,\mathrm{d}S, \qquad Y = p_y \,\mathrm{d}S$$

Hence

$$Y = Y_0 \sqrt{1-\beta^2} = 0, \qquad p_y = 0$$
$$p_x \,\mathrm{d}S = p_0 \,\mathrm{d}S_0, \qquad p_x = p_0$$

For the work done

$$\mathrm{d}\mathcal{T} = \frac{\mathrm{d}\mathcal{T}_0}{\sqrt{1-\beta^2}} = \frac{p_0 \,\mathrm{d}\mathcal{V}_0}{\sqrt{1-\beta^2}} = \frac{p \,\mathrm{d}\mathcal{V}_0}{\sqrt{1-\beta^2}}$$

We introduce the volume $\mathrm{d}\mathcal{V}$ which corresponds to $\mathrm{d}\mathcal{V}_0$ for simultaneous observation in K: $\mathrm{d}\mathcal{V} = \mathrm{d}\mathcal{V}_0 \sqrt{1-\beta^2}$. We then find

$$\mathrm{d}\mathcal{T} = \frac{p \,\mathrm{d}\mathcal{V}}{1-\beta^2}$$

Important remark. It is to be *stressed* that this $\mathrm{d}\mathcal{V}$ is not the volume swept out by $\mathrm{d}S$ in K (whereas $\mathrm{d}\mathcal{V}_0$ *is* the volume swept out by $\mathrm{d}S_0$ in K_0). If we were to use this volume swept out, $\delta\mathcal{V}$, we should have

volume swept out at A:

$$\delta\mathcal{V}_A = \mathrm{d}x \,\mathrm{d}S = \frac{\mathrm{d}l_0 + vT_0}{\sqrt{1-\beta^2}} \,\mathrm{d}S$$

volume swept out at B:

$$\delta\mathcal{V}_B = \frac{-\mathrm{d}l_0 + vT_0}{\sqrt{1-\beta^2}} \,\mathrm{d}S$$

total volume swept out:

$$\delta\mathcal{V} = \delta\mathcal{V}_A - \delta\mathcal{V}_B = \frac{2 \,\mathrm{d}l_0 \,\mathrm{d}S}{\sqrt{1-\beta^2}} = \frac{\mathrm{d}\mathcal{V}_0}{\sqrt{1-\beta^2}}$$

Thus as we might have expected, we find

$$\mathrm{d}\mathcal{T} = p \,\delta\mathcal{V}$$

(c) *Cylinder normal to the velocity*

In K_0, we have

$$X_0 = p_{0x} \,\mathrm{d}S_0 = 0, \qquad p_{0x} = 0$$
$$Y_0 = p_0 \,\mathrm{d}S_0 = p_{0y} \,\mathrm{d}S_0, \qquad p_{0y} = p_0$$

and in K

$$X = p_x \,\mathrm{d}S, \qquad Y = p_y \,\mathrm{d}S$$

so that

$$Y = Y_0 \sqrt{1-\beta^2} = p_0 \,\mathrm{d}S_0 \sqrt{1-\beta^2} = p_0 \,\mathrm{d}S$$

or

$$p_y \,\mathrm{d}S = p_0 \,\mathrm{d}S, \qquad p_y = p_0$$

Furthermore,

$$X = \frac{v}{c^2}\frac{dl_0}{T_0}Y_0 = \frac{v}{c^2}\frac{dl_0}{T_0}p_0\,dS_0$$

$$= \frac{v}{c^2}\frac{dl_0}{T_0}p_0\frac{dS}{\sqrt{1-\beta^2}} = p_x\,dS$$

$$p_x = \frac{vu_{0y}}{c^2}p_0$$

In K, therefore, there is a tangential component of pressure. The work is given by

$$d\mathcal{T} = p_x\,dS\,dx + p_y\,dS\,dx$$

and hence

$$d\mathcal{T} = \frac{p_0\,d\mathcal{V}_0}{\sqrt{1-\beta^2}}$$

When equilibrium is established afresh, we have a normal pressure $p = p_0$ (applying (a)); using the volume $d\mathcal{V}$ defined above,

$$d\mathcal{T} = \frac{p\,d\mathcal{V}}{1-\beta^2}$$

[185] The Principles

(a) *The equivalence principle*

We have shown in § [183] that heat transforms according to

$$dQ = \frac{dQ_0}{\sqrt{1-\beta^2}}$$

For a gas under pressure, work transforms according to

$$d\mathcal{T} = \frac{d\mathcal{T}_0}{\sqrt{1-\beta^2}}$$

These results are consistent with the invariant form of the equivalence principle

$$dQ = dU + d\mathcal{T}$$

provided we write

$$dU = \frac{dU_0}{\sqrt{1-\beta^2}}$$

This last formula is a consequence of the mass–energy equivalence:

$$dU = c^2\,dm, \qquad dU_0 = c^2\,dm_0$$

The transformation for dQ may also be demonstrated by setting out from the transformations for dU and $d\mathcal{T}$; the method matters little: the important thing is to bring out the consistency of the formulae.

(b) *Carnot's principle*

Because of its statistical interpretation, entropy is an invariant. If we require Carnot's principle,

$$\frac{dQ}{T} = dS$$

to be invariant, we must use the transformation

$$T = \frac{T_0}{\sqrt{1-\beta^2}}$$

for the temperature. Various other arguments also lead to this transformation (see ref. [4] of § [188]).

D. APPLICATION TO THE MECHANICS OF CONTINUOUS MEDIA

[185A] Fundamental Equations in Three-dimensional Form

(a) When the action that causes the proper mass to vary does not act on the velocity, we must use the equations of motion applicable to particles of variable proper mass (§ [174]):

$$\mathbf{F}\,dt = m_0\,d\left(\frac{\mathbf{v}}{\sqrt{1-\beta^2}}\right) \tag{1}$$

$$\mathbf{F}\,.\,\mathbf{v}\,dt = c^2 m_0\,d\left(\frac{1}{\sqrt{1-\beta^2}}\right) \tag{2}$$

in which the variable mass m_0 is *outside* the differential operator. This is the equation that I shall adopt here. The compressive forces which act on a drop of liquid (which we picture as cut out of the continuous medium) thus play two roles:

—they affect the velocity, and we shall allow for this together with the force due to the pressure;
—they compress the medium, even if the latter remains at rest, on average.

This postulate is of course open to generalization, to allow for heat flow, for example, but it covers all the cases considered by earlier authors (§ [188A]).

(b) Consider a drop of fluid of volume $d\bar{\omega}$, density ϱ and mass $dm = \varrho\, d\bar{\omega}$ with respect to a Galileian system of reference, K. This mass is the real, total mass including the potential elastic energy. Let $d\bar{\omega}_0$, ϱ_0 and dm_0 denote the corresponding proper values. Equations (1) and (2) transform into

$$d\mathbf{F} = dm_0 \frac{D}{Dt}\left(\frac{\mathbf{v}}{\sqrt{1-\beta^2}}\right) \tag{3}$$

$$d\mathbf{F}\,.\,\mathbf{v} = c^2\,dm_0 \frac{D}{Dt}\left(\frac{1}{\sqrt{1-\beta^2}}\right) \tag{4}$$

where

$$\frac{\mathrm{D}}{\mathrm{D}t} = \frac{\partial}{\partial t} + v^{\alpha}\frac{\partial}{\partial x^{\alpha}} \tag{5}$$

Defining a force density **f** by the relation $\mathbf{dF} = \mathbf{f}\,\mathrm{d}\varpi$, equations (3) and (4) become

$$\mathbf{f} = \frac{\varrho_0}{\sqrt{1-\beta^2}}\frac{\mathrm{D}}{\mathrm{D}t}\left(\frac{\mathbf{v}}{\sqrt{1-\beta^2}}\right) \tag{6}$$

$$\mathbf{f}\,.\,\mathbf{v} = \frac{c^2\varrho_0}{\sqrt{1-\beta^2}}\frac{\mathrm{D}}{\mathrm{D}t}\left(\frac{1}{\sqrt{1-\beta^2}}\right) \tag{7}$$

In indicial notation, using (5), we have

$$f^{\alpha} = \frac{\varrho_0}{\sqrt{1-\beta^2}}\left\{\partial_t\left(\frac{v^{\alpha}}{\sqrt{1-\beta^2}}\right) + v^{\beta}\,\partial_{\beta}\left(\frac{v^{\alpha}}{\sqrt{1-\beta^2}}\right)\right\} \tag{8}$$

$$f^{\alpha}v_{\alpha} = \frac{c^2\varrho_0}{\sqrt{1-\beta^2}}\left\{\partial_t\left(\frac{1}{\sqrt{1-\beta^2}}\right) + v^{\alpha}\,\partial_{\alpha}\left(\frac{1}{\sqrt{1-\beta^2}}\right)\right\} \tag{9}$$

The force f^{α} is the resultant of the mass forces and the pressure forces. If there are no mass forces, and in the general case of viscous liquids, f^{α} reduces to the familiar expression $\partial_{\beta}t^{\alpha\beta}$, where $t^{\alpha\beta}$ is the elastic tensor. For perfect fluids, this gives $\partial_{\beta}p_0\,\delta^{\alpha\beta}$. These are the expressions corresponding to the proper frame of reference. In an arbitrary Galileian coordinate system, it can be shown that (see the bibliography)

$$f^{\alpha} = -\partial^{\alpha}p_0 - \frac{V^{\alpha}V^{i}}{c^2}\,\partial_i p_0$$

in which V^i is the four-velocity ($\alpha = 1, 2, 3$; $i = 1, 2, 3, 4$).

(c) We must add an equation determining the variation of proper mass to the equations (8) and (9). If this variation is due wholly to the work done by a uniform pressure p_0, then

$$\mathrm{D}(\mathrm{d}m_0) = \frac{p_0}{c^2}\,\mathrm{D}(\mathrm{d}\varpi_0) \tag{10}$$

[185B] Four-dimensional Equations

We introduce the four-vector with components

$$\Phi^{\alpha} = f^{\alpha}, \qquad \Phi^4 = \frac{i}{c}\,\mathbf{v}\,.\,\mathbf{f}, \qquad \Phi^i v_i = 0 \tag{1}$$

and we shall use the four-velocity V^i and the four-acceleration Γ^i; equations (8) and (9) of § [185A] combine to give

$$\Phi^i = \varrho_0\Gamma^i = \varrho_0 V^j\,\partial_j V^i \tag{2}$$

This is a general equation, valid for viscous liquids granted the hypothesis about the variation of proper mass. In the case of perfect fluids, and using the value of f^α given above, we have

$$\Phi_i = -\partial_i p_0 - \frac{V_i V^k}{c^2}\partial_k p_0 \tag{3}$$

Furthermore,

$$\partial_i(\varrho_0 V^i) = -\frac{p_0}{c^2}\partial_i V^i \tag{4}$$

These equations differ from those obtained by earlier authors.

HISTORICAL AND BIBLIOGRAPHICAL NOTES

[186] On the Problem of the Right-angled Lever

(a) Previous to the appearance of my article [8], only the state of equilibrium under tension had been considered: the problem was to reconcile the existence of the couple C' with the fact that the lever does not rotate.

The problem was first stated by Lewis and Tolman [1], who solved it without bringing in the inertial energy; they derived incorrect formulae for the force transformation. In 1912, H. A. Lorentz [2] showed that the mass–energy equivalence should apply to all forms of energy, and in particular, to the interaction energy between particles. Acting on a suggestion by Sommerfeld who had noticed Lewis and Tolman's mistake, von Laue [3] reconsidered the question and displayed the role of the inertial energy in this kind of problem for the first time. He gave the following solution, which is the same as Epstein's [4]. The force F_B normal to the velocity does no work, but the force F_A does work at $F_0 v$ per unit time on the lever. Von Laue then considered that an energy current enters the lever through the face A. It runs along to the axis Ω where it flows out of the lever, since at Ω, the axis exerts on the lever a force which is equal and opposite to the force F_A. An energy current therefore flows through the arm ΩA but not through ΩB. In relativistic terms, this is equivalent to saying that the mass of the arm ΩA is greater than that of the arm ΩB.

Let us determine the additional momentum associated with this energy current. For the moment, we picture this energy flux as a liquid current; we denote the component of velocity of the energy along $A\Omega$ by v' and the area of cross-section of the lever-arm by dS. During time dt, energy $F_0 v\, dt$ enters. This energy is distributed along a length $v'\, dt$; so the energy density is

$$\frac{F_0 v\, dt}{v'\, dt\, dS} = \frac{F_0 v}{v'\, dS}$$

The density of the component of momentum parallel to $A\Omega$ is therefore given by

$$\text{energy density} \times \frac{v'}{c^2} = \frac{F_0 v}{c^2\, dS}$$

in which the mass-energy relation has been employed; the total component parallel to $A\Omega$ is thus

$$\frac{F_0 v}{c^2\, dS} l_0\, dS = \frac{F_0 v l_0}{c^2}$$

The increase of angular momentum per unit time, about a fixed point in K, is therefore

$$\frac{F_0 v l_0}{c^2} v = \beta^2 l_0 F_0$$

This is just the couple C'.

(b) This solution was reproduced in books by von Laue [5], Tolman [6], myself [7] and Yilmaz [10] The two concepts introduced—energy current and transverse impulse – are then used by various authors to establish the fundamental equations of continuous media. It becomes apparent that the problem of the right-angled lever is not a mere relativistic oddity: fundamental principles are involved.

(c) These two concepts had always struck me as mysterious, but the success of the calculations seemed to justify them. I was led to consider the problem afresh by my research in thermodynamics, however, and I became aware that serious relativistic mistakes appeared not only in the solution of the problem but even in the way in which it was stated. I have shown [8] that in a permanent state of tension, there is neither energy current nor transverse component of momentum; this is the solution given above in § [178](b), with some additional elucidation not given in the article. Furthermore, when we really think about it, it is absurd to speak about an energy current in a steady state. If energy is, in fact, moving along $A\Omega$ in K, then it must be moving similarly in K_0. In K_0, however, such a current of energy is inconceivable. This is a result of the very argument I am criticizing, since in K_0 the forces F_0 do not work in a state of tension. We thus arrive at an internal contradiction: an energy in K has no equivalent in K_0. Furthermore, we may notice that von Laue's energy density is indeterminate, as v' is arbitrary and is introduced into the calculation only for convenience. The total mass of the lever with respect to K is arbitrary, which is an absurd result.

We still have to explain why von Laue's calculation works. The explanation of the term $\beta^2 l_0 F_0$ lies in the relativity of simultaneity. From the relation

$$t = \frac{t_0 + \dfrac{v}{c^2} x_0}{\sqrt{1-\beta^2}}$$

it follows that, for a given value of t_0, the two ends of the bar ΩB are considered at times separated by

$$\Delta t = \frac{\dfrac{v}{c^2} l_0}{\sqrt{1-\beta^2}}$$

If we wish to transform to simultaneous observation in K, the lever must be allowed to move through $v\,\Delta t$. The corresponding additional moment of the force applied at B is therefore

$$v\,\Delta t\,F = \frac{\beta^2 l_0 F}{\sqrt{1-\beta^2}} = \beta^2 l_0 F_0$$

Remark 1. The problem assumes rest in K_0, and hence requires simultaneous forces in K_0. We could erect *another* problem, in which the forces were applied simultaneously in K. For the rod, these two possibilities have been considered.

Remark 2. The concept of an energy current has recently been defended by Kibble [9]; my comments and Kibble's replies are to be found at the end of the article. See Appendix VI for recent references.

[1] G. N. LEWIS and R. C. TOLMAN: The principle of relativity and non-Newtonian mechanics. *Phil. Mag.* **18** (1909) 510–523.
[2] H. A. LORENTZ: Over de Mass der Energie. *Verslag. Gew. Vergad. Wiss. Afdeeling Amsterdam* **20** (1911) 87–98.
[3] M. VON LAUE: Ein Beispiel zur Dynamik der Relativitätstheorie. *Verhandl. Deutsch. Phys. Ges.* (1911) 513–518.
[4] P. S. EPSTEIN: Ueber relativistische Statik. *Ann. Physik* **36** (1911) 779–795.
[5] M. VON LAUE: ref. [2] of Appendix I; see p. 258 of the French translation of 1926.
[6] R. C. TOLMAN: ref. [4] of Appendix I; see p. 79, 1950 ed.
[7] H. ARZELIÈS: ref. [32] of Appendix I; see pp. 29–31.
[8] H. ARZELIÈS: Sur le problème relativiste du levier coudé. *Nuovo Cimento* **35** (1965) 783–791.
[9] T. W. B. KIBBLE: ref. [5] of § [187].
[10] H. YILMAZ: ref. [36] of Appendix I, pp. 203–204.

[187] On the Compression of a Rod or a Gas

(a) This problem is discussed by Epstein [1]; his solution is then reproduced by von Laue [2] and myself (3). His argument employs the energy current, as in the case of the right-angled lever. The reasoning is now as follows. We disregard the forces F_y which do no work. At A, the force F_x does work $F_x v$ per unit time, and at B, it does work $-F_x v$. The rod is then assumed to carry an energy current of strength $F_x v$, giving a momentum density per unit length of $F_x v/c^2$. The total momentum is $F_x vl/c^2$ and the angular momentum about a fixed point in K increases by an amount

$$\frac{F_x vl}{c^2}\, v \sin\theta = \frac{F_{0x} v^2}{c^2}\, l_0 \sin\theta_0$$

per unit time. My criticisms are the same as those for the right-angled lever.

(b) I have reconsidered this problem in the course of research in thermodynamics [4], eliminating the concept of energy current. Kibble [5], however, believes the former theory to be correct. It would be helpful if other authors would examine the question.

[1] P. S. Epstein: ref. [4] of § [186].
[2] M. von Laue: ref. [5] of § [186].
[3] H. Arzeliès: ref. [17] of § [186].
[4] H. Arzeliès: Transformation relativiste de la température et de quelques autres grandeurs thermodynamiques. *Nuovo Cimento* **35** (1965) 792–804.
[5] T. W. B. Kibble: Relativistic transformation laws for thermodynamic variables. *Nuovo Cimento* **41** (1966) 72–85. This article is followed by comments by A. Gamba and H. Arzeliès.

[188] Variable Proper Mass and Thermodynamics

For half a century, the following formulae for the transformation of heat $\mathrm{d}Q$ and temperature T were accepted:

$$\mathrm{d}Q = \mathrm{d}Q_0\sqrt{1-\beta^2}, \qquad T = T_0\sqrt{1-\beta^2}$$

The new transformations have been connected with the dynamics of variable proper mass by Ott [1] and myself [2, 3] independently. The results of Ott and myself concur only for the transformations of Q and T. We differ over the work done (see above, § [185]) and the internal energy, and also over the parameters of black-body radiation. This topic is discussed at length in my thermodynamics textbook [4], both in the introduction and in the text. *See Appendix VI for recent references.*

[1] H. Ott: Lorentz-Transformation der Wärme und der Temperatur. *Z. Physik* **175** (1963) 70–104.
[2] H. Arzeliès: Nouvelles bases pour la thermodynamique relativiste. *Nucleus* **6** (1965) 250–252.
[3] H. Arzeliès: Transformation relativiste de la temperature. *Nuovo Cimento* **35** (1965) 792–802. (Article submitted 15 January 1964.)
[4] H. Arzeliès: *Thermodynamique relativiste et quantique*. Gauthier-Villars, Paris, 1967.

[188A] On the Equations for Continuous Media

(a) Almost all the earlier authors (the references to relativistic treatises that I give will provide a bibliographical starting-point, elaborated in my forthcoming book [12]) base the dynamics of perfect neutral fluids on the equation

$$\partial_i T^{ij} = 0 \tag{1}$$

where

$$T^{ij} = \left(\varrho_{00}+\frac{p_0}{c^2}\right)V^i V^j + p_0\delta^{ij} = \varrho_0 V^i V^j + p_0\delta^{ij} \tag{2}$$

I disregard any possible mass forces, which are irrelevant to our present objective, and take only the pressure forces into account. In equation (2), V^i is the four-velocity, p_0 the pressure (invariant) and δ^{ij}

the Kronecker delta. The symbol ϱ_{00} denotes the proper density at zero pressure, and ϱ_0 the proper density. Equation (1), or the corresponding three-dimensional equations, and in particular the expression for the tensor T^{ij}, have been known for half a century and derived by various methods. Von Laue [1] obtained them by transposing the equations of electromagnetism and the Maxwell–Minkowski tensor. I have shown elsewhere [2, 3] that the usual physical interpretation of this tensor is unacceptable in the general case, even with the modifications proposed by Kwal. The method of von Laue thus seems to me suspect, and moreover, a simple transposition can be justified only by agreement with experiment.

Other authors, Tolman [4], Møller [5] and Lichnérowicz [7] for example, arrive at the fundamental equations after setting out from the following two ideas:

(i) reasoning on a drop of fluid, and applying ordinary point dynamics (fixed proper mass);
(ii) the energy flux density due to the work done by pressure forces is introduced into the expression for the momentum; this in general produces a transverse component.

These are clear and precise hypotheses, which can be argued about on the theoretical plane and which I personally propose to reject. I believe that I have shown [8] that the concept of energy flux, in the sense that these authors employ it, must be abandoned and that all calculations on bodies in extension or compression must be made using the dynamics of variable proper mass. I pointed out in this article that the dynamics of continuous media must be re-examined on this new basis. This formed the object of §§ [185A] and [185B]. More details about these new formulae will be found in a series of *Notes* [11] and in an overall survey [12] to appear in due course.

We also mention a third method of obtaining equation (1): here, purely abstract definitions are set out, the only object of which is in fact to recover equation (1). Lichnérowicz [9], for example, asserts that the energy distribution in the medium is characterized by a tensor T^{ij}; he defines the concepts of density, pressure, . . ., in terms of the components of this tensor. Costa de Beauregard [10] proceeds by a series of successive inductions and by abstract postulates. Such a procedure, like that of von Laue, is not open to theoretical discussion (one can assert whatever one likes): it can be based only on experiment.

Some authors add the following equation to (1), in order to allow for variation of the proper mass (although still using the dynamics appropriate to invariant proper mass):

$$\partial_i(\varrho_{00}V^i) = -\frac{p_0}{c^2}\,\partial_i V^i \tag{3}$$

(b) Let us reconsider the equation of § [185B]; for the spatial components ($\alpha = 1, 2, 3$), we have

$$\varrho_0 V^j\,\partial_j V^i = -\partial^i p_0 - \frac{V^i V^j}{c^2}\,\partial_j p_0$$

Equation (1), which is the same as that obtained by earlier authors, gives

$$\left(\varrho_0+\frac{p_0}{c^2}\right)V^j\,\partial_j V^i = -\partial^i p_0 - \frac{V^i V^j}{c^2}\,\partial_j p_0$$

Their equation differs from mine by the term $(p_0/c^2)V^j\,\partial_j V^i$, corresponding to the energy current of von Laue. Furthermore, the expression for ϱ_0 is different in the two theories. According to the earlier authors, the mass of volume $\mathrm{d}\bar{\omega}_0$ is

$$\left(\varrho_{00}+\frac{p_0}{c^2}\right)\mathrm{d}\bar{\omega}_0$$

whereas for me, it is

$$\varrho_{00}\,\mathrm{d}\bar{\omega}_0+\int_{\mathrm{d}\bar{\omega}_{00}}^{\mathrm{d}\bar{\omega}_0}\frac{p_0}{c^2}\,\mathrm{d}(\mathrm{d}\bar{\omega}_0)$$

(c) My criticisms of equation (1) and the new equation that I advance imply that the tensor T^{ij} must be abandoned. This has important consequences in other domains: in electromagnetism (the Poynting vector, the energy density), in relativistic thermodynamics (§ [188]) and in gravitation (where the calculations in which T^{ij} is usually invoked—in my two books, for example—will have to be performed afresh). In view of these important implications, and not to seem immodest, let us say that I offer my equations as a working hypothesis. This task of critical revision is inevitable, however; it is now quite inadmissible to use the dynamics of invariant proper mass.

[1] M. von Laue: *Théorie de la relativité*. Gauthier-Villars, 1924, chapter VII; 5th German edition, Vieweg, Braunschweig, 1951.
[2] H. Arzeliès: *Electricité*. Gauthier-Villars, 1963, pp. 404–415.
[3] H. Arzeliès: *Rayonnement et dynamique du point chargé fortement accéléré*. Gauthier-Villars, 1966, pp. 28–52 and appendix III.

[4] R. C. Tolman: *Relativity, Thermodynamics and Cosmology*. Oxford, Clarendon Press, 1934 and 1950, chapter III.
[5] C. Møller: *Theory of Relativity*. Oxford, Clarendon Press, 1952, chapter VI.
[6] J. L. Synge: *Relativity, Special Theory*. North Holland, 1956, chapter VIII.
[7] A. Lichnérowicz: *Eléments de calcul tensoriel*. A. Colin, 1950, p. 191.
[8] H. Arzeliès: Sur le problème relativiste du levier coudé. *Nuovo Cimento* **35** (1965) 783–791.
[9] A. Lichnérowicz: *Théories relativistes de la gravitation et de l'électromagnétisme*. Masson, 1955, p. 11 and chapter IV.
[10] O. Costa de Beauregard: *La Théorie de la relativité restreinte*. Masson, 1949, chapter IV.
[11] H. Arzeliès: Sur l'équation fondamentale. *C. R. Acad. Sci. Paris*, **264** (1967) 161–162; Forces de tension dans un fluide en mouvement; retour sur l'équation fondamentale. *Ibid.*, **267A** (1968) 134–135; Sur les hypothèses minimales de l'hydrodynamique relativiste. *Ibid.*, **268A** (1969) 1355–1358; Equations de transfert et fluides relativistes; équations macroscopiques du mouvement. *Ibid.*, **271A** (1970) 442–444; Sur la mécanique des milieux continus. *Nuovo Cimento* **50** (1967) 287–291.
[12] H. Arzeliès: *Fluides relativistes*. Masson, Paris, 1971.

[189] On Currents in Translational Motion in an Electric Field

In the case of Fig. 46, von Laue [1] offered an analysis based on the same ideas as his theory of the right-angled lever (§ [186]). In the 1957 edition of this book, [2], I too adopted this standpoint, and proceeded as follows.

"In the system K_0, the loop tends to rotate about Oy_0, in such a way as to bring its axis towards the field $\mathfrak{B}_0$. In K, therefore, the plane of the circle is seen to move towards the plane formed by the electric field and the relative velocity: this is a paradoxical result, for there is no couple. As von Laue points out, this is the reverse of what happens with a right-angled lever where the couple produces no motion. Once again, the explanation involves the inertial energy. Let us picture a current of zero charge as a negatively charged ring rotating inside a stationary torus which is positively charged to the same charge density. The stationary ring will experience no couple arising from the field $\mathbf{E}$; for the moving ring, however, the velocities of half the charges have a component in the same direction as $\mathbf{E}$ while the velocities of the other half have a component in the opposite direction, at any given time. For the former, the charges supply energy to the ring, and conversely for the latter. From symmetry considerations, therefore, we have an energy current lying in the plane of the ring and perpendicular to $\mathbf{E}$, and hence a momentum in the same direction. This momentum is perpendicular to the velocity; the ring in uniform motion therefore possesses a transverse momentum, which requires a couple. Since this couple is non-existent, uniform motion is impossible and the ring begins to rotate."

I now reject this interpretation due to von Laue. We first notice that the statement "... a paradoxical result, for there is no couple" is somewhat hasty. There seems to be no couple because the usual laws giving the action of a field on a current have been applied to a moving current. According to these laws, the field $\mathbf{E}$ produces no couple, but they are valid only for stationary currents. The theory of von Laue employs the concept of energy currents, which I also reject. § [181] has shown that this tendency to rotate can be explained in a very different way.

[1] M. von Laue: ref. [2] of Appendix I, p. 261.
[2] H. Arzeliès: ref. [32] of Appendix I, pp. 241–2.

[190] On the Attempts to Display the Absolute Translational Motion of the Earth by Means of Electromagnetic Experiments

(a) *The experiments of Trouton and Noble*

The arrangement described in § [182], which we have treated as an exercise in relativistic dynamics, played a rather important historical role. We saw in *Kinematics* (§§ [28–30]) how the physicists at the end of the last century strove to display the translational motion of the earth with respect to the absolute system of the aether. We have described the fundamental second order experiment (of Michelson) involving light. It was natural that attempts of a similar nature, depending on electromagnetic phenomena, should be made. Fitzgerald [1] suggested that one might investigate experimentally whether electrical measurements are affected by the motion of the earth.

A second-order experiment was performed by Trouton and Noble [2, 3]; see Fig. 54. With the aether theory, the argument ran as follows. The condenser is at rest on the earth and hence in uniform translational motion through the aether. Thus in the aether, each plate creates an electric field and a magnetic field which act on the other plate; this yields the couple C', calculated in § [182]. As Newtonian dynamics was applied at that time, a rotation was anticipated as a consequence of this couple. The condenser, charged by a Wimshurst machine at about 2000 V, was suspended from a very fine bronze thread, 37 cm long. Observations were made at various times of day, and hence for various values of the angle α.

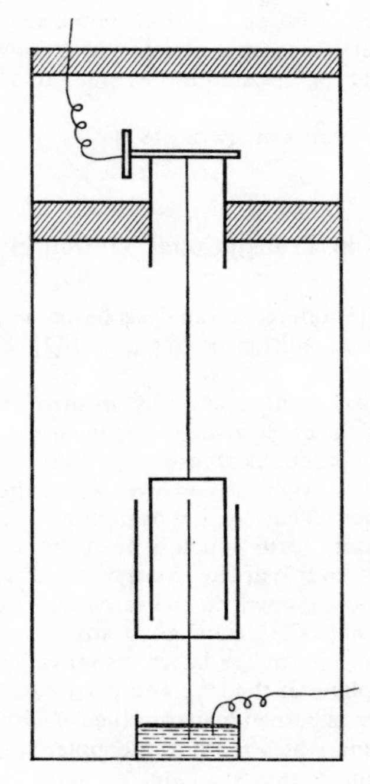

FIG. 54.

Calculation predicted a shift of the spot between 3 and 6 cm. The observed shifts never exceeded 0·35 cm, or 5%, and the authors therefore concluded that "there is no doubt that the result is a purely negative one". The relativistic interpretation is obvious. Within the approximation of the experiment, the Earth's frame of reference is a Galileian reference frame; the condenser is at rest in it, and hence experiences no couple. If we argue with respect to the solar system of reference (the old aether system), there should again be no rotation according to the principle of relativity. In relativistic dynamics, we must distinguish between the couples C and C'; there is certainly a couple C', but it produces no rotation.

A theoretical analysis of this negative result was made by H. A. Lorentz [4], and later, in relativistic terms, by von Laue [5, 6]; the latter gives a very full calculation of the couple, allowing for edge-effects at the plates. He only discusses the couple C', however, and introduces his concept of an energy current. Until the publication of the present book, all relativistic authors have adopted this interpretation (including myself, in the 1957 edition, p. 231).

(b) *Chase's experiment*

This author repeated the measurements [7, 9] more accurately, taking into account Epstein's remarks [8] about the influence of the angle between the vertical and the translational velocity on the sensitivity. He could have detected an aether wind of 3 km/s; the result was negative.

(c) *Tomaschek's experiments*

This physicist repeated the experiments at different altitudes: 120, 570 and 3457 metres (in the Jungfrau).

Apart from the two hypotheses mentioned in *Kinematics* (§ [28]) concerning the aether drag by the earth (total or not at all), some physicists (Lénard, for example) believed that the aether was wholly dragged only in the neighbourhood of the Earth. Further away they imagined it to be at rest (with respect to the distant stars), rather like a sphere travelling through a liquid. This explained the negative result at the surface; but one ought to obtain positive results at a great altitude. The results were negative, to within 3 km/s of aether wind. This could, of course, be explained by saying that the reduction of the drag was not great enough to be observable; this point of view was in contradiction with the positive optical results obtained by Miller at a moderate altitude (1734 m). This is certainly a proof that the weak positive results obtained in certain cases should be interpreted otherwise than as a denial of the principle of relativity.

[1] G. F. FITZGERALD: On electromagnetic effects due to the motion of the earth. *Sci. Trans. Royal Dublin Soc.* **1** (1882) 319–324; see also *Scientific Writings*, p. 557.

[2] F. T. TROUTON and H. R. NOBLE: The forces acting on a charged condenser moving through space. *Proc. Roy. Soc. London* **72** (1904) 132–133.

[3] F. T. TROUTON and H. R. NOBLE: The mechanical forces acting on a charged electric condenser moving through space. *Phil. Trans. Roy. Soc. London* **202A** (1904) 165–181.

[4] H. A. LORENTZ: Electromagnetic phenomena in a system moving with any velocity smaller than that of light. *Proc. Akad. Amsterdam* **6** (1904) 809–831.

[5] M. VON LAUE: Zur Theorie des Versuches von Trouton und Noble. *Ann. Physik* **38** (1912) 370–384.

[6] M. VON LAUE: ref. [2] of Appendix I.

[7] C. P. CHASE: A repetition of the Trouton–Noble ether drift experiment. *Phys. Rev.* **28** (1926) 378–383.

[8] P. S. EPSTEIN: Remark on the theory of the Trouton and Noble experiment. *Phys. Rev.* **29** (1927) 753.

[9] C. P. CHASE: The Trouton–Noble ether-drift experiment. *Phys. Rev.* **30** (1927) 516–519.

[10] R. TOMASCHEK: Ueber Versuche zur Auffindung elektrodynamischer Wirkungen der Erdbewegung in grossen Höhen. *Ann. Physik* **78** (1925) 743–756; *ibid.* **80** (1926) 509–514; *ibid.* **84** (1927) 161–162

CHAPTER XIV

RELATIVISTIC ROCKETS

A. GENERAL EQUATIONS; FUEL JET PROPELLED ROCKETS

[191] The Equations of Motion

(a) *No external force*. At time t, the rocket has mass m and velocity $\mathbf{v}$ and the ejected matter has velocity $\mathbf{w}$ ($\mathbf{w}$ is in the opposite direction to $\mathbf{v}$ for propulsion). At time $t+\mathrm{d}t$, the rocket has lost mass $\mathrm{d}m$; we count mass lost negative. Let $\mathrm{d}m'$ be the mass of the matter ejected when it is outside the rocket, and assume that there is no interaction between m and $\mathrm{d}m$. We write down the condition that *the total energy is the same at times t and $t+\mathrm{d}t$*:

$$\mathrm{d}m+\mathrm{d}m' = 0 \tag{1}$$

The total momentum is

$$m\mathbf{v}$$

at time t and

$$(m+\mathrm{d}m)(\mathbf{v}+\mathrm{d}\mathbf{v})+\mathbf{w}\,\mathrm{d}m'$$

at time $t+\mathrm{d}t$.

We write down the condition for *conservation of total momentum:*

$$m\mathbf{v} = (m+\mathrm{d}m)(\mathbf{v}+\mathrm{d}\mathbf{v})+\mathbf{w}\,\mathrm{d}m'$$

or

$$\begin{aligned} m\,\mathrm{d}\mathbf{v}+\mathbf{v}\,\mathrm{d}m+\mathbf{w}\,\mathrm{d}m' &= 0 \\ \mathrm{d}(m\mathbf{v})+\mathbf{w}\,\mathrm{d}m' &= 0 \end{aligned} \tag{2}$$

Using (1)

$$\begin{aligned} \mathrm{d}(m\mathbf{v}) &= \mathbf{w}\,\mathrm{d}m \\ \frac{\mathrm{d}}{\mathrm{d}t}(m\mathbf{v}) &= \mathbf{w}\frac{\mathrm{d}m}{\mathrm{d}t} \end{aligned} \tag{3}$$

We denote the proper mass of the rocket by m_0, so that

$$\mathrm{d}m = \mathrm{d}\left(\frac{m_0}{\sqrt{1-v^2/c^2}}\right)$$

We must be careful not to consider only the term $\mathrm{d}m_0\big/\sqrt{1-\beta^2}$; the total variation $\mathrm{d}m$

consists of the variation dm_0 and the variation dv. The equation of motion becomes

$$\frac{d}{dt}\left(\frac{m_0\mathbf{v}}{\sqrt{1-v^2/c^2}}\right)=\mathbf{w}\frac{d}{dt}\left(\frac{m_0}{\sqrt{1-v^2/c^2}}\right) \tag{4}$$

Expanding,

$$(\mathbf{v}-\mathbf{w})\frac{d}{dt}\left(\frac{m_0}{\sqrt{1-v^2/c^2}}\right)+\frac{m_0}{\sqrt{1-v^2/c^2}}\frac{d\mathbf{v}}{dt}=0$$

$$(\mathbf{v}-\mathbf{w})\left\{\frac{dm_0}{dt\sqrt{1-v^2/c^2}}+\frac{m_0 v\,dv}{c^2\,dt(1-v^2/c^2)^{3/2}}\right\}+\frac{m_0}{\sqrt{1-v^2/c^2}}\frac{d\mathbf{v}}{dt}=0 \tag{5}$$

We introduce the proper time of the rocket into the term in dm_0:

$$dt_0=dt\sqrt{1-\beta^2}, \qquad \beta=v/c$$

and the equation becomes

$$(\mathbf{v}-\mathbf{w})\left\{\frac{dm_0}{dt_0}+\frac{m_0 v}{c^2(1-\beta^2)^{3/2}}\frac{dv}{dt}\right\}+\frac{m_0}{\sqrt{1-\beta^2}}\frac{d\mathbf{v}}{dt}=0 \tag{6}$$

(b) *With external forces.* We now add a term $\mathbf{F}_e$ on the right-hand side:

$$\frac{d}{dt}\left(\frac{m_0\mathbf{v}}{\sqrt{1-\beta^2}}\right)=\mathbf{w}\frac{d}{dt}\left(\frac{m_0}{\sqrt{1-\beta^2}}\right)+\mathbf{F}_e$$

$$=\mathbf{w}\frac{dm}{dt}+\mathbf{F}_e \tag{7}$$

The force $\mathbf{F}_e$ may depend on the mass m (gravity, for example).

(c) Comparing (7) with the general equation of § [157], we obtain for the total force

$$\mathbf{F}=\mathbf{F}_e-\mathbf{w}\frac{dm'}{dt}=\mathbf{F}_e+\mathbf{w}\frac{dm}{dt}$$

Even in the absence of $\mathbf{F}_e$, there is now a force $\mathbf{F}$ in K_0. With the $\mathbf{E}$, $\mathfrak{B}$ notation, this gives

$$\mathbf{F}=\varepsilon\left(\mathbf{E}+\frac{\mathbf{v}}{c_0}\times\mathfrak{B}\right)+\mathbf{v}\frac{dm_0}{dt_0}$$

$$=\mathbf{F}_e+\mathbf{w}\frac{d}{dt}\left(\frac{m_0}{\sqrt{1-\beta^2}}\right)$$

and hence

$$\varepsilon\left(\mathbf{E}+\frac{\mathbf{v}}{c_0}\times\mathfrak{B}\right)=\mathbf{F}_e+\mathbf{w}\frac{d}{dt}\left(\frac{m_0}{\sqrt{1-\beta^2}}\right)-\mathbf{v}\frac{dm_0}{dt\sqrt{1-\beta^2}}$$

(d) For the rocket, and in the absence of an external force, the energy theorem is written (§ [159])

$$dW=\mathbf{F}\cdot\mathbf{v}\,dt+c^2\sqrt{1-\beta^2}\,dm_0$$

with

$$\mathbf{F} = \mathbf{w}\frac{\mathrm{d}m}{\mathrm{d}t}$$

whence

$$\frac{m_0\,\mathrm{d}v}{1-v^2/c^2}\left(1-\frac{\mathbf{v}\,.\,\mathbf{w}}{c^2}\right)+\mathrm{d}m_0\left(v-\frac{\mathbf{v}\,.\,\mathbf{w}}{v}\right) = 0$$

Two special cases are of interest, when $\mathbf{v}$ and $\mathbf{w}$ are perpendicular or parallel.

(e) *Four-dimensional formalism.* Let V_i be the four-velocity of the rocket and W_i that of the matter ejected. The four-impulses are thus

at time t:	rocket	$m_0 V_i$
at time $t+\mathrm{d}t$:	rocket	$(m_0+\mathrm{d}m_0)(V_i+\mathrm{d}V_i)$
	ejected matter $\mathrm{d}m_0' W_i$	

The change of impulse is therefore

$$\mathrm{d}m_0(V_i+\mathrm{d}V_i)+m_0\,\mathrm{d}V_i+\mathrm{d}m_0' W_i = \mathrm{d}(m_0 V_i)+\mathrm{d}m_0' W_i$$

since $\mathrm{d}m_0\,\mathrm{d}V_i$ is a higher order term. Setting this variation equal to zero, we recover the equations of motion and energy.

[192] The Energy Conversion Factor, ε

(a) In § [191] we have stressed the fact that the conservation of matter must be written

$$\mathrm{d}\left(\frac{m_0}{\sqrt{1-v^2/c^2}}\right)+\frac{\mathrm{d}m_0'}{\sqrt{1-w^2/c^2}} = 0 \tag{1}$$

with

$$\mathrm{d}m' = \frac{\mathrm{d}m_0'}{\sqrt{1-w^2/c^2}}$$

and *not*

$$\frac{\mathrm{d}m_0}{\sqrt{1-v^2/c^2}}+\frac{\mathrm{d}m_0'}{\sqrt{1-w^2/c^2}} = 0$$

Furthermore, we recall that $\mathrm{d}m_0$ is negative for a rocket, with our present conventions. Expanding (1), we find

$$\frac{-\mathrm{d}m_0}{\sqrt{1-v^2/c^2}} = m_0\,\mathrm{d}\left(\frac{1}{\sqrt{1-v^2/c^2}}\right)+\frac{\mathrm{d}m_0'}{\sqrt{1-w^2/c^2}} \tag{2}$$

In the system of reference K_0 in which $v = 0$, this equation simplifies to

$$-\mathrm{d}m_0 = \frac{\mathrm{d}m_0'}{\sqrt{1-w_0^2/c^2}} \tag{3}$$

for in this system, the second term vanishes, thus:

$$\mathrm{d}\left(\frac{1}{\sqrt{1-v^2/c^2}}\right) = \frac{v\,\mathrm{d}v}{c^2(1-v^2/c^2)^{3/2}} = 0$$

Equation (2) can be interpreted by saying that the energy corresponding to the proper mass lost (fuel burnt) reappears as an increase in the kinetic energy of the rocket and as the kinetic energy of the matter ejected with velocity w:

energy consumed = increase of kinetic energy of vehicle + energy ejected

In K_0 [equation (3)], the change of kinetic energy of the vehicle is zero. The energy $c^2\, dm_0$ is recovered wholly as the kinetic energy of the ejected gases.

The vital point to be made clear is that $dm_0' \neq dm_0$; *proper mass is not conserved.*

(b) Let us set

$$dm_0' = -dm_0(1-\varepsilon), \qquad -dm_0 = \frac{dm_0'}{1-\varepsilon}$$

We may then say that $-\varepsilon\, dm_0$ is the "useful" part of the fuel, the part that is transformed into kinetic energy of the vehicle. We shall call ε the *energy conversion factor (coefficient de conversion, Nutzfaktor).*

Substituting for dm_0 from (3), we obtain (with Sänger)

$$1 = \frac{1-\varepsilon}{\sqrt{1-w_0^2/c^2}}$$

or

$$\left|\frac{w_0}{c}\right| = \sqrt{1-(1-\varepsilon)^2} = \sqrt{\varepsilon(2-\varepsilon)}$$

[193] Mass Ratios

We shall use the following notation:

m_{01} = total proper mass (vehicle + fuel) initially;
m_0 = total proper mass at time t;
$\delta = \dfrac{m_{01}}{m_0}$ = mass ratio.

It is convenient in some applications to distinguish between the mass of the vehicle m_0^v (constant) and that of the fuel (m_0^f);

$$m_0 = m_0^v + m_0^f$$

We define a coefficient

$$\chi = \frac{m_{01}^v}{m_{01}^f}$$

We then have

$$m_{01} = m_{01}^f(1+\chi), \qquad \delta = \frac{1+\chi}{\chi}$$

[194] Extension of the Formalism of § [175] to the Present Case, Regarding the External Force as the Only Force

(a) *Longitudinal and transverse masses*

We denote the unit vectors along the tangent and principal normal to the trajectory by **t** and **n** respectively, and the radius of curvature by R; then

$$\mathbf{F}_e = \frac{\mathrm{d}}{\mathrm{d}t}\left(\frac{m_0 v\mathbf{t}}{\sqrt{1-\beta^2}}\right) - \mathbf{w}\frac{\mathrm{d}}{\mathrm{d}t}\left(\frac{m_0}{\sqrt{1-\beta^2}}\right)$$
$$= \frac{m_0\mathbf{t}}{(1-\beta^2)^{3/2}}\frac{\mathrm{d}v}{\mathrm{d}t} + \frac{m_0\mathbf{n}}{\sqrt{1-\beta^2}}\frac{v^2}{R} + \frac{\mathbf{v}\,\mathrm{d}m_0}{\mathrm{d}t\sqrt{1-\beta^2}} - \mathbf{w}\frac{\mathrm{d}}{\mathrm{d}t}\left(\frac{m_0}{\sqrt{1-\beta^2}}\right) \tag{1}$$

Writing

$$m_t = \frac{m_0}{(1-\beta^2)^{3/2}}, \qquad m_n = \frac{m_0}{\sqrt{1-\beta^2}}$$

we have

$$\mathbf{F}_e = m_t\boldsymbol{\gamma}_t + m_n\boldsymbol{\gamma}_n + \frac{\mathbf{v}\,\mathrm{d}m_0}{\mathrm{d}t\sqrt{1-\beta^2}} - \mathbf{w}\frac{\mathrm{d}}{\mathrm{d}t}\left(\frac{m_0}{\sqrt{1-\beta^2}}\right)$$

(b) *The energy theorem*

We scalar multiply the relation by $\mathbf{v}\,\mathrm{d}t$:

$$\mathbf{F}_e\,.\,\mathbf{v}\,\mathrm{d}t = \frac{m_0 v\,\mathrm{d}v}{(1-\beta^2)^{3/2}} + \frac{v^2\,\mathrm{d}m_0}{\sqrt{1-\beta^2}} - \mathbf{v}\,.\,\mathbf{w}\,\mathrm{d}\left(\frac{m_0}{\sqrt{1-\beta^2}}\right)$$
$$= \mathrm{d}\left(\frac{m_0c^2}{\sqrt{1-\beta^2}}\right) - c^2\frac{\mathrm{d}m_0}{\sqrt{1-\beta^2}} + \frac{v^2\,\mathrm{d}m_0}{\sqrt{1-\beta^2}} - \mathbf{v}\,.\,\mathbf{w}\,\mathrm{d}\left(\frac{m_0}{\sqrt{1-\beta^2}}\right)$$

Retaining the usual definition of the work $\mathrm{d}\mathcal{T}_e$ and of the energy $\mathrm{d}W$, the energy equation becomes

$$\mathrm{d}\mathcal{T}_e = \left(1-\frac{\mathbf{v}\,.\,\mathbf{w}}{c^2}\right)\mathrm{d}W - c^2\,\mathrm{d}m_0\sqrt{1-\beta^2}$$

(c) *Transformation of the force*

Passage from K_0 to K. Taking the axis Ox parallel to the velocity in K at the time considered, and writing the components of $\mathbf{F}_e$ as $X_e, \ldots$, we have

$$X_e = \frac{m_0\gamma_x}{(1-\beta^2)^{3/2}} + \frac{v\,\mathrm{d}m_0}{\mathrm{d}t\sqrt{1-\beta^2}} - w_x\frac{\mathrm{d}}{\mathrm{d}t}\left(\frac{m_0}{\sqrt{1-\beta^2}}\right)$$

$$Y_e = \frac{m_0\gamma_y}{\sqrt{1-\beta^2}} - w_y\frac{\mathrm{d}}{\mathrm{d}t}\left(\frac{m_0}{\sqrt{1-\beta^2}}\right)$$

In K_0

$$X_{0e} = m_0\gamma_{0x} - w_{0x}\frac{\mathrm{d}m_0}{\mathrm{d}t_0}$$

$$Y_{0e} = m_0\gamma_{0y} - w_{0y}\frac{\mathrm{d}m_0}{\mathrm{d}t_0}$$

From kinematics,

$$\gamma_{0x} = \frac{\gamma_x}{(1-\beta^2)^{3/2}}, \qquad \gamma_{0y} = \gamma_y$$

$$w_{0x} = \frac{w_x - v}{1-\dfrac{vw_x}{c^2}}, \qquad w_{0y} = \frac{w_y\sqrt{1-\beta^2}}{1-\dfrac{vw_x}{c^2}}$$

whence

$$X_{0e} = \frac{m_0\gamma_x}{(1-\beta^2)^{3/2}} - \frac{w_x - v}{1-\frac{vw_x}{c^2}} \frac{dm_0}{dt_0}$$

$$Y_{0e} = m_0\gamma_y - \frac{w_y\sqrt{1-\beta^2}}{1-\frac{vw_x}{c^2}} \frac{dm_0}{dt_0}$$

Hence

$$X_e \neq X_{0e}, \qquad Y_e \neq Y_{0e}\sqrt{1-\beta^2}$$

As we mentioned in § [177], the advantage of this formalism, which was that the usual transformations were retained, disappears here.

B. RECTILINEAR TRAJECTORIES (FUEL JET PROPELLED ROCKETS)

[195] Position and Velocity as Functions of the Proper Mass

a) *Ackerett's differential equation*

As an exercise, we perform the calculation of § [191] directly; we set out from

$$d(mv) + w\, dm' = 0, \qquad dm + dm' = 0 \tag{1}$$

and hence

$$m\, dv + (v-w)\, dm = 0 \tag{2}$$

We transform this so as to introduce the given quantities m_0 and w_0; first of all, we have

$$m = \frac{m_0}{\sqrt{1-v^2/c^2}}, \qquad dm = \frac{dm_0}{\sqrt{1-v^2/c^2}} + \frac{m_0}{c^2} \frac{v\, dv}{(1-v^2/c^2)^{3/2}}$$

Substituting into (2),

$$m_0\, dv + (v-w)\left(dm_0 + \frac{m_0}{c^2} \frac{v\, dv}{1-v^2/c^2}\right) = 0$$

or

$$\frac{m_0\, dv}{1-v^2/c^2}\left(1-\frac{vw}{c^2}\right) + (v-w)\, dm_0 = 0$$

This equation can also be extracted as a special case of equation (5) of § [191]. The velocity transformation gives

$$w_0 = \frac{w-v}{1-\frac{vw}{c^2}}$$

so that

$$m_0\, dv - \left(1-\frac{v^2}{c^2}\right) w_0\, dm_0 = 0 \tag{3}$$

Remark. w_0 is here taken positive in the same direction as v, which accounts for the difference in sign from the French edition, Vol. II, § [240].

(b) *The velocity of the rocket as a function of its mass for constant proper ejection velocity*

Let m_{01} and m_{02}, and v_1 and v_2, be the initial and final values of m_0 and v, respectively. integrating Ackerett's equation, we find

$$\frac{dv}{1-v^2/c^2} = w_0 \frac{dm_0}{m_0}$$

$$\frac{c}{2}\left[\log \frac{1+v/c}{1-v/c}\right]_{v_1}^{v_2} = w_0 \log \frac{m_{02}}{m_{01}}$$

$$\frac{c+v_2}{c-v_2}\,\frac{c-v_1}{c+v_1} = \left(\frac{m_{02}}{m_{01}}\right)^{2w_0/c}$$

Considering the propulsion case, and with zero initial velocity, we find

$$\frac{c+v_2}{c-v_2} = \left(\frac{m_{02}}{m_{01}}\right)^{2w_0/c}$$

$$\therefore \quad v_2 = c\,\frac{\left(\dfrac{m_{02}}{m_{01}}\right)^{2w_0/c} - 1}{\left(\dfrac{m_{02}}{m_{01}}\right)^{2w_0/c} + 1}$$

In this case, it would naturally be more suitable to count w_0 positively in the opposite direction to v_2; this would change the sign of the index. The final velocity depends only on the ratio of the masses; the behaviour of dm_0/dt_0 is not involved.

For low velocities,

$$v_2 \simeq w_0 \log \frac{m_{02}}{m_{01}}$$

We may also solve for the mass ratio:

$$\frac{m_{02}}{m_{01}} = \left(\frac{c+v_2}{c-v_2}\,\frac{c-v_1}{c+v_1}\right)^{c/2w_0}$$

and if v_1 is zero,

$$\frac{m_{02}}{m_{01}} = \left(\frac{c+v_2}{c-v_2}\right)^{c/2w_0}$$

Finally, if the rocket departs from rest with $m_0 = m_{01}$, its velocity v and proper mass m_0 at some arbitrary subsequent time are connected by the relations (where $w_0 < 0$)

$$\frac{m_0}{m_{01}} = \left(\frac{c+v}{c-v}\right)^{c/2w_0}$$

$$\frac{v}{c} = \frac{\left(\dfrac{m_0}{m_{01}}\right)^{2w_0/c} - 1}{\left(\dfrac{m_0}{m_{01}}\right)^{2w_0/c} + 1}$$

Figure 55 represents $\frac{v}{c}\left(\frac{m_0}{m_{01}}\right)$ for various ejection velocities. The continuous curves represent the relativistic formulae, the dotted curves, the non-relativistic formulae. The two forms of dynamics give indistinguishable results for ejection velocities smaller than $c/10$.

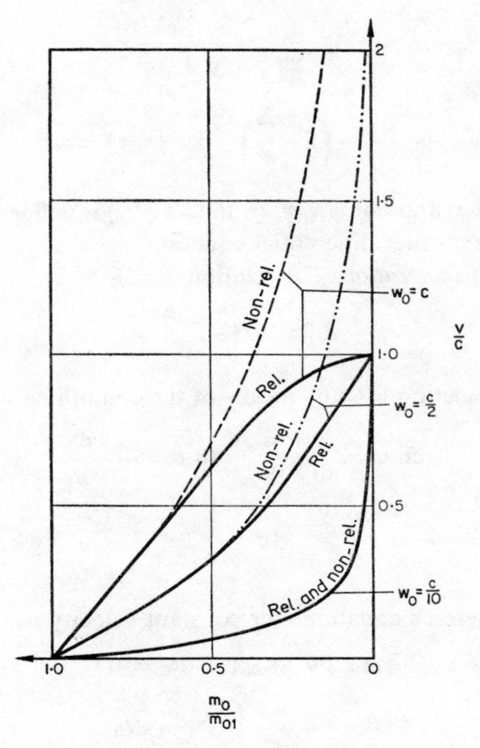

FIG. 55. (After Ackerett)

Using the notation of § [193], we have

$$\frac{v}{c} = \frac{\delta e^{-2w_0/c}-1}{\delta e^{-2w_0/c}+1} = \frac{\left(\dfrac{1+\chi}{\chi}\right)^{2\sqrt{\varepsilon(2-\varepsilon)}}-1}{\left(\dfrac{1+\chi}{\chi}\right)^{2\sqrt{\varepsilon(2-\varepsilon)}}+1}$$

[196] Position and Velocity as Functions of Time

(a) Let the gas be ejected according to

$$\frac{dm_0}{dt_0} = \varrho(t_0) \tag{1}$$

To integrate Ackerett's equation, we divide by dt:

$$m_0 \frac{dv}{dt} - \left(1 - \frac{v^2}{c^2}\right) w_0 \frac{dm_0}{dt_0} \frac{dt_0}{dt} = 0$$

where

$$t_0 = \frac{t - \frac{v}{c^2}x}{\sqrt{1-v^2/c^2}}; \qquad \frac{dt_0}{dt} = \frac{1 - \frac{v}{c^2}\frac{dx}{dt}}{\sqrt{1-v^2/c^2}} = \sqrt{1-v^2/c^2}$$

Hence

$$m_0 \frac{dv}{dt} - \left(1 - \frac{v^2}{c^2}\right)^{3/2} w_0 \varrho(v, t) = 0$$

We have a function $\varrho(v, t)$ and not $\varrho(x, v, t)$; in fact $\varrho(t_0)$ is defined with dt_0.

We thus obtain a first order differential equation.

(b) *Bade's method of integration.* The relation

$$dt_0^2 = dt^2 - \frac{dx^2}{c^2}$$

enables us to define a function $\theta(t_0)$ by means of the conditions

$$\text{ch}\,\theta = \frac{dt}{dt_0}, \qquad \text{sh}\,\theta = \frac{1}{c}\frac{dx}{dt_0}$$

so that

$$v = \frac{dx}{dt} = c\,\text{th}\,\theta$$

We substitute into Ackerett's equation; for constant velocity w_0,

$$m_0 c\, d\theta - w_0 \varrho(t_0)\, dt_0 = 0$$

$$\theta(t_0) = \frac{w_0}{c}\int_0^{t_0} \frac{\varrho(t_0)}{m_0}\, dt_0 \tag{2}$$

in which

$$m_0 = m_{01} + \int_0^{t_0} \varrho(t_0)\, dt_0$$

Hence

$$t = \int_0^{t_0} \text{ch}\,\theta\, dt_0, \qquad x = c\int_0^{t_0} \text{sh}\,\theta\, dt_0 \tag{3}$$

These are the equations of motion in terms of the parameter t_0.

(c) *Kooy's method of integration.* We start from equation (3) of § [195],

$$\frac{dv}{1 - v^2/c^2} = w_0 \frac{dm_0}{m_0} \tag{4}$$

With μ_0 denoting a *constant*, we write

$$m_0 = m_{01} - \mu_0 t_0, \qquad dm_0 = -\mu_0\, dt_0$$

$$\frac{dv}{1 - v^2/c^2} = -\frac{w_0 \mu_0\, dt_0}{m_{01} - \mu_0 t_0} \tag{5}$$

By integration, we have found

$$\frac{m_{01}-\mu_0 t_0}{m_{01}} = \left(\frac{c+v}{c-v}\right)^{c/w_0} \tag{6}$$

In (4), we replace dt_0 by $dt\sqrt{1-\beta^2}$ and $m_{01}-\mu_0 t_0$ by the expression for it obtained from (6); we obtain

$$\frac{dv}{dt}\frac{m_{01}-\mu_0 t_0}{(1-v^2/c^2)^{3/2}} = -\mu_0 w_0$$

and so

$$\frac{m_{01}}{(1-v^2/c^2)^{3/2}}\left(\frac{1+v/c}{1-v/c}\right)^{c/2w_0}\frac{dv}{dt} = -\mu_0 w_0$$

Thus, *if μ_0 and w_0 are constants,*

$$t = -\frac{m_{01}}{\mu_0 w_0}\int_0^v \left(\frac{1+v/c}{1-v/c}\right)^{c/2w_0}\frac{dv}{(1-v^2/c^2)^{3/2}}$$

This is the time required to attain a velocity v; $v(t)$ and the position $s(t)$ can thus be obtained numerically.

If μ_0 and w_0 are not constants, we must write

$$dm_0 = \mu_0(t_0)\,dt_0, \quad m_0 = m_{01} - \int_0^{t_0}\mu_0(t_0)\,dt_0$$

so that

$$\frac{dv}{1-v^2/c^2} = -\frac{\mu_0(t_0)w_0(t_0)\,dt_0}{m_{01}-\int_0^{t_0}\mu_0(t_0)\,dt_0}$$

Integrating,

$$c\int_0^v \frac{d(v/c)}{1-v^2/c^2} = -\int_0^{t_0}\frac{\mu_0(t_0)w_0(t_0)\,dt_0}{m_{01}-\int_0^{t_0}\mu_0(t_0)\,dt_0}$$

$$= \phi(t_0)$$

or

$$\frac{c}{2}\log\frac{1-v/c}{1+v/c} = \phi(t_0)$$

This relation gives $v(t_0)$. We then have

$$dt = \frac{dt_0}{\sqrt{1-\dfrac{v^2(t_0)}{c^2}}}, \qquad t = \int_0^{t_0}\frac{dt_0}{\sqrt{1-\dfrac{v^2(t_0)}{c^2}}}$$

This gives the dependence $t(t_0)$ or $t_0(t)$, and $v(t)$ and $s(t)$ can be derived.

If w_0 is constant,

$$\frac{c}{2}\log\frac{1-v/c}{1+v/c} = -w_0\int_0^{t_0}\frac{\mu_0(t_0)\,\mathrm{d}t_0}{m_{01}-\int_0^{t_0}\mu_0(t_0)\,\mathrm{d}t_0}$$

$$= +w_0\int_{m_{01}}^{m_0}\frac{\mathrm{d}m_0(t_0)}{m_0(t_0)} = w_0\log\frac{m_{01}}{m_0}$$

$$\left(\frac{1-v/c}{1+v/c}\right)^{c/w_0} = \frac{m_0}{m_{01}}$$

This is Ackerett's formula.

(d) *Marx's method of integration.* We employ the coefficient ε defined in § [192]; the conservation equations are of the form

$$\mathrm{d}\left(\frac{m_0}{\sqrt{1-\frac{v^2}{c^2}}}\right)-\frac{(1-\varepsilon)\,\mathrm{d}m_0}{\sqrt{1-\frac{w^2}{c^2}}} = 0$$

$$\mathrm{d}\left(\frac{m_0 v}{\sqrt{1-\frac{v^2}{c^2}}}\right)-\frac{(1-\varepsilon)\,\mathrm{d}m_0 w}{\sqrt{1-\frac{w^2}{c^2}}} = 0$$

or

$$\frac{\mathrm{d}}{\mathrm{d}m_0}\left(\frac{m_0}{\sqrt{1-\frac{v^2}{c^2}}}\right)-\frac{1-\varepsilon}{\sqrt{1-\frac{w^2}{c^2}}} = 0$$

$$\frac{\mathrm{d}}{\mathrm{d}m_0}\left(\frac{m_0 v}{\sqrt{1-\frac{v^2}{c^2}}}\right)-\frac{(1-\varepsilon)w}{\sqrt{1-\frac{w^2}{c^2}}} = 0$$

We write

$$x = \log m_0, \qquad \mathrm{d}x = \frac{\mathrm{d}m_0}{m_0}$$

$$\frac{1}{\sqrt{1-\frac{v^2}{c^2}}}+\frac{\mathrm{d}}{\mathrm{d}x}\frac{1}{\sqrt{1-\frac{v^2}{c^2}}}-\frac{1-\varepsilon}{\sqrt{1-\frac{w^2}{c^2}}} = 0$$

$$\frac{v}{\sqrt{1-\frac{v^2}{c^2}}}+\frac{\mathrm{d}}{\mathrm{d}x}\left(\frac{v}{\sqrt{1-\frac{v^2}{c^2}}}\right)-\frac{(1-\varepsilon)w}{\sqrt{1-\frac{w^2}{c^2}}} = 0$$

and then write

$$y = \frac{1}{\sqrt{1-\frac{v^2}{c^2}}}, \qquad 1-\frac{v^2}{c^2} = \frac{1}{y^2}$$

$$\frac{v}{c} = \frac{\sqrt{y^2-1}}{y}$$

We eliminate w between the two conservation equations; after some lengthy intermediate calculation, we obtain

$$\left(\frac{dy}{dx}\right)^2 = (y^2-1)\{1-(1-\varepsilon)^2\}$$

We write

$$z = \sqrt{1-(1-\varepsilon)^2}$$

and finally

$$z\,dx = \frac{dy}{\sqrt{y^2-1}}$$

Integrating,

$$\int_{\log m_{01}}^{\log m_0} z\,dx = \text{arc ch}\, y = \log\sqrt{\frac{1-v/c}{1+v/c}}$$

$$\frac{v}{c} = \frac{1-\exp 2\int_{\log m_{01}}^{\log m_0} z\,dx}{1+\exp 2\int_{\log m_{01}}^{\log m_0} z\,dx}$$

If z = constant,

$$\frac{v}{c} = \frac{1-\exp\left(2z\log\frac{m_0}{m_{01}}\right)}{1+\exp\left(2z\log\frac{m_0}{m_{01}}\right)} = \frac{1-\left(\frac{m_0}{m_{01}}\right)^{2z}}{1+\left(\frac{m_0}{m_{01}}\right)^{2z}}$$

Since $z = -w_0/c$, we thus recover Ackerett's formula.

[197] Energy Coefficients

(a) *Mechanical efficiency*

This is by definition the coefficient

$$\eta = \frac{\text{final kinetic energy of the useful mass}}{\text{energy supplied by the fuel}}$$

so that

$$\eta = \frac{\frac{m_0}{\sqrt{1-v^2/c^2}}-m_0}{\varepsilon(m_{01}-m_0)} = \frac{\frac{1}{\sqrt{1-v^2/c^2}}-1}{\varepsilon\left(\frac{m_{01}}{m_0}-1\right)}$$

$$= \frac{1}{\varepsilon}\,\frac{\frac{1}{\sqrt{1-v^2/c^2}}-1}{\left(\frac{c+v}{c-v}\right)^{c/2w_0}-1}$$

Writing

$$B = \frac{1}{\sqrt{1-v^2/c^2}} - 1$$

we obtain

$$\eta = \frac{1}{1-\sqrt{1-w_0^2/c^2}} \frac{B}{\{B+1+\sqrt{B(B+2)}\}^{-c/w_0}-1}$$

For low velocities,

$$\frac{m_0}{m_{01}} = e^{v/w_0}, \qquad \varepsilon = \frac{w_0^2}{2c^2}$$

The coefficient η then reduces to

$$\eta = \frac{\frac{1}{2}m_0v^2}{\varepsilon(m_{01}-m_0)} = \frac{\frac{1}{2}v^2}{\varepsilon(e^{-v/w_0}-1)} = \frac{\frac{v^2c^2}{w^2}}{e^{-v/w_0}-1}$$

For velocities in the neighbourhood of c (the so-called ultra-relativistic region),

$$B \gg 1$$

and

$$\eta = \frac{1}{\varepsilon} \frac{B}{(2B)^{-c/w_0}} = \frac{B^{\frac{w_0+c}{w_0}}}{2^{-c/w_0}\varepsilon}$$

(b) *The mass ratio for long journeys at constant velocity except for the initial launching and the braking on arrival; the Q coefficient*

Let $2x$ be the path for the round trip, t the corresponding time at constant velocity v and t_0 the proper time; we have

$$t = \frac{t_0}{\sqrt{1-v^2/c^2}} = \frac{2x}{v}$$

so that

$$\frac{v}{c} = \frac{1}{1+\left(\frac{ct_0}{2x}\right)^2}$$

Ackerett's mass ratio then becomes

$$\frac{m_{01}}{m_0} = \left(\frac{c-v}{c+v}\right)^{c/2w_0} = \left\{\frac{\sqrt{1+\left(\frac{ct_0}{2x}\right)^2}-1}{\sqrt{1+\left(\frac{ct_0}{2x}\right)^2}+1}\right\}^{2/cw_0}$$

If $v \simeq c$, $ct_0 \ll x$ and so

$$\frac{m_{01}}{m_0} \simeq \left(\frac{ct_0}{4x}\right)^{c/w_0}$$

Ackerett's coefficient corresponds to the change from zero velocity to velocity v. To allow for the braking on arrival, it must be squared. For the round trip, therefore, Ackerett's ratio must be raised to the fourth power.

It is convenient to use the coefficient

$$Q = \log_{10}\left(\frac{m_{01}}{m_0}\right)^4 \simeq \frac{4c}{w_0}\log_{10}\frac{ct_0}{4x}$$

We may also write

$$\frac{m_{01}}{m_0} = \frac{\text{mass at departure}}{\text{mass on return}} = 10^Q$$

C. PHOTON ROCKETS

[198] The Equations of Motion

The variation of proper mass is now a consequence of emission or absorption of photons, which are for the moment assumed to be monochromatic and of frequency ν. The velocity w used in this chapter is now equal to c in any reference system; we shall write it $\mathbf{c}$ to indicate direction. Let dn be the number of photons corresponding to the mass change dm. Conservation of energy gives

$$c^2\,dm + h\nu\,dn = 0 \tag{1}$$

and conservation of momentum

$$d(m\mathbf{v}) = -dn\frac{h\nu}{c^2}\mathbf{c} = \mathbf{c}\,dm \tag{2}$$

If there is an external force

$$\frac{d}{dt}(m\mathbf{v}) = -\frac{h\nu}{c^2}\frac{dn}{dt}\mathbf{c} + \mathbf{F}_e$$

In the system of reference of the rocket, the photon has frequency ν_0; we have

$$\nu = \frac{\sqrt{1-\beta^2}}{1-\beta\cos\delta}\nu_0, \qquad dt = \frac{dt_0}{\sqrt{1-\beta^2}}$$

If, as will be the case in practice, the photons are emitted in the opposite direction to the velocity v, we have

$$\delta = \pi, \qquad \cos\delta = -1$$

$$\nu = \frac{\sqrt{1-\beta^2}}{1+\beta}\nu_0 = \sqrt{\frac{1-\beta}{1+\beta}}\,\nu_0$$

The equation of motion then becomes

$$\frac{d}{dt}(m\mathbf{v}) = -\frac{h\mathbf{c}}{c^2}\frac{\sqrt{1-\beta^2}}{1+\beta}\frac{\nu_0\, dn\sqrt{1-\beta^2}}{dt_0}$$

$$= -\frac{dn}{dt_0}\frac{h\mathbf{c}}{c^2}\nu_0(1-\beta)+\mathbf{F}_e$$

Denoting the energy emitted per unit time by $\mathscr{E}_0 = dm_0/dt_0$, we may also write

$$\frac{d}{dt}(m\mathbf{v}) = -\frac{\mathscr{E}_0}{c^2}\mathbf{c}+\mathbf{F}_e$$

[199] Rectilinear Motion

(a) *The position and velocity as function of the proper mass*

We consider rectilinear motion directly. We may use Ackerett's equation (3), see §[195]:

$$\frac{dv}{1-v^2/c^2} = w_0\frac{dm_0}{m_0} \tag{1}$$

If we assume that *the energy emitted per unit time, $\mathscr{E}_0$, is constant*, we have

$$m_0 = m_{01}-\frac{\mathscr{E}_0}{c^2}t_0 \qquad dm_0 = -\frac{\mathscr{E}_0}{c^2}dt_0$$

Thus ($w_0 = c$)

$$\frac{dv}{1-v^2/c^2} = -\frac{\dfrac{\mathscr{E}_0}{c}dt_0}{m_{01}-\dfrac{\mathscr{E}_0}{c^2}t_0}$$

and

$$\left(\frac{1-v/c}{1+v/c}\right)^{1/2} = \frac{m_{01}}{m_{01}-\dfrac{\mathscr{E}_0}{c^2}t_0} = \frac{m_{01}}{m_0} \tag{2}$$

(b) *Position and velocity as a function of time* (Kooy)

We assume that $\mathscr{E}_0$ is constant. In equation (1), we replace dt_0 by $dt\sqrt{1-v^2/c^2}$:

$$\frac{m_{01}-\dfrac{\mathscr{E}_0}{c^2}t_0}{(1-v^2/c^2)^{3/2}}\frac{dv}{dt} = -\frac{\mathscr{E}_0}{c}$$

We replace $m_{01}-(\mathscr{E}_0/c^2)t_0$ by the value given by equation (2), and integrate:

$$t = \frac{cm_{01}}{\mathscr{E}_0}\int_0^v\left(\frac{1+v/c}{1-v/c}\right)^{1/2}\frac{dv}{(1-v^2/c^2)^{3/2}}$$

$$= \frac{cm_{01}}{\mathscr{E}_0}\int_0^v\frac{dv}{(1+v/c)(1-v/c)^2}$$

To complete the integration, we write

$$\frac{1}{(1+v/c)(1-v/c)^2} = \frac{A}{(1+v/c)} + \frac{B}{(1-v/c)} + \frac{C}{(1-v/c)^2}$$

$$A(1-v/c)^2 + B(1+v/c)(1-v/c) + C(1+v/c) = 1$$

$$A+B+C+(C-2A)\frac{v}{c}+(A-B)\frac{v^2}{c^2} = 1$$

This relation must be satisfied for all v, so that

$$A+B+C = 1, \qquad 2A-C = 0, \qquad A-B = 0$$

This gives

$$A = B = \tfrac{1}{4}, \qquad C = \tfrac{1}{2}$$

and hence

$$t = \frac{cm_{01}}{\mathcal{E}_0}\int_0^v \left\{\frac{1}{4(1+v/c)} + \frac{1}{4(1-v/c)} + \frac{1}{2(1-v/c)^2}\right\} dv$$

After integration, we obtain

$$t = \frac{cm_{01}}{\mathcal{E}_0}\left\{\frac{c}{4}\log\frac{1-v/c}{1+v/c} - \frac{v}{2(1-v/c)}\right\}$$

HISTORICAL AND BIBLIOGRAPHICAL NOTES

These are to be found at the end of Chapter XII.

PART THREE

Collision Theory

CHAPTER XV

ELASTIC COLLISIONS BETWEEN TWO CORPUSCLES

A. GENERAL REMARKS

[200] The Macroscopic Nature of the Theory

We consider two quasi-point particles of proper masses m_{01} and m_{02} in rectilinear motion which virtually coincide at time t_0. We assume that there is no interaction between the two particles except at the moment when they collide. As in pre-relativistic dynamics, of course, a "collision" does not necessarily imply "contact" in the everyday sense. For our arguments to be valid, we only require that the interaction last a time Δt and extend over a range Δr, negligible in comparison with macroscopic times and distances (occurring in the corresponding measurements).

This very simple limiting hypothesis is adequate to lead to important results. We shall, however, notice in passing that it is not capable of determining all the characteristics of the phenomena. To go further, we could retain the point nature of the corpuscles, but specify the dependence of the interaction on the distance; since the "collision" would no longer entail contact, we should also have to supplement the initial conditions. Notice that this method cannot be applied to microscopic particles, even if the law of force is correct, because it is impossible to specify the initial conditions. Quantum mechanics must be employed.

In many practical situations, the motion of the corpuscles is not uniform. With charged particles, for example, the trajectories are often curved because a magnetic field is necessary to take measurements. However, we shall always assume that the acceleration is small enough for the energy radiated to be negligible, in conformity with the approximations of the present work. For the case when this is not so, see my *Rayonnement et Dynamique*.

We shall of course also assume that the conditions which allow us to take the macroscopic point of view are fulfilled, that is, that the notion of trajectory can be retained.

[201] Elastic Collisions

(a) *Definition.* The collision is elastic if, after the collision, the corpuscles are separate, their proper masses are conserved and the total kinetic energy has remained constant. We thus adopt the classical definition. It follows immediately that the total energy in the relativistic sense also remains constant. We shall see that the conservation of total momentum is also a consequence of this definition.

Alternatively we could stipulate that proper mass and total momentum are conserved and deduce the other results.

During the interaction, the acceleration cannot be treated as negligible, and hence radiation is emitted at this moment. For the collision to be regarded as elastic, the energy lost in this way must be negligible.

(b) Let us demonstrate that conservation of kinetic energy implies conservation of momentum. Denoting the velocities of corpuscles 1 and 2 before and after the collision by a, b, α, β respectively (this notation is provisional, and is used only in this section), we write

$$m_{01}c^2\left(\frac{1}{\sqrt{1-a^2/c^2}}-1\right)+m_{02}c^2\left(\frac{1}{\sqrt{1-b^2/c^2}}-1\right)$$
$$= m_{01}c^2\left(\frac{1}{\sqrt{1-\alpha^2/c^2}}-1\right)+m_{02}c^2\left(\frac{1}{\sqrt{1-\beta^2/c^2}}-1\right)$$

and after simplification, we obtain the conservation relation for the total mass:

$$\frac{m_{01}}{\sqrt{1-a^2/c^2}}+\frac{m_{02}}{\sqrt{1-b^2/c^2}}=\frac{m_{01}}{\sqrt{1-\alpha^2/c^2}}+\frac{m_{02}}{\sqrt{1-\beta^2/c^2}}$$

We now write down the condition that energy is conserved with respect to another Galileian system K', travelling with velocity v with respect to K along the positive x-axis:

$$\frac{m_{01}}{\sqrt{1-a'^2/c^2}}+\frac{m_{02}}{\sqrt{1-b'^2/c^2}}=\frac{m_{01}}{\sqrt{1-\alpha'^2/c^2}}+\frac{m_{02}}{\sqrt{1-\beta'^2/c^2}}$$

In *Kinematics* (§ [80]), we obtained the velocity transformation for v_1

$$\frac{1}{\sqrt{1-v'^2/c^2}}=\frac{1-\dfrac{v}{c}\,\dfrac{v_{1x}}{c}}{\sqrt{1-v^2/c^2}\sqrt{1-v_1^2/c^2}}$$

whence

$$m_{01}\frac{1-va_x/c^2}{\sqrt{1-a^2/c^2}}+m_{02}\frac{1-vb_x/c^2}{\sqrt{1-b^2/c^2}}$$
$$= m_{01}\frac{1-v\alpha_x/c^2}{\sqrt{1-\alpha^2/c^2}}+m_{02}\frac{1-v\beta_x/c^2}{\sqrt{1-\beta^2/c^2}}$$

Simplifying with the aid of the equation in K, we obtain

$$\frac{m_{01}a_x}{\sqrt{1-a^2/c^2}}+\frac{m_{02}b_x}{\sqrt{1-b^2/c^2}}=\frac{m_{01}\alpha_x}{\sqrt{1-\alpha^2/c^2}}+\frac{m_{02}\beta_x}{\sqrt{1-\beta^2/c^2}}$$

This relation shows that the component of the sum of the momenta of the particles along Ox is unaffected by the collision. The proof clearly applies to the other components also; although we have considered a reference frame K' travelling along the x-axis of K, we could equally well have assumed that it was travelling along the y- or z-axes.

The calculations are thus similar to those of pre-relativistic dynamics; they are based on the two conservation theorems. Here, however, the conservation of energy can be more simply written as the conservation of relativistic mass.

[202] Inelastic Collisions

(a) In pre-relativistic dynamics, these collisions were characterized by a loss of kinetic energy, which was converted into some other form (heat, deformation, ...); total momentum and total mass were conserved. The present problem is more subtle.

To begin with, there is the possibility of energy loss by radiation at the moment of collision. This is the phenomenon usually known as Bremsstrahlung, or retardation radiation. In this case, so far as the particles are concerned, neither total energy nor total momentum is conserved; the proper masses may be conserved, however, and the corpuscles thus preserve their individuality.

At the moment of collision, there may also be creation or annihilation of particles, and the two incident particles may disappear.

In these two cases, we have to *establish an energy balance for the situation*, and write down that total momentum and total energy (or total mass) are conserved, including any possible radiation of course. The notions of proper mass and kinetic energy become secondary.

Pre-relativistic dynamics dealt only with a very special problem: it was assumed that after the collision, there were still two corpuscles, separate or united (the capture phenomenon), with unchanged proper masses; only temperature and shape could change. Mechanical energy disappeared; the energy lost was assumed to reappear in another form (heat, for example) but to calculate it was beyond the limits of dynamics.

We shall see that from the relativistic point of view, this case is no different from the preceding ones. The proper mass of a body that has been heated or deformed increases; because of the mass-energy equivalences there can be no question of mechanical energy lost. We shall always use conservation of energy.

The principal cases of inelastic collisions are studied in Chapter XVII.

(b) The conservation of total impulse is now also a consequence of the conservation of total energy (the kinetic energy is not conserved). For example, when two particles m_{01} and m_{02} collide and produce two different particles m_{03} and m_{04}, we write

$$\frac{m_{01}}{\sqrt{1-a^2/c^2}}+\frac{m_{02}}{\sqrt{1-b^2/c^2}}=\frac{m_{03}}{\sqrt{1-\alpha^2/c^2}}+\frac{m_{04}}{\sqrt{1-\beta^2/c^2}}$$

and the calculations resemble those of § [201] (b).

Finally, therefore, for all types of collision (elastic or inelastic), we write down that total energy and total momentum are conserved.

Remark. Everything therefore depends upon the conservation of total energy, which is the *fundamental postulate* of collision theory (and from which the conservation of impulse or momentum is deduced).

(c) *Characteristic axes and planes for the collision of two particles*. The momenta of the two incident particles define a plane Π (unless one of the particles is at rest). If there are two emergent particles, their velocities lie in Π; if more than two particles emerge, the resultant of their momenta lies in Π. The resultant of the incident momenta (and hence of the emergent momenta also) defines an axis and this, together with the axis normal to Π, forms a plane Π'. We have thus defined two axes and two characteristic planes.

B. TWO PARTICLES OF THE SAME MASS, ONE OF WHICH IS AT REST BEFORE THE COLLISION

[203] Fundamental Equations

A particle of proper mass m_0, travelling with velocity v_1 along the straight line 1 (Fig. 56) meets a particle at rest at A of the same mass m_0. This case has many important practical applications. In fact, no particle is ever rigorously at rest, but one of the initial velocities is assumed to be negligible in comparison with the other. After the collision, the first particle follows trajectory 1′ with velocity v_1'; the second particle follows 2′ with velocity v_2'.

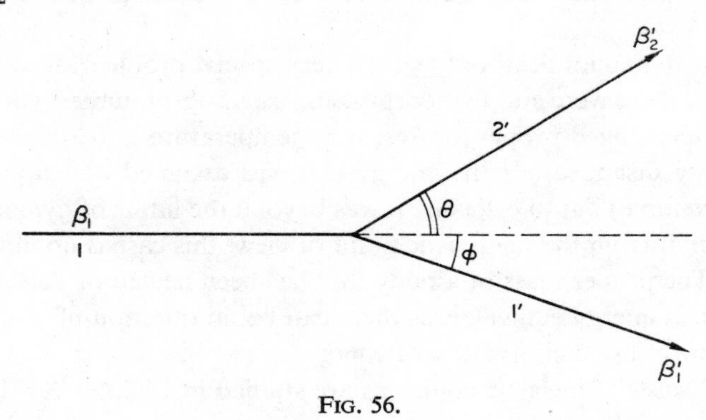

FIG. 56.

These three trajectories are coplanar. For, if we call the plane containing 1′ and 2′ P, 1 necessarily lies in P; if this were not so, momentum would not be conserved in the direction normal to P. In the present case, the two particles are *indistinguishable* in the quantum mechanical sense. This property should also have an effect here, because we are treating the collision problem quite schematically, and assuming that the point particles *coincide* at some given time. There is therefore no reason to ascribe the trajectory 1′ to the incident particle. The formulae must be unchanged if we reverse the trajectories and set m_1' on 2′ and m_2' on 1′, for example.

Conservation of energy gives

$$1+\frac{1}{\sqrt{1-\beta_1^2}} = \frac{1}{\sqrt{1-\beta_1'^2}}+\frac{1}{\sqrt{1-\beta_2'^2}} \tag{1}$$

Conservation of momentum along the line 1 and perpendicular to it gives

$$\frac{\beta_1}{\sqrt{1-\beta_1^2}} = \frac{\beta_1'}{\sqrt{1-\beta_1'^2}} \cos\phi + \frac{\beta_2'}{\sqrt{1-\beta_2'^2}} \cos\theta \tag{2}$$

$$\frac{\beta_1'}{\sqrt{1-\beta_1'^2}} \sin\phi = \frac{\beta_2'}{\sqrt{1-\beta_2'^2}} \sin\theta \tag{3}$$

We have only three equations and four unknowns:

$$\beta_1', \beta_2', \theta \quad \text{and} \quad \phi.$$

There is therefore some uncertainty in the problem; we shall return to this after establishing the formulae which will provide us with a basis for discussion.

[204] Total Velocities After the Collision as Functions of the Angles

Substituting for β_2' from (3) into (2),

$$\frac{\beta_1}{\sqrt{1-\beta_1^2}} = \frac{\beta_1'}{\sqrt{1-\beta_1'^2}} \cos\phi + \frac{\beta_1'}{\sqrt{1-\beta_1'^2}} \frac{\sin\phi}{\sin\theta} \cos\theta$$

$$= \frac{\beta_1'}{\sqrt{1-\beta_1'^2}} \frac{\sin(\phi+\theta)}{\sin\theta}$$

we obtain an expression for β_1':

$$\beta_1' = \frac{\beta_1 \sin\theta}{\sqrt{\beta_1^2 \sin^2\theta + (1-\beta_1^2)\sin^2(\phi+\theta)}}$$

Likewise, substituting for β_1' from (3) into (2), we find

$$\beta_2' = \frac{\beta_1 \sin\phi}{\sqrt{\beta_1^2 \sin^2\phi + (1-\beta_1^2)\sin^2(\phi+\theta)}}$$

These formulae allow us to calculate β_1' and β_2' from the values of ϕ and θ measured on photographs.

We notice that they satisfy the indistinguishability condition.

In the symmetrical case (Fig. 57),

$$\phi = \theta, \quad \beta_1' = \beta_2' = \frac{\beta_1}{\sqrt{\beta_1^2 + 4(1-\beta_1^2)\cos^2\phi}}$$

If one of the angles vanishes, θ say, β_1' is also equal to zero. Relation (2) then gives

$$\frac{\beta_1}{\sqrt{1-\beta_1^2}} = \frac{\beta_2'}{\sqrt{1-\beta_2'^2}}$$

The incident particle transfers all its kinetic energy to the particle originally at rest. In addition, conservation of momentum requires that ϕ should also vanish.

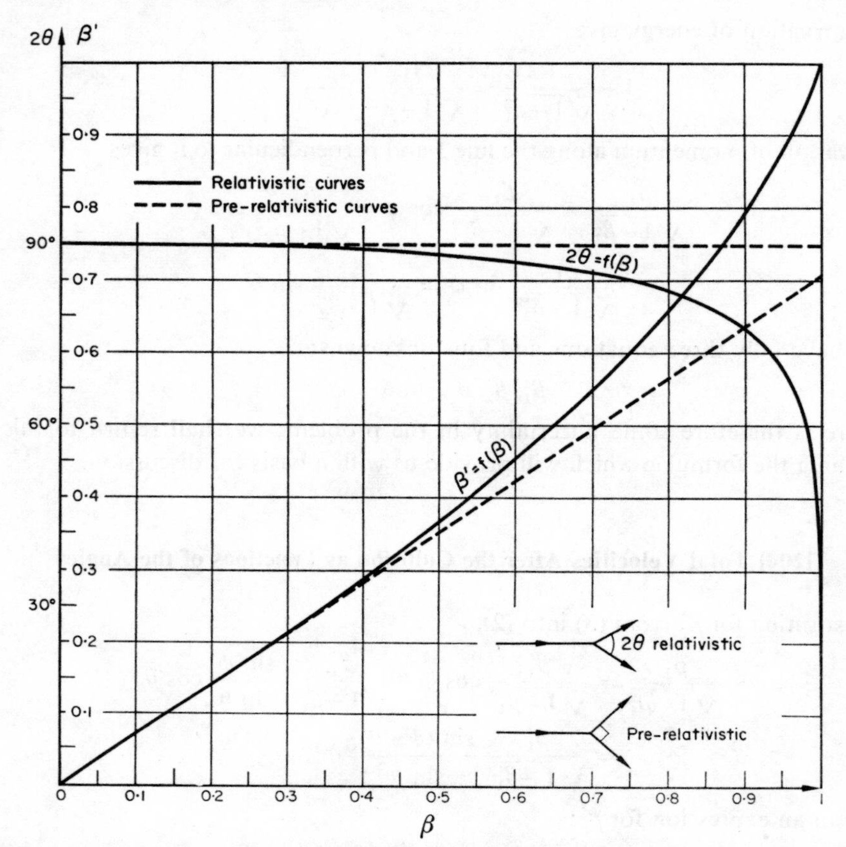

FIG. 57. Elastic collision between one particle and another identical particle, originally at rest; the symmetrical case.

[205] The Energy after Collision as a Function of the Angles

With the aid of the formulae derived,

$$\sqrt{1-\beta_1'^2} = \frac{\beta_1'}{\beta_1}\sqrt{1-\beta_1^2}\,\frac{\sin(\phi+\theta)}{\sin\theta}$$

so that

$$W_1' = \frac{m_0c^2}{\sqrt{1-\beta_1'^2}} = \frac{m_0c^2}{\sqrt{1-\beta_1^2}}\,\frac{\sqrt{\beta_1^2\sin^2\theta+(1-\beta_1^2)\sin^2(\phi+\theta)}}{\sin(\phi+\theta)}$$

or, introducing the initial energy,

$$W_1' = \frac{\sqrt{(W_1^2-m_0^2c^4)\sin^2\theta+m_0^2c^4\sin^2(\phi+\theta)}}{\sin(\phi+\theta)}$$

Similarly, of course,

$$W_2' = \frac{\sqrt{(W_1^2-m_0^2c^4)\sin^2\phi+m_0^2c^4\sin^2(\phi+\theta)}}{\sin(\phi+\theta)}$$

[206] The Angle Between the Two Trajectories After the Collision as a Function of the Angle θ or ϕ

We first use the above relations; a few simple calculations give

$$\cos(\phi+\theta) = \frac{\left(\frac{1}{\sqrt{1-\beta_1^2}}-1\right)\sin\theta\cos\theta}{\sqrt{\left(\frac{1}{\sqrt{1-\beta_1^2}}+1\right)^2 \sin^2\theta + 4\cos^2\theta}}$$

In the symmetrical case, the formula simplifies and gives the angle *as a function of the initial velocity alone* (Bothe):

$$\phi = 0, \quad \beta_1' = \beta_2'$$

$$\cos\theta = \cos\phi = \frac{\beta_1}{\sqrt{2\sqrt{1-\beta_1^2}+3\beta_1^2-2}}$$

The curve

$$2\theta = 2\phi = \theta+\phi = f(\beta_1)$$

is plotted in Fig. 57.

[207] The Angle Between the Two Trajectories After the Collision as a Function of the Masses

The foregoing formulae show that in the present circumstances, the phenomena are independent of the proper masses of the particles (equal in the present case). Nevertheless, on introducing the masses explicitly, the symmetry properties of the expressions thus obtained simplify certain arguments.

Let p_1, p_1', p_2' denote the momenta; we write down the conservation relations, projecting along and perpendicular to the direction $1'$:

$$p_1 \cos\phi = p_1' + p_2' \cos(\phi+\theta)$$
$$p_1 \sin\phi = p_2' \sin(\phi+\theta)$$

so that

$$\cos(\phi+\theta) = \frac{p_1^2 - (p_1'^2 + p_2'^2)}{2p_1'p_2'}$$

Denoting the masses m_1, m_1' and m_2' and using the relation

$$p^2 = (m^2 - m_0^2)c^2$$

we can write

$$\cos(\phi+\theta) = \frac{1}{2}\frac{(m_1^2+m_0^2)-(m_1'^2+m_2'^2)}{\sqrt{(m_1'^2-m_0^2)(m_2'^2-m_0^2)}}$$

We now use the conservation of energy in the form

$$m_1 + m_0 = m_1' + m_2'$$

and finally,

$$\cos(\phi+\theta) = \sqrt{\frac{(m_1'-m_0)(m_2'-m_0)}{(m_1'+m_0)(m_2'+m_0)}}$$

As indistinguishability requires, the masses m_1' and m_2' may be interchanged. Since

$$m_1' > m_0 \quad \text{and} \quad m_2' > m_0,$$

we have

$$0 < \theta+\phi < \pi/2$$

After the collision, the trajectories are inclined at an acute angle, unlike the pre-relativistic case. In the latter approximation, we have

$$m_1 = m_1' = m_2' = m_0, \quad \cos(\phi+\theta) = 0$$

$$\phi+\theta = \pi/2$$

Remark. The equations leave one of the angles arbitrary; this uncertainty is a consequence of the fact that we are assuming the particles to be points and we give no details about the type of interaction exerted during the collision. This is far from meaning that we may choose ϕ or θ arbitrarily. These angles obviously depend on the law of interaction and on the minimum separation of the corpuscles. We may choose ϕ (or θ) equal to zero (a case already examined); only experiment or a more complete theory can give the other values, however.

[208] Calculation Using the System in which the Total Momentum of the Two Particles is Zero ("Rest System" or Baryocentric System)

The following is a purely relativistic argument, depending on the formulae for a change of Galileian system.

Let K be the system in which the above calculations were performed (in K, particle 2 is at rest before the calculation).

Let K_0 be another Galileian system, travelling with velocity v with respect to K. With the axes as shown in Fig. 58, we have the transformation formulae.

$$(p_x)_0 = \frac{p_x - \frac{v}{c^2}W}{\sqrt{1-v^2/c^2}}, \quad (p_y)_0 = p_y \tag{1}$$

$$(W)_0 = \frac{W - v(p_x)_0}{\sqrt{1-v^2/c^2}} \tag{2}$$

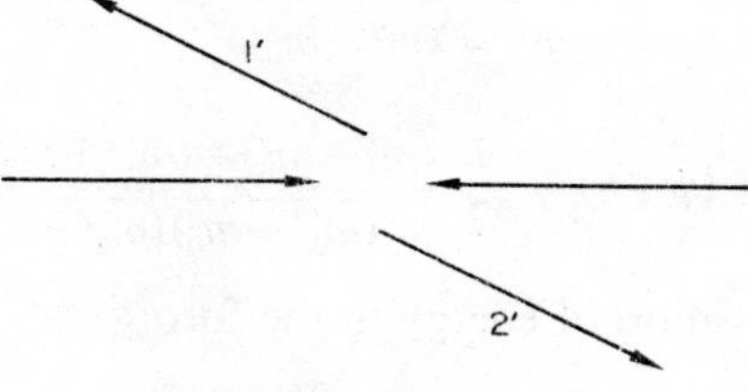

FIG. 58.

Let us obtain the value of v for which the total momentum in K_0 vanishes

$$(\mathbf{p}_1)_0+(\mathbf{p}_2)_0 = (\mathbf{p}_1')_0+(\mathbf{p}_2')_0 = 0 \tag{3}$$

Using (3) and (1), we have (before collision)

$$0 = (p_1)_0+(p_2)_0 = \frac{p_1-\frac{v}{c^2}W_1}{\sqrt{1-v^2/c^2}}+\frac{-\frac{v}{c^2}W_2}{\sqrt{1-v^2/c^2}}$$

so that

$$p_1-\frac{v}{c^2}(W_1+W_2) = 0$$

with

$$W_1 = \frac{m_0c^2}{\sqrt{1-\beta_1^2}}, \qquad W_2 = m_0c^2$$

$$p_1 = \frac{m_0\beta c}{\sqrt{1-\beta_1^2}}$$

Hence, for v, we find

$$v = \frac{\beta_1 c}{1+\sqrt{1-\beta_1^2}} \tag{4}$$

In pre-relativistic dynamics, we should have

$$v = \beta_1 c$$

We note that with equal proper masses, the two particles have equal and opposite velocities in K_0 before the collision. Since particle 2 was stationary in K, it had velocity $-v$ in K_0; particle 1 had velocity $+v$.

After the collision, the two particles again have equal and opposite velocities v along Ox, but they also have components of velocity along Oy:

$$(p_{1y}')_0 = -(p_{2y}')_0$$

After the collision, in K, we write

$$\tan\theta = \frac{p_{1y}'}{p_{1x}'}, \qquad \tan\phi = \frac{p_{2y}'}{p_{2x}'} = -\frac{p_{1y}'}{p_{2x}'}$$

$$\tan\theta\tan\phi = -\frac{(p_{1y}')^2}{p_{1x}'p_{2x}'} = -\frac{(p_{1y}')_0^2(1-v^2/c^2)}{\left(p_{1x}'+\frac{v}{c^2}W_1'\right)_0\left(p_{2x}'+\frac{v}{c^2}W_2'\right)_0}$$

where

$$(p_{2x}')_0 = -(p_{1x}')_0, \qquad (W_2')_0 = (W_1')_0$$

so that

$$\tan\theta\tan\phi = -\frac{(p_{1y}')_0^2(1-v^2/c^2)}{\left(\frac{v}{c^2}W_1'+p_{1x}'\right)_0\left(\frac{v}{c^2}W_1'-p_{1x}'\right)_0}$$

$$= -\frac{(p_{1y}')_0^2(1-v^2/c^2)}{\frac{v^2}{c^4}(W_1')_0^2-(p_{1x}')_0^2} = -(1-v^2/c^2)$$

Finally we substitute for v from (4),

$$\tan\theta\tan\phi = -\frac{2}{1+\dfrac{1}{\sqrt{1-\beta_1^2}}}$$

and obtain the formula given by Møller. This product depends only on the incident velocity, hence its usefulness.

Let us check the pre-relativistic limit, taking the absolute values of the angles:

$$\beta_1^2 \simeq 0 \quad \tan\theta\tan\phi \simeq 1$$

$$\tan(\theta+\phi) = \frac{\tan\theta+\tan\phi}{1-\tan\theta\tan\phi} \to \infty$$

$$\theta+\phi = \pi/2$$

In general,

$$0 < \tan\theta\tan\phi < 1, \quad 0 < \theta+\phi < \pi/2$$

In the symmetrical case,

$$\tan^2\theta = \tan^2\phi = \frac{2\sqrt{1-\beta_1^2}}{\beta_1^2}\left(1-\sqrt{1-\beta_1^2}\right)$$

which agrees satisfactorily with Bothe's formula for $\cos\theta$.

Remark. The above expression for v should be obtainable from a purely kinematic calculation. Thus, let q_0 be the velocity of the two particles in K_0; in K,

$$q_1 = \frac{q_0+v}{1+vq_0/c^2}, \quad q_2 = \frac{q_0-v}{1-vq_0/c^2}$$

If q_2 is to be zero (particle 2 at rest in K), we must take q_0 equal to v. Whence

$$q_1 = \frac{2v}{1+v^2/c^2}, \quad \frac{v}{c} = \frac{1-\sqrt{1-\beta_1^2}}{\beta_1}$$

The sign is chosen from the condition $v < c$. This expression can be shown to be identical with that of formula (4).

[209] Experimental Confirmation

(a) As we said in § [200], collisions between charged particles are generally very complex phenomena, and quantum mechanics and quantum field theory are required to study them. In particular, one must take into account the electromagnetic radiation produced by the collision.

Nevertheless, there are experimental cases, admittedly rather rare, in which this radiation is negligible and the interaction can be treated as an elastic collision. These cases offer striking confirmations of the relativistic formulae.

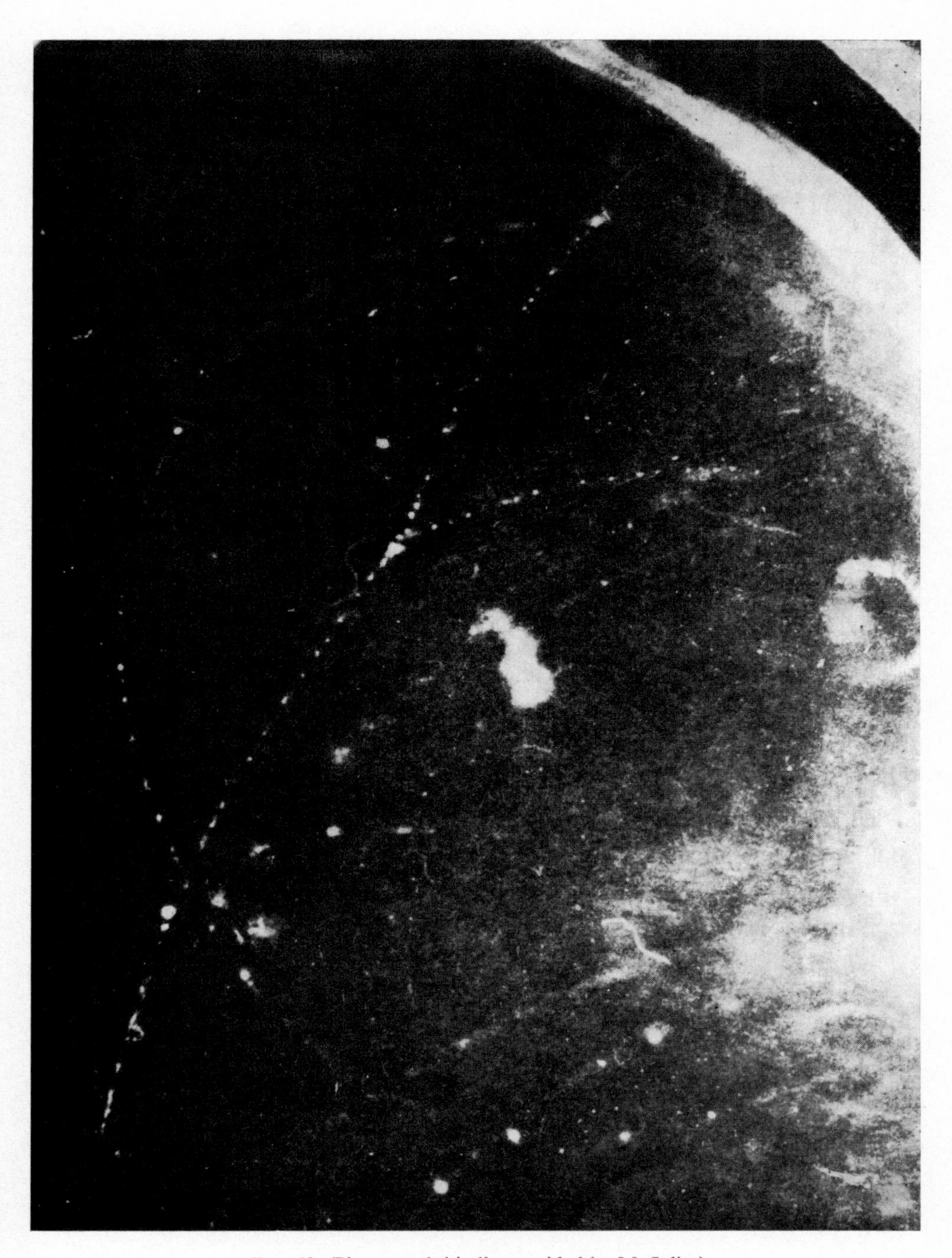

FIG. 59. (Photograph kindly provided by M. Joliot)

(b) *Champion's experiments (electron–electron)*

A table given by Champion (see the Bibliography) is reproduced below; it was obtained by observing electron–electron collisions (radium E as a source of β-rays). The agreement between theory and observation is striking. To illustrate the scale of the work involved, we mention that these fourteen elastic collisions, which are capable of yielding exact enough measurements, were extracted from among more than 30,000 tracks.

β_1 ± 1 or 2%	Energies (in eV)	θ measured $\pm 0{\cdot}5\%$	$\theta+\phi$ measured $\pm 1\%$	$\theta+\phi$ calculated $\pm 1\%$
85	457·1	20	83·6	82·7
83	403·4	26·6	81·2	81·7
83	403·4	31·4	81·0	81·0
82	380·2	22	84·1	83·2
85	457·1	22·2	82·2	81·7
83	403·4	22·4	82·1	82·3
84	428·9	23·4	82·7	81·7
90	658·5	24·5	79·6	77·4
88	562·4	35·4	76·8	77·3
85	457·1	21·1	82·7	81·8
91	700·0	36·9	75·2	75·2
93	885·0	29·6	72·5	72·6
85	457·1	21·6	82·4	81·4
82	380	36·9	80·6	81·0

(c) *The Joliot picture (electron–electron)*

The numerical values are as follows:

$$\beta_1 = 0{\cdot}968, \qquad \beta_1' = 0{\cdot}930$$
$$\beta_2' = 0{\cdot}910, \quad \theta+\phi = 60^\circ$$

The experimental confirmation involves measuring radii of curvature and deducing from them the angle between the trajectories by a calculation; this angle is then compared with that seen on the photograph (Fig. 59).

(d) *The Leprince–Ringuet picture (electron–electron)*

An electron (here a Compton recoil electron) collides with an electron in a Wilson cloud chamber. The numerical values are (Fig. 60):

$$\varrho_1 = 3{\cdot}15\text{ cm}, \quad \varrho_1' = 2{\cdot}03\text{ cm}, \quad \varrho_2' = 1{\cdot}82\text{ cm}$$
$$H = 1400\text{ gauss}, \quad \theta+\phi = 72^\circ$$

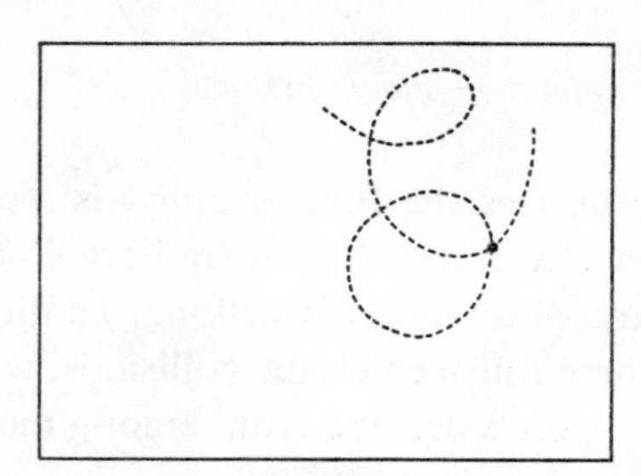

FIG. 60. The incident trajectory of the fast electron is on the extreme right of the diagram; the first break represents a collision between a fast electron and a slow electron. The two lower paths are the electron trajectories after the collision; they are inclined at an acute angle (72°).

C. TWO PARTICLES OF DIFFERENT MASSES, ONE OF WHICH IS STATIONARY BEFORE THE COLLISON

[210] The Fundamental Equations

Associating the suffix 2 with the particle initially at rest, we now write

$$\cos(\phi+\theta) = \frac{p_1^2-(p_1'^2+p_2'^2)}{2p_1'p_2'}$$

$$= \frac{1}{2}\frac{(m_1^2+m_{02}^2)-(m_1'^2+m_2'^2)}{\sqrt{(m_1'^2-m_{01}^2)(m_2'^2-m_{02}^2)}}$$

In the numerator of this equation, the difference between the sums of the squares of the masses before and after collision appears. Conservation of energy states that

$$m_1+m_{02} = m_1'+m_2'$$

or

$$m_1-m_1' = m_2'-m_{02}$$

During the collision, the two particles exchange the difference between their masses.

[211] Application to the Measurement of the Proper Mass of an Incident Particle (particle 2 is at rest)

(a) Conservation of energy gives

$$m_{02}c^2+m_1c^2 = m_2'c^2+m_1'c^2$$

Using the general relation

$$(mc^2)^2 = (m_0c^2)^2+p^2c^2,$$

we write

$$m_{02}c^2+\sqrt{(m_{01}c^2)^2+p_1^2c^2} = \sqrt{(m_{02}c^2)^2+p_2'^2c^2}+\sqrt{(m_{01}c^2)^2+p_1'^2c^2}$$

We set

$$m_{02}c^2 = p_{02}c, \quad \sqrt{(m_{02}c^2)^2+p_2'^2c^2}-m_{02}c^2 = cp_c$$

and the quantity cp_c represents the kinetic energy of the particle hit, after the collision.

We then have

$$\sqrt{(m_{01}c^2)^2+p_1^2c^2} = \sqrt{(m_{01}c^2)^2+p_1'^2c^2}+cp_c$$

or

$$\sqrt{(\mu p_{02})^2+p_1^2} = \sqrt{(\mu p_{02})^2+p_1'^2}+p_c, \qquad \mu = \frac{m_{01}}{m_{02}}$$

so that finally,

$$\mu^2 = \left(\frac{p_1'}{p_{02}}\right)^2\left\{\left(\frac{p_1^2-p_1'^2-p_c^2}{2p_1'p_c}\right)^2-1\right\} \tag{1}$$

This formula enables us in principle to deduce μ by measuring momenta. It requires the quantities p to be measured accurately, however, and this is not always possible. For example, Gorodetzky mentions the case where the incident particle is of high energy, and the difference $p_1^2-p_1'^2$ is inappreciable even though it is of the same order of magnitude as p_c^2.

By applying the principle of conservation of momentum, we have found above (§ [210])

$$p_1^2-p_1'^2 = 2p_1'p_2' \cos(\phi+\theta)+p_2'^2$$

In this way, the difference in question is obtained by measuring the angle $\phi+\theta$, and this can be performed more accurately. On making this substitution, we obtain

$$\mu^2 = \left(\frac{p_1'}{p_{02}}\right)^2\left\{\left(\frac{2p_1'p_2' \cos(\phi+\theta)+p_2'^2-p_c^2}{2p_1'p_c}\right)^2-1\right\}$$

Using the fact that

$$p_2'^2-p_c^2 = 2p_{02}p_c$$

we finally obtain

$$\boxed{\mu = \frac{p_1'}{p_{02}}\sqrt{\left\{\frac{p_2'}{p_c}\cos(\phi+\theta)+\frac{p_{02}}{p_1'}\right\}^2-1}} \tag{2}$$

This formula depends upon no suppositions about orders of magnitude; it is fundamental to the method.

The non-relativistic approximation is obtained by replacing p_c by its approximate value,

$$p_c = \sqrt{p_{02}^2+p_2'^2}-p_{02} \simeq \frac{1}{2}\frac{p_2'^2}{p_{02}}$$

whence

$$\mu = \frac{p_1^2-p_1'^2}{p_2'^2}$$

This could be obtained directly by a simple calculation. Hence

$$\mu = \frac{2p_1'}{p_2'}\cos(\phi+\theta)+1$$

We mention one other interesting expression derived from (1) and (2). It can easily be shown that

$$\frac{p'}{\mu p_{02}} = \frac{p_1'}{m_{01}c} = \frac{\beta_1'}{\sqrt{1-\beta_1'^2}}$$

so that using (1) and (2),

$$\frac{1}{\beta_1'} = \frac{p_1^2-p_1'^2-p_c^2}{2p_1'p_c} = \frac{p_2'}{p_c}\cos(\phi+\theta)+\frac{p_{02}}{p_1'}$$

$$\mu = \frac{p_1'}{p_{02}}\sqrt{\frac{1}{\beta_1'^2}-1}$$

This formula is convenient when the effect of errors of measurement of β_1', and p_1' is being examined.

(b) *The expression in terms of the radii of curvature and the inclination of the trajectories.* Let ϱ_1, ϱ_1' and ϱ_2' be the radii of curvature (at the moment of collision) of the projections of the trajectories on the principal plane; and let α_1, α_1' and α_2' be the angles between the trajectories and a plane Π perpendicular to the field H.

We define lengths R as follows:

$$R_1 = \varrho_1 \sec\alpha_1, \quad R_1' = \varrho_1' \sec\alpha_1'$$

$$R_2' = \varrho_2' \sec\alpha_2', \quad R_{02} = \frac{m_0c^2}{\varepsilon\mathcal{B}}$$

$$\sqrt{R_{02}^2+R_2'^2}-R_{02} = R_e$$

Notice that these R are not the radii of curvature of the trajectories, which are given by

$$\varrho_1\sec^2\alpha_1 = R_1\sec\alpha_1, \ \ldots$$

Substituting the corresponding R for the quantities p in (2), we have

$$\mu = \frac{R_1'}{R_{02}}\sqrt{\left\{\frac{R_2'}{R_e}\cos(\phi+\theta)+\frac{R_{02}}{R_1'}\right\}^2-1}$$

The quantities ϱ and α are obtained from the photographs by stereoscopic reconstruction, and hence the R are found.

To obtain θ, the angle θ' between the projections on the plane Π of the trajectories 1' and 2' is measured; whence,

$$\cos\theta = \cos\theta'\cos\alpha_1\cos\alpha_2+\sin\alpha_1'\sin\alpha_2'.$$

[212] A Direct Argument Giving μ in Terms of the Radii of Curvature of the Trajectories in a Magnetic Field in the Case where the Particles are Equally Charged and have Different Masses

(a) We use the formulae of § [211], assuming to begin with that the trajectories lie in the plane normal to $\mathfrak{B}$. In the present case, the conservation theorems can then be written in the following form:

conservation of momentum:

$$\Sigma \mathbf{R} = \text{const}$$

conservation of energy:

$$\Sigma \sqrt{\varrho_0^2 + R^2} = \text{const}$$

The quantities **R** denote the parallel and equal vectors along the radii of curvature at the point of impact.

Explicitly, the last relation is of the form

$$\sqrt{R_{01}^2 + R_1^2} + R_{02} = \sqrt{R_{01}^2 + R_1'^2} + \sqrt{R_{02}^2 + R_2'^2}$$

Since the charge ε is the same for the two particles, however, we have

$$R_{01} = \frac{m_{01}c^2}{\varepsilon \mathcal{B}}, \qquad R_{02} = \frac{m_{02}c^2}{\varepsilon \mathcal{B}}$$

$$\mu = \frac{m_{01}}{m_{02}}, \qquad R_{01} = \mu R_{02} = \mu R_0$$

whence

$$\sqrt{\mu^2 R_0^2 + R_1^2} + R_0 = \sqrt{\mu^2 R_0^2 + R_1'^2} + \sqrt{R_0^2 + R_2'^2}$$

To make the notation more compact, we set

$$\sqrt{R_0^2 + R_2'^2} = \varrho$$

and solving, we obtain

$$\mu^2 = \frac{(R_1 + R_1' + \varrho - R_0)(R_1 + R_1' - \varrho + R_0)(R_1 - R_1' + \varrho - R_0)(R_1 - R_1' - \varrho + R_0)}{4R_0^2(\varrho - R_0)^2}$$

(b) *High-energy incident trajectory*. This is the case for incident mesons, for example (Fig. 61). The radii R_1 and R_1' are large in comparison with R_0. If the term

$$R_1 - R_1' - \varrho + R_0$$

were employed as it stands, it would lead to an intolerable error, since the difference between R_1 and R_1' cannot be measured on the photographs with an accuracy of the order of R_0. We plot the triangle of rays (Fig. 61), using the momentum relation. Introducing the distance f of trajectory 2 from trajectory 1,

$$f = R_2'(1 - \cos\theta),$$

we can then write

$$R_1 - R_1' = R_2' - f\frac{2R_1}{(R_1 + R_1' + R_2') - 2R_2'}$$

It is this expression for $R_1 - R_1'$ that must be substituted into the general formula.

In practice, one of the following three approximate relations is adequate, chosen according to the relative orders of magnitude of R_1, R_2' and R_0:

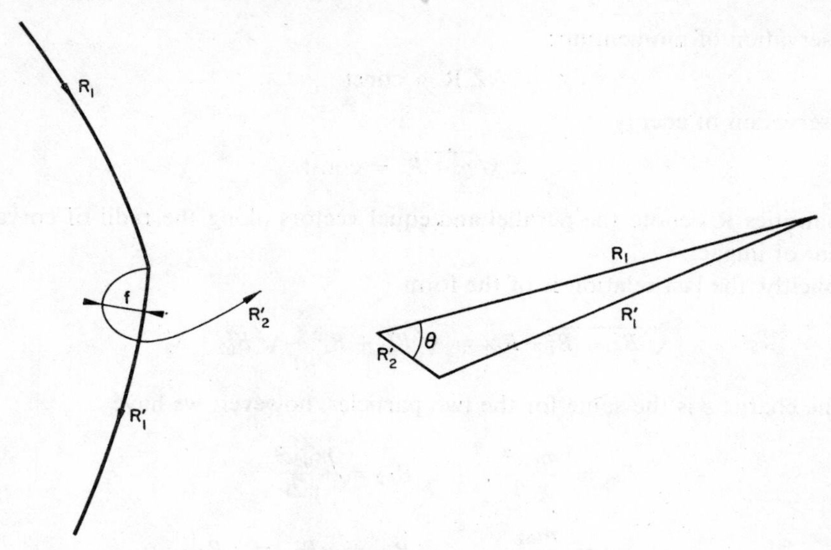

FIG. 61.

(i) $R_1 \gg R_0$ irrespective of R'_2:

$$\mu = \frac{R_1}{R_0}\sqrt{\frac{2(R_1-R'_2)\left(R_0-f\dfrac{R_1}{R_1-R'_2}\right)}{R_1R'_2}}$$

(ii) $R_1 \gg R'_2 \gg R_0$:

$$\mu = \frac{R_1}{R_0}\sqrt{\frac{2(R_0-f)}{R'_2}}$$

(iii) $R_1 \gg R_0$ and R'_2 the same order of magnitude as R_0:

$$\mu = \frac{R_1}{R_0(\varrho-R_0)}\sqrt{(R'_2+\varrho-R_0-f)(R'_2-\varrho+R_0-f)}$$

Remark. If the trajectories do not lie in a plane perpendicular to the field $\mathfrak{B}$, the formula must be modified. Denoting the angle between the trajectory and $\mathfrak{B}$ by

$$\frac{\pi}{2}-\xi$$

the radius of curvature is given by

$$mv = \varepsilon\mathcal{B}R\sec\xi$$

The calculations remain valid provided we make the substitution

$$R \rightarrow R\sec\xi$$

(c) *The measurements of Leprince-Ringuet, Nageotte, Gorodetzky and Richard-Foy.* These authors have obtained a photograph (Fig. 62) with which the method can be

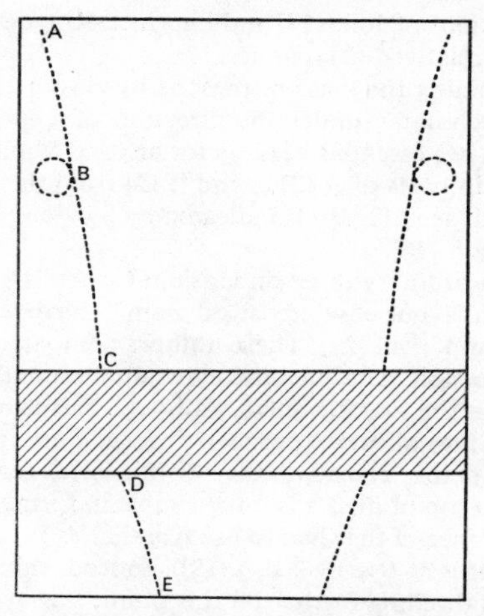

FIG. 62. Diagram of elastic collision between a meson and an electron. The fine trajectory *ABC* crosses a screen, from which it emerges along *DE* displaying increased ionization. At *B*, a remarkable phenomenon occurs: a collision with an electron, formerly at rest, which acquires a considerable energy during the collision. This is a very rare phenomenon, especially when the curvature of the primary ray *ABC* is measurable, as it is here. At *B*, we have all the details necessary for the application of the laws of mechanics.

applied to the determination of the mass of a meson. The numerical values are as follows:

$$\mathcal{B} = 2650 \text{ gauss},$$

$$115 \text{ cm} < R_1 < 120 \text{ cm}, \quad 1{\cdot}05 < R_2' < 1{\cdot}10 \text{ cm}$$

$$0 < f < 0{\cdot}03 \text{ cm}, \quad \cos \xi = 0{\cdot}97$$

The last of the formulae must be applied, making allowance for the changes arising from ξ. It is found that

$$\mu = 240$$

After discussing the various errors possible, the authors give

$$\mu = 240 \pm 22$$

HISTORICAL AND BIBLIOGRAPHICAL NOTES

[213] Collisions Between Charged Particles; Experimental Confirmations

(a) The first qualitative affirmations of the fact that the angle between the trajectories after an elastic collision of one particle with an identical stationary particle is acute appear to be due to Bothe [1] and Wilson [2]; they considered electron–electron collisions.

A paper by Champion [4] provides the first quantitative test (reproduced in § [209]). This article, which makes very pleasant reading, also contains a proof of the fundamental formulae. So far as I am aware, there are only the observations of Joliot [7] and Leprince-Ringuet after this. A mere glance at the photographs is enough to justify relativistic dynamics.

(b) The techniques for determining the mass of mesons by elastic collisions were developed in the laboratories of the Ecole Polytechnique, under the direction of Leprince-Ringuet [9, 12–17, 21–23]. These papers, and especially [22], are essential reading for anyone who wishes to learn how to apply the method. I have borrowed the main parts of §§ [211] and [212] from them.

The four-dimensional method (see § [214]) is indicated by Synge [6] and C. de Beauregard [24], and considerably elaborated by Romain [26].

A graphical representation in ordinary three-dimensional space is given by Blaton [11].

Champion's conclusions [4] in favour of relativistic dynamics have recently been contested by Faragó and Jánossy [27] and by Raboy and Trail [29]. These authors compare afresh the relativistic expression for the mass and Abraham's expression. After meticulous examination, they conclude that no experiment can enable one to choose between the two formulae. Although this is not the result they set out to establish, this does of course mean that all the experimental results are in agreement with the relativistic formula. For us, this is the vital point. The advantage of this latter formula is that it fits into a theory covering the whole of physics. On top of this, it is quite simple in form. I should like to see the expressions on the accelerator engineers' faces if they had to use Abraham's formula in calculating the properties of their machines (see *Rayonnement et Dynamique*, § [19]). Indeed, this is more or less what Champion replies [28]. I feel that it is worth dwelling further on this point, however, for as I see it, there is a lack of epistemological reflection in this. There is certainly no question of arguing about the intrinsic worth of these texts (they are genuinely sound); as they are set out, however—as a comparison of the relativistic expression with that of Abraham—their interest seems to me if not nil, at least purely historic. A healthier attitude towards physical theory would have saved these authors much wasted time. What in fact is the outcome of their research? Experiment confirms the relativistic formula (which we already knew); it also confirms Abraham's formula, within the accuracy of the measurements. Without doing any work at all, I can make the following statement: if a formula A is verified experimentally (the mass formula, for example), then an infinity of other formulae B can be obtained by adding to A terms smaller than the accuracy of the measurements, and all these formulae will be experimentally verified under the same conditions as A. This is self-evident, and the conclusion of the authors in question is no different from this if we specify that Abraham's formula is one of the formulae B. This would be a useful assertion, and would hence justify all this work, if it enabled us to choose between two parallel theories, one containing A and the other B. This is not the case, though; there is after all no question of returning to the old ideas of Abraham. Under these conditions, his mass formula, taken in isolation, can be of no use to us whatsoever. The comparison could be of practical value if the equivalent formula B led to simpler numerical calculations than A; here, the contrary is true. I do not see what remains, therefore, apart from the historical interest. In fact, all this is based on a false notion of physical theory. It is a delusion to believe that doubt is cast on the validity of a theory or a formula because others can be shown to be possible (see the Preface). I hasten to state that my remarks are directed only at the comparison I am considering. The article by Raboy and Trail contains other features (suggested experiments, . . .) which are interesting in themselves.

[1] W. Bothe: Untersuchungen an β-Strahlenbahnen. *Z. Physik* **12** (1922) 117–127. This paper contains the formula given in § [206] as Bothe's formula.

[2] C. T. R. Wilson: Investigations on X-rays and β-rays by the cloud methods. *Proc. Roy. Soc.* A **104** (1923) 1–24.

[3] H. C. Wolfe: Scattering of high velocity electrons in hydrogen as a test of the interaction energy of two electrons. *Phys. Rev.* **37** (1931) 591–601.

[4] F. G. Champion: On some close collisions of fast β particles with electrons photographed by the expansion method. *Proc. Roy. Soc.* A **136** (1932) 630–637.

[5] R. C. Tolman: *Relativity, Thermodynamics and Cosmology*, Oxford, Clarendon Press, 1st ed., 1934; pp. 4 2–47 of the 1950 printing.

[6] J. L. Synge: Collision problems and the conservation laws. *Phys. Rev.* **45** (1934) 500–501.

[7] Mme Curie: *Radioactivité*, Vol. I, Paris, 1935.

[8] W. H. McCrea: *Relativity Physics*, Methuen, London, 1st ed., 1935; p. 19 of the 3rd ed., 1949.

[9] L. Leprince-Ringuet: Thèse, Paris, 1936.

[10] H. R. Crane: An example of the relativity change of mass with speed. *Amer. Phys. Teacher* **6** (1938) 105.

[11] J. Blaton: Sur une interprétation géométrique de la conservation de l'impulsion et de l'énergie

dans les collisions atomiques et les phénomènes de désintégration. *Kgl. Danske Videnskab. Selsk. Mat.-Fys. Medd.* **24** (1950) No. 20.
[12] L. Leprince-Ringuet, S. Gorodetzky, E. Nageotte and R. Richard-Foy: Mesure de la masse d'un mésoton par choc élastique. *C. R. Acad. Sci. Paris* **211** (1940) 382–385.
[13] L. Leprince-Ringuet, E. Nageotte, S. Gorodetzky and R. Richard-Foy: Mesure de la masse d'un mésoton du rayonnement cosmique. *Cahiers de Physique* (1941) No. 3, 2–9.
[14] R. Richard-Foy: Mesure des masses des mésotons par choc élastique. Détermination du domaine d'application de la méthode. *C. R. Acad. Sci. Paris* **213** (1941) 724–726.
[15] L. Leprince-Ringuet and S. Gorodetzky: Mesure de la masse d'une particule par choc élastique, formule générale. Application à un cliché de choc permettant une vérification directe des formules de la relativité restreinte. *C. R. Acad. Sci. Paris* **213** (1941) 765–768.
[16] L. Leprince-Ringuet, E. Nageotte, S. Gorodetzky and R. Richard-Foy: Mesure directe de la masse d'un mésoton. *J. Phys. Rad.* (1941) 63–71.
[17] L. Leprince-Ringuet, E. Nageotte, S. Gorodetzky and R. Richard-Foy: Direct measurement of the mass of the mesoton. *Phys. Rev.* **59** (1941) 460–461.
[18] L. de Broglie: reference suppressed.
[19] P. G. Bergmann: *Introduction to the Theory of Relativity*. Prentice-Hall, New York, 1st ed., 1942; p. 85 of the 1950 printing.
[20] René Dugas: Sur le choc de deux particules relativistes. *C. R. Acad. Sci. Paris* **216** (1943) 287–288.
[21] S. Gorodetzky: Sur la formule permettant d'obtenir la masse d'une particule (par choc élastique ou par une autre méthode). Considerations générales sur les erreurs. *C. R. Acad. Sci. Paris* **219** (1944) 330–332.
[22] S. Gorodetzky: Sur la détermination de la masse des particules chargées du rayonnement cosmique. *Ann. Physique* **19** (1944) 5–70. Very interesting review of the theoretical and experimental results obtained by the team at the Ecole Polytechnique under the direction of Leprince-Ringuet.
[23] S. Gorodetzky: Sur une nouvelle forme de la relation permettant d'obtenir la masse d'une particule par choc élastique. *C. R. Acad. Sci. Paris* **220** (1945) 915–916.
[24] O. Costa de Beauregard: *La Théorie de la relativité restreinte*. Paris, Masson, 1949, pp. 107–110.
[25] C. Møller: *The Theory of Relativity*, Oxford, Clarendon Press, 1952, pp. 67 and 82.
[26] J. Romain: Théorie du choc élastique de particules dans l'espace-temps de Minkowski. *Bull. Acad. Roy. Belgium* **41,** No. 11 (1953) 1225–1241.
[27] P. S. Faragó and L. Jánossy: Review of the experimental evidence for the law of variation of the electron mass with velocity. *Nuovo Cimento* **5** (1957) 1411–1436.
[28] F. G. Champion: Variation of electron mass with velocity. *Nuovo Cimento* **7** (1958) 122.
[29] S. Raboy and C. C. Trail: On the relation between velocity and mass on the electron. *Nuovo Cimento* **10** (1958) 797–803.
[30] B. L. Robinson: Collisions of relativistic particles. *Amer. J. Phys.* **29** (1961) 369–370.
[31] K. G. Debrick: Kinematics of high energy particles. *Rev. Mod. Phys.* **34** (1962) 429–441.
[32] H. Bougrain, B. Perrier and J. P. Bailliard: *Recueil de problèmes de physique nucléaire*. Paris, Editions de la Revue d'Optique (1962).
[33] R. Hagedorn: *Relativistic Kinematics*. Benjamin, 1963.
[34] J. Babeki, T. Coghen and M. Miesowicz: The four-momentum transfer between groups of particles in cosmic ray jets of very high energies. *Acta Phys. Pol.* **26** (1964) 71–76.
[35] K. Bichteler: Bemerkungen über relativistischen Stossinvarianten. *Z. Physik* **182** (1965) 521–523.
[36] M. H. Cha and D. G. Simons: Geometrical methods for relativistic particle dynamics. *Amer. J. Phys.* **31** (1963) 280–284.
[37] J. G. Hegarty: Graphical method of illustrating relativistic conservation laws. *Amer. J. Phys.* **36** (1968) 270–272.

[214] The Theory of Collisions in the Minkowski Continuum

The two conservation theorems, of impulse and total energy, are expressed by the conservation of four-vector momentum (§§ [34] and [232]).

The theory of the indicatrix may be obtained as follows. Through a point O, we draw the four-vectors corresponding to all the possible energy states of a particle; the end-points M of these vectors describe a hyperbolic space with two surfaces of revolution; their equation is

$$p_x^2+p_y^2+p_z^2-p_u^2 = -m_0^2c^2$$

and they are asymptotic to the conical space (the light hypercone)

$$p_x^2+p_y^2+p_z^2-p_u^2 = 0$$

The two surfaces correspond to positive and negative energies respectively.

Suppose that we have two particles and the resultant OO' of their four-momenta. The point M, at the end of one of these four-vectors (originating at O) and at the origin of the other (ending at O'), therefore lies at the point of intersection of the positive surface of one of the hyperbolic spaces and the negative surface of the other hyperbolic space, for positive energies. This point M thus describes a quadric $\mathscr{E}$, the projection of which on the coordinate hyperplane is $\mathscr{E}_p$. The symmetry conditions and the requirement that the vectors be of finite length show that $\mathscr{E}_p$ must be an ellipsoid of revolution; we need only consider an elliptical section, therefore, and we thus obtain the indicatrix. The reader who wishes to go into this approach further should consult the work of C. de Beauregard and Romain, mentioned in § [213]. Direct calculation of the indicatrix (§ [232]) seems much simpler to me, and hence preferable for the user.

[215] On the Use of the Baryocentric Reference System

Collision phenomena and nuclear reactions are most simply explained in the baryocentric system of reference, which we may call the natural system of the phenomenon.

To see the essential features of the phenomenon, we must place ourselves in this system. The conservation laws are then particularly simple; if, for example, the collision is elastic, each particle conserves its momentum. The laws in the laboratory system are then obtained with the aid of the Lorentz transformation.

In the following example, which has already been treated in § [207], this method using the baryocentric system K_0 reduces the problem to a kinematical one.

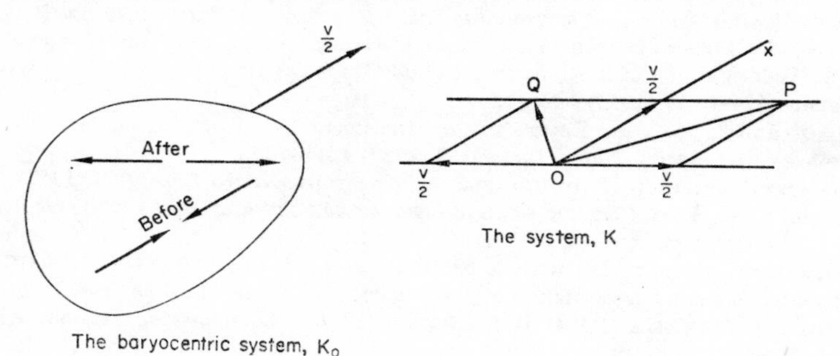

FIG. 62A.

Consider two electrons in K_0, travelling towards one another with velocities $v/2$; after collision, they travel away from one another at the same velocities. We now transfer to the system K, with respect to which K_0 has velocity $v/2$ parallel to one of the pre-collision velocities. In classical kinematics, we have Fig. 62A. In K, the electrons travel in directions OP and OQ. From the properties of the parallelogram we deduce that $\sphericalangle POQ$ is a right angle.

In relativistic kinematics, the resultants veer towards the translational velocity, that is towards Ox, whence the acute angle.

[216] On the Quantum Theory of Collisions

We have already said that a full study of the problem of collisions between "microscopic" particles involves quantum techniques. This is a fundamental problem for modern physics and its development at the present time is immense. The reader who wishes to supplement the point of view of the present work may obtain guidance, if he is a beginner, from the books by Heitler [2], Mott and Massey [1] and Massey and Burhop [3], for example.

Let a be the greatest distance (in order of magnitude) within which the interaction between the two particles is perceptible in practice. The macroscopic theory (relativistic or not) remains valid when the following conditions are satisfied:

— the particle trajectory must be well-defined, which requires that

$$mva \gg h$$

— the wavelength of the particles must be small compared with a;
— the exchange of momentum must be large compared with the quantum constant:

$$a \Delta p \gg h$$

[1] N. F. Mott and H. S. W. Massey: *Theory of Atomic Collisions.* Oxford, Clarendon Press, 1st ed. 1933; 2nd ed. 1949 (see especially p. 124); 3rd ed. 1965.
[2] W. Heitler: *The Quantum Theory of Radiation.* Oxford U.P., 1st ed., 1935; 2nd ed. 1944; 3rd ed. 1954.
[3] H. S. W. Massey and E. H. S. Burhop: *Electronic and Ionic Impact Phenomena.* Oxford: Clarendon Press, 1952.

CHAPTER XVI

THE COMPTON EFFECT

[217] General Remarks

This effect is a feature of X-ray scattering by certain bodies. In studying this scattering, the physicist Compton observed that the wavelength of the scattered rays was greater than that of the incident rays.

To a first approximation, the difference between the wavelengths obeys the law

$$\Delta\lambda = a \sin^2 \frac{\theta}{2}$$

n which a is a constant, equal to 0·048 Å and θ is the angle of scattering.

The phenomenon arises from the collisions between the incident photons and the electrons of the scattering medium. We shall see that the results obtained by applying relativistic dynamics are in excellent agreement with experiment.

[218] The Scattering Particles are Free and at Rest Before the Collision

(a) *The wavelength of the scattered light*

The incident beam of frequence ν_0 consists of photons, of which the energy W_0 and momentum p_0 are given by

$$W_0 = h\nu_0, \qquad p_0 = \frac{h\nu_0}{c}$$

We assume that the scattering medium contains particles of mass m_0, which scatter the photons by means of a mechanism that we describe below. We suppose that when a photon encounters a scattering particle, it is exactly as though two point particles collided. To begin with, we assume that the scattering particle is free and at rest. The incident photon arrives along AM (Fig. 63) with frequency ν_0 and velocity c. At M, it encounters a stationary free particle. After the collision, this particle has velocity v along

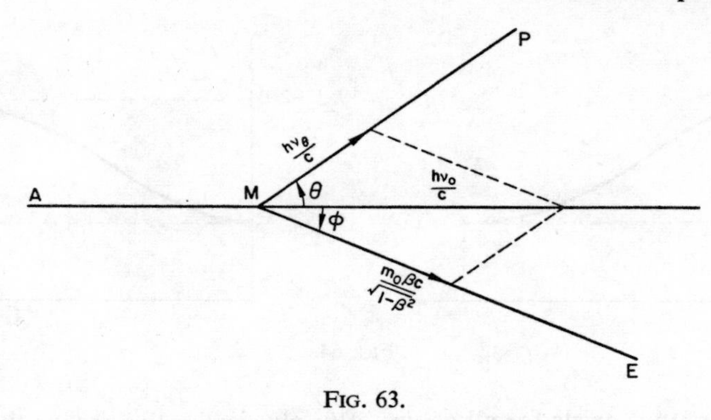

FIG. 63.

ME. The photon has velocity c (invariant by hypothesis) along *MP*; since the latter has lost energy, its frequency must have dropped. Let θ be the angle between *MP* and *AM* and ν_θ the frequency of the photon after the collision ($\nu_\theta < \nu_0$). Conservation of energy and momentum give

$$h\nu_0 + m_0c^2 = h\nu_\theta + \frac{m_0c^2}{\sqrt{1-\beta^2}}$$

$$\left(\frac{m_0\beta c}{\sqrt{1-\beta^2}}\right)^2 = \left(\frac{h\nu_0}{c}\right)^2 + \left(\frac{h\nu_\theta}{c}\right)^2 + 2\frac{h\nu_0}{c}\frac{h\nu_\theta}{c}\cos\theta$$

The collision is elastic in the sense of § [201]. Eliminating v between these two expressions (the first equation is solved for β^2 which is then substituted in the second), we have

$$\nu_\theta = \frac{\nu_0}{1+2\dfrac{h\nu_0}{m_0c^2}\sin^2\dfrac{\theta}{2}}$$

$$\lambda_\theta = \lambda_0 + 2\frac{h}{m_0c}\sin^2\frac{\theta}{2}$$

We write

$$\lambda_c = \frac{h}{m_0c}, \quad \alpha = \frac{h\nu_0}{m_0c^2} = \frac{\lambda_c}{\lambda_0}$$

The quantity λ_c is often known as the *Compton wavelength.*

With this notation, the formulae become

$$\nu_\theta = \frac{\nu_0}{1+2\alpha\sin^2\theta/2}, \quad \lambda_\theta = \lambda_0(1+2\alpha\sin^2\theta/2)$$

It is a characteristic of the phenomenon that the change of wavelength caused by the Compton effect depends only on the direction θ and not on the wavelength:

$$\Delta\lambda_\theta = \lambda_\theta - \lambda_0 = 2\alpha\lambda_0\sin^2\theta/2 = 2\lambda_c\sin^2\theta/2$$

When the scattering angle θ varies from 0 to π (Fig. 64), the frequency ν_θ falls from ν_0 to its minimum value

$$\nu_\theta = \frac{\nu_0}{1+2\alpha} \quad \text{or} \quad \frac{1}{\nu_\theta} = \frac{1}{\nu_0} + \frac{2h}{m_0c^2}$$

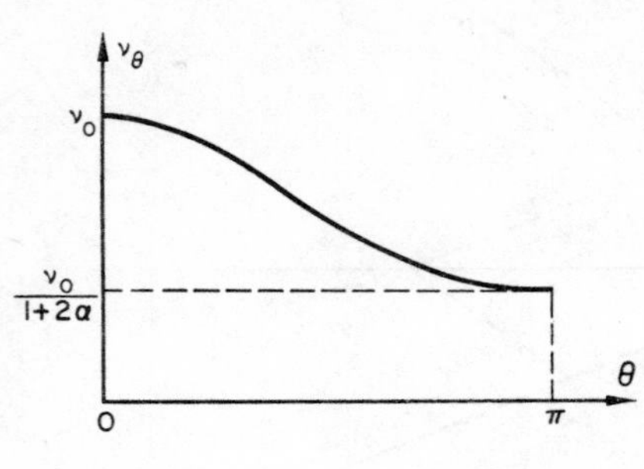

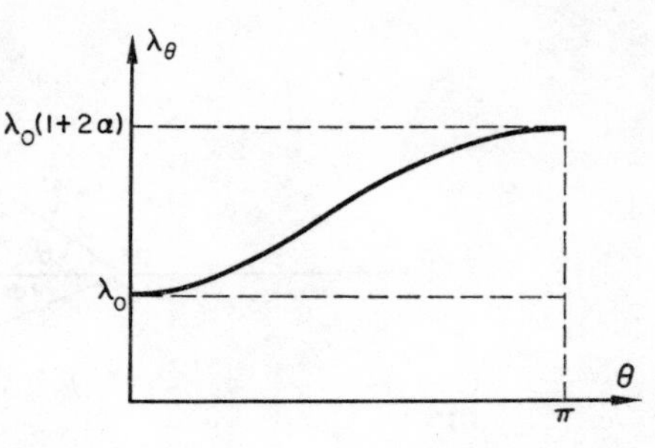

FIG. 64.

The frequency ν_θ thus exists for all values of θ; physically, this means that there always is a scattered photon. Indeed, if the collision equations were written down, assuming that there is no scattered photon, we should find that they were incompatible (except when ν_0 is zero).

(b) *The recoil electron*

To determine the direction ϕ of the recoil electron, we write down the equations for conservation of momentum in the incident direction and perpendicular to it:

$$\frac{h\nu_0}{c} = \frac{h\nu_\theta}{c}\cos\theta + \frac{m_0 v}{\sqrt{1-\beta^2}}\cos\phi$$

$$0 = \frac{h\nu_\theta}{c}\sin\theta + \frac{m_0 v}{\sqrt{1-\beta^2}}\sin\phi$$

so that

$$\tan\phi = \frac{\nu_\theta \sin\theta}{\nu_\theta\cos\theta - \nu_0}$$

Using the expression for ν_θ as a function of ν_0, we find

$$\tan\phi = -\frac{\cot\theta/2}{1+\alpha}$$

The energy of the recoil electron is given by

$$W_\phi = h(\nu_0 - \nu_\theta) = h\nu_0 \frac{2\alpha\sin^2\theta/2}{1+2\alpha\sin^2\theta/2} = \frac{2\alpha h\nu_0}{1+2\alpha+(1+\alpha)^2\tan^2\phi}$$

The present method leaves one of the angles θ, ϕ indeterminate, for lack of adequate information about the interaction mechanism. The results obtained are valid irrespective of this mechanism.

The angle ϕ can vary between 0 and $\pi/2$, and the angle θ between 0 and π; the electron is always sent forwards.

When θ is zero, ϕ vanishes, but the photon frequency does not change; since the electron receives no energy, it remains stationary. We must be careful about this, and not conclude that it is sent in the direction of the photon.

When θ is very small, the change in momentum is given by

$$\frac{h(\nu_0 - \nu_\theta)}{c} = \frac{h^2\nu_0^2\theta^2}{2m_0c^3}$$

since the frequency shift is small. It is under these conditions, for given ν_0, that the electron acquires least energy.

[219] The Scattering Particles are Free and in Motion before the Collision

(a) The foregoing calculation brings out the physical features fundamental to the phenomenon. We can now consider the general case.

Let the Ox axis be the direction of the incident photon and we add axes Oy and Oz. Before the collision, the scattering electron has velocity v_0 and direction cosines a_0, b_0 and c_0. After the collision, the photon and electron trajectories (electron velocity v) are characterized by the direction cosines a_p, b_p, c_p and a_e, b_e, c_e respectively. Conservation of energy and of the components of momentum gives

$$h\nu_0 + \frac{m_0c^2}{\sqrt{1-\beta_0^2}} = h\nu_\theta + \frac{m_0c^2}{\sqrt{1-\beta^2}}$$

$$\frac{h\nu_0}{c} + \frac{m_0v_0}{\sqrt{1-\beta_0^2}}a_0 = \frac{h\nu_\theta}{c}a_p + \frac{m_0v}{\sqrt{1-\beta^2}}a_e$$

$$\frac{m_0v_0}{\sqrt{1-\beta_0^2}}b_0 = \frac{h\nu_\theta}{c}b_p + \frac{m_0v}{\sqrt{1-\beta^2}}b_e$$

$$\frac{m_0v_0}{\sqrt{1-\beta_0^2}}c_0 = \frac{h\nu_\theta}{c}c_p + \frac{m_0v}{\sqrt{1-\beta^2}}c_e$$

We use the same angles θ and ϕ as in the preceding section. Let ϕ_0 denote the angle between v_0 and Ox, and ω the angle between the trajectory of the scattered photon and v_0:

$$\cos\phi_0 = a_0, \qquad \cos\phi = a_e, \qquad \cos\theta = a_p$$

$$\cos\omega = a_0a_p + b_0b_p + c_0c_p$$

$$a^2 + b^2 + c^2 = 1$$

Using the conservation of momentum and energy and eliminating β, we find

$$\nu_\theta = \frac{1-\beta_0\cos\phi_0}{1-\beta_0\cos\omega + 2\alpha\sqrt{1-\beta_0^2}\sin^2\theta/2}\nu_0$$

(b) In the general case, a supplementary effect arising from the terms in β_0 is added to the Compton effect. At ordinary temperatures, however, the velocities v_0 are small. The terms in β_0 are small and the Compton effect dominates over the supplementary effect; it is merely diminished. In the limit, when the terms in β_0 are negligible, we recover the formula of § [218].

(c) If the velocities v_0 are large, the supplementary effect may completely mask the Compton effect; the frequency is then seen to increase. To investigate this phenomenon, we assume that the Compton effect term (in α) is negligible in the formula. In this case, the electron velocity v_0 is virtually unaltered by the collision. The formula reduces to

$$\nu_\theta' = \frac{1-\beta_0 \cos \phi_0}{1-\beta_0 \cos \omega} \nu_0$$

This formula can be interpreted in an interesting way. A photon having frequency ν_0 in the system of the observer has frequency

$$\nu' = \frac{1-\beta_0 \cos \phi_0}{\sqrt{1-\beta_0^2}} \nu_0$$

in the proper system of the electron. In the latter system, therefore, the electron begins to oscillate with frequency ν'; classically, therefore, it emits light of the same frequency, ν'. In the system of the observer, at direction θ (and hence ω with v_0), the frequency of this light is given by

$$\nu_\theta' = \nu' \frac{\sqrt{1-\beta_0^2}}{1-\beta_0 \cos \omega} = \frac{1-\beta_0 \cos \phi_0}{1-\beta_0 \cos \omega}$$

We recover the formula to be interpreted. The phenomenon is therefore a scattering together with a Doppler effect.

HISTORICAL AND BIBLIOGRAPHICAL NOTES; COMPLEMENTS

[220] The Individual Compton Effect Considered as a Relativistic Elastic Collision; Experimental Confirmation

Compton's theory demonstrates that *macroscopic relativistic mechanics is valid for certain individual phenomena on a microscopic scale.* Its experimental confirmation is therefore of the utmost interest. There are two stages to this.

(a) *The existence of the individual phenomenon*

The observed Compton effect must be proved to be a consequence of elastic collisions, each of which involves only one photon and one electron. A recoil electron corresponds to each scattered photon.

Compton and Simon [16] were the first to show that this is indeed the case, by making a statistical study of the traces of recoil electrons and secondary electrons created by scattered photons (see below, (b)).

Bothe and Geiger [17] confirmed this result, and established that the scattered electron and the recoil electron appear simultaneously. For this, they used two point counters which recorded separately the recoil electrons and the scattered photons from a very small volume of hydrogen. Experiment shows that there are many coincidences between these different groups of events (one might expect to find only coincidences, but allowance must be made for the measuring conditions).

(b) *Quantitative confirmation of the relation between the directions of the scattered photon and the recoil electron*

The correctness of the formula

$$\tan \phi = \frac{1}{1+\alpha} \cot \frac{\theta}{2}$$

has to be established. Figure 65 shows how ϕ varies with θ for various values of the parameter α, that is for various values of the incident photon energy. The angle ϕ can be measured on the photograph on which the trajectory *ME* of the recoil electron appears. The trajectory of the corresponding scattered photon is not visible, but a trajectory *PQ* is sometimes seen on the photograph, starting from a point

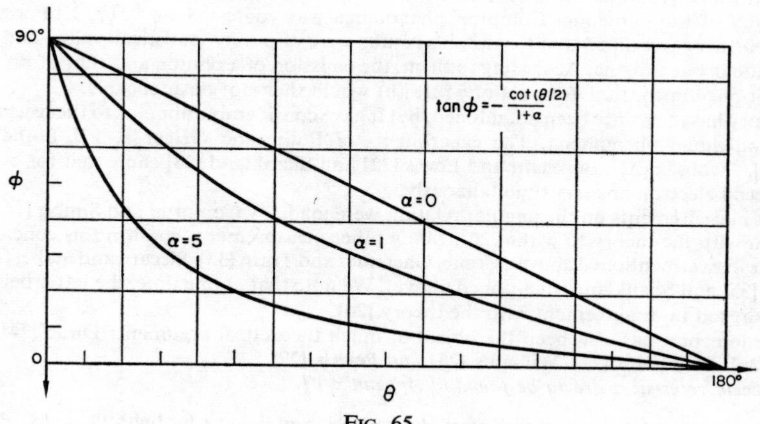

FIG. 65.

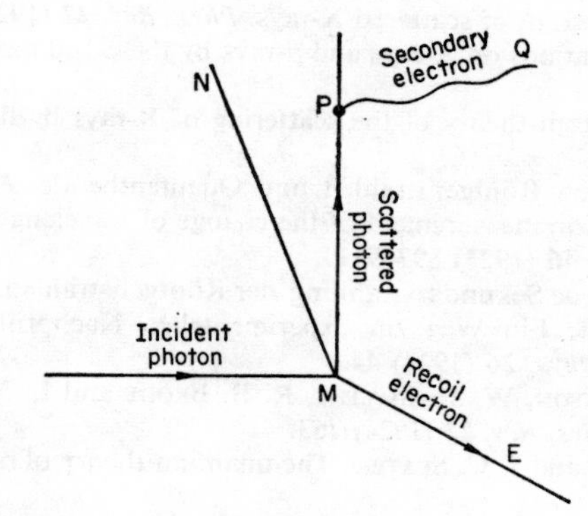

FIG. 66.

P out of line with the beam of incident X-rays (Fig. 66). This is the track of an electron released from an atom of the gas in the Wilson cloud-chamber by the photoelectric effect of the photon scattered at *M*. Thus *MP* is obviously the path of this scattered photon.

Using this method (Compton and Simon [14]), the angles ϕ and θ corresponding to a given collision can be measured. In fact, experiment reveals a difference between *MP* and the theoretical direction *MN*; this is characteristic of the approximate nature of the measurements.

(c) *General survey of the earlier work*

I give below only the bibliographical opening moves, to do justice to the early investigators. Since the subject is rather marginal in a book on dynamics, I refer the reader to specialized texts for an account of recent work.

Compton's fundamental article [1] contains the theory that has been set out in § [218] together with an experimental confirmation of the formula for the change of wavelength, $\Delta\lambda$. In the same year, Debye [5] gave the same theory, and set out the angular relation between the recoil electron and the scattered photon explicitly. These two articles are independent (Compton's appeared in May, Debye's on April 15th).

Much work was then devoted to verifying the expression for $\Delta\lambda$, and the range of measurements was extended; see in particular Compton [2, 4, 6], Woo and Compton [12], Becker, Watson *et al.* [9], Kallman and Mark [13] and Sharp [15].

The reality of the individual Compton phenomena was contested by Bohr, Kramers and Slater [10, 11]. Their theory assumed that Compton's results were of purely statistical value, and did not describe individual phenomena. According to them, the emission of a photon and that of an electron are two different phenomena; the formula of § [218](b) would then not be applicable.

This theory has long since been abandoned, but it gave considerable impetus to the research designed to display individual phenomena. The experiments of Bothe and Geiger [8, 17], Bothe and Maier-Leibnitz [21], Jacobsen [22], Burcham and Lewis [32] and Shankland [23] show that for each scattered photon, a recoil electron appears simultaneously.

The first measurements on the angular relation were made by Compton and Simon [14, 16] and are in agreement with the theory to within 20°. Many other measurements confirm this conclusion: Bothe and Maier-Leibnitz (mentioned above), Crane, Gaerttner and Turin [33], Piccard and Stahel [34], Williams and Pickup [35] and Shankland (mentioned above). We note that at one time, the latter believed that he had found a result in disagreement with the theory [20].

The phenomenon has also been the object of much theoretical argument: Dirac [24], Bohr [25], Cernuschi [26], Hoffmann [27], Williams [28] and Peierls [29].

More recent references are to be found in Appendix VI.

[1] A. H. Compton: A quantum theory of the scattering of X-rays by light elements. *Phys. Rev.* **21** (1923) 483–502.
[2] A. H. Compton: The spectrum of scattered X-rays. *Phys. Rev.* **22** (1923) 409–413.
[3] C. T. R. Wilson: Investigations on X-rays and β-rays by the cloud method. *Proc. Roy. Soc.* A **104** (1923) 1–24.
[4] A. H. Compton: A quantum theory of the scattering of X-rays by light elements. *Phys. Rev.* **21** (1923) 207.
[5] P. Debye: Zerstreuung von Röntgenstrahlen und Quantentheorie. *Phys. Z.* **24** (1923) 161–166.
[6] A. H. Compton: Absorption measurement of the change of wavelength accompanying the scattering of X-rays. *Phil. Mag.* **46** (1923) 897–911.
[7] W. Bothe: Ueber eine neue Sekundärstrahlung der Röntgenstrahlen. *Z. Phys.* **16** (1923) 319–321.
[8] W. Bothe and H. Geiger: Ein Weg zur experimentellen Nachprüfung der Theorie von Bohr, Kramers und Slater. *Z. Phys.* **26** (1924) 44.
[9] J. A. Becker, E. C. Watson, W. R. Smythe, R. B. Brode and L. M. Mott-Smith: The reality of the Compton-effect. *Phys. Rev.* **23** (1924) 763.
[10] N. Bohr, H. A. Kramers and J. C. Slater: The quantum theory of radiation. *Phil. Mag.* **47** (1924) 785–802.
[11] N. Bohr, H. A. Kramers and J. C. Slater: Ueber die Quantentheorie der Strahlung. *Z. Phys.* **24** (1924) 69–87.
[12] Y. H. Woo and H. A. Compton: The wavelength of molybdenum K rays when scattered by light elements. *Proc. Nat. Acad. Sci.* **10** (1924) 271–273.
[13] H. Kallman and H. Mark: Zur Grösse und Winkelabhängigkeit des Compton-effektes. *Naturwissenschaften* **13** (1925) 297–298.
[14] A. H. Compton and A. W. Simon: Directed quanta of scattered X-rays. *Phys. Rev.* **26** (1925) 289–299.
[15] H. M. Sharp: A precision measurement of the change of wavelength of scattered X-rays. *Phys. Rev.* **26** (1925) 692–696.
[16] A. H. Compton and A. W. Simon: Measurements of β-rays associated with scattered X-rays. *Phys. Rev.* **25** (1925) 306–313.

[17] W. BOTHE and H. GEIGER: Ueber das Wesen des Compton-effektes: ein experimenteller Beitrag zur Theorie der Strahlung. *Z. Phys.* **32** (1925) 639–663.
[18] F. KIRCHNER: Ueber den Compton-effekt an gebundenen Elektronen und einige andere Beobachtungen an Nebelkammer-Aufnahmen harter Röntgenstrahlen in Argongas. *Ann. Phys.* **83** (1927) 969–976.
[19] A. H. COMPTON and S. K. ALLISON: *X-rays in Theory and Experiment.* Macmillan, 1935. See especially pp. 199–262.
[20] R. S. SHANKLAND: An apparent failure of the photon theory of scattering. *Phys. Rev.* **49** (1936) 8–13.
[21] W. BOTHE and H. MAIER-LEIBNITZ: Eine neue experimentelle Prüfung der Photonenvorstellung. *Z. Phys.* **102** (1936) 143–155.
[22] J. C. JACOBSEN: Correlation between scattering and recoil in Compton-effect. *Nature* **138** (1936) 25.
[23] R. S. SHANKLAND: The scattering of γ-rays. *Phys. Rev.* **50** (1936) 571.
[24] P. A. M. DIRAC: Does conservation of energy hold in atomic processes? *Nature* **137** (1936) 298–299.
[25] N. BOHR: Conservation laws in quantum theory. *Nature* **138** (1936) 25–26.
[26] F. CERNUSCHI: Conservation de l'énergie et expérience de Shankland. *C. R. Acad. Sci. Paris* **203** (1936) 777–779.
[27] B. HOFFMANN, A. G. SHENSTONE and L. A. TURNER: Coincidences in time in Compton scattering. *Phys. Rev.* **50** (1936) 1092.
[28] E. J. WILLIAMS: Conservation of energy and momentum in atomic processes. *Nature* **137** (1936) 614–615.
[29] R. PEIERLS: Interpretation of Shankland's experiment. *Nature* **137** (1936) 904.
[30] W. BOTHE and H. MAIER-LEIBNITZ: Compton-Effekt und Photonentheorie. *Nachr... zu Göttingen* **2** (1936) 127–130.
[31] W. BOTHE and H. MAIER-LEIBNITZ: Photon theory and Compton-effect. *Phys. Rev.* **50** (1936) 187.
[32] W. E. BURCHAM and W. B. LEWIS: A repetition of the Bothe–Geiger experiment. *Proc. Camb. Phil. Soc.* **32** (1936) 637–642.
[33] H. R. CRANE, E. R. GAERTTNER and J. J. TURIN: A cloud chamber study of the Compton effect. *Phys. Rev.* **50** (1936) 302–308.
[34] A. PICCARD and E. STAHEL: Simultanéité de l'absorption du quantum primaire et de l'émission des rayons secondaires dans l'effet Compton et dans l'effet photoélectrique. *J. Phys. Rad.* **7** (1936) 326–328.
[35] E. J. WILLIAMS and E. PICKUP: Conservation of energy in radiation processes. *Nature* **138** (1936) 461–462.
[36] R. S. SHANKLAND: The Compton-effect with γ-rays. *Phys. Rev.* **51** (1937) 1024.
[37] R. S. SHANKLAND: The Compton-effect with γ-rays. *Phys. Rev.* **52** (1937) 414-418.
For complementary references, see Appendix IV.

[221] The Compton Effect in Quantum Mechanics

Only this theory can fully explain the intensity distribution; the reader will find it discussed in all books dealing with the interaction of matter and radiation. We mention the books of Louis de Broglie [2], W. Heitler [3] and A. Sommerfeld [4], for example.

In 1927 Schrödinger [1] gave an extremely simple treatment of the Compton problem, using the recently discovered de Broglie waves.

[1] E. SCHRÖDINGER: Ueber den Compton-effekt. *Ann. Phys.* **82** (1927) 257–264.
[2] L. DE BROGLIE: *Le Principe de correspondance et les interactions entre la matière et le rayonnement.* Hermann, Paris, 1938, pp. 107–147.
[3] W. HEITLER: *The Quantum Theory of Radiation.* Oxford, 1st ed. 1935; 3rd ed. 1953, pp. 211–231.
[4] A. SOMMERFELD: *Atombau und Spektrallinien.* Vieweg, Braunschweig, vol. **2** (1951) pp. 568–622.

[222] The Compton Effect and the Classical Theory of Electromagnetic Radiation (Application of Bohr's Correspondence Principle)

(a) Compton's theory was not immediately accepted, for the hypothesis concerning light quanta (photons) still startled traditional habits of thought. Before the recoil electrons were observed, some authors advanced the following interpretation which forms an interesting exercise in relativistic calculation.

They considered an incident electromagnetic wave of frequency ν, and they assumed that, under the pressure of the radiation, the electron is driven in the direction of the wave with velocity v. In the system K_0 of the electron, the incident wave has frequency

$$\nu_0 = \nu \frac{\sqrt{1-\beta^2}}{1+\beta}$$

The electron oscillates with this frequency ν_0, and radiates an electromagnetic wave of the same frequency. In the laboratory frame of reference K, the frequency of this wave varies with direction

$$\nu_\theta = \nu_0 \frac{\sqrt{1-\beta^2}}{1-\beta\cos\theta} = \nu \frac{1-\beta}{1-\beta\cos\theta}$$

We have then only to write

$$\beta = \frac{\alpha}{1+\alpha}, \qquad \alpha = \frac{h\nu}{m_0 c^2}$$

to recover the Compton law.

(b) To justify this value for the electron drag velocity, the constant h must be introduced. The authors of the theory devised the following procedure. Each electron absorbs a quantum of energy $h\nu$, acquires velocity v and radiates classically the energy difference W.

The conservation theorems give

$$h\nu = m_0 c^2\left(\frac{1}{\sqrt{1-\beta^2}}-1\right)+W$$

$$\frac{h\nu}{c} = \frac{m_0\,\beta c}{\sqrt{1-\beta^2}}+\frac{W\beta}{c}$$

hence the desired value for β.

(c) This theory was developed by Compton himself [1], by Bauer [2] and by Bauer, Auger and Perrin [3]. Assuming classical emission, Compton [1] calculated in this way the scattered intensity distribution —this problem was not dealt with in the elastic collision theory and had not yet been treated in quantum mechanics. This theory which is really rather too arbitrary a mixture, has now been abandoned; it cannot cope with the fact that the recoil electrons are observed *laterally*.

[1] A. H. COMPTON: refs. cited in § [220].

[2] E. BAUER: Sur le changement de longueur d'onde accompagnant la diffusion des rayons X. *C. R. Acad. Sci. Paris* **177** (1923) 1031–1033.

[3] E. BAUER, P. AUGER and F. PERRIN: Sur la théorie de la diffusion des rayons X. *C. R. Acad. Sci. Paris* **177** (1923) 1211–1212.

CHAPTER XVII

INELASTIC COLLISIONS (NUCLEAR REACTIONS)

A. CLASSIFICATION OF THE PHENOMENA

[223] Two Identical Corpuscles Remain in Contact (macroscopically speaking) after Colliding

The following arguments are also valid for elastic collisions during the collision; as we have neglected the duration of the collision, however, this does not correspond to anything that can be observed.

We can always consider a system K_0 in which the velocities before collision are equal and opposite (zero total momentum). Before the collision, in K_0, the total momentum and energy are

$$\mathbf{p}^0 = 0, \qquad W^0 = 2m_0c^2 + E^0$$

In another system K, with respect to which K_0 has velocity v,

$$\mathbf{p} = \frac{2m_0 + E^0/c^2}{\sqrt{1 - v^2/c^2}}\,\mathbf{v}, \quad W = \frac{2m_0c^2E^0}{\sqrt{1 - v^2/c^2}}$$

After the collision, in K_0,

$$(\mathbf{p}_0)' = 0, \qquad (E^0)' = 0$$

From the macroscopic point of view, the kinetic energy lost reappears as heat (internal kinetic energy) or as interaction energy. It is indeed pointless to be more precise here. Let M_0 be the total proper mass after collision; in K_0, conservation of total energy gives

$$W^0 = (W^0)', \quad 2m_0c^2 + E^0 = M_0c^2$$

$$M_0 = 2m_0 + \frac{E^0}{c^2}$$

The total mass M_0 after collision (or during the collision for the elastic case) is greater than the sum of the masses of the two corpuscles. All things being equal (the velocities, in particular), the mass of an elastically deformed body is greater than that of an undeformed body (§ [153]).

We thus confirm that $\mathbf{p}$ and W are conserved in every other frame of reference K. On calculating the difference between the kinetic energies before and after collision in

K we find

$$E-E' = \left(\frac{2m_0c^2+E^0}{\sqrt{1-v^2/c^2}}-2m_0c^2\right)-M_0c^2\left(\frac{1}{\sqrt{1-v^2/c^2}}-1\right) = E^0$$

This quantity is invariant.

Remark 1. The problem can be generalized by considering two different particles, m_{01} and m_{02}. When the particle 1 has velocity u_1 and particle 2 has velocity u_2, a particle of mass M_0 and velocity V is obtained after the collision, where

$$M_0^2 = m_{01}^2+m_{02}^2+2m_{01}m_{02}\frac{1-u_1u_2/c^2}{\sqrt{1-u_1^2/c^2}\sqrt{1-u_2^2/c^2}}$$

$$V = \frac{m_{01}u_1\sqrt{1-u_2^2/c_2^2}+m_{02}u_2\sqrt{1-u_1^2/c^2}}{m_{01}\sqrt{1-u_2^2/c^2}+m_{02}\sqrt{1-u_1^2/c^2}}$$

Remark 2. Inverse phenomena. If a stationary corpuscle of mass M_0 spontaneously disintegrates into two corpuscles m_{01} and m_{02}, then this means

$$M_0-(m_{01}+m_{02}) > 0$$

This difference ΔM is the *mass defect*, and the corresponding energy ΔMc^2 is the *binding energy*. With this definition, a particle is stable if the binding energy is negative, which is a perfectly natural statement. After disintegration, the energies of the two corpuscles are given by

$$W_1 = \frac{M_0^2+m_{01}^2-m_{02}^2}{2M_0}c^2$$

$$W_2 = \frac{M_0^2+m_{02}^2-m_{01}^2}{2M_0}c^2$$

[224] General Remarks Concerning the Transformation of "Elementary" Particles; Annihilation and Creation Phenomena

The distinction between elementary and composite particles is purely provisional. A particle is said to be elementary if, in the present state of knowledge, it forms an irreducible entity. A particle C is said to be composite if the constituent elementary particles, A and B say, are known.

Relativistic dynamics dictates the stability condition

$$m_C < m_A+m_B$$

Experiment shows that elementary particles are liable to transform into one another. When one or more particles disappear and radiation (as photons) is emitted, we speak of annihilation; conversely, we have creation when radiation becomes particles. Some examples will be given in the following sections.

[225] A Free Electron can Neither Absorb Nor Emit a Photon

(a) This is not obvious. For, since the photon is characterized by zero rest mass, its energy and momentum could be entirely transferred to the electron. The result follows directly from § [218], however (existence of a scattered photon), and from the fact that this collision procedure is reversible.

(b) *Jeans' direct argument*. Suppose that after the electron–photon interaction, only the electron remains. Now consider the proper system K_0 in which the electron is at rest after the collision. In K_0, the electron has lost energy since it has passed from a state of motion to a state of rest; in addition, the energy of the photon has disappeared. The interaction is thus associated with a loss of energy which is in contradiction to relativistic dynamics. An analogous argument explains why emission is impossible.

[226] Two Particles (Corpuscle or Photon) Interact to give a Single Particle; Possible Cases

(a) *We shall seek the transformations that are permitted by the conservation laws; this is not, of course, to say that these possible cases do in fact occur in nature*

For convenience, we shall use the word corpuscle to mean a particle of non-zero proper mass.

The relation

$$\frac{W^2}{c^2} - p^2 = m_0^2 c^2$$

enables us to characterize the particles.

There are two possible cases if we agree that the proper mass m_0 is positive (for negative masses and energies, see § [240]).

(i) For a photon

$$m_0 = 0, \qquad W = cp$$

(ii) For a positive energy corpuscle

$$m_0 \neq 0, \qquad W > cp$$

We write down the conservation laws, labelling the incident particles with suffices 1 and 2 and the resultant particle, 3.

$$W_1 + W_2 = W_3, \qquad \mathbf{p}_1 + \mathbf{p}_2 = \mathbf{p}_3$$

An elementary property of the triangle gives the relation

$$p_3 \leqslant p_1 + p_2$$

between the amplitudes; the equality holds only when $\mathbf{p}_1$ and $\mathbf{p}_2$ are parallel. Hence

$$\frac{W_3}{cp_3} \geqslant \frac{W_1 + W_2}{cp_1 + cp_2}$$

(b) If the incident particles are two corpuscles having positive energy or one such corpuscle and a photon, at least one of the two inequalities

$$W_1 > cp_1, \quad W_2 > cp_2$$

holds, and so

$$W_1 + W_2 > cp_1 + cp_2 \quad \text{or} \quad W_3 > cp_3$$

If two corpuscles with positive energy, or one such corpuscle and a photon, interact to give a single particle, then the latter is a corpuscle with positive energy.

(c) *Two incident photons*

We have

$$W_1 = cp_1, \qquad W_2 = cp_2$$

$$W_1 + W_2 = cp_1 + cp_2$$

If $\mathbf{p}_1$ and $\mathbf{p}_2$ are not parallel,

$$p_3 < p_1 + p_2, \qquad W_3 > cp_3$$

If $\mathbf{p}_1$ and $\mathbf{p}_2$ are parallel and in the same sense,

$$p_3 = p_1 + p_2, \qquad W_3 = cp_3$$

When two photons interact to give a single particle, the latter is

— a corpuscle with positive energy if the directions of the incident photons are different;
— a photon if they are moving in the same direction.

Remark. The first possibility is known as *creation*. The second is an unattainable limiting case, using the elementary concept of the photon. For two photons travelling in the same direction can never collide since they have the same velocity. It seems, therefore, that if such events did occur, we should have to explain them in terms of dispersion *in vacuo*, and hence ascribe a proper mass to the photon (see *Kinematics*, § [59]).

[227] The Inverses of the Foregoing Phenomena

As the procedures studied above are reversible, we come to the following conclusions.

(a) A corpuscle may disintegrate and give rise to:
- — two corpuscles (four unknowns);
- — one corpuscle and a photon (three unknowns);
- — two photons with non-parallel momenta (two unknowns).

(b) A photon may disintegrate and give rise to:
- — two photons having parallel momenta;
- — two corpuscles of the same proper mass.

Remark. The energy and the three components of momentum of each corpuscle are not all fixed by the conservation relation; hence the "unknowns".

[228] Two Corpuscles or a Corpuscle and a Photon (a Pair) Interact to Give Another Pair

With suitable initial conditions, all combinations are possible. In particular, *two corpuscles of the same sign may be annihilated and yield two photons although, as we showed in* § [226], *the result of such annihilation can never be one photon.*

B. QUANTITATIVE EXAMINATION OF VARIOUS CASES

[229] Two Particles of Different Masses, One of which is Stationary Prior to Collision, Give Two Other Particles

(a) *The practical interest of this problem; basic equations*

A particle of proper energy $W_{01} = m_{01}c^2$, to which kinetic energy E_1 is added, strikes a stationary particle of proper energy W_{02}. After the collision, we have two different particles by hypothesis (nuclear reaction), of proper energies W_{03} and W_{04} and kinetic energies E_3 and E_4. The formulae are also valid when particle 1 is replaced by a photon ($W_{01} = 0$, $E_1 = h\nu$).

Expressing the magnitude of the momentum in terms of the kinetic energy E and the proper energy W_0, we have

$$p = \frac{1}{c}\sqrt{E(E+2W_0)}$$

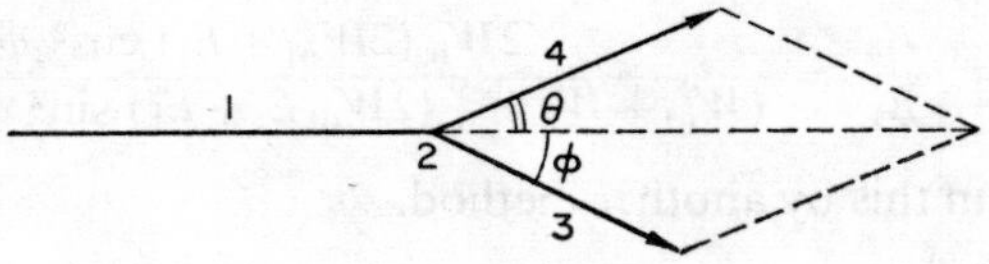

FIG. 67.

Using the angles shown in Fig. 67, the conservation theorems give

$$W_{01}+W_{02}+E_1 = W_{03}+W_{04}+E_3+E_4 \quad (1)$$

$$\sqrt{E_1(E_1+2W_{01})} = \sqrt{E_3(E_3+2W_{03})}\cos\phi+\sqrt{E_4(E_4+2W_{04})}\cos\theta \quad (2)$$

$$0 = \sqrt{E_3(E_3+2W_{03})}\sin\phi-\sqrt{E_4(E_4+2W_{04})}\sin\theta \quad (3)$$

(b) *The angles expressed in terms of the energies or momenta*

To obtain the angle between the two trajectories after collision, we square (2) and (3) and add them:

$$\cos(\phi+\theta) = \frac{E_1(E_1+2W_{01})-\{E_3(E_3+2W_{03})+E_4(E_4+2W_{04})\}}{2\sqrt{E_3(E_3+2W_{03})\,E_4(E_4+2W_{04})}}$$

Using the quantities W_0 and E, the formula for the special case of § [207] becomes

$$\cos(\phi+\theta) = \sqrt{\frac{E_3E_4}{(2W_0+E_3)(2W_0+E_4)}}$$

If we use the momenta, we obtain the same expression as in § [207]:

$$\cos(\phi+\theta) = \frac{p_1^2-(p_3^2+p_4^2)}{2p_3p_4}$$

To obtain the angle between one of the emergent trajectories and the incident trajectory, ϕ for example, we eliminate θ between (2) and (3), giving

$$\cos\phi = \frac{E_4(E_4+2W_{04})-\{E_1(E_1+2W_{01})+E_3(E_3+2W_{03})\}}{2\sqrt{E_1(E_1+2W_{01})\,E_3(E_3+2W_{03})}} = \frac{p_4^2-(p_1^2+p_3^2)}{2p_1p_3}$$

with a similar formula for $\cos\theta$. We notice, incidentally, that these formulae for $\cos\phi$ and $\cos\theta$ can be read immediately off the triangle of momenta. Even if the velocity of 2 is not zero, they remain valid; we have only to replace $\mathbf{p}_1$ by the resultant of $\mathbf{p}_1+\mathbf{p}_2$.

(c) *The efficiency of an elastic collision*

If we wish to give an uncharged particle (particle 2) a certain velocity, we can use the present procedure. The efficiency of the operation is defined as the ratio of the useful kinetic energy and the incident kinetic energy E_1. Since the collision is elastic, there is no transmutation of particles; we thus write

$$W_{02} = W_{03}, \quad W_{01} = W_{04}, \quad E_1 = E_3+E_4$$

Substituting into the formula of (b), we have

$$\text{Efficiency} = \frac{E_3}{E_1} = \frac{2W_{02}(2W_{01}+E_1)\cos^2\phi}{(W_{01}+W_{02})^2+(2W_{01}E_1+E_1^2)\sin^2\phi+2E_1W_{02}}$$

We shall later obtain this by another method.

[230] The Heat of Reaction Liberated in an Inelastic Collision

(a) By extension of the classical terminology, this is the difference between the proper energies before and after the collision. If this energy, Q, is negative, it remains stored in the particles, and hence kinetic energy is lost. In the opposite case, kinetic energy is gained.

By definition,

$$Q = (W_{01}+W_{02})-(W_{03}+W_{04})$$

(b) *Lithium nuclei bombarded with protons.* When a proton ${}_1^1\text{H}$ penetrates into a lithium nucleus ${}_3^7\text{Li}$, the composite particle which results is unstable and divides into two helium nuclei ${}_2^4\text{He}$:

$${}_3^7\text{Li}+{}_1^1\text{H} \rightarrow {}_2^4\text{He}+{}_2^4\text{He}$$

Taking the mass of oxygen to be 16, the masses are as follows:

$$^{7}_{3}\text{Li} = 7{\cdot}0166, \qquad ^{1}_{1}\text{H} = 1{\cdot}0076, \qquad ^{4}_{2}\text{He} = 4{\cdot}0028$$

The mass difference is thus 0·0186 atomic units, or $0{\cdot}309 \times 10^{-25}$ g. After the collision, there is therefore excess kinetic energy

$$Q = 27{\cdot}7 \times 10^{-6} \text{ erg}$$

The difference between the kinetic energies before and after collision has been measured directly by N. M. Smith, who finds

$$17{\cdot}28 \pm 0{\cdot}03 \text{ MeV} = (27{\cdot}6 \pm 0{\cdot}05) + 10^{-6} \text{ erg}$$

which is in excellent agreement with the theoretical value for Q.

[231] Calculation Using the "Rest" System (the Baryocentric System)

(a) *The velocity v of this system*

In the system K used above, the particles 1 and 2 with proper masses m_{01} and m_{02} have velocities v_1 and 0; in the baryocentric system K_0, which has velocity v with respect to K, we denote the velocities of the particles by v_{01} and v_{02}. The velocity transformation then gives

$$v_{02} = -v$$

$$v_{01} = \frac{v_1 - v}{1 - vv_1/c^2} \cdot \frac{v_{01}}{\sqrt{1 - v_{01}^2/c^2}} = \frac{v_1 - v}{\sqrt{1 - v_1^2/c^2}\sqrt{1 - v^2/c^2}}$$

The system $\boldsymbol{K_0}$ is defined by

$$\frac{m_{01}v_{01}}{\sqrt{1 - v_{01}^2/c^2}} + \frac{m_{02}v_{02}}{\sqrt{1 - v_{02}^2/c^2}} = 0$$

or

$$\frac{m_{01}v_1}{\sqrt{1 - v_1^2/c^2}} = \left(m_{02} + \frac{m_{01}}{\sqrt{1 - v_1^2/c^2}}\right) v$$

From this relation, v may be calculated. To obtain v, we may also use the following argument, based on the relation

$$\mathbf{p} = \frac{W}{c^2}\mathbf{v}$$

We consider the system of two particles 1 and 2 as a composite particle at rest in K_0, and hence travelling with velocity v in K; we immediately find

$$\mathbf{v} = c^2 \frac{\mathbf{p}_1 + \mathbf{p}_2}{W_1 + W_2} \qquad (\text{here, } p_2 = 0)$$

This formula is in everyday use in cosmic ray laboratories.

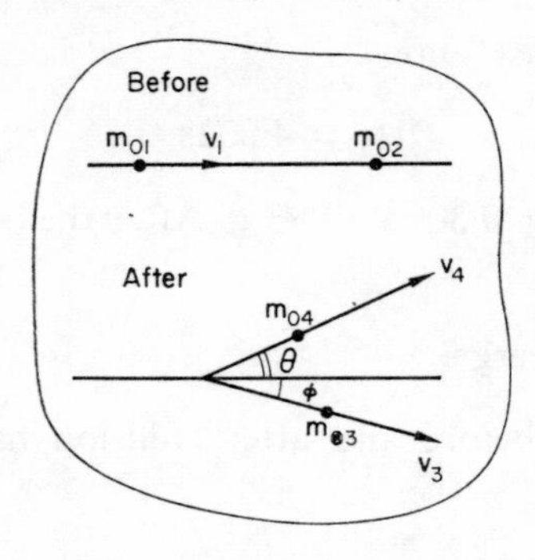

The laboratory system, K

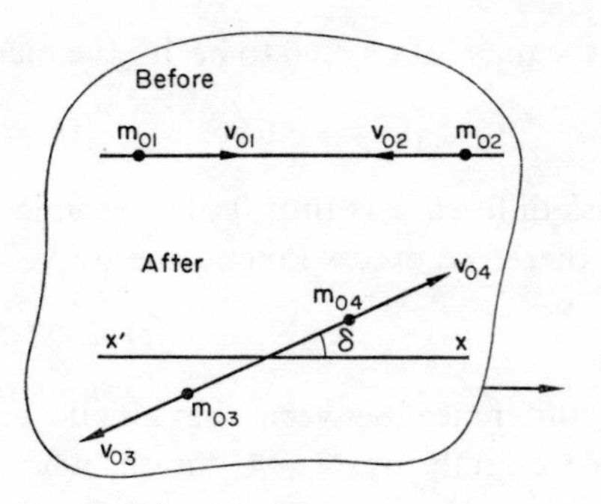

The baryocentric system, K_0

FIG. 68.

Remark. Figure 68 illustrates the phenomena in systems K and K_0 before and after the collision. One of the angles θ, ϕ in K, and the angle δ in K_0, are arbitrary (§ [207]).

(b) *Total energy in K_0 before the collision*

In K_0, the energy is given by

$$W_{K_0} = \frac{m_{01}c^2}{\sqrt{1-v_{01}^2/c^2}} + \frac{m_{02}c^2}{\sqrt{1-v_{02}^2/c^2}}$$

with

$$\sqrt{1-v_{01}^2/c^2} = \frac{\sqrt{1-v_1^2/c^2}\sqrt{1-v^2/c^2}}{1-vv_1/c^2}$$

In this expression, we substitute for v in terms of v_1 and then for v_1 in terms of m_{01} and E_1:

$$v_1 = c\sqrt{1-\left(\frac{m_{01}c^2}{m_{01}c^2+E_1}\right)^2}$$

Finally, we obtain an expression for W_{K_0}, in terms of the known quantities m_{01}, m_{02} and E_1:

$$W_{K_0} = c^2\sqrt{m_{01}^2+m_{02}^2+2m_{01}m_{02}\left(1+\frac{E_1}{m_{01}c^2}\right)}$$

we shall use this formula in § [233].

(c) *The energy distribution after an elastic collision (in K_0 and in K)*

Landau and Lifshitz have made some very significant comments on this. From the relation

$$\mathbf{p}_{01}+\mathbf{p}_{02} = 0$$

it is clear that in K_0, the two momenta remain equal and opposite after the collision. Furthermore, conservation of energy shows that the absolute value of each momentum remains the same. Hence the only effect of the collision on the momentum vectors is to rotate them through an angle δ about their initial direction, $x'x$ (Fig. 68).

Before the collision, particle 2 has momentum

$$-\frac{m_{02}v}{\sqrt{1-v^2/c^2}}$$

parallel to $x'x$. After the collision, the component of the momentum of 2 along $x'x$ is

$$-\frac{m_{02}v}{\sqrt{1-v^2/c^2}}\cos\delta$$

and the energy of 2 is

$$(W_{K_0})_2 = \frac{m_{02}c^2}{\sqrt{1-v^2/c^2}}$$

Transforming back into the original system K, we have

$$W_2' = \frac{m_{02}c^2\{1-(v^2/c^2)\cos\delta\}}{1-v^2/c^2}$$

and after eliminating v,

$$W_2' = m_{02}c^2 + \frac{m_{02}(W_1^2 - m_{01}^2c^4)}{m_{01}^2c^2 + m_{02}^2c^2 + 2m_{02}W_1}(1-\cos\delta)$$

The second term is the energy acquired by particle 2 during the collision. Using the conservation of energy,

$$W_1' = W_1 - \frac{m_{02}(W_1^2 - m_{01}^2c^4)}{m_{01}^2c^2 + m_{02}^2c^2 + 2m_{02}W_1}(1-\cos\delta)$$

W_2' will be greatest and W_1' least (corresponding to the largest transfer of energy from 1 to 2 and hence maximum efficiency in the sense of § [229](c)) when δ is equal to π

$$(W_2')_{\max} = m_{02}c^2 + \frac{2m_{02}(W_1^2 - m_{01}^2c^4)}{m_{01}^2c^2 + m_{02}^2c^2 + 2m_{02}W_1}$$

$$(W_1')_{\min} = m_{01}c^2 + \frac{(W_1 - m_{01}c^2)(m_{02}-m_{01})^2c^2}{m_{01}^2c^2 + m_{02}^2c^2 + 2m_{02}W_1}$$

To obtain the pre-relativistic approximation, we write

$$\frac{(W_1')_{\min} - m_{01}c^2}{W_1 - m_{01}c^2} = \frac{(m_{02}-m_{01})^2}{m_{01}^2 + m_{02}^2 + \dfrac{2m_{02}W_1}{c^2}}$$

and for low velocities, we recover the pre-relativistic result:

$$\frac{E_1'}{E_1} = \left(\frac{m_{01}-m_{02}}{m_{01}+m_{02}}\right)^2$$

For velocities in the neighbourhood of c, W_1 is very large; the ratio given above tends to zero and $(W_1')_{\min}$ tends towards

$$\frac{m_{02}^2+m_{01}^2}{2m_{02}}c^2$$

If $m_{01} \ll m_{02}$ *(a light particle falls on a heavy stationary particle)*, pre-relativistic dynamics predicts negligible energy transfer:

$$E_1' \simeq E_1$$

At relativistic velocities, on the contrary, we have

$$\frac{E_1'}{E_1} \simeq \frac{m_{02}^2}{m_{02}^2+\dfrac{2m_{02}W_1}{c^2}} = \frac{1}{1+\dfrac{2W_1}{m_{02}c^2}}$$

The light particle may thus transfer almost all its energy to the heavy particle if W_1 is large enough in comparison with $m_{02}c^2$.

The results for the case $m_{02} \ll m_{01}$ *(a heavy particle falls on a light stationary particle)* are analogous; at relativistic velocities,

$$\frac{E_1'}{E_1} = \frac{1}{1+\dfrac{2m_{02}W_1}{m_{01}^2c^2}}$$

For high efficiency, we must have

$$2m_{02}W_1 \gg m_{01}^2c^2$$

[232] The Elliptic Indicatrix of the Momenta for Two Emergent Particles

(a) *The equation of the indicatrix*

In § [229](b), we found

$$\cos\phi = \frac{p_4^2-(p^2+p_3^2)}{2pp_3} \tag{1}$$

We write the conservation of energy in the form (§ [17])

$$\sqrt{p_3^2+m_{03}^2c^2}+\sqrt{p_4^2+m_{04}^2c^2} = \frac{W}{c} \tag{2}$$

In these formulae, p and W denote the resultant of the momenta and the total energy respectively, which are known quantities. By hypothesis, we also know what the nature of the emergent particles is (m_{03} and m_{04}). Since one of the angles θ, ϕ is arbitrary, an infinite number of solutions is possible. Let OO' be the resultant $\mathbf{p}$ and $\overline{OM}$, $\overline{MO'}$ the momenta $\mathbf{p}_3$ and $\mathbf{p}_4$ (Fig. 69). We shall see that, because of the supplementary condition

(2), the point M traces out a curve when ϕ or θ is varied. Eliminating p_4 from (1) and (2), we have

$$4p_3^2(p^2c^2\cos^2\phi - W^2) - 4(p^2c^2 - W^2 + m_{04}^2c^4 - m_{03}^2c^4)p_3\cos\phi$$
$$+\frac{1}{c^2}(p^2c^2 - W^2 + m_{04}^2c^2 - m_{03}^2c^2)^2 - 4m_{03}^2c^4W^2 = 0$$

By definition, this curve $f(p_3, \phi)$ is the momentum indicatrix; a point M corresponds to each value of ϕ, and hence so does a value of p_3 and, in consequence, of p_4. We could of course have derived an indicatrix $f(p_4, \theta)$.

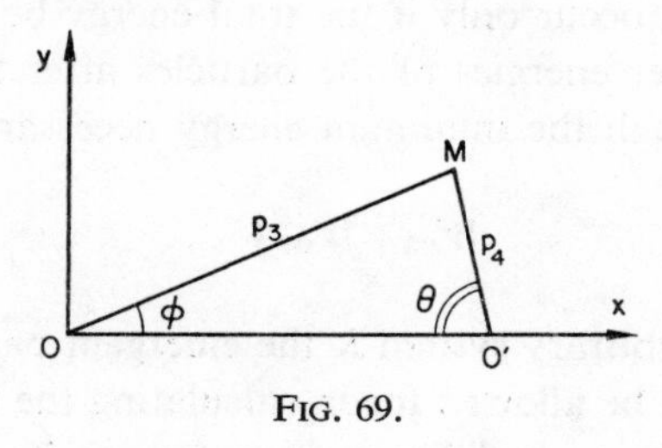

FIG. 69.

In Cartesian coordinates (Fig. 69),

$$x = p_3\cos\phi, \qquad y = p_3\sin\phi$$

we have the equation

$$4p^2c^2x^2 - 4W^2(x^2+y^2) - 4p(p^2c^2 - W^2 + m_{04}^2c^4 - m_{03}^2c^4)x + \ldots = 0$$

which represents a conic. Since the momenta p_3 and p_4 must remain finite, this conic can only be an ellipse; the analytical condition

$$W^2 - p^2c^2 > 0$$

is indeed satisfied.

We notice that the energy condition also gives us

$$W^2 - p^2c^2 > (m_{03} + m_{04})^2c^4$$

In the baryocentric system ($p = 0$) the indicatrix is a circle

$$-4W^2(x^2+y^2) + \frac{1}{c^2}(-W^2 + m_{04}^2 - m_{03}^2c^4)^2 - 4m_{03}^2c^4W^2 = 0$$

The modulus of the momentum of each particle, and hence its energy, is independent of ϕ. If one of the particles is a photon, we set m_{03} or m_{04} equal to zero.

(b) *An application*

Let us obtain the law for the Compton effect by means of the indicatrix. Using the notation of § [218], and replacing ϕ by θ in the indicatrix equation, we have

$$W = h\nu_0 + m_0c^2, \quad p = \frac{h\nu_0}{c}$$

$$p_3 = \frac{h\nu_\theta}{c}, \qquad m_{03} = 0, \qquad m_{04} = m_0$$

A simple calculation yields the expression $\nu_\theta = f(\theta)$ of § [218].

[233] The Energy Required for a Given Reaction (the Energy Threshold) in the Case of Two Incident Particles

(a) We again consider two particles, one of which is stationary before the collision; we now assume that there is an arbitrary number of particles after the collision:

$$W_{01}+E_1+W_{02} = W_{03}+E_3+W_{04}+E_4+ \dots$$

A nuclear reaction can occur only if the total energy before the collision is greater than the sum of the proper energies of the particles after the reaction. At first sight, therefore, it seems as though the minimum energy necessary (the energy threshold) is given by

$$W_{03}+W_{04}+ \dots$$

This is wrong, for in an arbitrary system K the emergent particles will necessarily have kinetic energy, which must be allowed for in calculating the threshold. In fact, the total momentum before the collision is different from zero in K, and hence the same is true after the collision. We obtain the threshold by considering the baryocentric system in which the total momentum is zero and where, in consequence, the kinetic energy after the collision *may* also be zero.

Let us consider the case of a reaction which gives three particles labelled 3, 4 and 5, for example. Using the formula of § [231](b), the energy threshold is defined by

$$W_{K_0} \geqslant (m_{03}+m_{04}+m_{05})c^2$$

or, substituting for W_{K_0} from § [231] and solving for E_1,

$$E_1 \geqslant \frac{c^2}{2m_{02}}\{(m_{03}+m_{04}+m_{05})^2-(m_{01}+m_{02})^2\}$$

(b) *Example*. Suppose that we wish to provoke the nuclear reaction

$$p+p \rightarrow p+n+\pi^+$$

(this example is borrowed from a course by Bernard). A proton p of kinetic energy E_1 strikes a stationary proton and produces a proton, a neutron n and a meson π^+. Denoting the common mass of proton and neutron by m_0, and that of the meson by m_π, we have

$$E_1 \geqslant \frac{c^2}{2m_0}\{(2m_0+m_\pi)^2-4m_0^2\}$$

or

$$E_1 \geqslant 2m_\pi c^2\left(1+\frac{m_\pi}{4m_0}\right)$$

Numerically,

$$m_0c^2 = 938 \text{ MeV}, \qquad m_\pi c^2 = 141 \text{ MeV}$$

$$E_1 \geqslant 293 \text{ MeV}$$

[234] The Energy Threshold when Photons are Involved in the Reaction

(a) *The energy in the baryocentric system*

We assume that particle 1, say, is replaced by a photon, frequency ν, and particle 2 is at rest. The baryocentric system is defined by

$$\frac{h\nu_0}{c}+\frac{m_{02}v_{02}}{\sqrt{1-v_{02}^2/c^2}}=0$$

with (§ [63])

$$v_{02}=-v, \qquad \nu_0=\sqrt{\frac{c-v}{c+v}}\,\nu$$

so that

$$\sqrt{\frac{c-v}{c+v}}\,\frac{h\nu}{c}-\frac{m_{02}v}{\sqrt{1-v^2/c^2}}=0$$

$$\left\{\left(\frac{h\nu}{c}\right)^2-m_{02}^2c^2\right\}\beta^2-2\beta\left(\frac{h\nu}{c}\right)^2+\left(\frac{h\nu}{c}\right)^2=0$$

$$\beta=\frac{1}{1+\dfrac{m_{02}c^2}{h\nu}}$$

To obtain $\beta<1$ we take the positive sign, and the energy is therefore

$$W_{K_0}=\sqrt{\frac{1-\beta}{1+\beta}}\,h\nu+\frac{m_{02}c^2}{\sqrt{1-\beta^2}}=\frac{h\nu}{\sqrt{1-\beta^2}}\left(1-\beta+\frac{m_{02}c^2}{h\nu}\right)$$

Remark. If particle 2 has velocity v_2 parallel to that of the photon, we write

$$\frac{h\nu_0}{c}+\frac{m_{02}v_{02}}{\sqrt{1-v_{02}^2/c^2}}=0$$

and now

$$\frac{v_{02}}{\sqrt{1-v_{02}^2/c^2}}=\frac{v_2-v}{\sqrt{1-v_2^2/c^2}\sqrt{1-v^2/c^2}}$$

so that

$$\sqrt{\frac{c-v}{c+v}}\,\frac{h\nu}{c}+m_{02}\frac{v_2-v}{\sqrt{1-v_2^2/c^2}\sqrt{1-v^2/c^2}}=0$$

and hence we obtain v by means of a second-degree equation.

(b) *The energy threshold and the frequency threshold*

If there is no photon after the collision, the calculation is the same as that of § [233], using the above expression for W_{K_0}. If v_2 is zero, a frequency threshold for the incident photon can then be determined.

When there are emergent photons, the energies of these photons are added to the proper energies of the corpuscles. For example,

$$W_{K_0} \geqslant m_{03}c^2 + m_{04}c^2 + h\nu_5$$

[235] The Two Particles are in Motion before Colliding

(a) *The energy available in the frame of reference of each particle*

In the foregoing calculations, it was always assumed that one particle was practically stationary before the collision. This does in fact correspond to the usual case in which a low density beam of particles from an accelerator impinges on a target; the collision probabilities are appreciable because of the density of the target.

There is no doubt that with present-day accelerators, much higher reaction energies would be obtained by firing two beams, travelling in opposite directions with velocities v_1 and v_2, at one another. Hitherto, this idea had not been taken very seriously, because the probability of collision was virtually zero with such low beam current densities. Now, however, this is a practical possibility, for beam current densities have been considerably raised; this is one of the ways in which the energy is being currently increased.

In the laboratory system K, consider the special case of two particles 1 and 2, travelling parallel to Ox towards one another with velocities v_1 and v_2; in K, particle 2 has energy W. In the system K_1 of particle 1, the latter is initially stationary and experiences the impact of particle 2, with energy (§ [29])

$$W_1 = W\frac{1+v_1v_2/c^2}{\sqrt{1-v_1^2/c^2}}$$

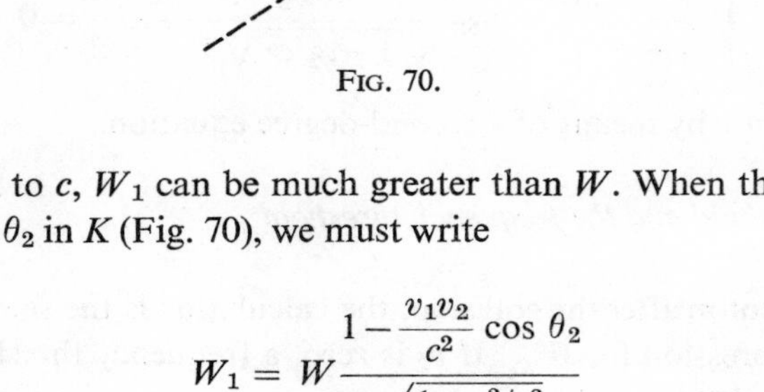

FIG. 70.

If v_1 is nearly equal to c, W_1 can be much greater than W. When the two velocities are inclined at an angle θ_2 in K (Fig. 70), we must write

$$W_1 = W\frac{1-\dfrac{v_1v_2}{c^2}\cos\theta_2}{\sqrt{1-v_1^2/c^2}}$$

(b) *Determination of the baryocentric system of two particles*

In K, we have the momenta $m_1\mathbf{v}_1$ and $m_2\mathbf{v}_2$ inclined at an angle ϕ (Fig. 71). The baryocentric system K_0 is moving at velocity v with respect to K along the direction $x'x$, inclined at angle θ_1 to v_1 (for example); the unknowns are v and θ_1.

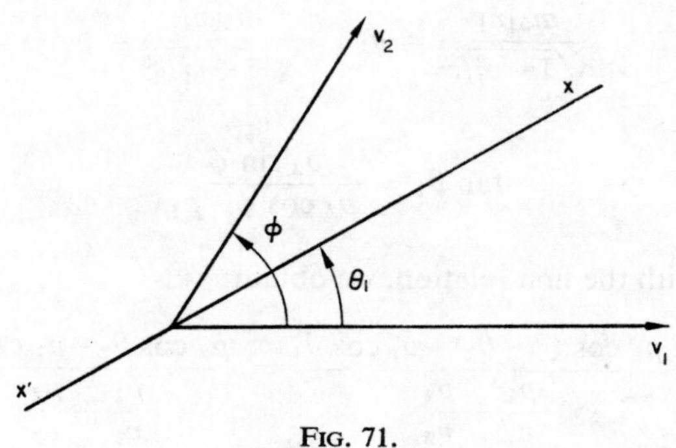

FIG. 71.

Let the velocity components of 1 and 2 be

$$v_{1x},\quad v_{1y},\quad v_{2x},\quad v_{2y}\quad \text{in}\quad K$$

$$v_{01x},\quad v_{01y},\quad v_{02x},\quad v_{02y}\quad \text{in}\quad K_0$$

From the definition of K_0,

$$\frac{m_{01}v_{01x}}{\sqrt{1-v_{01}^2/c^2}}+\frac{m_{02}v_{02x}}{\sqrt{1-v_{02}^2/c^2}}=0$$

$$\frac{m_{01}v_{01y}}{\sqrt{1-v_{01}^2/c^2}}+\frac{m_{02}v_{02y}}{\sqrt{1-v_{02}^2/c^2}}=0$$

with

$$v_{01x}=\frac{v_1\cos\theta_1-v}{1-\dfrac{vv_1\cos\theta_1}{c^2}},\qquad v_{01y}=v_1\sin\theta_1\frac{\sqrt{1-v^2/c^2}}{1-\dfrac{vv_1\cos\theta_1}{c^2}}$$

$$\sqrt{1-\frac{v_{01}^2}{c^2}}=\frac{\sqrt{1-v_1^2/c^2}\sqrt{1-v^2/c^2}}{1-\dfrac{vv_1\cos\theta_1}{c^2}}$$

$$v_{02x}=\frac{v_2\cos(\phi-\theta_1)-v}{1-\dfrac{vv_2\cos(\phi-\theta_1)}{c^2}},\qquad v_{02y}=v_2\sin(\phi-\theta_1)\frac{\sqrt{1-v^2/c^2}}{1-\dfrac{vv_2\cos(\phi-\theta_1)}{c^2}}$$

$$\sqrt{1-\frac{v_{02}^2}{c^2}}=\frac{\sqrt{1-v^2/c^2}\sqrt{1-v^2/c^2}}{1-\dfrac{vv_2\cos(\phi-\theta_1)}{c^2}}$$

Hence

$$m_{02}\frac{v_2 \cos(\phi-\theta_1)-v}{\sqrt{1-v_2^2/c^2}}+m_{01}\frac{v_1 \cos\theta_1-v}{\sqrt{1-v_1^2/c^2}}=0$$

$$m_{02}\frac{v_2 \sin(\phi-\theta_1)}{\sqrt{1-v_2^2/c^2}}+m_{01}\frac{v_1 \sin\theta_1}{\sqrt{1-v_1^2/c^2}}=0$$

We write

$$\frac{m_{01}v_1}{\sqrt{1-v_1^2/c^2}}=p_1, \qquad \frac{m_{02}v_2}{\sqrt{1-v_2^2/c^2}}=p_2$$

The last relation gives

$$\tan\theta_1=\frac{p_2 \sin\phi}{p_2 \cos\phi-p_1}$$

and hence θ; now, with the first relation, we obtain

$$v=\frac{p_2 \cos(\phi-\theta_1)+p_1 \cos\theta_1}{\dfrac{p_1}{v_1}+\dfrac{p_2}{v_2}}=\frac{p_2 \cos\theta_2+p_1 \cos\theta_1}{\dfrac{p_1}{v_1}+\dfrac{p_2}{v_2}}$$

We recover the usual special case of § [231] by writing

$$v_2=0, \qquad \phi=0, \qquad \frac{p_2}{v_2}=m_{02}$$

so that

$$\theta_1=0, \qquad v=\frac{p_1}{\dfrac{p_1}{v_1}+m_{02}}$$

(c) *The baryocentric system of two photons*

The formulae already proved for two corpuscles may be used, written in terms of p_1 and p_2. The direct calculation, which I give below, is an interesting relativistic exercise (application of the formulae for the Doppler effect). In the system K, consider two photons of frequencies ν_1 and ν_2, the velocities of which are inclined at an angle ϕ. We are required to find the velocity v of the baryocentric system K_0; θ_1 and θ_2 denote the angles between $\mathbf{v}$ and the velocities of the photons 1 and 2 respectively.

We first decide the signs in the Doppler effect formulae. With the notation of Fig. 72,

$$\cos\theta_0=\frac{\cos\theta-\beta}{1-\beta\cos\theta}$$

If θ is zero, so too is θ_0. To obtain $\theta_0>\pi/2$, we must have

$$\cos\theta<\beta$$

We return to the two photons. So that the sum of their momenta can be zero, we must have the arrangement of Fig. 72.

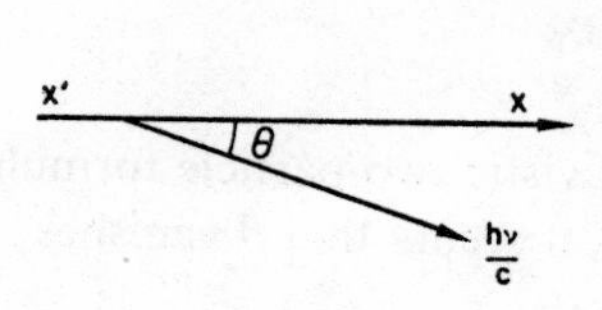

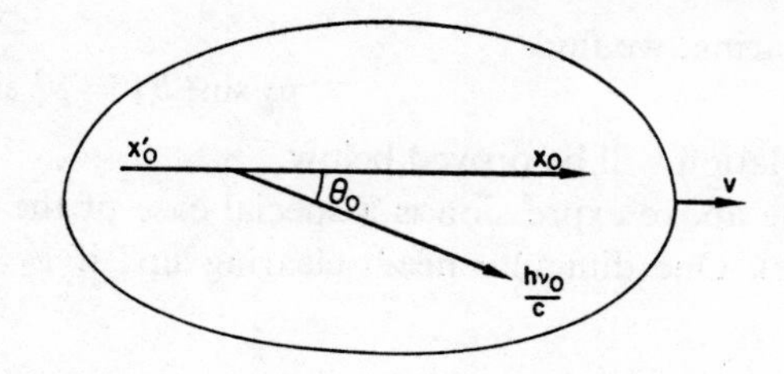

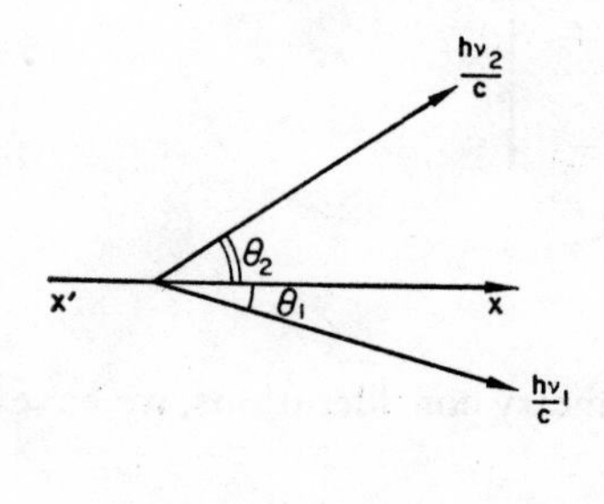

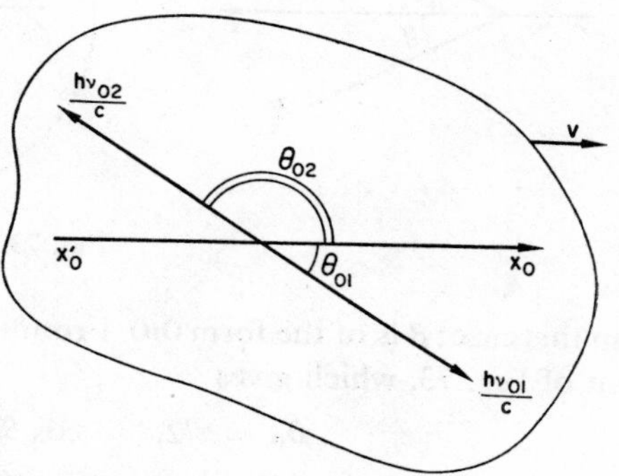

FIG. 72.

In K_0:

$$\nu_{01} = \frac{1-\beta \cos \theta_1}{\sqrt{1-\beta^2}}\, \nu_1, \qquad \cos \theta_{01} = \frac{\cos \theta_1 - \beta}{1-\beta \cos \theta_1}$$

$$\nu_{02} = \frac{1-\beta \cos \theta_2}{\sqrt{1-\beta^2}}\, \nu_2 = \frac{1-\beta \cos (\phi-\theta_1)}{\sqrt{1-\beta^2}}\, \nu_2$$

$$\cos \theta_{02} = \frac{\cos \theta_2 - \beta}{1-\beta \cos \theta_2} = \frac{\cos (\phi-\theta_1)-\beta}{1-\beta \cos (\phi-\theta_1)}$$

The system K_0 will be baryocentric if

$$\nu_{01} = \nu_{02}, \qquad \cos \theta_{01} = -\cos \theta_{02}$$

so that

$$(1-\beta \cos \theta_1)\, \nu_1 = \{1-\beta \cos (\phi-\theta_1)\}\nu_2$$

$$\frac{\cos \theta_1 - \beta}{1-\beta \cos \theta_1} = \frac{\beta - \cos (\phi-\theta_1)}{1-\beta \cos (\phi-\theta_1)}$$

The first relation gives us

$$\beta = \frac{\nu_2 - \nu_1}{\nu_2 \cos (\phi-\theta_1) - \nu_1 \cos \theta_1} = \frac{p_2 - p_1}{p_2 \cos (\phi-\theta_1) - p_1 \cos \theta_1}$$

which can also be written

$$\beta = \frac{p_2 \cos (\phi-\theta_1) + p_1 \cos \theta_1}{p_1 + p_2}$$

Comparing, we find

$$p_2^2 \sin^2 \theta_2 = p_1^2 \sin^2 \theta_1$$

this relation will be proved below.

The above expression is a special case of the relativistic two-particle formula ($v_1 = v_2 = c$). One difficulty needs clearing up: if $\nu_2 = \nu_1$, it seems that β vanishes. In fact,

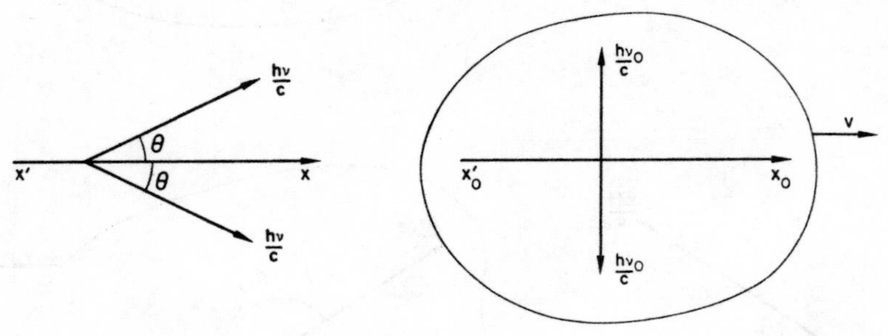

FIG. 73.

$\theta_1 = \theta_2$ in this case; β is of the form 0/0. From symmetry considerations, we have the arrangement of Fig. 73, which gives

$$\theta_0 = \pi/2, \qquad \cos\theta_0 = 0$$

$$\beta = \cos\theta, \qquad \nu_0 = \sqrt{1-\beta^2}\,\nu$$

and has two special cases:

—parallel photons

$$\theta = 0, \qquad \beta = 1, \qquad \nu_0 = 0$$

—photons in opposite directions

$$\theta = \pi/2, \qquad \beta = 0, \qquad \nu_0 = \nu$$

We now return to the case, $\nu_2 \neq \nu_1$. For the two special cases just mentioned, we have

—parallel photons

$$\theta_1 = \theta_2 = 0, \qquad \beta = 1, \qquad \nu_0 = \sqrt{\frac{1-\beta}{1+\beta}}\,\nu$$

—photons in opposite directions

$$\theta_1 = 0, \qquad \theta_2 = \pi$$

$$\beta = \frac{\nu_1 - \nu_2}{\nu_1 + \nu_2}, \qquad \nu_0 = \sqrt{\frac{1-\beta}{1+\beta}}\,\nu_1 = \sqrt{\frac{1+\beta}{1-\beta}}\,\nu_2$$

The velocity of K_0 is in the same direction as that of the higher frequency photon.

In the general case ($\nu_2 \neq \nu_1$, $\theta_2 \neq \theta_1$), the formula giving β contains the unknown angle θ_1. To determine this angle, we use the second baryocentric condition, in which we substitute for β in terms of θ_1:

$$\left\{\cos\theta_1 - \frac{\nu_2 - \nu_1}{\nu_2 \cos(\phi - \theta_1) - \nu_1 \cos\theta_1}\right\}\left\{1 - \frac{\nu_2 - \nu_1}{\nu_2 \cos(\phi - \theta_1) - \nu_1 \cos\theta_1}\cos(\phi - \theta_1)\right\}$$

$$= \left\{1 - \frac{\nu_2 - \nu_1}{\nu_2 \cos(\phi - \theta_1) - \nu_1 \cos\theta_1}\cos\theta_1\right\}\left\{\frac{\nu_2 - \nu_1}{\nu_2 \cos(\phi - \theta_1) - \nu_1 \cos\theta_1} - \cos(\phi - \theta_1)\right\}$$

We find

$$\nu_1^2 \sin^2 \theta_1 = \nu_2^2 \sin^2 (\phi - \theta_1) = \nu_2^2 \sin^2 \theta_2$$

and finally

$$\tan \theta_1 = \frac{\nu_2 \sin \phi}{\nu_2 \cos \phi - \nu_1} = \frac{p_2 \sin \phi}{p_2 \cos \phi - p_1}$$

which is the same as the corpuscle expression.

[236] Annihilation Phenomena

(a) *A neutral particle produces two photons*

We consider the system of the particle (the baryocentric system), and we denote the mass of the particle by m_0 and the frequencies of the two resultant photons by ν_{01} and ν_{02}; the latter obviously travel in opposite directions along an arbitrary straight line (θ_0 is arbitrary). We have

$$\frac{h\nu_{01}}{c} - \frac{h\nu_{02}}{c} = 0, \quad h\nu_{01} + h\nu_{02} = m_0 c^2$$

so that

$$\nu_{01} = \nu_{02} = \nu_0 = \frac{m_0 c^2}{2h}$$

It is obvious that the two frequencies will be equal, on symmetry grounds.

We transfer to the system K in which m_0 has velocity $\mathbf{v}$, using the Doppler effect formulae (§ [63]). The emergent photons are inclined to $\mathbf{v}$ at angles (Fig. 74)

$$\theta_0, \quad \pi - \theta_0 \quad \text{(baryocentric system)}$$

$$\theta_1, \quad \theta_2 \quad \text{(system } K\text{)}$$

where θ_0 is taken smaller than $\pi/2$. Denoting the frequencies and corpuscle mass in K by ν_1, ν_2 and m respectively, we have

$$\nu_1 = \frac{1+\beta \cos \theta_0}{\sqrt{1-\beta^2}} \nu_0 = \frac{mc^2}{2h}(1+\beta \cos \theta_0)$$

$$\nu_2 = \frac{1-\beta \cos \theta_0}{\sqrt{1-\beta^2}} \nu_0 = \frac{mc^2}{2h}(1-\beta \cos \theta_0)$$

$$\cos \theta_1 = \frac{\cos \theta_0 + \beta}{1+\beta \cos \theta_0}, \qquad \cos \theta_2 = \frac{-\cos \theta_0 + \beta}{1-\beta \cos \theta_0}$$

One of the angles θ_1, θ_2 is arbitrary. The three cases illustrated in Fig. 75 are possible, depending upon the relative values of β and $\cos \theta_0$.

In the special case where θ_0 vanishes,

$$\nu_1 = \frac{mc^2}{2h}(1+\beta), \qquad \nu_2 = \frac{mc^2}{2h}(1-\beta)$$

$$\theta_1 = 0, \qquad \theta_2 = \pi$$

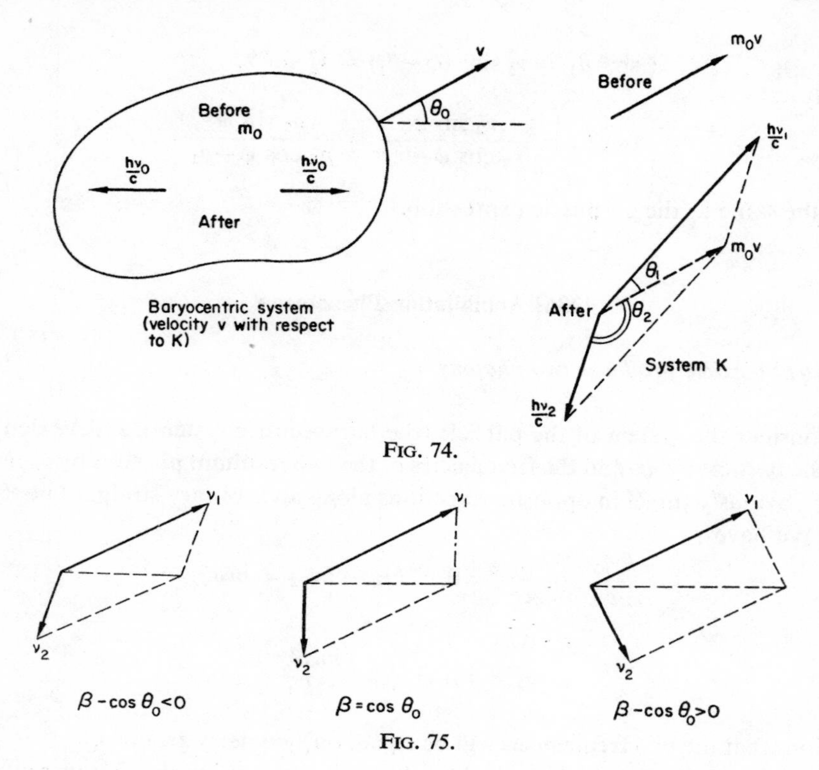

FIG. 74.

FIG. 75.

We could calculate in K directly; the equations are now

$$\frac{h\nu_1}{c}\cos\theta_1 + \frac{h\nu_2}{c}\cos\theta_2 = mv$$

$$\frac{h\nu_1}{c}\sin\theta_1 + \frac{h\nu_2}{c}\sin\theta_2 = 0$$

$$h\nu_1 + h\nu_2 = mc^2$$

but the previous argument displays the special features of the phenomenon much better.

(b) *An electron-positron pair produces two photons*

In the baryocentric system, the two corpuscles (of mass m_0) are travelling towards one another with velocities $\mathbf{v}$. The emergent photons are travelling in opposite directions along a straight line arbitrarily inclined to $\mathbf{v}$; their common frequency is

$$\nu = \frac{m_0c^2}{h\sqrt{1-v^2/c^2}}$$

We now consider the system in which the electron is at rest, which is virtually the same as the laboratory system. The conservation laws give

$$\frac{m_0 v}{\sqrt{1-\beta^2}} = \frac{h\nu_1}{c}\cos\theta_1 + \frac{h\nu_2}{c}\cos\theta_2 \tag{1}$$

$$0 = \frac{h\nu_1}{c}\sin\theta_1 - \frac{h\nu_2}{c}\sin\theta_2 \tag{2}$$

$$m_0c^2 + \frac{m_0c^2}{\sqrt{1-\beta^2}} = h\nu_1 + h\nu_2 \tag{3}$$

Notice that the mass and energy of the positron are counted positive (see § [241]).

From (1) and (2), we obtain

$$\frac{m_0^2\beta^2}{1-\beta^2} = \frac{h^2}{c^4}\{\nu_1^2+\nu_2^2+2\nu_1\nu_2\cos(\theta_1+\theta_2)\} \tag{4}$$

The velocity v is given (by the conditions prevailing before the collision); we can deduce ν_1 and ν_2 from (3) and (4) in terms of the angle $\theta_1+\theta_2$ between the velocities of the emergent photons. Squaring (3) and subtracting from (4), we have

$$1+\frac{1}{\sqrt{1-\beta^2}} = \frac{2h^2}{m_0^2c^4}\nu_1\nu_2\sin^2\frac{\theta_1+\theta_2}{2}$$

Combining this relation, which gives $\nu_1\nu_2$, with (3), we obtain

$$(\nu_1-\nu_2)^2 = (\nu_1+\nu_2)^2 - 4\nu_1\nu_2$$

$$= \frac{m_0^2c^4}{h^2}\left(1+\frac{1}{\sqrt{1-\beta^2}}\right)\left(1+\frac{1}{\sqrt{1-\beta^2}} - \frac{2}{\sin^2\dfrac{\theta_1+\theta_2}{2}}\right) \tag{5}$$

This expression shows that the angle $\theta_1+\theta_2$ is not wholly arbitrary:

$$\tan\frac{\theta_1+\theta_2}{2} \geqslant \sqrt{\frac{2}{\dfrac{1}{\sqrt{1-\beta^2}}-1}} \tag{6}$$

Finally, we solve (3) and (5) for ν_1 and ν_2:

$$\begin{matrix}\nu_1\\ \nu_2\end{matrix} = \frac{m_0c^2}{2h}\sqrt{1+\frac{1}{\sqrt{1-\beta^2}}}\left(\sqrt{1+\frac{1}{\sqrt{1-\beta^2}}} \pm \sqrt{\frac{1}{\sqrt{1-\beta^2}}-1-2\cot^2\frac{\theta_1+\theta_2}{2}}\right)$$

For $\theta_1+\theta_2 = \pi$, ν_1 is greatest and ν_2 least.

[237] Creation Phenomena

(a) *Two photons give a neutral particle*

This is the inverse of the phenomenon discussed in § [236] (a). In the baryocentric system, the two photons have the common frequency ν_0; the proper mass of the emergent particle is therefore

$$m_0 = \frac{2h\nu_0}{c^2}$$

In the laboratory system, we set out from the equations at the end of § [236] which are also valid for this inverse case, and we argue as in case (b) for (3) and (4):

$$m_0 = \frac{2h\sqrt{\nu_1\nu_2}}{c^2}\sin\frac{\theta_1+\theta_2}{2}$$

where $\theta_1+\theta_2$ is the angle between the two photons.

(b) *Two photons give an electron–positron pair*

This is the inverse of the phenomenon considered in § [236] (b). The mass m_0 is given by formula (4) of the latter section.

Remark 1. Two photons can give rise to an unlimited number of particles, subject of course to certain conditions.

Remark 2. Pair creation is illustrated in the photograph of Fig. 76.

C. PARTICLES AND ANTIPARTICLES

[238] General Remarks

I have already stressed many times that macroscopic relativistic dynamics is valid when studying individual interaction phenomena between microscopic particles. The Compton effect is the best-known example of this. Pursuing this idea, we shall examine the concept of antiparticle, which is usually considered to be essentially quantum mechanical in nature. We shall see that the ideas at the heart of this concept spring from macroscopic dynamics.

Generally speaking, the too all-pervasive tendency to adopt the "quantum mechanical" attitude is to be resisted, for it leads one to set aside real problems, wrongly regarding them as pseudo-problems (some physicists deny the objectivity of an individual microscopic phenomenon); or, although the existence of such phenomena may be recognized, all macroscopic concepts are rejected as unworthy of study.

These are dogmatic attitudes incompatible with the normal development of physics.

[239] The Concept of Antiparticle†

(a) Consider a particle (in the ordinary sense of the word) of charge ε and proper mass m_0, acted on by a field $\mathbf{E}$, $\mathfrak{B}$ and hence by a force

$$\mathbf{F} = \varepsilon\left(\mathbf{E}+\frac{\mathbf{v}}{c_0}\times\mathfrak{B}\right)$$

† We assume the particle has no spin; see § [241].

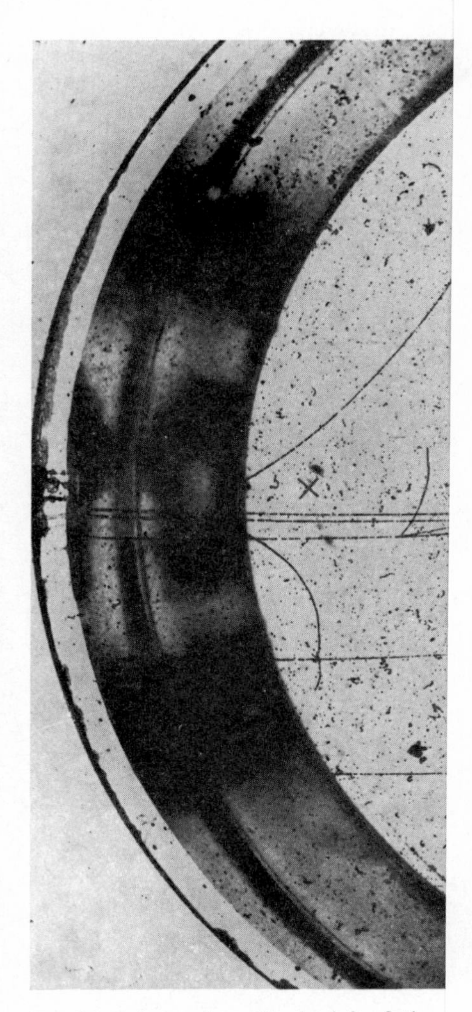

FIG. 76. Pair creation. On the left of the
chamber and interacts to give a neutral xi an
and the latter immediately decays into two
and a negative pion. The neutral kaon dec
(neutral xi) event where all the elements re
decay products are visible in the chamber. T
ratories, including CERN, in an experiment
the

At time t it passes through a point A with velocity $\mathbf{v}$; at an infinitesimally close subsequent time $t+\mathrm{d}t$, it passes through a point B, such that

$$AB = \mathrm{d}s = v\,\mathrm{d}t$$

The momentum changes by an amount

$$\mathbf{dp} = \mathbf{p}_B - \mathbf{p}_A = \mathbf{F}\,\mathrm{d}t$$

We say that the element of trajectory is described in the sense $\overline{AB}$; let ab be the corresponding world-line. This is illustrated in Fig. 77; since an element of arc-length is involved, we have taken the spatial axis Ox along this arc for simplicity.

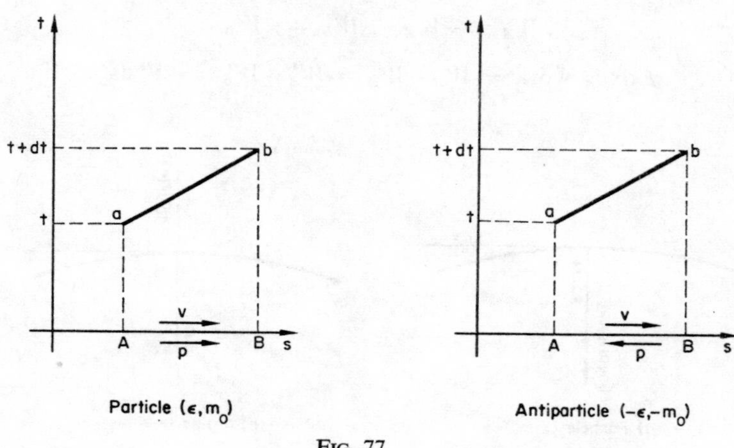

FIG. 77.

(b) Let us now make the transformations

$$\varepsilon' = -\varepsilon, \qquad m_0' = -m_0$$
$$\mathrm{d}t' = -\mathrm{d}t, \qquad \mathrm{d}x' = -\mathrm{d}x$$

To consider negative masses is compatible with relativistic dynamics (§ [17]); a particle of negative mass is *by definition* an antiparticle. The velocity is unaltered:

$$v' = \frac{\mathrm{d}x'}{\mathrm{d}t'} = \frac{-\mathrm{d}x}{-\mathrm{d}t} = v$$

The acceleration is changed:

$$\boldsymbol{\gamma} = \frac{\mathbf{dv}}{\mathrm{d}t} = \frac{\mathbf{dv}'}{-\mathrm{d}t'} = -\boldsymbol{\gamma}'$$

Since the mass is negative, the momentum $\mathbf{p}'$ is in the opposite direction to the velocity, and is hence in the direction $\overline{BA}$ (Fig. 77):

$$\mathbf{p}' = -\mathbf{p}, \qquad \mathbf{dp}' = -\mathbf{dp}$$

The force is not altered, since we must of course make the transformations for the charges which create the field $\mathbf{E}, \mathcal{B}$

$$\mathbf{F}' = \mathbf{F}$$

Hence the laws

$$\mathbf{F}\, \mathrm{d}t = \mathbf{dp} \qquad \text{or} \qquad \mathbf{F} = m\boldsymbol{\gamma}$$

are not altered.

This is, of course, true for a succession of arcs $\overline{AB}$, and hence for a trajectory of arbitrary length. To make the nature of the problem quite clear, let us assume that the velocity is small; force and acceleration are then parallel. If we are concerned with a particle, therefore, we have Fig. 78(a), and if with an antiparticle, we have Fig. 78(b). In addition, the energy of an antiparticle is negative, equal to $-m_0c^2$; hence

$$W'_A = -W_A, \quad W'_B = -W_B$$

$$F\,\mathrm{d}s = \text{work} = W_B - W_A = W'_A - W'_B = -\mathbf{F}'\,\mathbf{ds}'$$

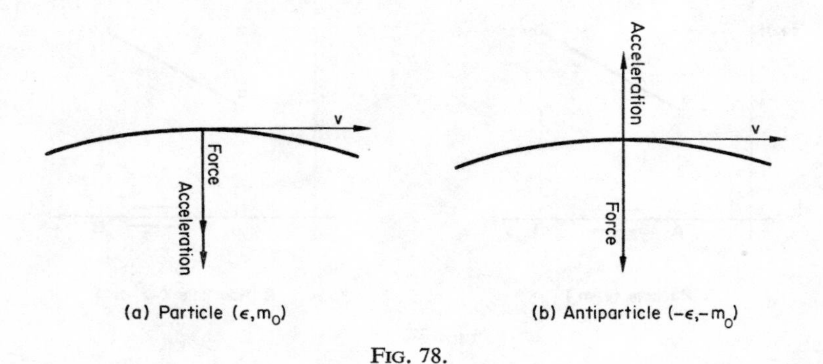

FIG. 78.

The observable effects, that is the succession of positions in space at each instant of time—the world-line—can be interpreted in two ways. In ordinary terms, the particle of charge ε and positive mass describes a trajectory as time increases, that is to say, in the sense past–future. We may, however, also say that we have to do with an anti-particle, defined by charge $-\varepsilon$ and negative mass $-m_0$. For the sense of travel, two terminologies are possible: the kinematic path (in the same sense as the velocity) as time increases and the dynamic path (in the same sense as the momentum). In the preceding formulae, we have arbitrarily chosen the sense $\overline{AB}$. But it seems more natural to select, for the antiparticle, the sense $\overline{BA}$, past–future in time t', future–past in time t. In the following section, we shall encounter much stronger reasons which also militate in favour of the future–past path.

(c) *The four-dimensional argument*

We take the dynamical equation in the form (§ [49])

$$\mathrm{d}p_i = \frac{\varepsilon}{c_0}\mathcal{M}_{ij}\,\mathrm{d}x^j, \quad p_i = m_0 V_i$$

This expression is invariant under the substitution

$$\varepsilon' = -\varepsilon, \quad m_0' = -m_0$$
$$(\mathrm{d}x^j)' = -\mathrm{d}x^j \quad \mathcal{M}'_{ij} = -\mathcal{M}_{ij}$$

We can therefore say that in a given field $\mathcal{M}_{ij}$, the world-lines of a particle and its antiparticle are superposable. Furthermore, the homogeneity of the laws leads us to say that the trajectory of an antiparticle is described in the sense of decreasing time, and hence in the sense future–past.

[240] Application to Creation and Annihilation Phenomena

(a) *Electron and positron*

In the creation phenomenon, a γ-ray produces an electron and a positron (pair) at a point M at time t. The positron is an unstable particle. At a later time $t+\Delta t$, it strikes an electron at N, the pair is annihilated and a γ-ray appears. The trajectories are shown in Fig. 79, together with the direction of travel. If we imagine that these trajectories lie in a plane, the xOy plane for example (there is nothing in principle to forbid this), then we can represent the world-lines on Fig. 79 (they have the same appearance as the trajectories). In ordinary terms, the points on all these lines are ranged in order of increasing time.

We can, however, also imagine that we have the antiparticle of the positron on MN, that is an electron of negative mass, going backwards in time (Fig. 80). In these terms, there is no longer creation or disappearance of particles. We have an electron–particle

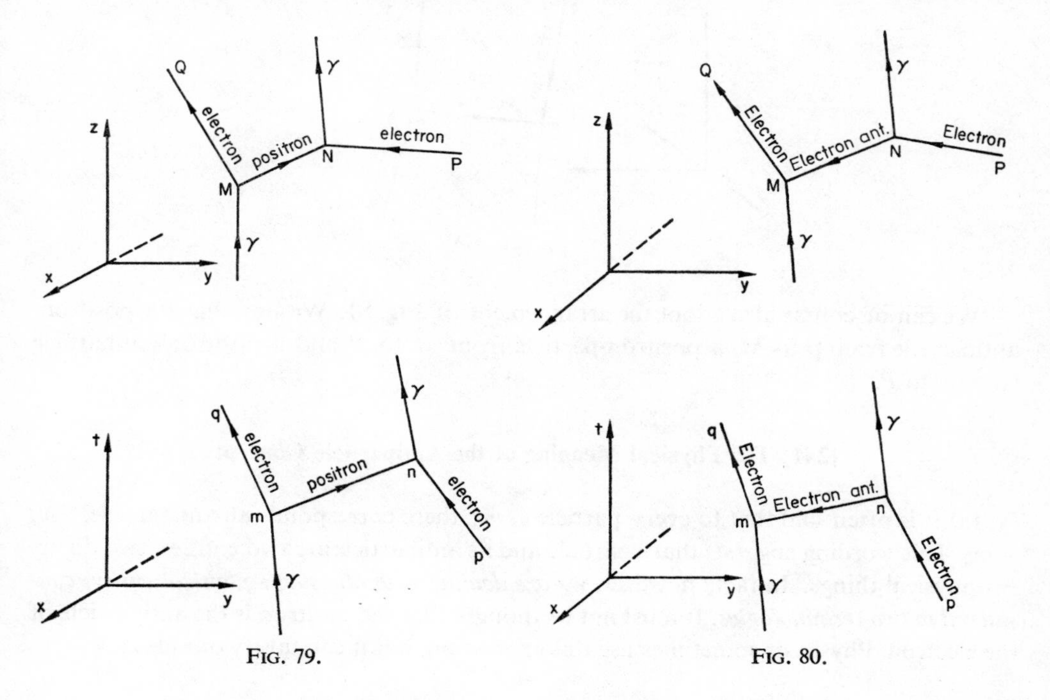

FIG. 79.

FIG. 80.

travelling from P to N, an electron–antiparticle from N to M and an electron–particle from M to Q. Only one particle is involved, therefore, in different energy states. If, therefore, we consider n particles at a given time, this number n will be conserved in time (past or future) even if there are phenomena of the kind called creation or annihilation; this offers major advantages in certain theories.

The conservation of number of particles like the existence of a rest mass thus becomes a characteristic of particles, as opposed to photons.

From this point of view, we connect particles and radiation in the same way as in the theory of the atom, and this is highly satisfactory. We know that when an atom changes its energy level or state, radiation is emitted or absorbed. Here again, when the particle emits radiation it changes state and becomes an antiparticle; the passage from the negative energy state to the positive energy state corresponds to absorption of radiation.

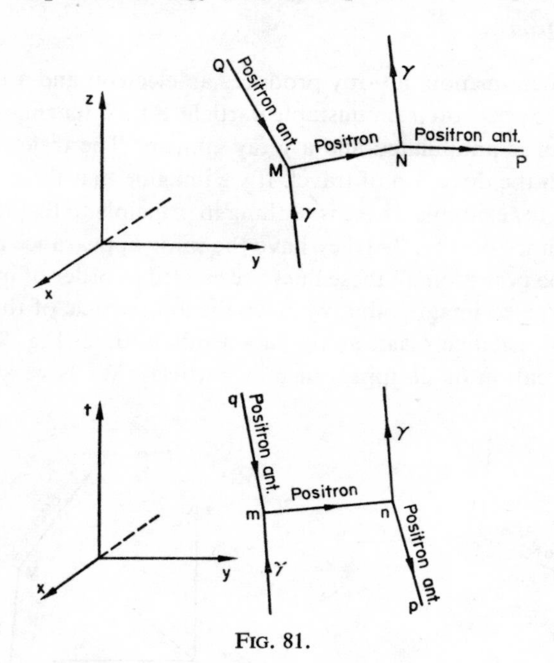

FIG. 81.

We can of course also adopt the arrangement of Fig. 81. We now have a positron-antiparticle from Q to M, a positron-particle from M to N and a positron-antiparticle from N to P.

[241] The Physical Meaning of the Antiparticle Concept

(a) It is often said that to every particle ε, m_0, there corresponds an antiparticle, $-\varepsilon$, $-m_0$. This wording suggests that a particle and its antiparticle are two entities, two different physical things. In fact, however, *we are dealing with one single physical entity, considered in two terminologies.* It must not be thought that the positron is the antiparticle of the electron. Physicists sometimes use this expression, but it can falsify our ideas.

In ordinary terms, the positron is a particle in the same sense as is the electron (positive mass and energy, . . .). The positron-particle (sometimes called the physical positron) behaves like an electron-antiparticle, however (which moves backwards through time); we can also say that an electron-particle behaves like a positron-antiparticle.

To make this vocabulary perfectly comprehensible, let us reconsider the creation of a pair at the point M. If we assumed that a particle and an antiparticle were created at M (the antiparticle having negative energy), the total energy of positron + electron would be zero, which is not the case. In fact, two particles are created at M, or in the new vocabulary, there is a change of state at M.

(b) *Another remark*

For a given field and initial velocity, particle and antiparticle follow the same trajectory; this would not be the case if we assumed that a particle and an antiparticle were created at M.

(c) *Matter and antimatter*

The statements concerning the double-terminology of particle-antiparticle must be carefully distinguished from the following law: to every particle of mass m_0 and charge ε there corresponds another particle of the *same mass* and charge $-\varepsilon$.

There is naturally a close connection between the two questions. To a particle of charge ε, relativistic dynamics attributes two states: positive energy and negative energy. This is the first point. It shows that the negative energy state gives the same observable effects as a charge $-\varepsilon$ having positive mass, whence the law about the existence of particle $+\varepsilon$ and $-\varepsilon$ for a given positive mass.

Among the known pairs, one of the particles is much stabler than the other; this explains why only one member of each pair of particles is commonly observed. It has become habitual to use the antiparticle concept for the unstable particle—for the positron and not for the electron, for the negative proton and not for the proton—but this is purely conventional.

The vocabulary is undecided, and is a source of confusion. The prefix "anti" is used with two different meanings, sometimes to label the antiparticle (negative mass) and at other times to label the unstable particle (positive mass) corresponding to the common stable particle.

This terminology can be extended to clusters of particles, hence the expression, antimatter. The antihydrogen atom, for example, would consist of a negative proton (the nucleus) and a positive planetary electron, if we use the prefix in the second sense.

We can, however, also call antihydrogen the antiparticle that gives the same observable effects as hydrogen. The nucleus would be a proton-antiparticle (that is, an antiproton of negative mass going backwards through time) and there would be a planetary electron-antiparticle (that is, a positron of negative mass).

(d) *The need to generalize the macroscopic theory of antiparticles to include particles with spin*

If we limit ourselves to characterizing a particle by its mass and its charge, we see that a neutral particle has no corresponding particle. For the neutron, for example, we could use the concept of an antineutron in the sense we have adopted (negative mass), but such an antineutron would be indistinguishable from an ordinary neutron. So far as the experimentalists are concerned, however, the antineutron does exist.

We must therefore anticipate a generalization of the macroscopic antiparticle concept.

[242] Remarks About the Interactions

(a) *General comments*

In § [226], we studied the various possibilities when two or more particles are made to interact. We have only spoken of particles; the conclusions reached are complete, for as we have just seen, antiparticles are not distinct entities different from particles.

With the new terminology, however, considering the example of § [226] (a), we can describe the event in three ways: the γ-ray creates the particles electron and positron; the positron-antiparticle interacts with the γ-ray to give an electron; the electron-antiparticle interacts with the γ-ray to give a positron. It is therefore of interest to extend the conclusions of § [226].

Notice to begin with that the conservation laws are identical in both terminologies; they take a similar form to Kirchhoff's laws for electric circuits:

$$\Sigma W = 0, \quad \Sigma \varepsilon = 0, \quad \Sigma \mathbf{p} = 0$$

(b) *Reconsideration of the discussion of* § [226]

Let us consider the case of two incident corpuscles having negative energy, or one such corpuscle and a photon.

For a negative energy corpuscle,

$$W = -c\sqrt{p^2+m_0^2c^2}, \quad W = -cp$$

With the notation of § [226], the conservation laws again give

$$\frac{W_3}{cp_3} \geqslant \frac{W_1+W_2}{cp_1+cp_2} \tag{1}$$

In the present situation (two incident negative energy corpuscles or one such corpuscle and a photon), at least one of the two inequalities

$$W_1 < -cp_1, \quad W_2 < -cp_2$$

is satisfied, and hence

$$W_1 + W_2 < -cp_1 - cp_2, \qquad -\frac{W_1 + W_2}{cp_1 + cp_2} > 1$$

Relation (1) can be written

$$-\frac{W_3}{cp_3} \leqslant \frac{W_1 + W_2}{cp_1 + cp_2} \quad \text{or} \quad N \leqslant M$$

N₄ —— −1 —— N₃ —— 0 —— N₂ —— +1 —— N₁ —— M

Fig. 82.

We therefore have four possibilities (Fig. 82):

$$N_1 \to -\frac{W_3}{cp_3} \geqslant 1, \qquad N_2 \to 0 < -\frac{W_3}{cp_3} \leqslant 1$$

$$N_3 \to -1 \leqslant -\frac{W_3}{cp_3} < 0, \quad N_4 \to -\frac{W_3}{cp_3} \leqslant -1$$

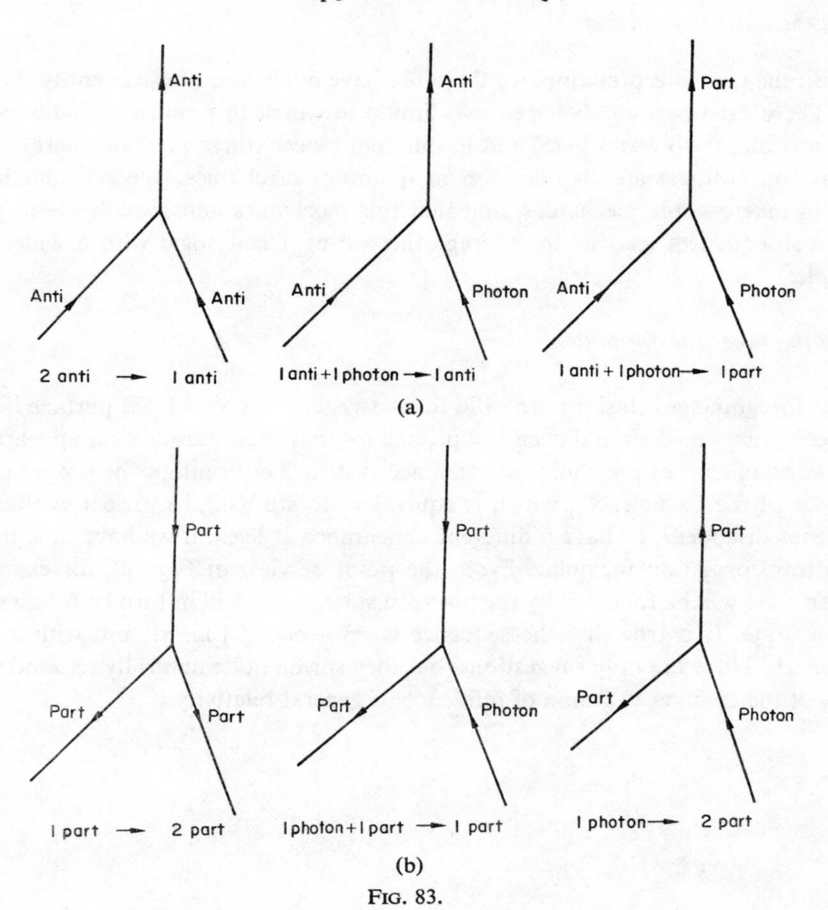

Fig. 83.

For the cases N_1 and N_2, W_3 must be negative; they give

$$W_3 \geqslant -cp_3, \quad W_3 \leqslant -cp_3$$

respectively. Only the case N_2 is possible therefore.

For the cases N_3 and N_4, W_3 must be positive, and they give

$$W_3 \leqslant cp_3, \quad W_3 \geqslant cp_3$$

respectively. Only N_4 is possible.

Applying the conservation of energy in addition, we can conclude that the interaction of two antiparticles gives an antiparticle, and that the interaction of an antiparticle and a photon gives a particle or an antiparticle. We thus obtain Fig. 83(a). If we return to particle language, we have Fig. 83(b).

[243] Consequences of an Epistemological Nature

(a) *On the identity principle*

With the new interpretation, we therefore have one single physical entity throughout time. There exist periods (between two limits) in which this entity is *simultaneously* in different states (two states here) and in different places (three positions here). The conclusion concerning states is common in quantum mechanics. We see that in fact it arises in macroscopic mechanics, and that this mechanics adds details about position. These consequences lead us to envisage the use of a new logic with a wider identity principle.

(b) *On the time transformation*

The foregoing conclusions are valid for the system K in which the particle is moving. It is very possible—I should even say probable—that their paradoxical appearance is a direct consequence of the choice of reference system. Let us adopt the proper system of reference of the particle, K_0, which is equivalent to studying its proper evolution. The paradoxes disappear, or have a different appearance at least, if we have time inversions in the transformation formulae. From the point of view of Fig. 80, for example, the electron state will be followed by the positron state which will in turn be followed by the electron state. It is true that the sequence is reversed on Fig. 81, but with a different vocabulary. These are only suggestions, but they spring quite naturally to mind when we carry out the changes of frame of reference of general relativity.

HISTORICAL AND BIBLIOGRAPHICAL NOTES

[244] On the Antiparticle Concept

I only wish to indicate briefly the three stages in the development of this concept, without claiming to give a bibliography, properly speaking.

(a) *Dirac's theory; negative energies; prediction of the positron*

Since the beginnings of relativistic dynamics, the relation

$$W = \pm c\sqrt{p^2 + m_0^2c^2}$$

has of course been known (§ [17]) but the solution with the negative sign was rejected as having no physical meaning. When developing his quantum theory of the relativistic electron, however, Dirac noticed that the negative energy states are necessarily conserved. He showed, in somewhat different terms, that a negative electron in such a state produces the same effects as a particle (with positive energy) and positive charge equal in magnitude to that of the electron.

b) *The experimental discovery of the positron*

The experimental discovery of the positron verified this standpoint very strikingly. The first photographs were obtained in 1932 by Anderson [1, 3, 5], Blackett and Occhialini [2], Curie and Joliot [4, 6], Thibaud [7], Perrin [8, 9] and others. The bibliography has now grown enormous; the reader is referred in particular to books on cosmic rays.

These results drew attention to Dirac's theory; considerably generalized, it now forms the corpus of knowledge known as the quantum theory of fields.

Dirac's prediction concerning the electron can be extended to every corpuscle. We mention, as an example of a more recent confirmation, the discovery of the negative proton by Chamberlain, Segrè and Wiegand [13].

(c) *Time inversion*

Adopting either the macroscopic or quantum point of view, Stückelberg [10, 11], Feynmann [12] and Costa de Beauregard [14] have brought out the antiparticle concept, together with the new idea of time inversion. The future of this idea cannot as yet be foretold; it seems to open up very interesting new theoretical perspectives. If it becomes definitely incorporated into physics, it will form as important a step as the introduction of the relativity of simultaneity by Einstein.

[1] C. D. ANDERSON: The apparent existence of easily deflectable positives. *Science* **76** (1932) 238–239.
[2] P. M. S. BLACKETT and G. P. S. OCCHIALINI: Photographs of tracks of penetrating radiation. *Proc. Roy. Soc.* **139** (1933) 699–727.
[3] C. D. ANDERSON: The positive electron. *Phys. Rev.* **43** (1933) 491–494.
[4] I. CURIE and F. JOLIOT: Sur l'origine des électrons positifs. *C. R. Acad. Sci. Paris* **196** (1933) 1581–1583.
[5] C. D. ANDERSON and S. H. NEDDERMEYER: Positrons from γ-rays. *Phys. Rev.* **43** (1933) 1034.
[6] F. JOLIOT: Preuve expérimentale de l'annihilation des électrons positifs. *C. R. Acad. Sci. Paris* **197** (1933) 1622–1625.
[7] J. THIBAUD: L'annihilation des positons au contact de la matière et la radiation qui en résulte. *C. R. Acad. Sci. Paris* **197** (1933) 1619–1633.
[8] F. PERRIN: Possibilité de matérialisation par interaction d'un photon et d'un électron. *C. R. Acad. Sci. Paris* **197** (1933) 1100–1102.

[9] F. Perrin: Matérialisation d'électrons lors du choc de deux électrons positifs. Processus divers d'annihilation des électrons positifs. *C. R. Acad. Sci. Paris* **197** (1933) 1302–1304.
[10] E. C. G. Stückelberg: Remarque à propos de la création de paires de particules en théorie de relativité. *Helv. Phys. Acta* **14** (1941) 588–594.
[11] E. C. G. Stückelberg: La mécanique du point matériel en théorie de relativité et en théorie des quanta. *Helv. Phys. Acta* **15**, fasc. I (1942) 23–27.
[12] R. P. Feynmann: The theory of positrons. *Phys. Rev.* **76** (1949) 749–759.
[13] O. Chamberlain, E. Segrè, C. Wiegand and T. Ypsilantis: Observation of antiprotons. *Phys. Rev.* **100** (1955) 947–950.
[14] O. Costa de Beauregard: *Théorie synthétique de la relativité restreinte et des quanta.* Gauthier-Villars (1957) pp. 37–38.
[15] E. Amaldi: Le antiparticelle. *Suppl. Nuovo Cimento* **19** (1961) 101–131.

APPENDIX I

SUPPLEMENTARY MATERIAL

[245] Some General Works†

References 1–27 are from Appendix III of my *Relativistic Kinematics*.

[1] A. Einstein: Zur Elektrodynamik bewegter Körper. *Ann. Physik.* **17**, 17 June 1905, pp. 891–921 English translation: On the electrodynamics of moving bodies; published in H. A. Lorentz, A. Einstein, . . . : *The Principle of Relativity* (translated into English by Perret and Jeffrey), Methuen, London, 1923. Recently republished by Dover, U.S.A.

[2] M. von Laue: *Die Relativitätstheorie*; 5th ed. Vieweg, Braunschweig, 1952. The first edition appeared in 1911. Translated into French by G. Letang, Gauthier-Villars, 1926.

[3] A. Einstein: *The Meaning of Relativity*. Methuen, London, 5th ed., 1951. The first edition, which appeared in 1923, contained the texts of his lectures at the University of Princeton, in May 1921.

[4] R. C. Tolman: *Relativity, Thermodynamics and Cosmology*. Oxford, Clarendon Press, 1st ed., 1934.

[5] W. H. McCrea: *Relativity Physics*. Methuen, London, 3rd ed., 1949 (1st ed. 1935).

[6] P. G. Bergmann: *Introduction to the Theory of Relativity*. Prentice Hall, New York, 7th ed., 1955 (the first ed. appeared in 1942).

[7] O. Costa de Beauregard: *La Théorie de la relativité restreinte*. Masson, Paris, 1949.

[8] G. Møller: *The Theory of Relativity*. Oxford, Clarendon Press, 1952, reprinted in 1955.

[9] L. Landau and E. Lifshitz: *The Classical Theory of Fields*. Addison-Wesley Press, Cambridge, Massachusetts, 1951; the original work appeared in Russian in 1948.

[10] K. Jellinek: *Weltsystem, Weltäther und die Relativitätstheorie*. Wepf, Basel, 1949, 450 pages. Pages 1–142 deal with special relativity.

[11] P. Couderc: *La Relativité*. Presses Universitaires, Paris, 1952.

[11a] E. Terradas and R. Ortiz: *Relatividad*. Espasa-Calpe, Madrid, 1952.

[12] A. Papapetrou: *Spezielle Relativitätstheorie*. Deutsch. Verlag Wiss., Berlin, 1955, 170 pages.

[13] J. L. Synge: *Relativity, the Special Theory*. North Holland, Amsterdam, 1956.

[14] G. Stephenson and G. W. Kilmister: *Special Theory for Physicists*. Longmans, London, 1958.

[15] A. Tonnelat: *Les Principes de la théorie électromagnétique et de la relativité*. Masson, Paris, 1959.

[16] Louis de Broglie: *Eléments de théorie des quanta et de mécanique ondulatoire*. Gauthier-Villars, Paris, 1953.

[17] J. Aharoni: *The Special Theory of Relativity*. Oxford, Clarendon Press, 1959.

[18] W. Rindler: *Special Relativity*. Oliver and Boyd, Edinburgh and New York, 1960.

[19] V. A. Fock: *The Theory of Space Time and Gravitation*. Translated by Kemmer, Pergamon Press, 1959.

[20] W. Pauli: *Theory of Relativity*. Pergamon Press, 1958 (translated from the German text of 1921).

[21] S. Becker: Divertissements astronautico-relativistes. *Perspectives* X (1961), 59–116.

---

† This list gives a general indication, but has no pretensions to completeness.

[22] G. HOLTON: Resource letter SRT-1 on special relativity theory. *Amer. J. Phys.* **30** (1962) 462–469. This is not a discussion but a bibliographical review; nevertheless, I quote this text here, for its intention is excellent. It is intended to put at the disposal of non-specialist teachers publications which are both sound and easily understood, dealing with special relativity.

I shall take this occasion to mention that the "Commission on College Physics" of the U.S.A. also publishes Newsletters which often contain interesting information (articles, lectures, films ...) about relativity and which demonstrate the increasing importance of this discipline in the teaching of the physical sciences.

[23] P. W. BRIDGMAN: *A Sophisticate's Primer of Relativity*. Wesleyan University Press, Middletown, Conn., 1962.

[24] R. HAGEDORN: *Relativistic Kinematics*. Benjamin, New York, 1964. Despite its title, only 20 pages of this book are devoted to kinematics; the remainder (about 140 pages) is concerned with dynamics.

[25] R. KATZ: *An Introduction to the Special Theory of Relativity*. Van Nostrand, New York, 1964.

[26] P. G. BERGMANN: The special theory of relativity. *Handbuch der Physik*, **4** (1962) 109–202.

[27] D. BOHM: *The Special Theory of Relativity*. Benjamin, New York, 1965.

[28] W. G. V. ROSSER: *An Introduction to the Theory of Relativity*. Butterworth, London, 1964.

[29] E. WHITTAKER: *History of the Theories of Aether and Electricity*. Nelson, London, 1951 (the first edition was published in 1910).

This book is very interesting and most useful. None the less, I must address a grave reproach to the author: one of the most important chapters is entitled "The relativity theory of Poincaré and Lorentz". The text has to be read quite carefully to detect Einstein's name. According to Whittaker, Einstein merely deduced the Lorentz transformation from a highly debatable postulate (about the velocity of light) and applied this new theory to particular problems, such as aberration and the Doppler effect. The historical perspective is completely falsified: the role of Einstein in special relativity is comparable with that of Lavoisier in chemistry. Both were confronted with a mass of facts and theories—their genius lay in their discovery of the guiding idea that converted a shapeless mass into a satisfying pattern, transforming a closed horizon into a dawn of major research. For Einstein, this guiding idea was the new interpretation of the concepts of space and time. To understand this, however, *one must rise above the simple mathematical formalism and extract the physical meaning and the philosophical implications from the symbols; in short, one must have the spirit of the true physicist*. Moreover, the theory of gravitation is wholly due to Einstein, formulae included; but Chapter V of Whittaker's book is just called "Gravitation". This reflects a regrettable bias against Einstein which, we must admit, was not uncommon among his contemporaries. A prophet is not without honour ....

I must insist on this point. Lavoisier's name is linked with the idea of conservation of mass. Through the relation expressing the equivalence of mass and energy, which owes much to him, Einstein has played a major role in this aspect of chemistry, which has given rise to nuclear chemistry (or physics). The Einsteinian use of the concepts of space and time is equivalent to choice of a standard for time-intervals and lengths. It was also by a choice of standard, or rather, of a reference system, that Copernicus transformed astronomy and, at the same time, our attitude towards the Earth. Einstein's role is thus identical, in this sense, with that of Copernicus, and is no less important. Finally, Newton's name is associated with the law of gravitation, which governs the motion of the planets. The law stated by Einstein gives a closer description of the planetary phenomena and enables us to begin to understand the world of the galaxies.

Remark. Recent work on the historical role of Einstein is to be found in Appendix V of *Relativistic Kinematics*.

[30] R. DUGAS: *Histoire de la Mécanique*. Dunod, Paris, 1950; *La Mécanique au XVIIe siècle*. Dunod, Paris, 1954.

[31] H. ARZELIÈS: *La Cinématique relativiste*. Gauthier-Villars, Paris, 1955; *Relativistic Kinematics*. Pergamon, Oxford, 1966.

[32] H. ARZELIÈS: *La Dynamique relativiste et ses applications*. Gauthier-Villars, Paris, vol. I 1957, vol. II 1958.

[33] H. ARZELIÈS: *Electricité*. Gauthier-Villars, Paris, 1963.

[34] H. ARZELIÈS: *Rayonnement et dynamique du corpuscle chargé fortement accéléré*. Gauthier-Villars, Paris, 1966.

[35] O. COSTA DE BEAUREGARD: *Précis of Special Relativity*. Academic Press, New York and London, 1966.

[36] H. YILMAZ: *Theory of Relativity and the Principles of Modern Physics*. Blaisdell, Waltham, Mass.

[37] R. K. Pathria: *The Theory of Relativity*. Hind. Publ. Corp., Delhi, 1963.
[38] J. A. Smith: *Introduction to Special Relativity*. Benjamin, New York, 1965.
[39] C. Kacser: *Introduction to the Special Theory of Relativity*. Prentice Hall, Englewood Cliffs, 1967.
[40] A. P. French: *Special Relativity*. Nelson, New York, 1968.
[41] H. M. Schwartz: *Introduction to Special Relativity*. McGraw-Hill, New York, 1968.
[42] A. Shadowitz: *Special Relativity*. Saunders, Philadelphia, 1968.
[43] F. W. Sears and R. W. Brehme: *Introduction to the Theory of Relativity*. Addison-Wesley, Reading, (Mass.), 1968.
[44] R. Dutheil: *Initiation à la physique relativiste*. Gauthier-Villars, Paris, 1969.
[45] N. D. Mermin: *Space and Time in Special Relativity*. McGraw-Hill, New York, 1968.
[46] R. Resnik: *Introduction to Special Relativity*. Wiley, New York, 1968.

APPENDIX II

(Complementary to Chapter IX)

[246] Quantum Definition of Mass and Force

In the present work, we are considering only the macroscopic scale (pre-relativistic or relativistic); this must be borne in mind in connection with the definitions of force and mass that we have advanced. In § [112], we saw that it would be possible to define mass from the equations of quantum theory. The parameter that is introduced in this way can only be interpreted if force has been defined beforehand. More generally, we may define mass as the parameter that occurs in Schrödinger's equation,

$$-\frac{\hbar^2}{2m}\nabla^2\Psi+U\Psi=-i\hbar\frac{\partial\Psi}{\partial t}$$

Thus Roubaud-Valette [1], for example, writes: "Considérons une particule de nom propre m_0, liée à un système de référence S..."† This assumes that the potential function U has already been defined, however, which requires the concept of force.

It seems, therefore, from the operational point of view, that when we write down the basic quantum equations, we are assuming that macroscopic definitions of force and mass are available (just as we assume the macroscopic concepts of space and time). Taylor [2] advances the idea of a quantum force, regarding the basic equations as established. He defines such a force in two equivalent ways, and shows how it is related to the macroscopic force. This article is very illuminating.

[1] J. Roubaud-Valette: La transformation de Lorentz et la mécanique ondulatoire. *C. R. Acad. Sci. Paris* **213** (1941) 563–566.

[2] P. L. Taylor: Wave mechanics and the concept of force. *Amer. J. Phys.* **37** (1969) 29–33.

† Let us consider a particle of proper "name" m_0, connected with a system of reference S.

APPENDIX III

[247] Complementary References to § [172]

[63] N. S. KALITZIN: Relativistic mechanics of the particle of variable mass (in Russian). *C. R. Acad. Sci. Bulg.* **7** (1954) 9–11.
[64] N. S. KALITZIN: Relativistic mechanics of the particle of variable mass (in Russian). *Zh. Eksp. Teor. Fiz.* **23** (1955) 631.
[65] N. S. KALITZIN: Verallgemeinerung der Grundformel der relativistischen Mechanik für einige praktisch wichtige Nichtinertialsysteme. *Nuovo Cimento* **13** (1959) 173–185.
[66] C. MØLLER: ref. [8] of Appendix I, § 38, pp. 106–107.
[67] L. D. LANDAU and E. M. LIFSHITZ: *The Classical Theory of Fields*. Pergamon, Oxford, 1962 (2nd ed.), p. 28.
[68] G. MARX: The mechanical efficiency of interstellar vehicles. *Astronautica Acta* **9** (1963), part 3, 131–138.
[69] D. F. SPENCER and L. D. JAFFE: Feasibility of interstellar travel. *Astronautica Acta* **9** (1963), part 2, 49–58.
[70] C. P. GADSEN: Newton's law with variable mass. *Amer. J. Phys.* **32** (1964) 61.
[71] A. M. BORK and A. B. ARONS: Newton's laws of motion and the 17th century laws of impact. *Amer. J. Phys.* **32** (1964) 313–317.
[72] K. B. POMERANZ: Newton's law for systems with a variable rest mass. *Amer. J. Phys.* **32** (1964) 386.
[73] A. M. BORK and A. B. ARONS: Newton's law and variable mass. *Amer. J. Phys.* **32** (1964) 646.
[74] H. L. ARMSTRONG: Rotation and angular momentum. *Amer. J. Phys.* **33** (1965) 507.
[75] G. CARINI: Relativistic dynamics of a body with a variable rest mass. *Atti Soc. Peloritana Sc. Fis. Mat. Nat.* **11** (1965) 401.
[76] K. B. POMERANZ: The relativistic rocket. *Amer. J. Phys.* **34** (1966) 565–566.
[77] G. L. TRIGG: Law of motion. *Amer. J. Phys.* **34** (1966) 71 and 988–989.
[78] H. L. ARMSTRONG: On the form in which Newton's law is expressed. *Amer. J. Phys.* **34** (1966) 982.
[79] C. P. GADSEN: Laws of motion for variable mass systems. *Amer. J. Phys.* **34** (1966) 987–988.
[80] M. S. TIERSTEN: Force, momentum change and motion. *Amer. J. Phys.* **37** (1969) 82–87.
[81] K. B. POMERANZ: Some relativistic effects in systems with a variable rest mass. *Amer. J. Phys.* **37** (1969) 741–744. Erratum, *Amer. J. Phys.* **38** (1970) 119.
[82] C. LEIBOVITZ: Rest mass in special relativity. *Amer. J. Phys.* **37** (1969) 834–835.

Among other definitions, this author adopts the following formulae (see my Sections [169] and [157]:

$$K^i = \frac{\mathrm{d}\Pi^i}{\mathrm{d}T} \qquad \mathbf{F} = \frac{\mathrm{d}\mathbf{p}}{\mathrm{d}t}$$

but he believes that F is the ordinary force (electrical force for example). Since the preceding equation gives

$$\frac{\mathrm{d}m_0}{\mathrm{d}T} = -\frac{K_i V^i}{c^2}$$

he concludes that the variation of proper mass is a function of the applied force, and hence that m_0 cannot remain constant in any field at all ... *This is completely wrong.* It must be stressed that in these formulae, **F** is *not* the applied force. Other authors use these formulae, but without giving the *operational* physical meaning of **F**. It is virtually certain that they are making the same mistake; if we are to use the term "force" without further qualification, we must refer to the ordinary force (which is the only kind that is defined).

[83] M. S. Tiesten: Erratum to ref. [80], *Amer. J. Phys.* **37** (1969) 1285.
[84] C. Barrabes: *Mouvement d'un objet ponctuel de masse propre variable*. D.E.A. (directed by J. Henry). Institut Henri Poincaré, Paris, May 1970.
[85] M. Carrassi: Heat and fictitious forces in variable rest mass relativistic dynamics. University of Genoa, Italy (personal communication), 1970.

APPENDIX IV

[248] Some Recent References on the Compton Effect (complementary to § [220])

[1] J. OLSEN: Elastic scattering as a relativistic Doppler effect. *Risö Rept.* no. 82 (1964).
[2] G. FARMER: Derivation of Compton scattering relation in covariant notation. *Amer. J. Phys.* **34** (1966) 614.
[3] R. E. SIEMON and D. R. SNIDER: Elastic collisions as Lorentz transformations with application to Compton scattering. *Amer. J. Phys.* **34** (1966) 614–615.
[4] A. NIELSON and J. OLSEN: Formal analogy between Compton scattering and Doppler effect. *Amer. J. Phys.* **34** (1966) 621–622.
[5] W. M. J. WEIGELE, P. T. TRACY and E. M. HENRY: Compton effect and electron binding. *Amer. J. Phys.* **34** (1966) 1116–1121.
[6] J. OLSEN: Note on elastic scattering. *Amer. J. Phys.* **36** (1968) 366–367.
[7] W. M. J. WEIGELE, P. T. TRACY and E. M. HENRY: Generalized Compton equation. *Amer. J. Phys.* **37** (1969) 806–808.

APPENDIX V

Errata to Relativistic Kinematics

p. 65, 1. 10: *In* $y_2 \div m(v)$, *the symbol* $\div$ *denotes divided by.*

p. 107, 1. 23: *For* The transverse Doppler effect *Read* The radial Doppler effect.

p. 224, 1. 1: *For* when the latter *Read* when H.

p. 236, 1. 4: *After "relativistic* physics" *and before "Return to the ..." Insert the following passage:* "In the system of the disc, space is metrically non-Euclidean. These two statements are well established. We must conclude, therefore, that for the inhabitants of the disc the nature of the geometry changes during the transitory phase. The final result is the same, however the disc is started."

APPENDIX VI

Complementary References to §§ [40], [186] and [187] (the force transformation, the jointed lever and the rod)

[1] H. M. SCHWARTZ: ref. [41] of Appendix I, p. 217. Draws attention to the rod problem and appears shaken by my arguments against the von Laue current but remains on the fence.

[2] R. G. NEWBURGH: The relativistic problem of the right-angled lever: the correctness of the Laue solution. *Nuovo Cimento* **61 B** (1969) 201–209. This author argues in favour of the energy current.

[3] G. CAVALLERI and G. SALGARELLI: Revision of the relativistic dynamics with variable rest mass and applications to relativistic thermodynamics. *Nuovo Cimento* **52 A** (1969) 722–754. These authors reject the von Laue current, as I do. Nevertheless, an exchange with Cavalleri reveals that they do not draw the same consequences as I do for continuous media (§ [188A]), for which they retain the usual formulae (which I reject).

[3 *bis*] H. ARZELIÈS: On the heat current and other fundamental concepts in relativistic thermodynamics. Pittsburgh Symposium, April 1969. In this paper, the basis for a relativistic study of irreversible changes is established.

[4] G. B. BROWN: What is wrong with relativity? *Bull. Inst. Phys. Phys. Soc. (London)* **18** (1967) 71.

[5] S. ARANOFF: Torques and angular momentum on a system at equilibrium in special relativity. *Amer. J. Phys.* **37** (1969) 453–455. This author rejects the notion of energy current.

[6] L. KARLOV: On the transformation of force in relativistic statics. *Lett. Nuovo Cimento* **3** (1970) 37–39. I agree with this author over the rejection of the energy current but *disagree entirely* over the force transformation. Karlov notices that the direction of a force does not transform like the direction of a straight line (the wire exerting the force). This has been known *for a long time* (see, for example, p. 37 of my *Dynamique relativiste* of 1957). The reason for this is very simple. The orientation of a line depends on the distribution of its component points in time (the convention to be chosen). Force, on the contrary, is defined at a point and its direction is independent of any conventions about time. This absolutely standard explanation appears to have escaped the author. He finds the statement absurd (for him, the force must "obviously" be in the same direction as the wire) and he concludes that the force transformation formulae must be modified. This *false problem* gave rise to articles by Ray, Rindler, Johns and others.

[7] J. R. RAY: Reply to Karlov's article on the transformation of force. *Lett. Nuovo Cimento* **3** (1970) 739–741.

[8] W. RINDLER: On the transformation of force in relativistic statics. *Lett. Nuovo Cimento* **3** (1970) 742–743.

[9] K. A. JOHNS: The paradoxical behaviour of force in special relativity. *Lett. Nuovo Cimento* **4** (1970) 351–354.

[10] G. CAVALLERI and G. SPINELLI: Does a rod, pushed by a force, accelerate less than the same rod pulled by the same force? *Nuovo Cimento* **66 B** (1970) 11–20.

[11] J. PAVAGEAU: Définition relativiste, dans un référentiel d'inertie quelconque, du moment retardé d'une force. Résolution du paradoxe du levier coudé. *C. R. Acad. Sci. Paris* **270 A** (1970) 1074.

[12] R. G. NEWBURGH: The Thomas precession and the relativistic right-angled lever. *Amer. J. Phys.* **38** (1970) 1158.

[13] J. W. BUTLER: The Lewis–Tolman lever paradox. *Amer. J. Phys.* **38** (1970) 360–368.

[14] F. L. MARKLEY: On rigid bodies and special relativity theory. Personal communication, April 1970.

[15] H. VOS and C. C. JONKER: On the concept of force in special relativity. Personal communication.

A LIST OF WORKS BY THE AUTHOR

Published by Gauthier-Villars, Paris

(1) *La cinématique relativiste*, 1955, 228 pp.
(2 and 3) *La dynamique relativiste et ses applications*, Fasc. I, 1957, 304 pp.; Fasc. II, 1958, 451 pp. (in collaboration with R. Mendez).
(4) *Milieux conducteurs et polarisables en mouvement*, 1959, 357 pp. (in collaboration with J. Henry).
(5 and 6) *Relativité généralisée. Gravitation*, Fasc. I, 1961, 377 pp.; Fasc. II, 1963, 274 pp. (in collaboration with J. Moulis).
(7) *Electricité macroscopique et relativiste*, 1963, 727 pp.
(8) *Rayonnement et dynamique du corpuscule chargé fortement accéléré*, 1966, 426 pp. (in collaboration with I. Tordjman).
(9) *Thermodynamique relativiste et quantique*, 1967, 704 pp.

Published by Pergamon Press

(10) *Relativistic Kinematics*, 1966, 298 pp. A translation of (1), with revisions and additions to bring the text up to date.
(11) *Relativistic Point Dynamics*, 1971, 382 pp. A translation of (2 and 3), thoroughly revised ad reorganized and containing new chapters on the dynamics of variable proper mass.

Published by Masson, Paris

(12) *Fluides relativistes*, 1971, 209 pp.
In preparation
(13) *Thermohydrodynamique irréversible* (complementing 9).
(14) *Dynamique relativiste des particules multipolaires*.

I should like to thank in advance any authors of articles on relativistic topics who are kind enough to send me reprints of their work. I shall also be interested to learn of work which is in progress at the present time. My permanent address is

Henri Arzelies, Le Truel, Lozere (48), France.

AUTHOR INDEX

Note: The numbers refer to sections and not to pages.